U.S. CUSTOMARY UNITS AND SI EQUIVALENTS

Quantity	U.S customary unit	SI equivalent
Area	mi^2	[illegible]
	acre	[illegible]
	ft^2	[illegible]
	in^2	[illegible]
Concentration	lb/million gal	0.1200 mg/L
Energy	ft·lb	1.356 J
Force	lb	4.448 N
Flow	ft^3/s	0.0283 m^3/s
	gal/min	0.003785 m^3/min
Length	ft	0.3048 m
	in	25.40 mm
	mi	1.609 km
Mass	grain	64.80 mg
	oz	28.35 g
	lb	0.4536 kg
	ton	907.2 kg
Power	ft·lb/s	1.356 W
	hp	745.7 W
Pressure	lb/ft^2	47.88 Pa
	lb/in^2	6.895 kPa
	ft (of water)	2.988 kPa
Velocity	ft/s	0.3048 m/s
	in/s	0.0254 m/s
	gal/ft^2 per min	0.0407 m/min
		58.678 m/day
	gal/ft^2 per day	0.0407 m/day
Volume	ft^3	0.02832 m^3
	yd^3	0.7646 m^3
	gal	0.003785 m^3
	acre·ft	1233.5 m^3

WATER SUPPLY AND SEWERAGE

McGraw-Hill Series in Water Resources and Environmental Engineering

Consulting Editors

Rolf Eliasen, Paul H. King, and Ray K. Linsley

Bailey and Ollis: *Biochemical Engineering Fundamentals*
Biswas: *Models for Water Quality Management*
Bouwer: *Groundwater Hydrology*
Canter: *Environmental Impact Assessment*
Chanlett: *Environmental Protection*
Chow, Maidment and Mays: *Applied Hydrology*
Eckenfelder: *Industrial Water Pollution Control*
Linsley and Franzini: *Water Resources Engineering*
Metcalf and Eddy, Inc.: *Wastewater Engineering: Collection and Pumping of Wastewater*
Metcalf and Eddy, Inc.: *Wastewater Engineering: Treatment, Disposal, Reuse*
McGhee: *Water Supply and Sewerage*
Peavy, Rowe, and Tchobanoglous: *Environmental Engineering*
Rich: *Low-Maintenance, Mechanically-Simple Wastewater Treatment Systems*
Sawyer and McCarty: *Chemistry for Environmental Engineers*
Tchobanoglous, Theisen, and Eliassen: *Solid Wastes: Engineering Principles and Management Issues*

WATER SUPPLY AND SEWERAGE

Sixth Edition

Terence J. McGhee
Lafayette College

McGraw-Hill, Inc.
New York St. Louis San Francisco Auckland Bogotá
Caracas Lisbon London Madrid Mexico City Milan
Montreal New Delhi San Juan Singapore
Sydney Tokyo Toronto

This book is printed on acid-free paper.

This book was set in Times Roman by Better Graphics, Inc.
The editors were B. J. Clark and David A. Damstra;
the production supervisor was Kathryn Porzio.
The cover was designed by Carla Bauer.
R. R. Donnelley & Sons Company was printer and binder.

WATER SUPPLY AND SEWERAGE

8 9 0 DOC/DOC 9 9 8 7

ISBN 0-07-060938-1

Library of Congress Cataloging-in-Publication Data

McGhee, Terence J.
Water supply and sewerage engineering / Terence J. McGhee. —6th ed.
p. cm.—(McGraw-Hill series in water resources and environmental engineering)
ISBN 0-07-060938-1
1. Water-supply engineering. 2. Sewerage. I. Title.
II. Series.
TD345.M33 1991
628—dc20 90-35930

ABOUT THE AUTHOR

Terence J. McGhee is professor and head of civil engineering at Lafayette College in Easton, Pennsylvania. He has taught and conducted research since 1968 at Lafayette, Tulane, the University of Nebraska, and the University of Louisville.

He is a graduate of Newark College of Engineering, Virginia Polytechnic Institute, and the University of Kansas and is registered as a professional engineer in Nebraska and Louisiana.

In addition to his academic activities, he has served as a consultant on both local and international environmental projects, and has worked and traveled in Latin America. He is a member of the American Society of Civil Engineers and the Interamerican Association of Sanitary and Environmental Engineering (AIDIS).

CONTENTS

PREFACE

This edition, like the former, is intended to introduce the design of water and wastewater treatment systems to the undergraduate student in civil engineering. The text has been revised to incorporate recent improvements in our understanding of fundamental phenomena, applications of new technologies and materials, and new computational techniques. The material has also been reorganized to reduce repetition.

A significant change will be found in the expansion of the treatment of hydraulic principles and their concentration in a single chapter. Since many engineering students may never have a formal course in unconfined or open-channel flow, and since design and control of water and wastewater treatment processes requires knowledge of this material, a concise treatment is presented in addition to pipeline flow principles. Practicing engineers will also find this chapter useful, since it summarizes a great deal of useful theoretical and practical information which is usually scattered among many sources.

In recent years there has been a tendency in some environmental engineering curricula to focus on theoretical and laboratory analyses of water and wastewater problems. While such research can lead to improved techniques and better understanding of fundamental concepts, practicing engineers must still be concerned with designing treatment, distribution, and collection systems that work—and that work despite any variations in flow and quality which may occur. This book is intended to be used in the training of undergraduate engineers for this design task.

Numerical calculations have been rounded to the nearest significant figure. Similarly, in conversions between S.I. and English units, numbers have been rounded, particularly when presenting a range within which values typically fall. The purpose in both cases is the same: to avoid giving a false impression of accuracy.

The text may be used in a comprehensive course covering all aspects of water and wastewater treatment or in separate courses dealing with design of hydraulic networks (Chaps. 1, 2, 3, 4, 6, 7, 12, 13, 14, 15, 16, 17, and 26) and design of treatment systems (Chaps. 1, 2, 5, 8, 9, 10, 11, 18, 19, 20, 21, 22, 23, 24, 25, and 26).

McGraw-Hill and I would like to thank the following reviewers of this edition for their many helpful comments and suggestions: Anthony G. Collins, Clarkson University; David W. Hubly, University of Colorado; R. A. Minear, Institute for Environmental Studies; and Paul D. Trotta, Northern Arizona University.

As in the past, the comments and suggestions of users of the text will be greatly appreciated.

Terence J. McGhee

WATER SUPPLY AND SEWERAGE

CHAPTER 1

INTRODUCTION

1-1 Environmental Engineering

Environmental engineering involves the application of technology to minimize unfavorable impacts of both humans on the environment and of the environment on humans. Although technology is often painted as a villain warring with nature, we must recognize that the use of tools is, in fact, an intrinsic characteristic of humanity and that life in a state of nature is " . . . nasty, brutish and short."[1] Technology has controlled the spread of many communicable diseases, expanded agricultural production, and greatly improved the quality and length of our lives. At the same time it has produced air and water pollution, radioactive and hazardous wastes, and perhaps even modified the climate of the world. The task of the environmental engineer is to reduce or eliminate the unfavorable impacts of our activities while enhancing, or at least preserving, their favorable results.

This book deals with the design, construction, and operation of systems for the treatment and supply of potable water and for the collection, treatment, and disposal of wastewater. These constitute an important part of environmental engineering but cannot be considered in a vacuum, since these activities themselves produce wastes, require energy and raw materials, and can be adversely affected by air pollution, hazardous and nuclear waste management, and industrial activity.

1-2 Sources of Environmental Contaminants

Contaminants, to the environmental engineer, can be defined as constituents of the air, water, or soil which render them unsuitable for their intended use. Such agents may be chemical or biological in nature and may result from natural forces, life processes of other species, or our own activities.

Naturally occurring contaminants of water include viruses, bacteria, and higher life forms; dissolved mineral species; soluble organic by-products of life processes; and organic and inorganic suspended solids. These natural contaminants may be increased in concentration and supplemented by other materials produced by industrial or agricultural technology.

Although most natural waters are unsuitable for consumption, the common perception that pollution results from human activity is justified in the sense that many once-safe supplies have been rendered unsafe by our actions.

1-3 Water Supply

Provision of an adequate quantity of water has been a matter of concern since the beginning of civilization. Even in ancient cities, local supplies were often inadequate and aqueducts were built to convey water from distant sources. Such supply systems did not distribute water to individual residences, but rather brought it to a few central locations from which the citizens could carry it to their homes.

Until the middle of the seventeenth century, pipes which could withstand significant pressures were not available. Pipe made of wood, clay, or lead was used, but generally was laid at the hydraulic grade line. The development of cast iron pipe and the gradual reduction in its cost, together with the development of improved pumps driven by steam, made it possible for even small communities to provide public supplies and deliver the water to individual residences.

The provision of an adequate quantity of water responded to only a part of the need since, as noted above, most natural waters are not suitable for consumption. Additionally, as cities grew, their wastes contaminated their own or other supplies. Treatment methods were thus required in order to protect the health of the consumers.

Coagulants and filtration have been used in water treatment since at least 2000 B.C., although their application in municipal treatment in the United States was not common until about 1900. Figure 1-1 illustrates the effect of various treatment techniques on the incidence of typhoid fever in Philadelphia. The city's supply was untreated until 1906, when slow sand filters were placed in service. An immediate reduction in the number of cases occurred, and further improvement was observed when disinfection with chlorine was introduced in 1913. The discovery of the existence of carriers of the disease and their control after 1920 contributed to a still greater decrease in the number of cases.

Outbreaks of waterborne disease still occur in the United States and other countries with generally modern treatment systems. The average number of such incidents in the United States in the period 1976 to 1980 was 38 per year.[2] Most of these outbreaks were associated with obvious deficiencies in treatment or distribution systems.

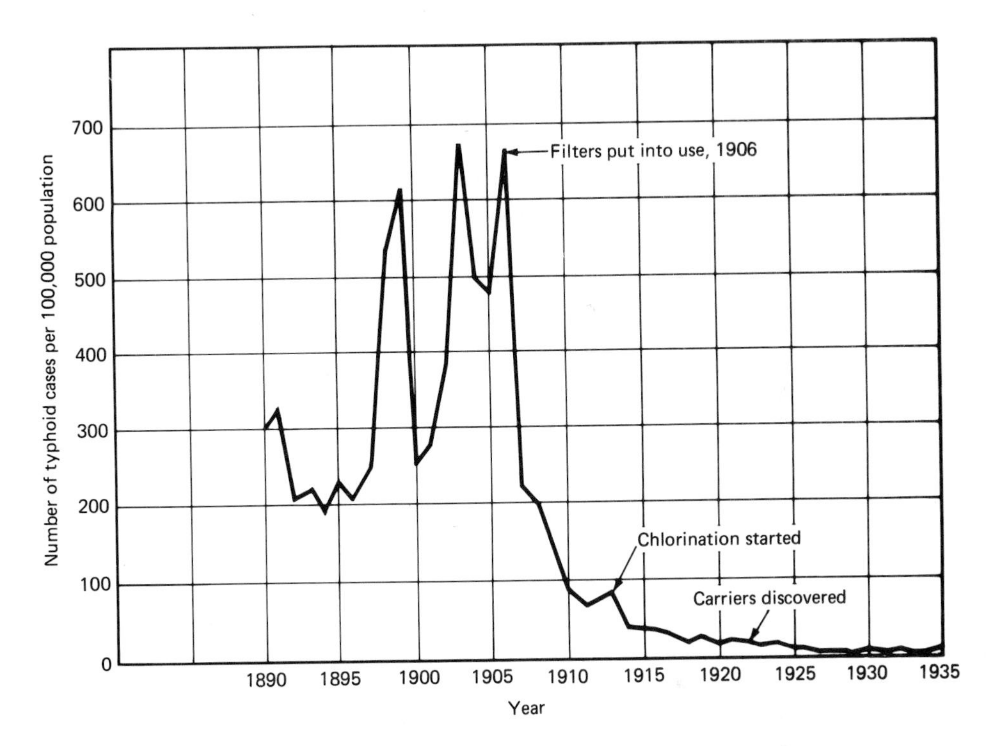

FIGURE 1-1
Typhoid fever cases per 100,000 population from 1890 to 1935 in Philadelphia.

1-4 Sewerage

Remains of sanitary sewers have been found in the ruins of the ancient cities of Crete and Assyria. The sewers of Rome were primarily intended to carry storm water, but, since refuse was deposited in the streets, these sewers often carried organic material as well. Drainage of medieval cities was generally effected by flow over the surface of the streets, which, as in Rome, were the common site for disposal of offal of all sorts, as well as animal and human excrement. The emptying of chamber pots from upper story windows with the warning "Gare l'eau" presented a hazard to the clothes of the passers by as well as to their health. When storm sewers were built, it was not unusual to prohibit their use for household wastes. However, as adequate water supplies were provided and flush toilets began to be utilized, it became evident that prompt removal of waterborne wastes was an important factor in public health. For this reason the storm sewers were also used to carry sanitary sewage.

Provision of sewers did not provide a complete solution to the health and environmental problems associated with high population density. The contaminated flows were generally discharged to the nearest surface water, where their subsequent decomposition often created nuisance conditions and where they became a source of disease to downstream users. The impact on downstream users of the water resource has been the impetus for development and application of waste treatment techniques.

It is possible to treat wastewater to any degree that may be desired—to make it suitable for any purpose whatsoever. The degree of treatment that should be provided has often been a controversial issue, since it is the waste producer who must pay for the treatment but it is the downstream users who receive the benefits. In the United States this issue has been resolved by establishing stream standards. These standards are based on an analysis of stream uses, including waste disposal, and selection of a best possible use. This best use then establishes the required quality of the water—highest for shellfish harvesting and water-contact sports, somewhat lower for public water supply, and lower still for such uses as transportation, fishing, and non-water-contact sports. Each waste discharge must then be treated to the degree necessary to prevent degradation of the stream quality below the established level.

Present regulations in the United States establish which bodies of water are *quality-limited* and which are *effluent-limited*. Those waters which are of a quality suitable for their highest intended use are defined as *effluent-limited*. Wastes discharged to such waters must, in general, be treated to the degree obtained in secondary systems (Chap. 22). Waters which are not suitable for their highest intended use under such effluent limitations are governed by water quality and are said to be *quality-limited*. Such streams are analyzed in order to determine the total pollutional load which can be assimilated without degradation. This allowable waste load is allocated to present and future discharges. Treatment at each discharge point is then tailored to meet this *waste load allocation*. Waste discharges are regulated by the National Pollutional Discharge Elimination System (NPDES), under which permits are issued to each source of wastewater. These permits specify the required waste quality and allowable flow and require the submission of regular reports as well as immediate notification in the event of significant permit violations.

1-5 Interrelationship of Environmental Problems

From the discussion above, it is evident that the disposal of wastewater from one community may have an impact upon the degree of water treatment required at another community located downstream. The situation is in fact still more complicated since, in the treatment of either water or wastewater, certain sludges are produced. Unless these are carefully managed, they may contaminate either ground or surface waters. In general, the removal of a contaminant from one region of the environment will result in its being put somewhere else—where its effects may be as undesirable as in the original location.

The thoughtless disposal of solid and hazardous wastes in the past has created a large number of long-term sources of environmental contamination which will continue to cause problems for many years. Incineration of solid wastes can cause air pollution and subsequent contamination of soil and water.

Cleaning the exhaust gases of an incinerator produces liquid and solid wastes which must, in turn, be handled somehow. Environmental engineers must thus consider very carefully all the effects of their actions so that pollutants are not simply hidden or transferred from one place to another.

REFERENCES

1. Thomas Hobbes, *Leviathan*, 1651.
2. Edwin C. Lippy and Stephen C. Waltrip, "Waterborne Disease Outbreaks—1946–1980: A Thirty-five-year Perspective," *Journal of American Water Works Association*, **76**:2:60, 1984.

CHAPTER 2

QUANTITY OF WATER AND SEWAGE

WATER

2-1 Relation of Quantity and Population

It is self-evident that a large population will use more water than a small one and that water use must be, in some measure, related to population. While this is certainly true, and while water consumption estimates have been historically based on population projections, such techniques are not always satisfactory.[1] As is noted below, water consumption is also influenced by factors such as climate, economic level, population density, degree of industrialization, cost, pressure, and quality of the supply. A number of multivariate projection techniques have been developed which relate water use to one or more of these factors in addition to population.[2,3] When such methods can be shown to be applicable to a particular community, they should be used in preference to the procedures presented in this chapter.

An analysis of the future demand of a particular community should always begin by considering present use. To the extent possible, consumption should be broken down by classes of users (domestic, commercial, industrial, public), area of the city, economic level of the users, season of the year, etc. The rather common procedure of dividing total use by total population to derive a per capita consumption should be applied only with great care, since (1) the entire population may not be served by the municipal system, (2) there may be large industrial users which will not change with population, and (3) the characteristics as well as the size of the population may be changing.

2-2 Population Estimation

Since population is always a relevant factor in estimating future water use, it is necessary to predict, in some manner, what the future population will be. The date in the future for which the projection is made depends on the particular component of the system which is being designed. Elements of the system which are relatively easy to expand tend to have shorter design lives, hence population projection periods may range from as little as 5 to as many as 50 years (Art. 2-7).

Population data may be obtained from the records of the U.S. Bureau of the Census, which conducts the census every 10 years and publishes reports of its findings. Estimating population in years between censuses (1992, for example) is sometimes done by relating the population in the census year to some readily available datum—such as the number of telephone or electrical connections or number of school children—and assuming that the same ratio will be maintained in the interim year.

Estimating population in the future is another matter. It is certain that our estimate will be wrong in some degree and we can only try to be as reasonable as possible in selecting an appropriate technique. Thorough knowledge of the community and external factors which may affect its growth are very important in such analyses.

ARITHMETIC METHOD. The assumption in this method is that the rate of growth is constant. The validity of the method may be tested by examining the growth of the community to determine if approximately equal incremental increases have occurred between recent censuses. Mathematically the hypothesis may be expressed as

$$\frac{dP}{dt} = K \tag{2-1}$$

in which dP/dt is the rate of change of population and K is a constant. K is determined graphically or from populations in successive censuses as

$$K = \frac{\Delta P}{\Delta t} \tag{2-2}$$

The population in the future is then estimated from

$$P_t = P_0 + Kt \tag{2-3}$$

where P_t is the population at some time in the future, P_0 is the present population, and t is the period of the projection.

UNIFORM PERCENTAGE METHOD. The hypothesis of geometric or uniform percentage growth assumes a rate of increase which is proportional to population:

$$\frac{dP}{dt} = K'P \tag{2-4}$$

Integration of this equation yields

$$\ln P = \ln P_0 + K' \Delta t \tag{2-5}$$

This hypothesis is best tested by plotting recorded population growth on semilog paper. If a straight line can be fitted to the data, the value of K' can be determined from its slope. Computerized least squares techniques may be used to fit the line, but a graphical presentation is valuable in that it permits evaluation of how well the data fit the assumed function.

CURVILINEAR METHOD. This technique involves the graphical projection of the past population growth curve, continuing whatever trends the historical data indicate. A commonly used variant of this method includes comparison of the projected growth to the recorded growth of other cities of larger size. The cities chosen for the comparison should be as similar as possible to the city being studied. Geographical proximity, likeness of economic base, access to similar transportation systems, and other such factors should be considered. As an example, in Fig. 2-1, city A, the city being studied, is plotted up to 1990, the year in which its population was 51,000. City B reached 51,000 in 1950, and its growth is plotted from 1950 to 1990. Similarly, curves are drawn for cities C,

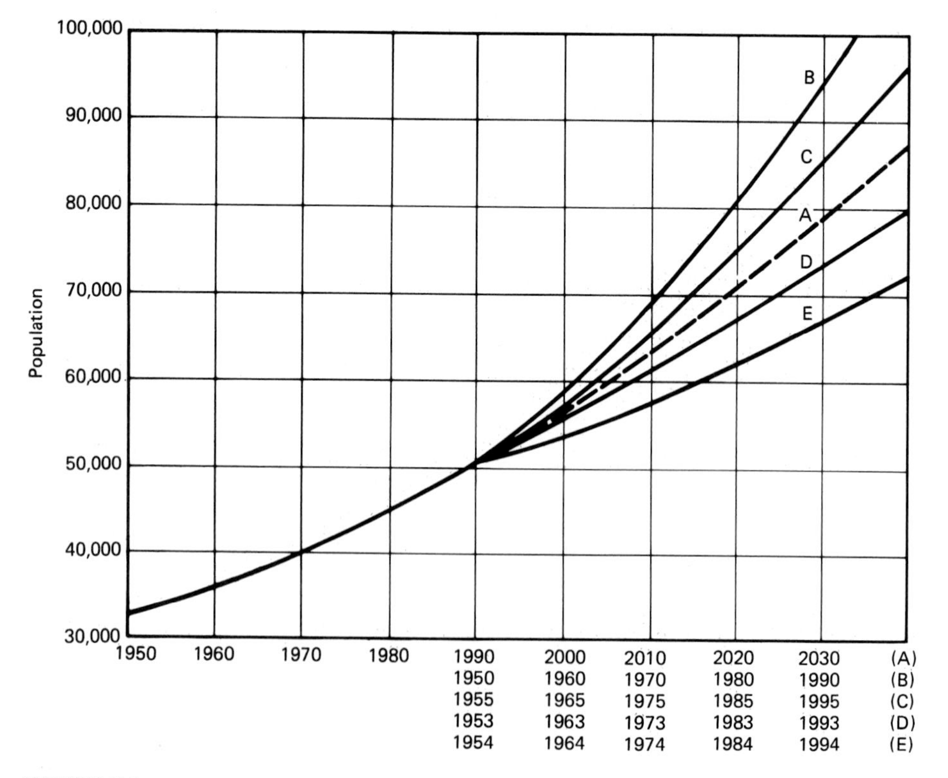

FIGURE 2-1
Curvilinear method of population prediction. The dashed line is the forecast for city A. Scales A, B, C, and D refer to the corresponding cities.

D, and E from the years in which they reached A's 1990 population. A's growth curve is then projected by considering the recorded growth of the comparison cities.

LOGISTIC METHOD. The logistic curve used in modeling population growth has an S shape—combining a geometric rate of growth at low population with a declining growth rate as the city approaches some limiting population. The hypothesis of logistic growth may be tested by plotting the census data on logistic paper—on which it will appear as a straight line if the hypothesis is valid. In the short term a logistic projection can be based on the equation

$$P = \frac{P_{sat}}{1 + e^{a + b\,\Delta t}} \tag{2-6}$$

in which P_{sat} is the saturation population of the community and a and b are constants. P_{sat}, a, and b may be determined from three successive census populations and the equations

$$P_{sat} = \frac{2P_0P_1P_2 - P_1^2(P_0 + P_2)}{P_0P_2 - P_1^2} \tag{2-7}$$

$$a = \ln \frac{P_{sat} - P_2}{P_2} \tag{2-8}$$

$$b = \frac{1}{n} \ln \frac{P_0(P_{sat} - P_1)}{P_1(P_{sat} - P_0)} \tag{2-9}$$

where n is the time interval between succeeding censuses. Substitution of these values in Eq. (2-6) permits the estimation of population for any period Δt beyond the base year corresponding to P_0.

DECLINING GROWTH METHOD. This technique, like the logistic method, assumes that the city has some limiting saturation population, and that its rate of growth is a function of its population deficit:

$$\frac{dP}{dt} = K''(P_{sat} - P) \tag{2-10}$$

After estimation of the saturation population according to some rational basis such as land available and existing population density, K'' may be determined from successive censuses and the equation

$$K'' = -\frac{1}{n} \ln \frac{P_{sat} - P}{P_{sat} - P_0} \tag{2-11}$$

where P and P_0 are populations recorded n years apart. Future population can then be estimated from this value and

$$P = P_0 + (P_{sat} - P_0)(1 - e^{K''\,\Delta t}) \tag{2-12}$$

RATIO METHOD. The ratio method of forecasting relies on the population projections made by professional demographers for the state or the nation. The method is based on the assumption that the ratio of the population of the city being studied to that of the larger group will continue to change in the future in the same manner that has occurred in the past. The ratio is calculated for a series of censuses, the trend line is projected into the future, and the projected ratio is multiplied by the forecast regional population to obtain the city's population in the year of interest.

Use of good judgment in population estimation is important since, if the estimate is too low, the system will soon be inadequate and redesign, reconstruction, and refinancing will be necessary. On the other hand, overestimation of population results in excess capacity which must be financed by a smaller population at a higher unit cost and which may never be used as a result of deterioration or technological obsolescence.

The selection of an appropriate technique is not always easy, and many engineers will test all methods against the recorded growth and eliminate those methods which are clearly inapplicable. The growth of a community with limited land area for future expansion might be modeled by the declining growth or logistic technique, while another, with large resources of land, power, and water and good transportation might be best predicted by the geometric or uniform percentage growth model. In nearly all cases, comparison is made to the recorded growth patterns of similar cities.

2-3 Water Use for Different Purposes

Municipal water demand is commonly classified according to the nature of the user. The ordinary classifications are:

Domestic. Water furnished to houses, hotels, etc. for sanitary, culinary, and other purposes. Use varies with the economic level of the consumers, the range being 75 to 380 L (20 to 100 gal) per capita per day. These figures include water used for air conditioning and watering of lawns and gardens—practices which may have a substantial effect upon total use in some parts of the country. Domestic consumption is typically about 50 percent of the total, but represents a larger fraction where the total consumption is small.

Commercial and industrial. Water furnished to industrial and commercial establishments such as factories, offices, and stores. The importance of this varies locally, depending on whether there are large industries and whether the industries obtain their water from the municipal system. Self-supplied industrial water nationwide is estimated to be more than 200 percent of municipal demand.

The quantity of water required for commercial and industrial purposes has been related to such factors as units produced, number of persons employed, or floor area of the establishment. Such factors, if they are used, should either be derived locally or be checked by comparison with recorded consumption. In cities of over 25,000 persons, commercial consumption is about 15 percent of the total.

Public use. Water furnished to public buildings and used for public services. This includes water for city buildings, schools, flushing streets, and fire protection, for which the municipal supplier usually is not paid. Water used for such purposes amounts to 50 to 75 L per capita per day. Fire supply (Art. 2-6) does not affect the average consumption greatly, but has a major effect upon peak rates (Art. 2-5).

Loss and waste. Water which is "unaccounted for" in the sense that it is not assigned to a specific user. Unaccounted-for water is attributable to errors in meter readings, unauthorized connections, and leaks in the distribution system. Loss and waste can be reduced significantly by careful maintenance of the system and a regular program of meter recalibration and replacement.[4,5]

The total consumption is the sum of the individual elements listed above. In the United States in 1980, the total consumption on a per capita basis averaged 535 L (142 gal) per day for privately owned utilities and 640 L (171 gal) per day for publicly owned supplies.[6] Consumption in the year 2000 has been estimated to be distributed as shown in Table 2-1.[7] Such average figures, while they give a general idea of the probable flow required, should not be used for design since actual values vary widely. Table 2-2 presents reported figures from a number of cities in the United States. It is clear that each city must be studied carefully, particularly with regard to industrial and commercial use and loss and waste. Reported per capita use may be based on the number of persons actually served or on the census population of the community. This factor can have a substantial effect on reported values.

2-4 Factors Affecting Water Use

The average daily per capita water consumption in American cities varies from 130 to 2000 L (35 to 530 gal). Local use depends upon such factors as the size of the community, presence of industries, quality of the water, its cost, its pres-

TABLE 2-1
Projected consumption of water for various purposes in the year 2000

Use	Liters per capita/day*	Percentage of total
Domestic	300	44
Industrial	160	24
Commercial	100	15
Public	60	9
Loss and waste	50	8
Total	670	100

* Gallons = liters × 0.264

TABLE 2-2
Recorded rates of water consumption in some American cities

City	Average daily per capita consumption, L*	Maximum 1-day consumption in a 3-year period, L	Maximum in proportion to average, %
Rochester, N.Y.	451	637	141
Syracuse, N.Y.	728	917	126
Hartford, Conn.	671	887	132
Albany, N.Y.	671	860	128
El Paso, Tex.	447	739	165
Portland, Me.	572	773	135
Camden, N.J.	641	963	150
Albuquerque, N.M.	402	766	190
Winston-Salem, N.C.	447	580	130
Waterloo, Iowa	383	625	163
Passaic, N.J.	807	1016	126
South Gate, Calif.	550	891	162
Fort Smith, Ark.	474	652	138
Poughkeepsie, N.Y.	569	728	128
Tyler, Tex.	371	743	200
Monroe, La.	584	875	150
Spartanburg, S.C.	754	955	126
Pomona, Calif.	629	1092	173
St. Cloud, Minn.	277	701	254
Salina, Kan.	603	1357	225
Alliance, Ohio	796	1114	140
Ashtabula, Ohio	766	985	129

* Gallons = liters × 0.264

sure, the climate, characteristics of the population, whether supplies are metered, and the efficiency with which the system is maintained. The more important of these factors are treated separately below.

Size of the city has an indirect effect in that small communities tend to have more limited uses for water. On the other hand, the presence of an important water-using industry may result in high per capita use. Small communities are more likely to contain areas which are inadequately served by both water and sewerage systems. In unsewered homes, water consumption seldom exceeds 40 L (10 gal) per capita per day. Extension of sewers may thus produce increased use of water.

Industry and commerce have a pronounced effect upon total consumption. Since industrial use has no direct relation to population, great care must be taken in estimating present or future water use in a restricted portion of a city. One must study the existing industries of the area, their actual use of water, and assess the probability of establishment of more industrial facilities.

Industries frequently use auxiliary water supplies for some purposes—a factor which reduces their consumption of the municipal supply.

Commercial consumption is largely dependent on the number of people employed in business districts and cannot be estimated on the basis of the number of residents. Sanitary use in business facilities amounts to about 55 L (15 gal) per person per 8-h day. Estimates are sometimes made upon the basis of floor area [10 to 15 L/(m^2 · day)] or ground area [up to 95 L/(m^2 · day) in highly developed districts].

Water is sometimes used for air conditioning, either directly, if it is sufficiently cool, or as a heat sink in conjunction with a mechanical system. Such use of water is generally discouraged, but in cities in which these systems are permitted water use may be substantially higher than average.

Characteristics of the population, particularly economic level, can produce substantial variations from the average use of about 300 L (80 gal) per capita per day. In high-value districts of a city or in suburban communities with a similar population, per capita use will be high—perhaps as much as 380 L (100 gal) per day for domestic use alone. Watering lawns will increase this figure still more. In slum districts consumption rates of 100 L (25 gal) per capita per day are typical, although rates as low as 50 L (13 gal) per capita per day have been reported. Still lower use is found in low-value districts where sewerage is not provided and water supplies are inadequate.

Metering of water supplied to individual users has been shown to reduce consumption substantially—perhaps by as much as 50 percent.[8] In the absence of meters, users have no incentive to conserve water and waste is much more common. Metering is also desirable in that it permits analysis of use patterns of different classes of users, thereby providing data which is useful in planning expansion of facilities and in assessing the magnitude of loss due to leaks in the distribution system.

Miscellaneous factors include climate, quality, pressure, system maintenance, and conservation programs. During hot, dry weather, water consumption will increase as a result of lawn watering and more bathing. During cold weather, particularly in the south, where pipes are often exposed, consumption is increased by the practice of letting taps run to prevent freezing.

Water which is of poor quality (colored, odorous, or otherwise objectionable) will be used less than water which is satisfactory to consumers. In this connection it should be recognized that improvements in the quality of a public supply are likely to result in increased consumption.

High pressure in the system also results in greater use; in addition, it increases losses from leaks. Reducing water pressure can reduce per capita use by up to 6 percent.[8]

A well-designed program of maintenance will reduce loss and waste in the system. Leak surveys (Art. 7-9) aid in detecting both leaks in the pipes and the presence of unauthorized connections.

Conservation programs may be either short-term (during a period of drought) or permanent. Such programs may limit or prohibit lawn watering; encourage the use of drought-resistant yard plantings; require or encourage the use of flow-restricting showers, water-saving toilets, and similar devices; and apply rate schedules in which the unit cost to the consumer increases with total use. In some communities, flow-restricting devices may be installed at the meters of those individuals who refuse to voluntarily reduce their consumption.

2-5 Variations in Water Use

The values given above are average annual consumption rates which, while useful, are not sufficient for the design of water systems. Water consumption varies during the day (being lower at night), from day to day during the week, from week to week during the month, and from month to month during the year.

Pumping records, that is, the flows measured at the pumping station or water source, are extremely important in evaluating variations in demand. As may be noted from Table 2-2, there is no clearly defined relationship between average and peak flow which is applicable in all communities. For this reason each community should be carefully studied to determine variations in rate with time and location.

In the absence of data it may be necessary to estimate the maximum rates. The maximum daily consumption is likely to be 180 percent of the annual average and may reach 200 percent. The Goodrich formula is sometimes used for estimating consumption:

$$p = 180t^{-0.10} \tag{2-13}$$

In this equation p is the percentage of the annual average rate which occurs during shorter periods and t is the length of the period in days from $\frac{1}{12}$ to 360. The formula predicts the maximum daily rate to be 180 percent of the annual average rate, the weekly maximum to be 148 percent of the annual average, and the monthly maximum to be 128 percent of the annual average.

The maximum hourly rate is likely to be about 150 percent of the average for that day. The maximum hourly rate for a community having an average annual rate of 670 L per capita per day could be estimated as being 670 × 1.80 × 1.50, or 1800 L per capita per day—to which must be added the fire demand (Art. 2-6). The values of the Goodrich formula are based on data from residential communities of moderate size. Larger cities will have smaller ratios of peak to average flow. Minimum rates are also important, particularly in the design of pumping stations. The minimum rate depends on leakage, night industrial use, and the portion of the peak demand which is provided from storage. Typical minima range from 25 to 50 percent of the daily average.

The design of the distribution system is affected by peak consumption in specific areas. Residential areas have particularly high ratios of peak to average flow because of lawn watering, air conditioning, and major water-using appliances such as washing machines and dishwashers. Hourly peaks as high as 1000 percent of the annual average have been recorded in suburban areas and peaks of 300 to 400 percent are not uncommon in residential areas of large cities. The ratio of peak to average flow increases with decreasing population density.

2-6 Fire Demand

Although the actual amount of water used for fire fighting in a year is small, the rate of use is high. The Insurance Services Office[9] calculates required fire flow from the formula

$$F = 18C(A)^{0.5} \tag{2-14}$$

in which F is the required flow in gal/min [(L/min) ÷ 3.78], C is a coefficient related to the type of construction, and A is the total floor area in ft^2 ($m^2 \times 10.76$) excluding the basement.

A variety of special factors which can affect the required fire flow are presented in Ref. 9. In general, C is 1.5 for wood frame construction, 1.0 for ordinary construction, 0.8 for noncombustible construction, and 0.6 for fire-resistive construction. The fire flow calculated from the formula is not to exceed 8000 gal/min (32,400 L/min) in general, nor 6000 gal/min (22,680 L/min) for one-story construction. The minimum fire flow is not to be less than 500 gal/min (1890 L/min). Additional flow may be required to protect nearby buildings. The total for all purposes for a single fire is not to exceed 12,000 gal/min (45,360 L/min) nor be less than 500 gal/min (1890 L/min).

Table 2-3 may be used to determine the required fire flow for groups of single- and two-family residences. The fire flow must be maintained for a minimum of 4 h as shown in Table 2-4. Most communities will require a duration of 10 h.

In order to determine the maximum water demand during a fire, the fire flow must be added to the maximum daily consumption rate. If it is assumed that a community with a population of 22,000 has an average consumption of 600 L per capita per day and a fire flow dictated by a building of ordinary

TABLE 2-3
Residential fire flows

Distance between adjacent units		Required fire flow	
ft	m	gal/min	L/min
> 100	> 30.5	500	1890
31–100	9.5–30.5	750–1000	2835–3780
11–30	3.4–9.2	1000–1500	3780–5670
≤ 10	≤ 3.0	1500–2000*	5670–7560*

* For continuous construction use 2500 gal/min (9450 L/min).

TABLE 2-4
Fire flow duration

Required fire flow		
gal/min	L/min	Duration, h
< 1000	< 3780	4
1000–1250	3780–4725	5
1250–1500	4725–5670	6
1500–1750	5670–6615	7
1750–2000	6615–7560	8
2000–2250	7560–8505	9
> 2250	> 8505	10

construction with a floor area of 1000 m^2 and a height of six stories, the calculation is as follows:

$$\text{Average domestic demand} = 22{,}000 \times 600 = 13.2 \times 10^6 \text{ L/day}$$

$$\text{Maximum daily demand} = 1.8 \times \text{average} = 23.76 \times 10^6 \text{ L/day}$$

$$F = 18(1)(1000 \times 10.76 \times 6)^{0.5} = 4574 \text{ gal/min } (24.89 \times 10^6 \text{ L/day})$$

$$\text{Maximum rate} = 23.76 \times 10^6 + 24.89 \times 10^6 = 48.65 \times 10^6 \text{ L/day}$$

From Table 2-4, the fire flow must be maintained for 10 h, hence the total flow required during this day would be

$$23.76 + 24.89 \times (10/24) = 34.13 \times 10^6 \text{ L}$$

This represents an average per capita rate of 1551 L/day. The difference between the maximum daily domestic rate and the higher values above is frequently provided from elevated storage tanks which are filled during periods of lower demand. Article 7-3 discusses this matter further.

2-7 Design Periods for Water Supply Components

The economic design period of the components of a water supply system depends on their life, first cost, the ease with which they can be expanded, and the likelihood that they will be rendered obsolete by technological advances. In order to design the parts of a water system, the flow at the end of the design period must be estimated. "Conservatism," that is, overestimation of the design flow, must be avoided since this can burden a relatively small community with the cost of extravagant works designed for a far larger population. The different elements of the treatment and distribution systems may appropriately be designed for different periods and their design may be based upon different flow criteria.

Development of source will be based upon a design period which depends on the nature of the source. Groundwater supplies are typically easily expanded by construction of additional wells and design periods may be as short as 5 years. On the other hand, surface supplies which require the construction of impoundments to meet demand during periods of low river flow are designed for much longer periods, perhaps as much as 50 years. Ordinary river intakes without impoundments are designed for intermediate periods, depending on the difficulty of expansion. Design lives of about 20 years are often used for such structures. The design capacity of the source is normally based upon meeting the maximum daily demand rate expected during the design period, but not necessarily upon a continuous basis.

Pipe lines from the source are generally designed for a long life, since the life of pipe is very long and the cost of the material is only a small portion of the cost of construction. A design period of 25 years or more would not be unusual. The design itself is based on provision of economical conveyance at average daily flow at the end of the design period with suitable velocities under all anticipated flow conditions.

Water treatment plant components are commonly designed for a period of 10 to 15 years since expansion is generally simple if it is considered in the original design. Most treatment units will be designed on the basis of average daily flow at the end of the design period, since overloads do not result in major losses of treatment efficiency. Hydraulic design should be based on the maximum anticipated flow through the plant (which is not necessarily the same as the maximum anticipated water use), but must consider velocities under all potential flow conditions.

Pumping plant facilities are generally designed for a period of about 10 years, since modification and expansion are easy if provision for change is made initially. Pump selection and design of the pump control system require knowledge of the maximum flow including fire demand, the average flow, and the minimum flow expected during the period. Total installed pumping capacity will exceed the maximum flow expected to be pumped.

Storage within the distribution system normally consists of elevated steel tanks which are relatively inexpensive and easy to construct. Their life, however, is potentially quite long, hence they are seldom replaced. Design of such structures is closely linked to design of the pumping plant and requires knowledge of average consumption, fire demand, maximum hour, maximum day, maximum week, and maximum month, as well as the capacity of the source and the pipe lines from the source.

Distribution system elements are normally installed below the streets. Their life is very long and their replacement is very expensive, thus the design period is indefinite and the capacity is based on maximum anticipated development of the area served. One must consider anticipated population densities (which may range from 3500 to 250,000 persons/km^2), zoning regulation (which will aid in predicting future population density and industrial demand), and the factors discussed above which can affect per capita flow. Design is based on provision of adequate pressure for fire protection at maximum hourly flow including fire demand.

SEWAGE

2-8 Sources of Sewage

Sewage consists of liquid wastes produced in residences, commercial establishments, and institutions; liquid wastes discharged from industries; and any subsurface, surface, or storm water which enters the sewers. The first of these is commonly called *sanitary* or *domestic sewage*, the second *industrial waste*, while the third includes *infiltration*, *inflow*, and *storm sewage*.

Sewers are often classified according to their use. *Sanitary sewers* carry domestic sewage, industrial waste, and whatever ground, surface, and storm water enters through joints, manhole covers, and defects in the system. *Storm sewers* are designed to carry the surface and storm water passing through or generated in the area which they serve. *Combined sewers* carry all types of sewage in the same conduits.

This chapter deals with the estimation of flows produced by residences, industries, and defects in the system. Estimation of storm water flow is treated in Chap. 13.

2-9 Relation to Water Use

Sanitary sewage and industrial wastes are derived principally from the water supply. Very little water is actually "consumed" in the sense that it is permanently removed from the community's environment. For this reason, estimation of sewage flows should be prefaced by a study of both present water consumption and that expected in the future. The proportion of the water supplied which will reach the sewers depends in large part upon local conditions. Water used in steam generation in industry, air conditioning, and watering lawns and gardens may or may not reach the sewers. On the other hand, industries may have their own sources of water, but may discharge their wastes to the municipal sewers. In individual communities the sanitary sewage flow may vary from 70 to 130 percent of the water consumed. It is fairly common to assume that the average rate of sewage flow is equal to the average rate of water consumption, but this should be done only after careful consideration of the actual nature of the community. Such an estimate would be too high for a residential community in an area with hot, dry summers, and too low for a community containing industries with private water supplies.

2-10 Infiltration and Inflow

Infiltration is the water which enters sewers through poor joints, cracked pipes, and the walls of manholes. Inflow enters through perforated manhole covers, roof drains connected to the sewers, and drains from flooded cellars. Inflow is associated with runoff events (rainfall or snowmelt), while infiltration is drawn from the soil and may occur even in dry weather.

Sewage treatment is an expensive process and the cost of the facilities is closely associated with flow. For this reason, efforts to reduce inflow and infiltration are generally economically justified. Surveys employing flow measurement in the sewers late at night (which gives a reasonable estimate of infiltration), flow measurement during storms, smoke testing to reveal sources of inflow, and surface flooding are often conducted.

The amount of infiltration depends upon the care with which the sewer system is constructed, the height of the groundwater table, and the character of the soil. An expansive soil will tend to pull joints apart and permit more leakage, while granular soils permit easy travel of water to the sewers where it may enter through joints or breaks. Since construction conditions and soil characteristics vary widely, infiltration is difficult to predict without actual flow measurements. Sewer size apparently has little effect since, although large sewers have greater joint length, the workmanship is likely to be better than in small sewers. Infiltration rates in old systems have been measured to be from 35 to 115 $m^3/(km \cdot day)$ [15,000 to 50,000 gal/(mi · day)], with even higher rates being found in localized regions where sewers are below the groundwater table or construction is particularly poor. Specifications for new sewer projects now limit infiltration to 45 L/(km · day) per mm of diameter [500 gal/(mi · day) per in].

Since sewers deteriorate with age, estimates of infiltration, even for new systems, should be reasonably generous. The values given above are those which have been measured for public sewers and do not include infiltration or inflow in the house sewers which connect the buildings to the system. These are often poorly installed, and their construction should be carefully controlled and inspected.

2-11 Fluctuations in Sewage Flow

Wherever possible, gaugings of flow in existing sewers should be made in order to determine actual variations. Recording gauges are available or can be devised that will give the depth in the outfall sewer or in the main leading from a section of the community. To design a system for a previously unsewered town or section of a city, an estimate must be made of the fluctuations to be expected in the flow.

Sewage flow, like water consumption, will vary with the time of day, day of the week, season of the year, and weather conditions. The variations from the mean are less than those observed in water supply because the sewers do not flow full and thus provide a degree of equalization. Data have been collected in many communities which permit estimation of the ratio of peak and minimum flows to the average as a function of either average flow or population.[10] Figures 2-2 and 2-3 present compilations of such data and may be useful in communities where flow measurements are not obtainable.

2-12 Design Periods for Sewerage System Components

As in water system design, the engineer must select appropriate periods of design and determine which flow rate is to be used for different components of the sewerage system.

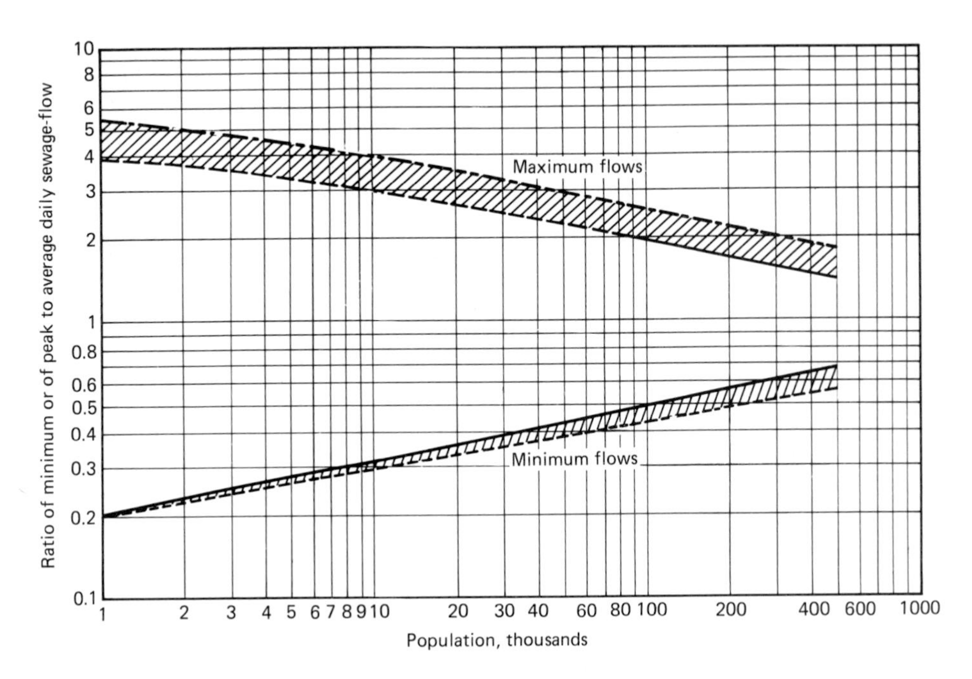

FIGURE 2-2
Ratio of extreme flows to average daily flow in various areas of the United States. (*After Ref. 10.*)

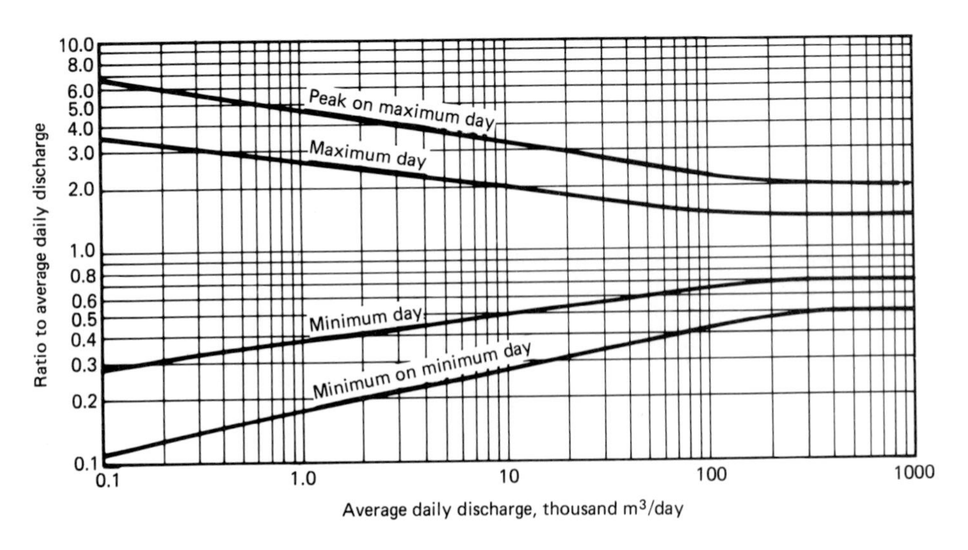

FIGURE 2-3
Ratio of extreme flows to average daily flow in New England. (*After Ref. 10.*)

Sewers are designed for an indefinite period since, like the water distribution system, they are long-lived and expensive to replace. For this reason they are designed to provide for the ultimate development of the district which they serve. This requires estimation of maximum population and maximum industrial development in the various areas of the community. The design is based upon maximum anticipated flow rate with the sewers flowing partially full.

Sewage pumping may be required at the treatment plant and at intermediate locations within the collection system. Pumping facilities are reasonably easy to expand and are relatively short-lived, hence the design period seldom exceeds 10 years. The rates of flow which are required are the average, maximum, and minimum expected during the design period.

Sewage treatment facilities are generally designed to be adequate for a period of 15 to 20 years. Design of individual components may be based on either average or peak flow rate expected. Hydraulic design requires an estimate of minimum flow as well, since minimum velocities must be maintained in some portions of the plant.

PROBLEMS

2-1. Using the methods of Art. 2-2, estimate the population of a nearby city 5, 10, 15, and 20 years in the future. Explain why certain of the techniques are less likely to be correct for this particular city. Select a single value of population which you would use for design purposes in each year.

2-2. For the community of Prob. 2-1 determine the design flow which you would use for:

(*a*) A groundwater source
(*b*) Pipe lines from source to community
(*c*) Water treatment plant
(*d*) Pumping plant

2-3. A community has experienced the growth in population and water use shown. Estimate the population, per capita water use, and total average daily water demand in the year 2010.

Year	1950	1960	1970	1980	1990
Population	8000	8990	11,300	14,600	18,400
Average daily flow, m^3	2270	2720	3630	4970	6600

2-4. Estimate the maximum daily rate, maximum hourly rate, and minimum hourly rate of water use for the community of Prob. 2-3 in the year 2010. What volume of water would be used in the maximum week? In the maximum month?

2-5. A city with a present population of 58,000 persons used a total of 9,526,500 m^3 of water during the last 12 months. On the maximum day during that period 42,000,000 L were used. Estimate the average and maximum daily flows to be expected in 10 years, when the population is estimated to be 72,500.

2-6. A community's population is estimated to be 35,000 20 years from now. The present population is 28,000 and the present average water consumption is 16,000

m^3/day. The existing water treatment plant has a design capacity of 19,000 m^3/day. Assuming an arithmetic rate of population growth, determine in what year the plant will reach design capacity.

2-7. Determine the fire flow required for a three-story building with a plan area of 700 m^2 and wood frame construction. If this building dictates the fire flow for the community of Prob. 2-3, what is the maximum daily demand plus fire flow? What percentage of the *average* demand does the fire flow constitute?

2-8. Determine the fire flow required for an area which consists of single family residences from 1.5 to 3 m (5 to 10 ft) apart. For how long must this flow be provided? What total volume of water must be available for fire protection?

2-9. Considering the community of Prob. 2-3, estimate the average wastewater flow in the year 2010 assuming that the average sewage flow is equal to the average water consumption. Estimate the maximum hourly wastewater flow and compare this to the maximum hourly water consumption. Explain the difference between these values.

2-10. A residential community has an estimated ultimate population density of 15,000 per km^2 and an area of 120,000 m^2. The average wastewater flow is presently 300 L per capita per day. Estimate the maximum sewage flow rate to be expected from this area.

2-11. From the following table of recorded average monthly flows at a community's water and sewage treatment plants estimate the percentage of sewage flow contributed by infiltration and inflow.

	Flow in 10^5 m^3/month	
Month	**Total water pumped**	**Total sewage treated**
January	6.3	6.6
February	6.0	6.3
March	6.4	7.2
April	6.2	9.2
May	6.8	9.3
June	7.0	8.7
July	8.6	7.2
August	8.9	7.5
September	7.3	7.3
October	6.3	7.6
November	6.0	7.7
December	6.2	6.5

REFERENCES

1. John J. Boland et al., "A Research Agenda for Municipal Water Demand Forecasting," *Journal of American Water Works Association*, **75**:1:20, 1983.
2. T. R. Burke, "A Municipal Water Demand Model for the Coterminous United States," *Water Resources Bulletin*, **6**:4, 1970.
3. M. B. Sonnen and D. E. Evenson, "Demand Projection Considering Conservation," *Water Resources Bulletin*, **15**:2, 1979.

4. John E. Pilzer, "Leak Detection—Case Histories," *Journal of American Water Works Association*, **73**:565, 1981.
5. Ellen E. Moyer et al., "The Economics of Leak Detection and Repair: A Case Study," *Journal of American Water Works Association*, **75**:1:28, 1983.
6. Harris F. Seidel, "Water Utility Operating Data: An Analysis," *Journal of American Water Works Association*, **77**:5:50, 1985.
7. "The Nation's Water Resources," United States Water Resources Council, 1968.
8. William O. Maddaus, "The Effectiveness of Residential Water Conservation Measures," *Journal of American Water Works Association*, **79**:3:68, 1987.
9. Kenneth J. Carl et al., "Guidelines for Determining Fire-Flow Requirements" *Journal of American Water Works Association*, **65**:335, 1973.
10. "Design and Construction of Sanitary and Storm Sewers," American Society of Civil Engineers, 1969.

CHAPTER 3

HYDRAULICS

3-1 Scope

The usual undergraduate engineering courses in fluid mechanics cover both theoretical and practical matters dealing with hydrostatics and hydrodynamics.[1,2,3] No attempt will be made to repeat that subject matter in this chapter. Rather, the emphasis will be on application of the fundamental principles to problems of particular interest to engineers involved in water and wastewater conveyance and treatment. Basic equations are presented without derivation. The reader is directed to the references for more detailed information.

3-2 Flow in Pipes

The fundamental equation of flow in pipes is Bernoulli's equation, which is written between two points as

$$\frac{P_1}{\gamma} + \alpha_1 \frac{V_1^2}{2g} + z_1 = \frac{P_2}{\gamma} + \alpha_2 \frac{V_2^2}{2g} + z_2 + \mathrm{HL}_{1-2} \tag{3-1}$$

The coefficients on the velocity head terms arise from the integration of Euler's differential equation over the cross-sectional area of flow and reflect the nonuniform velocity distribution which exists in the flow of real fluids. α will never be equal to 1 and lies between 1 and 2 for all flow conditions in circular pipes. In general, as the velocity increases, α decreases; thus, as the velocity becomes more significant, its departure from ideality decreases. Additionally, one may observe that the velocity head terms on either side of the equation will tend to cancel one another and that velocity head in most (but not all) pipeline problems is negligible with respect to other terms. For these reasons, α is commonly taken as being equal to 1.

The terms grouped under HL_{1-2} include energy losses due to pipe friction and momentum changes in fittings. These are, in part, a function of fluid properties, pipe characteristics, and the flow regime which exists in the pipe.

3-3 Flow Regimes

Reynolds, in the latter part of the nineteenth century, identified two different flow regimes which are characterized as *laminar* and *turbulent*. In laminar flow, streamlines remain parallel to one another and no mixing occurs between adjacent *laminae* or layers. In turbulent flow, mixing occurs across the pipe. The distinction between these two regimes lies in the fact that the shear stress in laminar flow results from viscosity, while that in turbulent flow results from momentum exchanges occurring as a result of motion of fluid particles from one layer to another. Reynolds found that the transition from laminar to turbulent flow is related to a dimensionless parameter

$$N_R = \frac{Vd\rho}{\mu} = \frac{Vd}{\nu} \tag{3-2}$$

which is now known as the Reynolds number. In Eq. (3-2), V is the velocity in the pipe, d the pipe diameter, ρ the density of the fluid, μ its viscosity, and ν its kinematic viscosity (μ/ρ). The actual value of N_R at which the transition from laminar to turbulent flow occurs depends on whether velocity is increasing or decreasing and on physical characteristics of the pipe. For practical purposes, however, the upper limit of laminar flow in pipes may be defined as $N_R = 2100$ to 4000.

In most circumstances the flow of water in pipes involves turbulent conditions. The maximum velocities at which laminar flow could exist in pipes of different size conveying water are summarized in Table 3-1. Since velocities of water in pipes are generally in the range of 0.6 m/s (2 ft/s) or greater, it is clear that there will be few occasions in which laminar flow will occur in the conveyance of water. This is not to say, however, that laminar flow conditions never exist in water or wastewater systems, since water is not the only fluid which must be handled. Polymers and concentrated solutions of metallic coagulants may have viscosities ranging from 20 to 1000 times that of water and will exhibit laminar flow at velocities 20 to 1000 times greater than those in Table 3-1.

3-4 Bernoulli's Equation

Bernoulli's equation, as noted in Art. 3-2, is commonly written between two points as

$$\frac{P_1}{\gamma} + \frac{V_1^2}{2g} + z_1 = \frac{P_2}{\gamma} + \frac{V_2^2}{2g} + z_2 + \mathrm{HL}_{1-2} \tag{3-1a}$$

The terms in the equation are commonly called pressure head P/γ, velocity head $V^2/2g$, static head z, and head loss HL_{1-2}. There are no circumstances of practical interest in which head losses are negligible, thus it is necessary to define the magnitude of these losses and their variation before considering any actual applications. The energy (or head) losses result from shear stress along

TABLE 3-1
Maximum velocity of laminar flow of water in circular pipes (N_R = 4000)

Diameter		Maximum velocity of laminar flow	
mm	in	m/s	ft/s
6	$\frac{1}{4}$	0.71	2.34
12	$\frac{1}{2}$	0.36	1.17
25	1	0.18	0.59
50	2	0.09	0.29
150	6	0.03	0.10
300	12	0.01	0.05

the walls of the pipe and within the fluid and from momentum changes at entrances, exits, changes in cross section or direction, and fittings such as valves.

3-5 The Darcy-Weisbach Equation

Darcy, Weisbach, and others proposed, on the basis of experiment, that the energy loss resulting from friction varies as

$$\text{HL} = f\frac{L}{d}\left(\frac{V^2}{2g}\right) \tag{3-3}$$

in which f is a friction factor, L is the pipe length, d its internal diameter, and $V^2/2g$ the velocity head. It can be shown by dimensional analysis that f depends only and uniquely upon the Reynolds number N_R and another dimensionless parameter, e/d, called relative roughness, where e is the height of surface roughness on the wall of the pipe and depends on the pipe material.

Both theoretical and experimental considerations lead to the conclusion that the dependence of f on N_R and e/d is different in laminar and turbulent flow regimes. In fact, in laminar flow, f is dependent only on N_R and may be calculated from

$$f = \frac{64}{N_R} \tag{3-4}$$

which, if substituted in Eq. (3-3) yields, for laminar flow,

$$\text{HL} = \frac{32\mu LV}{\gamma d^2} \tag{3-5}$$

Example 3-1 A polymeric coagulant, undiluted, has an absolute viscosity of 0.48 kg/(m · s) [0.01 lb · s/ft²] and a specific gravity of 1.15. This fluid is to be pumped at a rate of 3.78 L/min (1 gal/min) through 15.25 m (50 ft) of ½-in-diameter schedule 40 pipe. What is the head loss due to friction?

Solution

½-in-diameter schedule 40 pipe has an internal diameter of 0.622 in (15.8 mm).

$$V = \frac{Q}{A} = \frac{3.78 \times 10^{-3}}{60 \times \pi\,(0.0158)^2/4} = 0.321 \text{ m/s } (1.05 \text{ ft/s})$$

$$N_R = \frac{Vd\rho}{\mu} = \frac{(0.321)(0.0158)(1000)(1.15)}{0.48} = 12.1 < 2100$$

Therefore flow is laminar and Eqs. (3-4) and (3-5) apply.

$$\text{HL} = \frac{32\mu LV}{\gamma d^2} = \frac{32(0.48)(15.25)(0.321)}{9806(1.15)(0.0158)^2} = 26.7 \text{ m } (87.6 \text{ ft})$$

or

$$f = \frac{64}{N_R} = \frac{64}{12.1} = 5.29$$

$$\text{HL} = f\frac{L}{d}\frac{V^2}{2g} = 5.29\,\frac{15.25}{0.0158}\,\frac{(0.321)^2}{9.806 \times 2} = 26.7 \text{ m } (87.6 \text{ ft})$$

No matter what flow regime exists within a pipe, a laminar film will exist next to the wall. If this film is substantially thicker than the roughness e of the pipe, the pipe will be hydraulically smooth. The actual film thickness is given by

$$\delta = \frac{32.8d}{N_R f^{0.5}} \tag{3-6}$$

If e/δ is less than 0.25, the pipe may be considered smooth, while if e/δ is greater than 6, the pipe is wholly rough.

Within the turbulent flow regime, as velocity and N_R increase, it is evident that the thickness of the laminar film will decrease and the effect of viscous friction will decrease while roughness will become more important. In the region described as "completely" turbulent, f depends only upon e/d. Nikuradse developed an experimental equation,

$$f^{-0.5} = 1.14 + 2.0 \log\frac{d}{e} \tag{3-7}$$

which reflects this phenomenon. In the region between laminar flow and complete turbulence, both viscous effects (N_R) and roughness (e/d) affect f. The variation of f with these parameters is shown on the Moody diagram[4] of

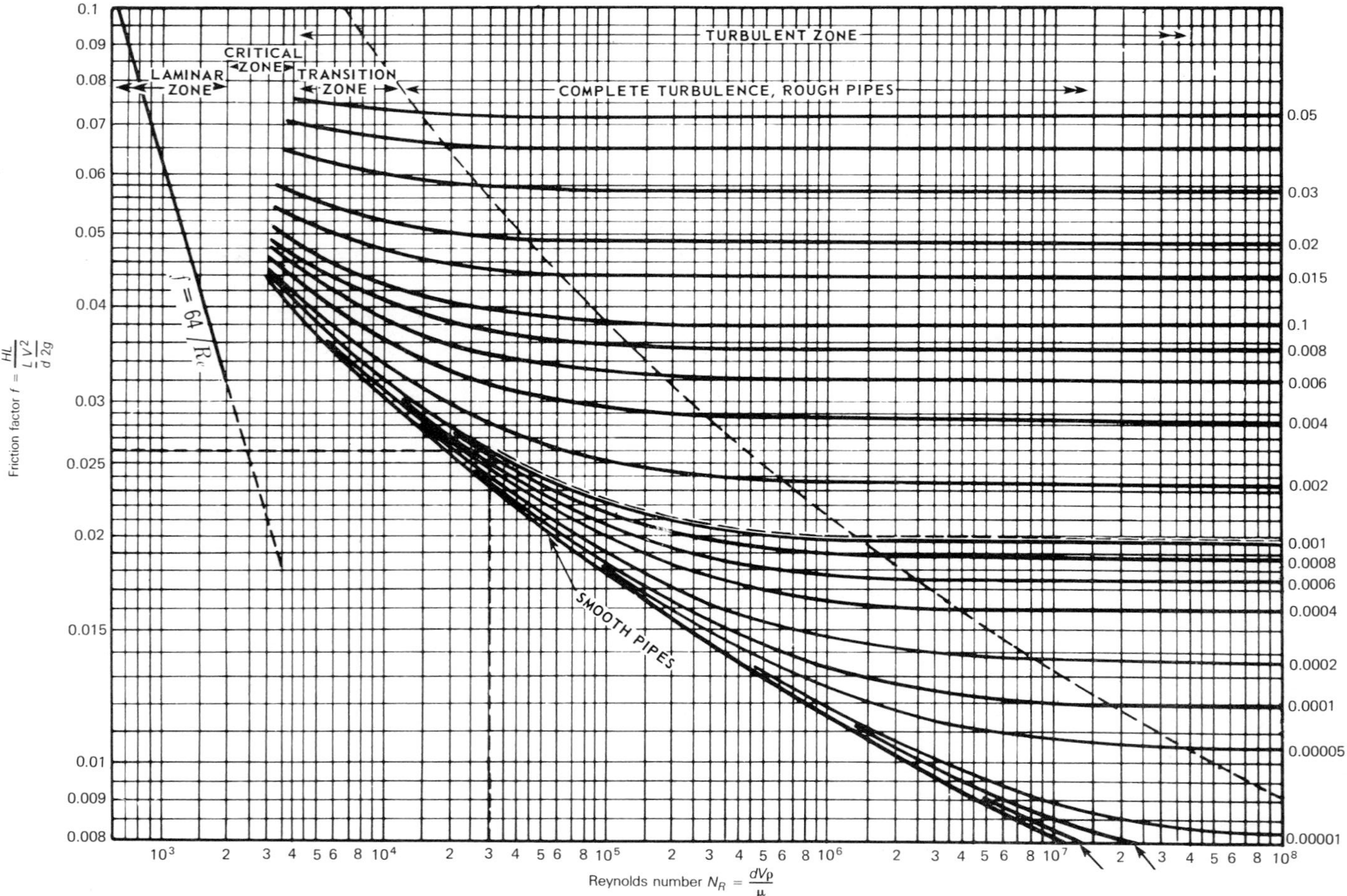
TURBULENT ZONE
CRITICAL ZONE
LAMINAR ZONE
TRANSITION ZONE
COMPLETE TURBULENCE, ROUGH PIPES
$f = 64/Re$
SMOOTH PIPES
0.1
0.09
0.08
0.07
0.06
0.05
0.04
0.03
0.025
0.02
0.015
0.01
0.009
0.008
Friction factor $f = \frac{HL}{\frac{L}{d}\frac{V^2}{2g}}$
0.05
0.03
0.02
0.015
0.1
0.008
0.006
0.004
0.002
0.001
0.0008
0.0006
0.0004
0.0002
0.0001
0.00005
0.00001
10^3 2 3 4 5 6 8 10^4 2 3 4 5 6 8 10^5 2 3 4 5 6 8 10^6 2 3 4 5 6 8 10^7 2 3 4 5 6 8 10^8
Reynolds number $N_R = \frac{dV\rho}{\mu}$

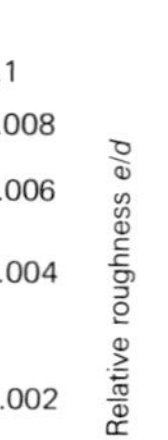
Relative roughness e/d

Fig. 3-1. Figure 3-2 presents the variation of e/d with d for various commercial pipe materials.

Example 3-2 A 24-in class 1 ductile iron pipe 90 m long with a neat cement lining carries a flow of 1.5 m^3/s (52.9 ft^3/s). What is the friction loss in the pipe?

Solution
The internal diameter of the pipe is 24.79 in (0.630 m). From Fig. 3-2, e is 0.000005 and e/d is 2.02×10^{-7}. This pipe is hydraulically smooth.

$$N_R = \frac{Vd}{\nu} = \frac{1.5 \times 0.630}{0.312 \times 1.007 \times 10^{-6}} = 3.008 \times 10^6$$

From Fig. 3-1, $f = 0.0098$. Checking to see if the pipe is, in fact, hydraulically smooth gives

$$\delta = \frac{32.8 \times 0.630}{3.008 \times 10^6 \times (0.0098)^{0.5}} = 0.00007$$

$$\frac{e}{\delta} = 0.07 < 0.25$$

therefore the pipe is smooth, as assumed and $f = 0.0098$.

$$HL = f\frac{L}{d}\frac{V^2}{2g} = 0.0098\frac{90}{0.63}\frac{4.81^2}{9.806 \times 2} = 1.65 \text{ m } (5.41 \text{ ft})$$

The determination of f for a given pipe at a given flow rate is completely straightforward. It is more difficult, however, to determine the flow which would be produced at some limiting head loss, since we cannot determine f with certainty until the flow is known and the flow cannot be found without f.

Example 3-3 A 14-in-diameter schedule 80 steel pipe has an inside diameter (ID) of 12.50 in (317.5 mm). How much flow can this pipe carry if the allowable head loss is 3.5 m in a length of 200 m?

Solution
From Fig. 3-2, e for this pipe is 0.00015 in (0.00381 mm).

$$\frac{e}{d} = \frac{0.00381}{317.5} = 0.000012$$

The friction factor from Fig. 3–1, can vary from 0.0084 to as much as 0.04, depending on the flow. Assuming $f = 0.01$,

$$V = \left[\frac{HL \times 2g \times d}{L \times f}\right]^{0.5} = \left[\frac{3.5 \times 9.806 \times 2 \times 0.3175}{200 \times 0.01}\right]^{0.5} = 3.30 \text{ m/s}$$

$$N_R = \frac{Vd}{\nu} = \frac{3.3 \times 0.3175}{1.007 \times 10^{-6}} = 1.04 \times 10^6$$

FIGURE 3-1
Moody diagram for circular pipe.[4] (*Used with permission of Crane Co.*)

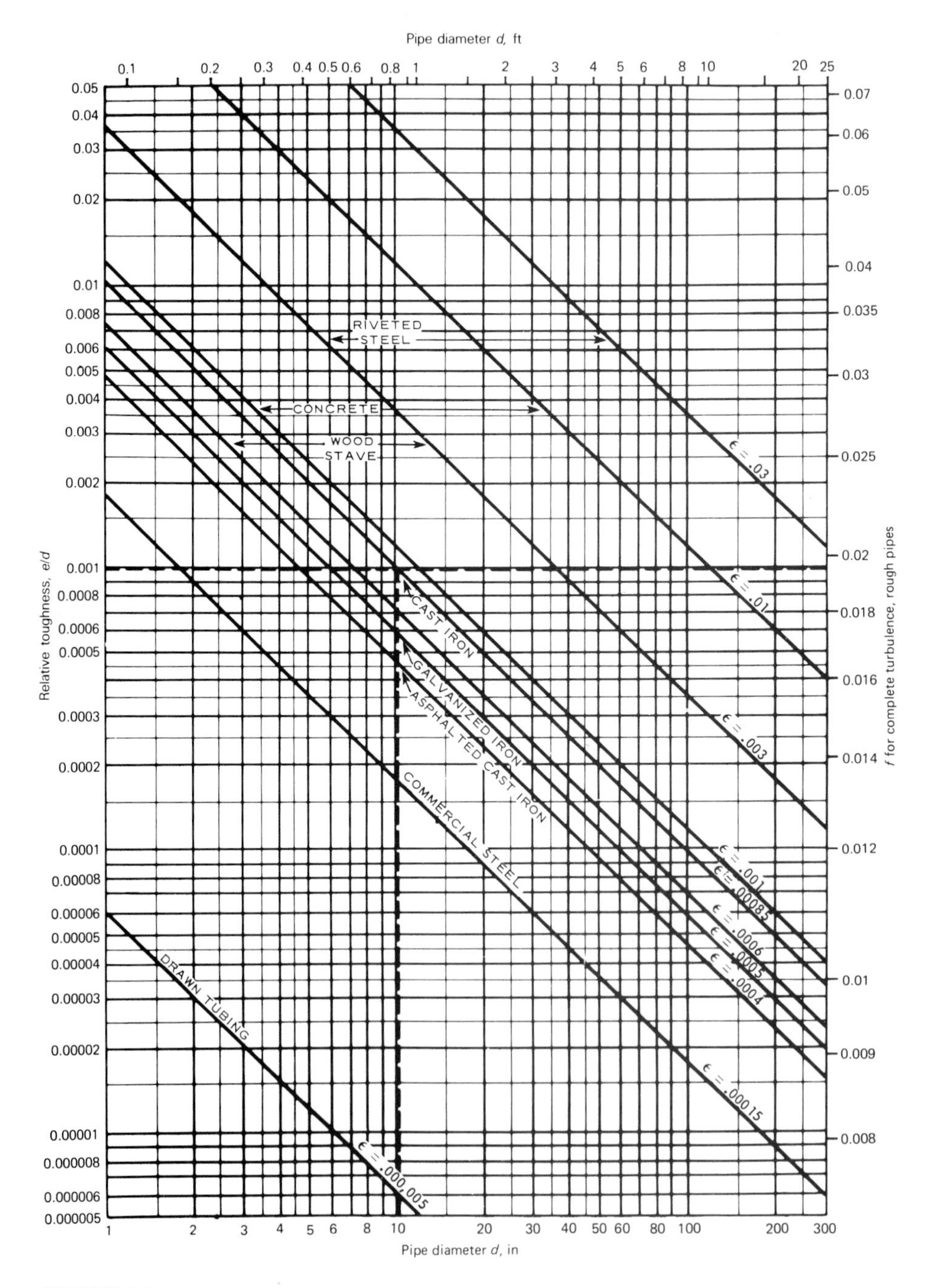

FIGURE 3-2
Relative roughness versus diameter for circular pipe.[4] (*Used with permission of Crane Co.*)

From Fig. 3-1, at $N_R = 10^6$, $f = 0.012$. Recalculating gives $V = 3.01$ m/s, $N_R =$ 949,000, $f \approx 0.012$. Therefore the pipe will carry

$$Q = 3.01 \times \pi \frac{0.3175^2}{4} = 0.238 \text{ m}^3/\text{s } (8.40 \text{ ft}^3/\text{s})$$

3.6 The Hazen-Williams Equation

In addition to the Darcy-Weisbach equation, a number of more or less empirical equations which are more easily solved have been developed for use in pipe-flow problems. The Hazen-Williams equation is such a relation and is

$$V = kCR^{0.63}S^{0.54} \tag{3-8}$$

in which C is a factor dependent on relative roughness, R is the hydraulic radius (the area of flow divided by the wetted perimeter), S is the slope of the energy grade line (HL/L), and k is a factor dependent on units (0.849 for m/s and m; 1.318 for ft/s and ft). The equivalent friction factor represented by C is modified by N_R, and the equation is therefore not totally divorced from the actual phenomena which occur in pipeline flow. The Darcy-Weisbach and Hazen-Williams equations can be shown to give comparable results at moderately high Reynolds numbers when appropriate values are chosen for C. The equation is particularly useful in that it can be solved directly and thus can be used to give a first approximation to N_R when the Darcy-Weisbach equation is used.

Example 3-4 Use the Hazen-Williams equation with $C = 120$ to obtain an estimate of f in Example 3-3.

Solution

$$V = 0.849(120)\left(\frac{0.3175}{4}\right)^{0.63}\left(\frac{3.5}{200}\right)^{0.54} = 2.32 \text{ m/s}$$

$$N_R = \frac{2.32 \times 0.3175}{1.007 \times 10^{-6}} = 733{,}000$$

From Fig. 3-1, $f \approx 0.0125$.

$$V = \left[\frac{3.5 \times 9.806 \times 2 \times 0.3175}{200 \times 0.0125}\right]^{0.5} = 2.95 \text{ m/s}$$

$$Q = 0.233 \text{ m}^3/\text{s } (8.22 \text{ ft}^3/\text{s})$$

A better initial answer would have been obtained with a higher value of C; however, C for steel pipe is generally taken as 120. Accepted values of C for pipes of different materials are presented in Table 3-2. These values may not be representative of actual conditions in real pipes and should be used with caution in the absence of considerable experience. A nomogram which permits graphical solution of the Hazen-Williams equation is presented in Fig. 3-3. The graph will yield either discharge, pipe size, or energy slope given the other two

TABLE 3-2
Hazen-Williams coefficients for various pipe materials

Description of pipe	Value of C
Cast iron:	
New	130
5 years old	120
10 years old	110
20 years old	90–100
30 years old	75–90
Concrete	120
Cement lined	140
Welded steel, as for cast iron 5 years older	
Riveted steel, as for cast iron 10 years older	
Plastic	150
Asbestos cement	140

variables. The figure is constructed for $C = 100$; however, the results obtained can be readily corrected for other values of C by using Eqs. (3-9) through (3-11).

Given flow and diameter, find S from nomogram

$$S_C = S_{100}\left(\frac{100}{C}\right)^{1.85} \qquad (3\text{-}9)$$

Given flow and S, find diameter from nomogram

$$d_C = d_{100}\left(\frac{100}{C}\right)^{0.38} \qquad (3\text{-}10)$$

Given diameter and S, find flow from nomogram

$$Q_C = Q_{100}\left(\frac{C}{100}\right) \qquad (3\text{-}11)$$

The most common errors associated with the use of the Hazen-Williams equation and the nomogram are incorrect selection of C; failure to use actual pipe diameter, rather than nominal diameter; and use in circumstances in which it is not applicable (low N_R, fluids other than water).

The Manning equation, which is discussed in Art. 3-13, can also be used in the solution of pipeline problems. Difficulties with its use are analogous to those presented for the Hazen-Williams equation.

3-7 Fittings and Transitions in Pipe

Losses in fittings and transitions in pipes are sometimes called *minor losses*, but are minor only to the extent that other losses are major. In long pipelines, fitting losses may be negligible, but in situations such as those which exist

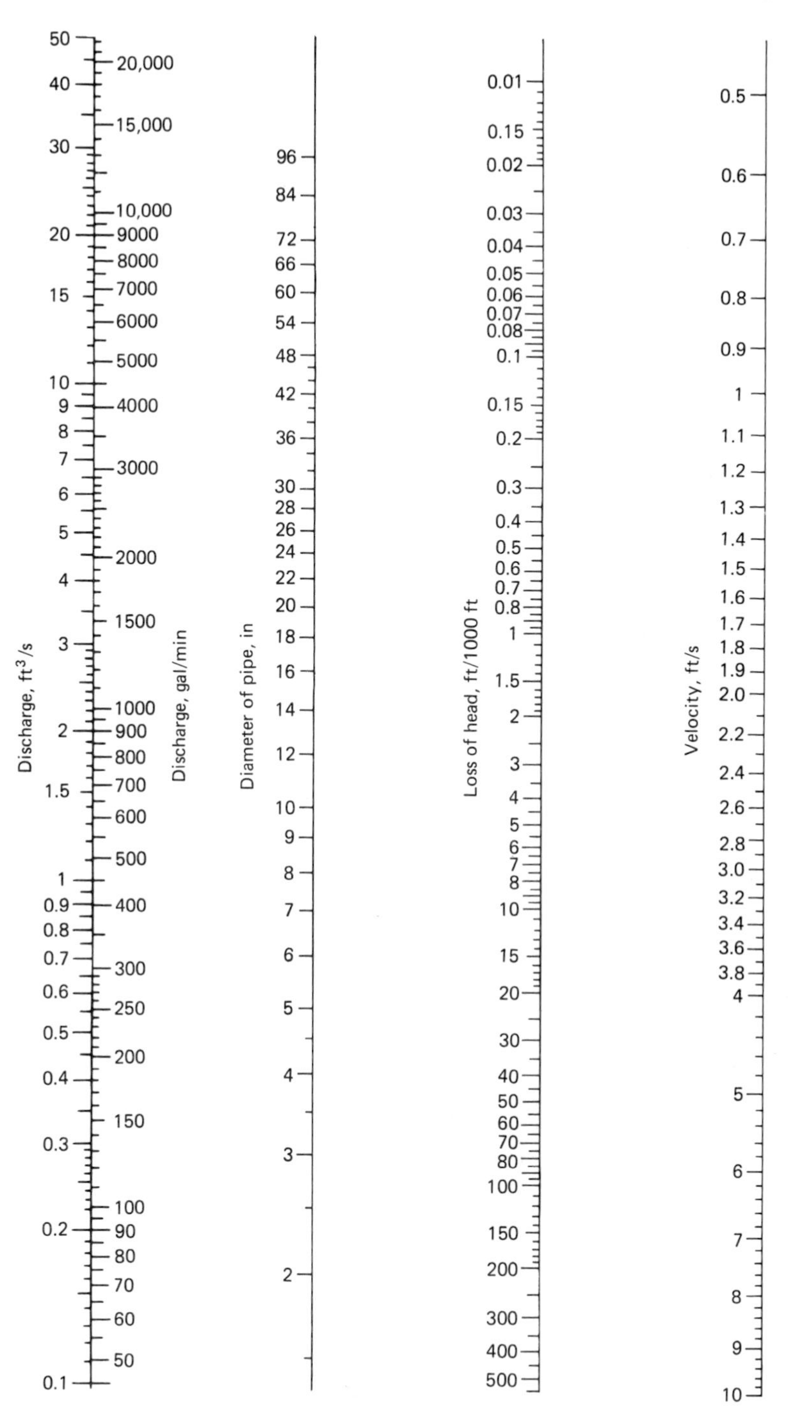

FIGURE 3-3
Nomogram for Hazen-Williams formula for $C = 100$.

within water and wastewater plants, these losses may be far greater than those due to friction.

When water in a pipeline moves from a region of lower to higher velocity, that is, through a contraction, losses are relatively small. On the other hand, passage through an expansion is inefficient, and energy losses are greater. The losses in expansions are attributable primarily to boundary layer effects which cause flow separation and extreme turbulence as the velocity decreases. Boundary layer effects also contribute to losses at changes in direction by establishing secondary flow patterns which produce a double spiral motion in a plane normal to the main flow path.

Losses in fittings have been found, in general, to vary as the velocity head. The head loss may thus be expressed as

$$\mathrm{HL} = K\frac{V^2}{2g} \tag{3-12}$$

where K is a function of changes in direction, obstructions, or changes in velocity. Since K is not a "friction" factor, it is reasonable that it not vary with N_R, and this has been found to be the case. K is constant for a given fitting, but is not constant for all fittings of a given type because of the lack of geometric similarity among different sizes. It has been found that K varies with fitting size in the same way that f varies with pipe size in wholly rough or completely turbulent flow. In that range of flow, f is constant for a given pipe size, as noted in Art. 3-5 and as may be seen in Fig. 3-1. The similarity in the way in which K and f_{turb} vary is thought to be more a matter of coincidence than an indication that relative roughness is important in fitting losses.

Expansions in pipelines, as noted above, produce substantial energy losses. At *abrupt* enlargements, application of the momentum equation yields an energy loss, sometimes called the *Borda-Carnot loss*, equal to

$$\mathrm{HL} = \frac{(V_1 - V_2)^2}{2g} \tag{3-13}$$

In *gradual* enlargements the head loss depends upon the shape. In cone-shaped enlargements there is a combination of frictional loss and eddying loss resulting from the expansion. When the central angle is small, friction predominates. When it is large, eddying is most significant. Some investigators have found that there is both a minimum and maximum value for K (at about 6° and 60° respectively). It is certain that beyond a central angle of about 45° the head loss is at least as great as that in an abrupt expansion. For such enlargements

$$\mathrm{HL} = K\frac{(V_1 - V_2)^2}{2g} \tag{3-14}$$

in which K varies with central angle from about 0.1 to at least 1. One may calculate the energy loss in terms of the upstream velocity alone by defining the contraction ratio β as $\beta = d_1/d_2$. Then $V_1 = V_2/\beta^2$ and

$$HL = K(1 - \beta^2)^2 \frac{V_1^2}{2g} \tag{3-15}$$

where V_1 is the velocity in the smaller pipe.

The value of the entire coefficient $K(1 - \beta^2)^2$ has been found[5] to vary with central angle in the following manner:

$$K_1 = 2.6 \sin \frac{\theta}{2} (1 - \beta^2)^2$$

for $\theta \leq 45°$, and

$$K_1 = (1 - \beta^2)^2$$

for $45° \leq \theta \leq 180°$.

The discharge from a pipeline into a relatively large reservoir may be considered a special case of Borda-Carnot expansion in which $V_2 = 0$. For this circumstance Eq. (3-13) may be rewritten as

$$HL = \frac{V_1^2}{2g} \tag{3-16}$$

Gradual contractions in pipelines produce insignificant head loss in most cases. A contraction will, of course, increase the velocity and, to the extent the velocity distribution and α are changed, the pressure downstream will be less than indicated by Eq. (3-1*a*). If the contraction is particularly long (small central angle), the net loss will be greater as a result of wall friction. Reference 5 recommends use of

$$K_2 = 0.8 \sin \frac{\theta}{2} (1 - \beta^2)$$

for $\theta \leq 45°$ and

$$K_2 = 0.5 \left(\sin \frac{\theta}{2}\right)^{0.5} (1 - \beta^2)$$

for $45° \leq \theta \leq 180°$, where K_2 is to be multiplied by the velocity head in the smaller pipe. The *sudden contraction* represented by large values of θ produces a *vena contracta* and a consequent expansion in which energy losses are high, reaching a maximum loss of $0.5\ (1 - \beta^2)V_2^2/2g$. The equations above correspond very closely to Weisbach's original data except for large values of β, in which case the loss is small in any event.

The square-edged pipe entrance from a large tank is a special case of sudden contraction in which $\beta = 0\ (V_1 = 0)$. For this situation

$$HL = 0.5 \frac{V_2^2}{2g} \tag{3-17}$$

If the entrance is rounded, the loss can be substantially reduced. Any radius of rounding greater than $0.14d$ will prevent formation of the *vena contracta*. For such a "bell-mouth" entrance, K is commonly taken as 0.04. Actual values range up to 0.5 depending upon the radius of rounding.

Example 3-5 Water flows from one basin with a water surface at 100-m elevation to another with water surface at 99.5 m. Flow is through a 10-ft-long, class 2, 8-in-diameter ductile iron pipe with a square-edged entrance and submerged discharge. What is the flow?

Solution
The total energy loss is the sum of the friction, entrance, and exit losses. Bernoulli's equation between the two basins is

$$\frac{V_1^2}{2g} + \frac{P_1}{\gamma} + z_1 = \frac{V_2^2}{2g} + \frac{P_2}{\gamma} + z_2 + \mathrm{HL}_{1-2}$$

$$\mathrm{HL}_{1-2} = 100 - 99.5 = 0.5 \text{ m}$$

$$\mathrm{HL}_{1-2} = K_{\text{entrance}}\frac{V^2}{2g} + K_{\text{exit}}\frac{V^2}{2g} + f\frac{L}{d}\frac{V^2}{2g}$$

Class 2 8-in ductile iron pipe has an ID of 8.39 in (213.1 mm) and an area of 0.0357 m². $V = 28.04Q$. Since the Reynolds number is unknown, f is unknown. If the flow is wholly turbulent, $f = 0.014$ (from Figs. 3-1 and 3-2). Then

$$\mathrm{HL}_{1-2} = \frac{V^2}{2g} + 0.5\frac{V^2}{2g} + 0.014\frac{3.05}{0.213}\frac{V^2}{2g} = 0.5$$

Then $V = 2.40$ m/s and

$$N_R = \frac{Vd}{\nu} = \frac{2.40(0.213)}{1.007 \times 10^{-6}} = 508{,}000$$

From Fig. 3-1, $f = 0.0155$. Repeating the calculation, $V = 2.38$ m/s. The total flow is thus 0.0581 m³/s (3 ft³/s). It is interesting to note that, of the total head loss, only about 13 percent is attributable to pipe friction.

Example 3-6 Estimate the head loss factor K for a venturi which has an entrance angle of 20° and an exit angle of 5°. Express K in terms of both the velocity in the pipe and that in the throat of the venturi.

Solution
The loss in the contraction is given by

$$K = 0.8 \sin\frac{20}{2}(1 - \beta^2) = 0.139(1 - \beta^2)$$

The loss in the expansion is given by

$$K = 2.6 \sin\frac{5}{2}(1 - \beta^2)^2 = 0.113(1 - \beta^2)^2$$

Both losses are in terms of the velocity in the throat. To express this loss in terms of the velocity in the pipe, we note that $V_1A_1 = V_2A_2$, whence

$$V_2 = V_1 \frac{A_1}{A_2} = V_1 \frac{d_1^2}{d_2^2} = V_1 \left(\frac{1}{\beta}\right)^2$$

and

$$V_2^2 = V_1^2 \left(\frac{1}{\beta}\right)^4$$

Substituting, we obtain

$$\mathrm{HL} = \frac{0.139(1 - \beta^2) + 0.113(1 - \beta^2)^2}{\beta^4} \frac{V_1^2}{2g}$$

where V_1 is the velocity in the pipe.

Values of K for fittings of various characteristics are presented in App. 2. In using this information it must be remembered that the losses in the fittings have been determined from the friction factor for steel pipe in the wholly rough zone and the velocity in steel pipe of the schedule which would be used with the particular fitting class. In nearly all cases of interest in water and wastewater, that implies schedule 40 pipe. The internal diameter and area of various sizes of schedule 40 pipe are listed in Table 3-3.

TABLE 3-3
Dimensions for schedule 40 pipe

Nominal	Internal diameter		Area	
in	in	mm	ft^2	m^2
$\frac{1}{8}$	0.269	6.83	0.0004	0.00004
$\frac{1}{4}$	0.364	9.25	0.0007	0.00007
$\frac{3}{8}$	0.493	12.52	0.0013	0.00012
$\frac{1}{2}$	0.622	15.80	0.0021	0.00020
$\frac{3}{4}$	0.824	20.93	0.0037	0.00034
1	1.049	26.64	0.0060	0.00056
$1\frac{1}{4}$	1.380	35.05	0.0104	0.00097
$1\frac{1}{2}$	1.610	40.89	0.0141	0.00131
2	2.067	52.50	0.0233	0.00217
$2\frac{1}{2}$	2.469	62.71	0.0332	0.00309
3	3.068	77.93	0.0513	0.00477
4	4.026	102.26	0.0884	0.00821
5	5.047	128.19	0.1390	0.01291
6	6.065	154.05	0.2006	0.01864
8	7.981	202.72	0.3474	0.03228
10	10.020	254.51	0.5475	0.05087
12	12.000	304.80	0.7854	0.07297
14	13.124	333.35	0.9394	0.08728
16	15.000	381.00	1.2272	0.11401
18	16.876	428.65	1.5533	0.14431
20	18.812	477.82	1.9305	0.17932
24	22.624	574.65	2.7921	0.25936
32	30.624	777.85	5.1151	0.47521
36	34.500	876.30	6.4918	0.60311

Example 3-7 Find the head loss through three 90° elbows, two gate valves, and a swing check valve in a 12-in ductile iron pipe carrying a flow of 0.3 m^3/s.

Solution
The values of K for the fittings, from App. 2, are

$$K_{\text{elbows}} = 0.39$$

$$K_{\text{gate valve}} = 0.10$$

$$K_{\text{check valve}} = 0.65$$

These values are to be multiplied by the velocity head produced by the stipulated flow in a schedule 40 pipe.

$$\frac{V^2}{2g} \text{ (for the fittings)} = \frac{(0.3/0.07297)^2}{2(9.806)} = 0.862 \text{ m}$$

$$\text{HL}_{\text{fittings}} = [3 \times 0.39 + 2 \times 0.10 + 0.65]\, 0.862 = 1.74 \text{ m (5.7 ft)}$$

Note that a ductile iron pipe might have an ID of 12.22 to 12.46 in (310.4 to 316.5 mm), depending on its class, but this diameter is not used in finding the head loss in the fittings. Were the wrong value to be used, the error in head loss could be as much as 16 percent.

Example 3-8 Determine the flow through the pipe system shown in Fig. 3-4. All pipe is schedule 40.

Solution
From App. 2, for the various fittings,

Fitting	K	β	K	$K(V^2/2g)$
Entrance	0.5		0.5	$4.79Q^2$
Contraction	Art. 3-7	0.505	0.143	$20.98Q^2$
90° bend	App. 2		0.180	$26.42Q^2$
90° bend	App. 2		0.180	$26.42Q^2$
Gate value	App. 2		0.120	$17.61Q^2$
Check valve	App. 2		0.750	$110.06Q^2$
90° bend	App. 2		0.180	$26.42Q^2$
90° bend	App. 2		0.180	$26.42Q^2$
Expansion	Art. 3-7	0.505	0.277	$40.65Q^2$
Exit	1.0		1.0	$9.58Q^2$
Total				$309.35Q^2$

The losses in the straight pipe will be

$$\text{HL}_{12} = f_{12}\frac{50}{0.305}(9.58Q^2) = 1569.8f_{12}Q^2$$

$$\text{HL}_6 = f_6\frac{350}{0.154}(146.75Q^2) = 333{,}553f_6Q^2$$

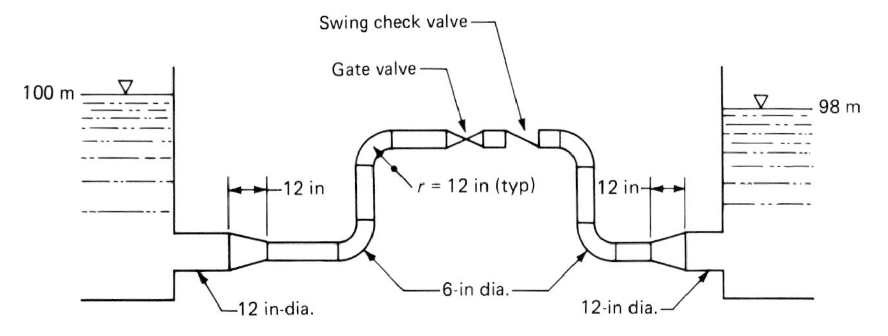

FIGURE 3-4
Pipe system of Example 3-8.

Since the Reynolds number will vary in the two pipes by a factor of about 2, trial values of f should reflect that difference. Assuming N_R in the small pipe is approximately 200,000, $f_6 = 0.0175$ and $f_{12} = 0.0188$. The total loss is thus

$$\text{HL} = 1569.8(0.0188)Q^2 + 333{,}553(0.0175)Q^2 + 309.35Q^2 = 2$$

whence $Q = 0.0180\ \text{m}^3/\text{s}$, $V_6 = 0.965$ m/s, and $V_{12} = 0.247$ m/s. From these velocities, N_R in the 6-in line is 148,600 and in the 12-in line is 75,300. With new values of $f_6 = 0.0181$ and $f_{12} = 0.0196$, the flow is recalculated and found to be 0.0177 m^3/s.

This solution assumes the velocity in the pipe is sufficiently large to completely open the check valve. If this is not the case, the losses in the valve will be substantially higher. The velocity required is given in App. 2 as

$$V_{min} = 48\overline{V}^{0.5}$$

where $\overline{V}$ is the specific volume (that is, the reciprocal of the density of the fluid). For water, this is 1/62.4 in English units.

$$V_{min} = 48\left(\frac{1}{62.4}\right)^{0.5} = 6.08\ \text{ft/s}\ (1.85\ \text{m/s})$$

Since the calculated velocity in the valve is only about one-half the required value, the valve will not open completely. Correct design requires that the valve size be reduced.

A 4-in valve will have a velocity of 2.16 m/s at a flow of 0.0177 m^3/s. The flow will be somewhat reduced by this change, but the velocity should still be above 1.85 m/s. From App. 2 the loss produced by a 4-in check valve in a 6-in line will be

$$\text{6-in to 4-in contraction} \approx 0.08\,\frac{V_2^2}{2g}$$

$$\text{4-in valve} \approx 0.80\,\frac{V_4^2}{2g}$$

$$\text{6-in to 4-in expansion} \approx 0.14\,\frac{V_4^2}{2g}$$

or approximately $771Q^2$ instead of the $110.06Q^2$ used initially. This change gives a flow of 0.0168 m^3/s or 0.59 ft^3/s. The reduction in velocity and N_R is not sufficient to make a substantial change in the friction factors.

3-8 Equivalent Lengths

Since the head loss in fittings may be expressed as

$$\text{HL} = K\frac{V^2}{2g} \tag{3-12}$$

and this expression bears an evident similarity to that for the head loss in pipes

$$\text{HL} = f\frac{L}{d}\left(\frac{V^2}{2g}\right) \tag{3-3}$$

one could consider that

$$K = f\frac{L}{d} \tag{3-18}$$

where f is the friction factor for the pipe and L/d is an equivalent length, expressed in pipe diameters, which would yield the same loss at a given flow. Since f is a function of velocity and K is not, it follows that the equivalent length must vary inversely with f. There is no reason why one could not express K in terms of equivalent length at some stipulated velocity in order to simplify tabulation of data. Such values should not, however, be added to other real lengths of pipe and used to calculate head loss at some other flow condition.

3-9 Flow Coefficients

Some manufacturers of valves, particularly control valves and metering valves, express the capacity and hydraulic characteristics in terms of a flow coefficient C_v. C_v is usually defined in the United States as the flow in gallons per minute at a pressure loss of 1 lb/in^2. If we substitute, with appropriate unit conversions, in Eq. (3-12), we obtain

$$C_v = \frac{29.84d^2}{K^{0.5}} \tag{3-19}$$

where d is in inches.

Example 3-9 A manufacturer offers a 12-in globe valve which is reported to have $C_v = 5000$. Find the value of K for this valve.

Solution
$d = 12.00$ in for schedule 40 pipe.

$$K = \left(\frac{29.84d^2}{C_v}\right)^2 = \left(\frac{29.84 \times 12^2}{5000}\right)^2 = 0.74$$

3-10 Flow Measurement in Pipelines

Orifice plates are the simplest and cheapest flow measurement device. Since they develop a substantial head loss, orifices may also be used to regulate flow. Orifices, like flow nozzles and venturis, produce a differential pressure in the pipe by producing a differential velocity. If one writes Bernoulli's equation from point 1 to 2 in Fig. 3-5, neglecting head loss, one obtains

$$\frac{P_1}{\gamma} + \frac{V_1^2}{2g} = \frac{P_2}{\gamma} + \frac{V_2^2}{2g}$$

Since $V_1 = Q/A_1$ and $V_2 = Q/A_2$ and $A_2/A_1 = d_2^2/d_1^2 = \beta^2$,

$$Q = \frac{A_2}{(1 + \beta^4)^{0.5}}\left(2g\,\frac{P_2 - P_1}{\gamma}\right)^{0.5} \tag{3-20}$$

Since there will be, in fact, some head loss, this equation is modified by an experimental coefficient,

$$Q = \frac{C_d A_2}{(1 + \beta^4)^{0.5}}\left(2g\,\frac{P_2 - P_1}{\gamma}\right)^{0.5} \tag{3-21}$$

C_d has been found to vary with Reynolds number and with β, thus some manufacturers combine C_d and the factor $1/(1 + \beta^4)^{0.5}$ into a single coefficient,

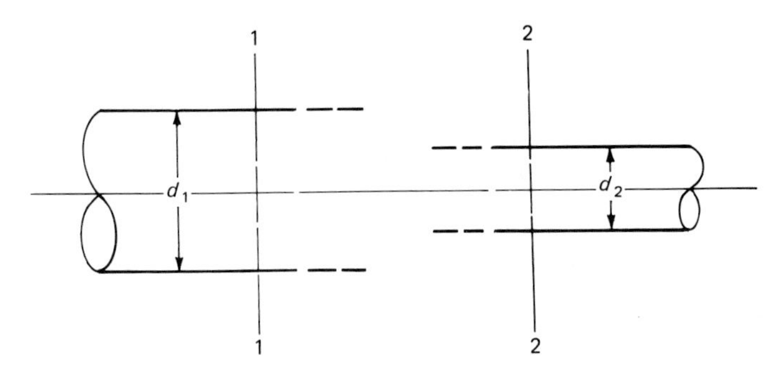

FIGURE 3-5
Flow measurement by velocity differential.

$$Q = CA_2\left(2g\frac{P_2 - P_1}{\gamma}\right)^{0.5} \tag{3-22}$$

The coefficient C can also be written in terms of A_1, the pipe area.

The value of C is dependent upon the location at which the two pressures are measured. Original data[6] collected for orifice plates indicated that the maximum pressure differential would be measured if the taps were located 0.4 pipe diameters upstream and 0.8 pipe diameters downstream of the face of the plate. These locations were thought to correspond to the location of undisturbed flow and the *vena contracta* respectively. More recent investigations[7] have indicated that the actual location of the *vena contracta* depends on β and varies from 0.3 to 0.9 pipe diameters downstream of the plate. A great deal of data have been collected for orifice plates with so-called *radius taps* located 0.5 pipe diameters upstream and 1 pipe diameter downstream of the plate. Figure 3-6 shows the variation of C with N_R and β.

The orifice plate must be constructed carefully if it is to function as intended. The plate thickness must not exceed the lesser of 1/30 of the pipe ID, 1/8 of the orifice bore, or 1/8 of the difference between the two. The plate must be flat and β should generally be between 0.1 and 0.8 if radius taps are used.

Example 3-10 A square-edged orifice with radius taps is installed in a 6-in schedule 40 pipe. The orifice is 3.00 in in diameter. Find the flow rate when a mercury manometer connected across the taps measures 122 mm.

Solution

For a 6-in pipe, $d = 6.056$ in and $\beta = 3.00/6.056 = 0.495$. N_R is unknown. Assuming the flow will occur in the range in which C is constant, $C \approx 0.62$. The differential pressure in meters of water is given by $(13.57 - 1) \times 0.122 = 1.533$ m.

$$Q = CA(2gh)^{0.5} = 0.62(0.00456)(2 \times 9.806 \times 1.533)^{0.5} = 0.0155 \text{ m}^3/\text{s}$$

The Reynolds number in the pipe is

$$N_R = \frac{Vd}{\nu} = \frac{0.832(0.154)}{1.007 \times 10^{-6}} = 127{,}260$$

Thus, from Fig. 3-6, a better value for C would be 0.622, which gives a flow not significantly different from that calculated.

The head loss across an orifice will be less than the differential pressure at the taps, since a portion of the velocity head is recovered in the subsequent expansion. The value of K for the orifice is given by

$$K = \frac{1 - \beta^2}{C^2\beta^4} \tag{3-23}$$

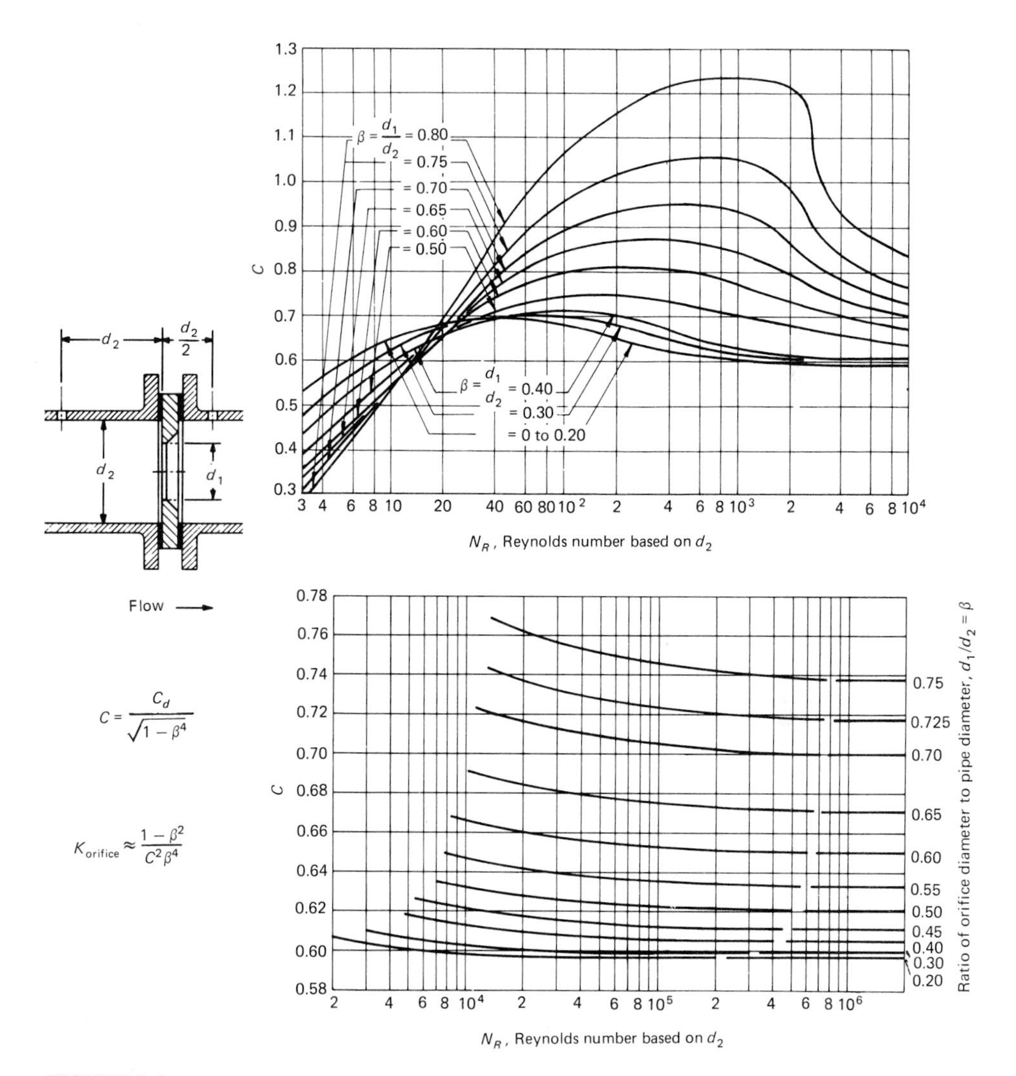

FIGURE 3-6
Flow coefficients for square-edge orifices. (*Used with permission of Crane Co.*)

which, in the case of Example 3-10, yields

$$K = \frac{1 - 0.495^2}{0.622^2(0.495)^4} = 32.5$$

This value is multiplied by the velocity head in the pipe to give a loss of

$$\text{HL} = 32.5 \frac{0.832^2}{2(9.806)} = 1.15 \text{ m}$$

or approximately 75 percent of the differential pressure.

Flow nozzles and *venturis* provide flow measurement in a similar fashion by producing a differential head proportional to flow. The advantage of these devices lies in their lower head loss. Since the entrances to both are smooth, no *vena contracta* is formed and the differential head for a given value of β will be less than in an orifice plate. The head loss in a flow nozzle may be considered to be primarily that which results from an abrupt expansion, while that in a venturi may be calculated as shown in Example 3-6. Values for C for these devices must generally be obtained from the manufacturers, who may present their data in different forms. A curve showing the variation of C for one type of flow nozzle is presented in Fig. 3-7.

Other flow meters include various mechanical (propeller) devices, magnetic meters, and sonic meters. *Mechanical meters* are essentially pumps run in reverse in which the flow produces a rotational velocity. The speed of rotation is related to propeller characteristics, velocity, and drag in the meter. If the drag is kept relatively low, the meter will be linear. Propeller meters are easily fouled and may be damaged by abrasive suspensions. Their use should be limited to clear water.

Magnetic meters require that the fluid be a conductor, which is generally the case in water and wastewater systems. The meters create a magnetic field through which the water flows. In accordance with Faraday's law, a voltage is produced in the fluid which is proportional to velocity. This voltage is mea-

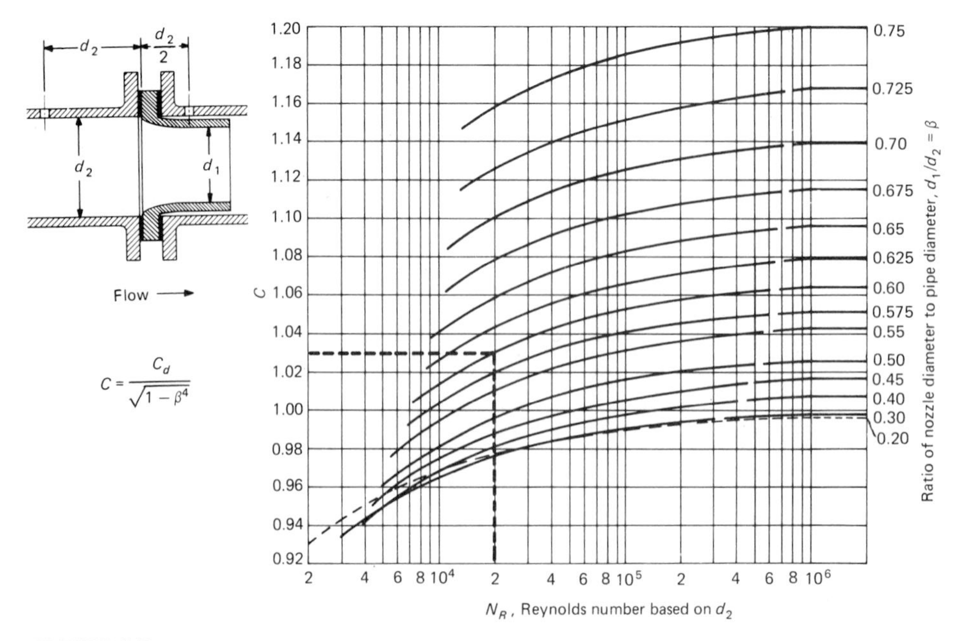

FIGURE 3-7

Flow coefficients for flow nozzles. As an example, the flow coefficient C for a diameter ratio β of 0.60 at a Reynolds number of 20,000 (2×10^4) equals 1.03. (*Used with permission of Crane Co.*)

sured by the meter. Magnetic meters are somewhat less affected than others by variations in velocity profile produced by fittings (see below). In order to provide acceptable accuracy, the velocity may need to be kept relatively high, which can require using a meter smaller than the pipe. The subsequent expansion can provide a substantial head loss. Solids and gas bubbles do not cause major losses in accuracy.

Sonic meters utilize a pressure wave transmitted through the fluid at an angle to the direction of flow. Such pressure waves move at the speed of sound in the medium. Since the fluid has a velocity component in the direction of the pressure wave, a Doppler effect will be observed in which the frequency of the imposed pulse is altered in proportion to velocity. Sonic meters are particularly sensitive to upstream fittings and cannot be used with flows which are high in solids or entrained gases. Pulsating flows, like those produced by plunger pumps, cannot be measured satisfactorily by these devices.

Meter location is dictated by the need to ensure a relatively well-developed and uniform velocity distribution across the pipe. Fittings, such as bends, will produce asymmetric profiles and rotational flow, while reducers will produce "jetting." Some manufacturers claim their meters will work in close proximity to fittings, but good practice dictates that a substantial distance should be provided between fittings and meters. It is particularly important to note that the required distance is greater when the contraction at the meter is less. Thus a meter which offers no restriction to the flow will be more sensitive to upstream disturbances than a venturi or orifice plate. Typical recommended spacings range up to 30 pipe diameters, depending upon the meter and type of upstream disturbance. These distances can be substantially reduced by installation of straightening vanes.[7]

3-11 Open-Channel Flow

Open-channel flow is flow in which the surface of the liquid is at atmospheric pressure. Such flow may be laminar or turbulent, uniform or varied, and subcritical, critical, or supercritical.

Laminar and *turbulent* flow in open channels is analogous to that in pipes. There is, however, no important circumstance in which laminar flow occurs in open channels in water or wastewater structures or processes.

Uniform flow is flow in which the depth, width, and velocity remain constant along a channel. *Varied* flow involves a change in these, with a change in one producing changes in the others. Most circumstances of open-channel flow in water and wastewater systems involve varied flow. The concept of uniform flow is valuable, however, in that it defines a limit which the varied flow may be considered to be approaching in many cases.

Critical flow defines a state of flow between two flow regimes. Critical flow coincides with minimum specific energy for a given discharge and maximum discharge for a given specific energy. Critical flow occurs in flow measurement devices, at or near free discharges, and establishes "controls" in

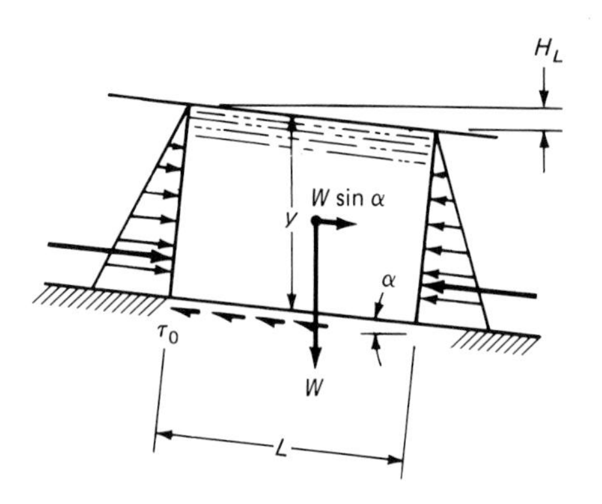

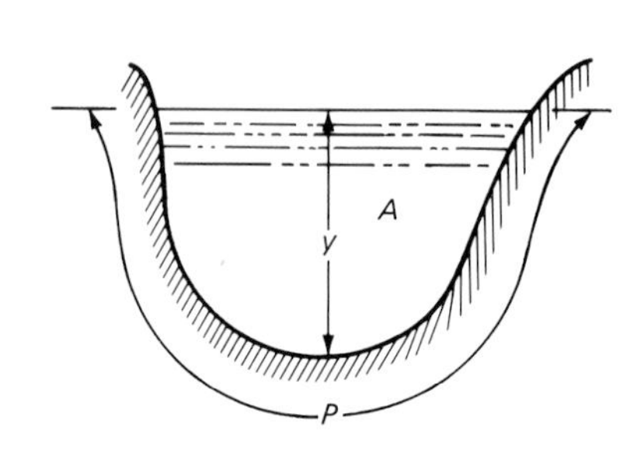

FIGURE 3-8
Uniform flow.

open-channel flow. Critical flow occurs frequently in water and wastewater systems and is very important in their operation and design.

3-12 The Chézy Equation

The condition of uniform flow may be analyzed by considering certain developments from pipeline flow together with Fig. 3-8. Although there are certain variations in velocity profile and shear stress produced by the free surface in open-channel flow, some useful conclusions may be drawn from consideration of the mean shear stress. Summing forces in the direction of flow in Fig. 3-8, one obtains

$$W \sin \alpha = \tau_0 LP \tag{3-24}$$

For small angles $\sin \alpha = \tan \alpha = S$, from which

$$A\gamma LS = \tau_0 LP$$

$$\tau_0 = \gamma \frac{A}{P} S = \gamma RS \tag{3-25}$$

where R is the hydraulic radius as defined in Art. 3-6. In pipeline flow it may be shown that $\tau_0 = f\rho V^2/8$. If it is assumed that this fundamental relationship is valid for open channels as well, one may write

$$f\rho \frac{V^2}{8} = \gamma RS$$

from which

$$V = \left(\frac{8g}{f} RS\right)^{0.5} = C(RS)^{0.5} \tag{3-26}$$

which is commonly called the Chézy equation.

The Chézy coefficient varies with roughness and with N_R inversely as does f in the Darcy-Weisbach equation. Modified Moody diagrams of the sort

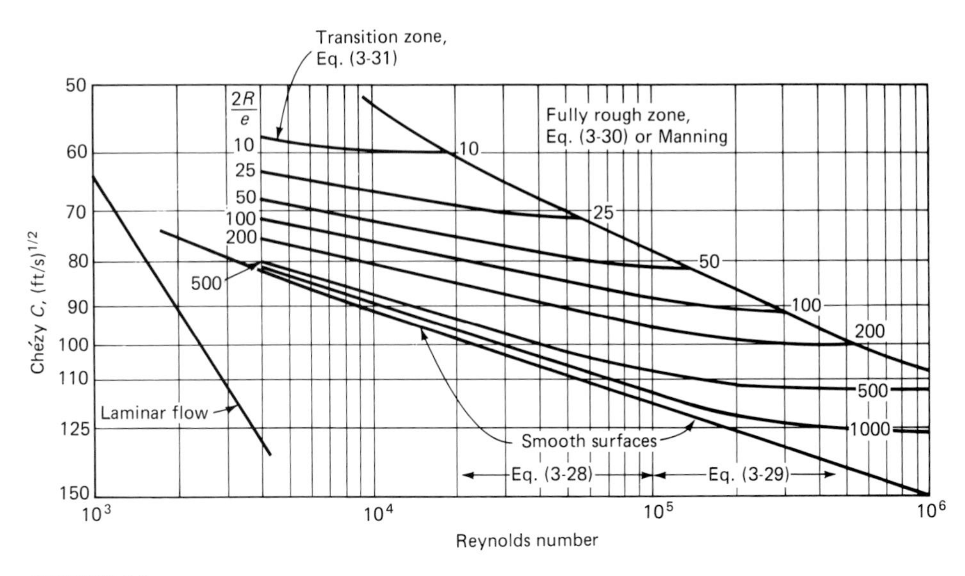

FIGURE 3-9
Modified Moody diagram for Chézy C.

shown in Fig. 3-9 reflect the variation of C with N_R and relative roughness, which, in this case, is expressed as $2R/e$. Typical values of e are presented in Table 3-4 and can be used in conjunction with Fig. 3-9 to find appropriate values of C. It must be observed that Chézy's C is not identical in English and SI units. Since $C = (8g/f)^{0.5}$, where f is dimensionless,

$$\frac{C_{\mathrm{SI}}}{C_{\mathrm{English}}} = \left(\frac{9.8}{32.2}\right)^{0.5} = 0.522 \tag{3-27}$$

Example 3-11 A very smooth cement plastered surface ($e = 0.0003$ m) forms an open channel 0.6 m wide which carries a flow of 0.25 m³/s at a depth of 0.5 m. Find the values of C and S.

Solution

$$e = 0.0003$$

$$R = \frac{A}{P} = \frac{0.5 \times 0.6}{2 \times 0.5 + 0.6} = 0.1875$$

$$\frac{2R}{e} = \frac{2(0.1875)}{0.0003} = 1250$$

$$N_R = \frac{4VR}{\nu} = \frac{4(0.25/0.3)(0.1875)}{1 \times 10^{-6}} = 625{,}000$$

From Fig. 3-9, $C \approx 120$ in English units. In SI units, $C_{\mathrm{SI}} = 0.552(120) = 66.2\ \mathrm{m^{0.5}/s}$. Then, since $V = C(RS)^{0.5}$,

$$S = \frac{V^2}{C^2R} = \frac{(0.25/0.3)^2}{66.2^2 \times 0.1875} = 0.0008$$

TABLE 3-4
Values of *e* for concrete and masonry surfaces

Description	*e* m	*e* ft
Concrete class 4 (monolithic construction, cast against oiled steel forms with no surface irregularities).	0.0002	0.0005
Very smooth cement-plastered surfaces, all joints and seams hand-finished flush with surface.	0.0003	0.001
Concrete cast in lubricated steel molds, with carefully smoothed or pointed seams and joints.	0.0005	0.0016
Wood-stave pipes, planed-wood flumes, and concrete class 3 (cast against steel forms or spun-precast pipe). Smooth troweled surfaces. Glazed sewer pipe.	0.0006	0.002
Concrete class 2 (monolithic construction against rough forms or smooth-finished cement-gun surface, the latter often called *Gunite* or *shot concrete*). Glazed brickwork.	0.0015	0.005
Short lengths of concrete pipe of small diameter without special facing of butt joints.	0.0024	0.008
Concrete class 1 (precast pipes with mortar squeeze at the joints). Straight, uniform earth channels.	0.003	0.01
Roughly made concrete conduits.	0.004	0.014
Rubble masonry.	0.006	0.02
Untreated Gunite.	0.003–0.009	0.01–0.03

It is worth noting that this flow condition, which is not atypical, is not in the fully rough zone of Fig. 3-9. This circumstance is analogous to that illustrated for pipe flow in Examples 3-3 and 3-4 and demonstrates the need for care in the use of empirical equations. The several zones of Fig. 3-9 may be described mathematically as shown in Eqs. (3-28) through (3-31).

$$C = 28.6N_R^{1/8} \tag{3-28}$$

$$C = 4(2g)^{0.5} \log \left[\frac{N_R(8g)^{0.5}}{2.51C} \right] \tag{3-29}$$

$$C = 4(2g)^{0.5} \log \left[\frac{12R}{e} \right] \tag{3-30}$$

$$C = -2(8g)^{0.5} \log \left[\frac{e}{12R} + \frac{2.5C}{N_R(8g)^{0.5}} \right] \tag{3-31}$$

In these equations, C is in English units. The result, as above, must be corrected by Eq. (3-27) if SI units are used.

Example 3-12 Determine C in Example 3-11 using the appropriate equation.

Solution
From $2R/e = 1250$, $N_R = 625{,}000$, and Fig. 3-9, the appropriate equation is Eq. (3-31).

$$C - 2(8g)^{0.5} \log \left[\frac{e}{12R} + \frac{2.5C}{N_R(8g)^{0.5}} \right]$$

$$C = -32.1 \log (1.333 \times 10^{-4} + 2.492 \times 10^{-7} C)$$

Solving by trial gives

$$C = 121$$

3-13 Manning's Equation

Manning and others attempted to relate the Chézy coefficient to roughness and channel geometry in a simpler fashion. The relationship took the form, in English units,

$$C = \frac{1.49}{n} R^{1/6} \tag{3-32}$$

where n is a friction factor of sorts. Substitution in Chézy's equation leads to

$$V = \frac{1.49}{n} R^{2/3} S^{1/2} \tag{3-33}$$

or

$$V = \frac{1}{n} R^{2/3} S^{1/2} \tag{3-34}$$

in SI units. This equation is properly applied only in the fully rough zone of Fig. 3-9. Typical values of Manning's n are presented in Table 3-5.

Example 3-13 Calculate S in Example 3-11 using Manning's equation.

Solution
From Table 3-5, $n = 0.011$. From Example 3-11, $R = 0.1875$ m, $V = 0.833$ m/s.

$$S = \left(\frac{0.833(0.011)}{0.1875^{2/3}} \right)^2$$

$$S = 0.0008$$

In this case Manning's equation gives a value for S which is nearly identical to that obtained from Chézy's equation. Using Manning's equation with a fixed value of n gives a more or less fixed value for C. This, as may be seen from Fig. 3-9, is correct only in the fully rough zone. Use of Manning's equation without consideration of N_R and relative roughness is not a conservative design procedure, since in the transition zone of Fig. 3-9, C decreases and calculated slopes will be less than those actually required.

TABLE 3-5
Values of Manning's roughness coefficient

Material	Value
Glass, plastic, machined metal	0.010
Dressed timber, joints flush	0.011
Sawn timber, joints uneven	0.014
Cement plaster	0.011
Concrete, steel-troweled	0.012
Concrete, timber forms, unfinished	0.014
Untreated Gunite	0.016
Brickwork or dressed masonry	0.014
Rubble set in cement	0.017
Earth, smooth, no weeds	0.020
Earth, some stones and weeds	0.025

3-14 Best Hydraulic Cross Section

If the Manning equation is written in terms of flow rather than velocity, one obtains

$$Q = \frac{1}{n} A R^{2/3} S^{1/2} \tag{3-35}$$

It is evident that for a channel of given area and slope the discharge will be a maximum for maximum hydraulic radius. Since maximum hydraulic radius for a channel of fixed area also requires minimum wetted perimeter, finding R_{max} will not only maximize discharge, but also will tend to minimize the cost of construction.

Example 3-14 For the trapezoidal section shown in Fig. 3-10, find the proportions which will maximize R.

Solution

$$P = b + \frac{2y}{\cos(90 - \alpha)}$$

$$A = by + \frac{y^2}{\tan\alpha}$$

whence

$$b = \frac{A}{y} - \frac{y}{\tan\alpha}$$

$$R = \frac{A}{p} = \frac{A}{A/y - y/\tan\alpha + 2y/\cos(90 - \alpha)}$$

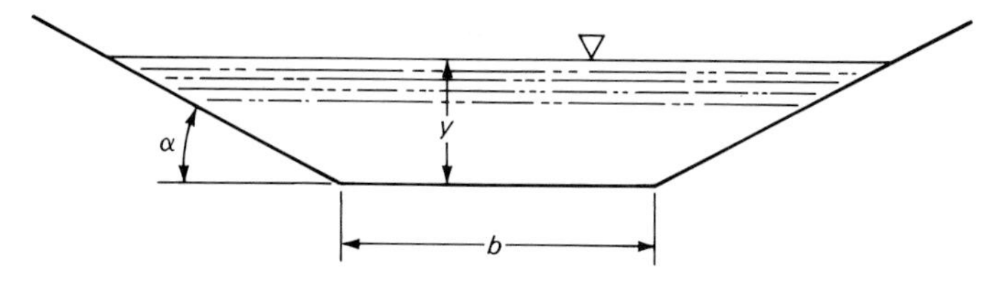

FIGURE 3-10
Trapezoidal section of Example 3-14.

for R_{max}, $dR/dy = 0$. Then

$$\frac{dR}{dy} = \frac{-A[-A/y^2 - 1/\tan\alpha + 2/\cos(90 - \alpha)]}{[A/y - y/\tan\alpha + 2y/\cos(90 - \alpha)]^2} = 0$$

$$A = y^2\left[\frac{-1}{\tan\alpha} + \frac{2}{\cos(90 - \alpha)}\right]$$

$$R = \frac{y^2[-1/\tan\alpha + 2/\cos(90 - \alpha)]}{2[-y/\tan\alpha + 2y/\cos(90 - \alpha)]} = \frac{y}{2}$$

For maximum flow a trapezoidal channel should thus be proportioned so that the hydraulic radius equals one-half the depth. A rectangle is a special case of the trapezoid for which $\alpha = 90°$, hence $R = y/2$ for best hydraulic section. R for a rectangular section, however, is also given by

$$R = \frac{by}{b + 2y}$$

$$\frac{y}{2} = \frac{by}{b + 2y}$$

$$b = 2y$$

Thus for best hydraulic section a rectangular channel should be proportioned so the depth of flow is one-half the width.

Example 3-15 Find the best dimensions for a rectangular channel intended to carry a flow of 0.5 m^3/s at a velocity of 1.2 m/s.

Solution

$$A = \frac{Q}{V} = \frac{0.5}{1.2} = 0.4166 \text{ m}^2$$

$$y \times b = y \times 2y = 0.4166$$

$$y = 0.456 \text{ m}$$

$$b = 0.912 \text{ m}$$

The actual dimensions would be modified somewhat for ease of construction and to provide some freeboard. If the section were dimensioned in English units the width would be 3 ft and the depth perhaps 1 ft 9 in.

3-15 Controls in Open-Channel Flow

Manning's equation permits calculation of S given V, n, and R; V given n, S, and R; and R given V, n, and S. While these values may be calculated, there is no requirement that they will, in fact, occur. Unlike a pipe flowing full, an open channel can carry water at any depth. The depth which actually occurs depends on which channel feature is acting as a control.

A control may be defined as a device which creates a unique relationship between depth and discharge. In this sense, a long sloping channel is a control since, if it is sufficiently long, uniform depth will be achieved and, from Manning's equation, $y = f(Q)$. Other devices which produce such a unique relationship include weirs, orifices, flumes, sluice gates, etc. These all produce critical flow somewhere in their immediate vicinity and will control the depth for some distance upstream and/or downstream.

If we consider a rectangular channel 1 m wide and 1 m deep carrying a flow of 0.62 m^3/s with $n = 0.012$ and $S = 0.001$, the depth of flow could be 0.34 m (at a free discharge), 0.569 m (if the channel is long), or any other depth between 0.34 and 1 m. The effect of controls and details of the calculation of the depths cited above are discussed further in the following articles.

3-16 Specific Energy

The specific energy equation is written with respect to the channel bottom rather than an arbitrary fixed datum. The energy at a channel section is thus

$$E = y + \frac{V^2}{2g} \tag{3-36}$$

For a rectangular section, $V = Q/by = q/y$, where q is the discharge per unit width. This substitution gives

$$E = y + \frac{q^2}{2gy^2} \tag{3-37}$$

Equation (3-37) may be rearranged as

$$(E - y)y^2 = \frac{q^2}{2g}$$

from which it may be observed that $(E - y)y^2$ is a constant for constant discharge and that a graph of y versus E must be asymptotic to $E = y$ and $y = 0$. A plot of specific energy versus depth for constant q is shown in Fig. 3-11. The point of minimum specific energy, E_{min}, may be determined by taking the

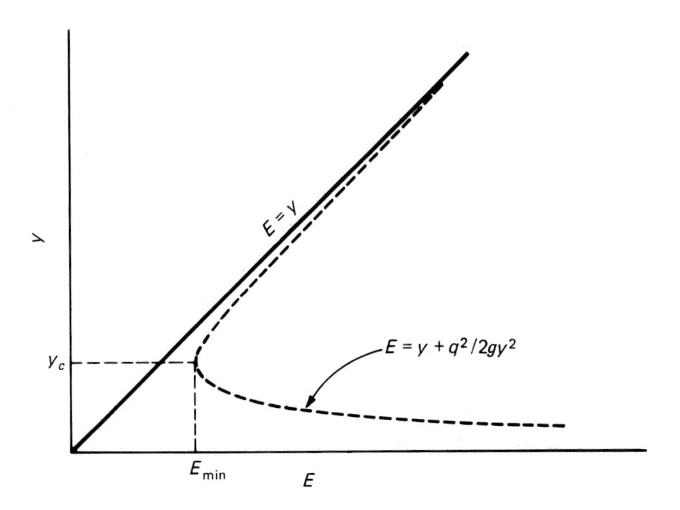

FIGURE 3-11
Specific energy versus depth at constant discharge.

derivative of Eq. (3-37) with respect to y, setting this equal to zero, and solving for y. From this one obtains

$$y_c = \left(\frac{q^2}{g}\right)^{1/3} \tag{3-38}$$

In similar manner, if one considers how q varies with y for constant E by rewriting Eq. (3-37) as

$$q = 2gy^2(E - y)$$

differentiating, equating to zero, and solving for y, one obtains

$$y_c = \tfrac{2}{3}E \tag{3-39}$$

The graph of y versus q for constant E is shown in Fig. 3-12. The two conditions illustrated in Figs. 3-11 and 3-12 (minimum E and maximum q) are, in fact, different representations of the same circumstance. This may be readily shown by substitution of Eq. (3-37) in Eq. (3-38) to yield

$$y_c = \tfrac{2}{3}E$$

The point of minimum specific energy and maximum discharge is called *critical flow*. The velocity corresponding to this condition is called *critical velocity*, and, for a rectangular channel, is given by

$$V_c = (gy_c)^{0.5} \tag{3-40}$$

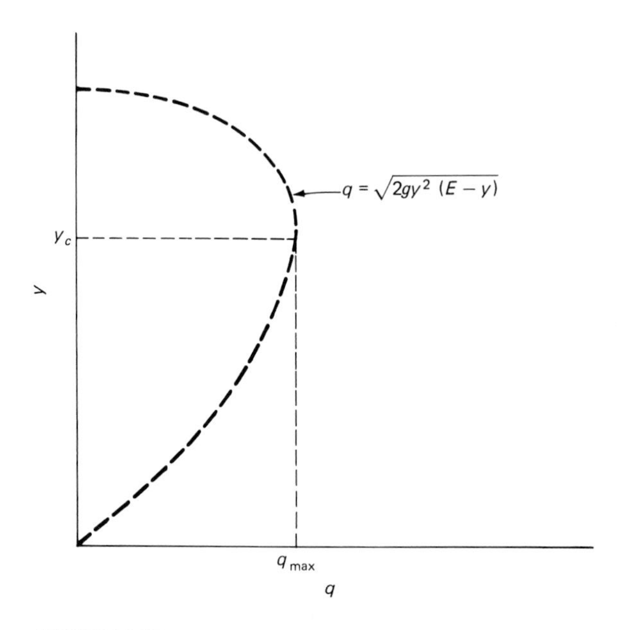

FIGURE 3-12
Discharge versus depth for constant specific energy.

For nonrectangular channels, the depth corresponding to critical flow may be found by solving

$$\frac{Q^2}{g} = \frac{A^3}{b} \tag{3-41}$$

where b is the width of the free water surface.

Example 3-16 Determine the critical depth in a trapezoidal channel with a base width of 2 m and side slopes of 1 on 1 which carries a flow of 5 m³/s.

Solution
For this channel

$$A = 2y + y^2$$

$$b = 2 + 2y$$

$$\frac{5^2}{9.806} = \frac{(2 + y^2)^3}{2 + 2y}$$

$$y = 0.754 \text{ m} \qquad \text{(by trial)}$$

Since critical flow defines a unique relationship between depth and discharge, a point of critical flow is a control, and, in fact, location of such points is the first step in analyzing the flow profile in open-channel problems.

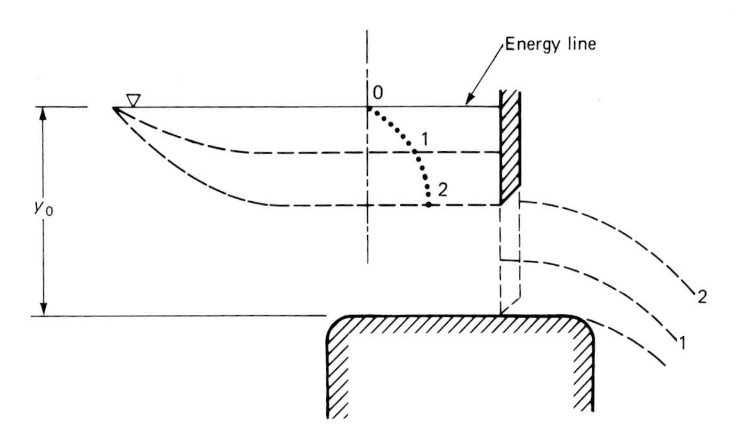

FIGURE 3-13
Broad-crested weir.

3-17 Location of Critical Flow Points

Critical flow will occur in certain definable circumstances, including transitions from normal subcritical flow to normal supercritical flow and vice versa. Neither of these circumstances is likely in sanitary engineering structures. More important locations of critical flow include overflow structures, similar free discharges, and flumes.

Overflow structures are particularly important, since all weirs are special cases of this condition. Considering a broad-crested weir controlled by a sluice gate (Fig. 3-13), one may observe that, beginning with the gate closed, $q = 0$, $E = y_0$. As the gate is raised the discharge will increase and the water surface above the gate will drop, since $y_0 = y + q^2/2gy^2$. Since E is constant the points 0, 1, and 2 describe the upper portion of the q versus y curve (Fig. 3-12). When the gate is raised above point 2 it no longer contacts the surface and cannot affect the discharge. Since q increases no further, point 2 corresponds to q_{max}, and thus the depth at some point on the weir must correspond to critical depth. It may also be shown[3] that curved free discharges where streamlines are not parallel also require that critical depth occur. In general, we may say that for a free discharge (no backwater), critical flow will occur at or near the discharge point.

Flumes are built in various configurations, but all are designed with the intent of producing critical depth in the flume throat and thereby creating a unique depth-discharge relationship. Considering Eq. (3-37) and neglecting any energy losses between the upstream channel and the throat, one obtains

$$E_1 = E_2$$

$$y_1 + \frac{q_1^2}{2gy_1^2} = y^2 + \frac{q_2^2}{2gy_2^2}$$

If we require that $y_2 = y_c$, this can be achieved by either raising the channel bottom so $y_2 = y_c$ or by reducing the channel width so that q_2 increases, thereby causing y_2 to be equal to y_c. Both of these things are done in most flumes. It is important to note that raising the bottom or reducing the width will cause the water surface in the constriction to drop if the upstream flow is subcritical.

Example 3-17 Water is flowing in a rectangular channel which is 2 m wide and carries a flow of 2 m³/s. The channel is constricted to a width of 0.5 m. Determine the depth in the constriction and the depth upstream.

Solution
The constriction may or may not produce critical flow. If critical flow occurs,

$$y_c = \left(\frac{q^2}{g}\right)^{1/3} = \left[\frac{(2/0.5)^2}{9.806}\right]^{1/3} = 1.177 \text{ m}$$

Since $E = 1.5y_c$, $E = 1.766$ m. If losses are neglected, the upstream energy is

$$y + \frac{q^2}{2gy^2} = 1.766$$

$$y + \frac{1^2}{19.616y^2} = 1.766$$

$$y = 1.748 \text{ m}$$

The depth in the throat might also be affected by other controls further downstream which could cause $y > y_c$.

Example 3-18 A channel 1 m wide and 1 m deep carries a flow of 0.62 m³/s on a slope of 0.001 with $n = 0.012$. The channel terminates in a free discharge. Find the normal depth and the depth in the vicinity of the discharge.

Solution
The normal depth is calculated from Manning's equation. $Q = 0.62$, $A = 1 \times y$, $P = 1 + 2y$, $R = y/(1 + 2y)$.

$$0.62 = \frac{1}{0.012} y \left(\frac{y}{1 + 2y}\right)^{2/3} (0.001)^{1/2}$$

$$y = 0.569 \text{ m} \qquad \text{(by trial)}$$

At a free discharge, $y = y_c$.

$$y = \left(\frac{q^2}{g}\right)^{1/3} = \left(\frac{0.62^2}{9.808}\right)^{1/3}$$

$$= 0.34 \text{ m}$$

The depth in the channel will be 0.34 m near the discharge and will increase upstream, approaching 0.569 m if the channel is sufficiently long. These cal-

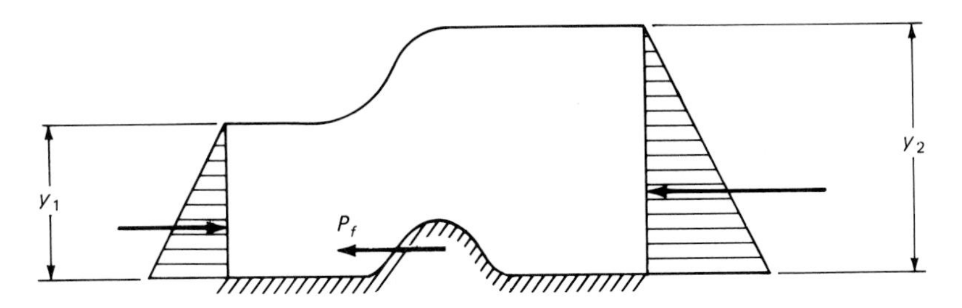

FIGURE 3-14
Definition sketch for momentum equation.

culations confirm the statements made in Art. 3-15 concerning depths in such a channel.

3-18 The Momentum Function

The energy equation permits solution of a large number of open-channel flow problems. However, there are some situations which require use of the momentum function. In general, if one considers water moving in a rectangular open channel and subjected to some force as shown in Fig. 3-14, one may write

$$\Sigma F = \Delta MV$$

$$\gamma \frac{y_1^2}{2} \cdot b - P_f - \gamma \frac{y_2^2}{2} \cdot b = Q\rho \, \Delta V$$

$$\gamma \frac{y_1^2}{2} \cdot b - P_f - \gamma \frac{y_2^2}{2} \cdot b = qb\rho\left(\frac{q}{y_2} - \frac{q}{y_1}\right)$$

$$\frac{P_f}{b} = \gamma\left[\left(\frac{q^2}{y_2} + \frac{y_2^2}{2}\right) - \left(\frac{q^2}{y_1} + \frac{y_1^2}{2}\right)\right] \tag{3-42}$$

where Pf/b is the force per unit width of channel. If $Pf/b = 0$, then the right-hand side of the equation equals zero and

$$\frac{q^2}{y_1} + \frac{y_1^2}{2} = \frac{q^2}{y_2} + \frac{y_2^2}{2} \tag{3-43}$$

The quality $(q^2/y + y^2/2)$ is called the *momentum function*:

$$M = \frac{q^2}{y} + \frac{y^2}{2} \tag{3-44}$$

If M is plotted versus y for constant q, one obtains a curve of the form shown in Fig. 3-15. As with the specific energy function (Art. 3-16), one may determine the value of y corresponding to M_{min}. When this is done one obtains

$$y = \left(\frac{q^2}{y}\right)^{1/3}$$

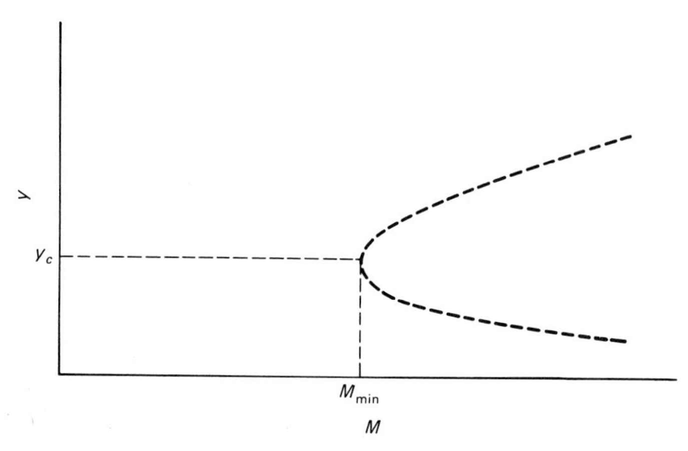

FIGURE 3-15
Momentum function versus depth at constant discharge.

indicating that the condition of minimum value of the momentum function corresponds to critical depth. The momentum function can often be used to advantage to determine the depth of flow across a transition in which the energy loss is unknown. The application of this relation to hydraulic jumps is a typical example of such problems.

Example 3-19 1 m^3/s discharges under a sluice gate at a velocity of 3 m/s in a channel 1 m wide. Determine whether the flow is supercritical and, if so, the conjugate depth to which the water will rise following a hydraulic jump.

Solution

$$y_c = \left(\frac{q^2}{g}\right)^{1/3} = \left(\frac{1}{9.808}\right)^{1/3} = 0.467 \text{ m}$$

$$V_c = \frac{1}{0.467} = 2.14 \text{ m/s}$$

Therefore $V > V_c$ and the flow is supercritical. To find the conjugate depth, use

$$\frac{q^2}{gy_1} + \frac{y_1^2}{2} = \frac{q^2}{gy_2} + \frac{y_2^2}{2}$$

$$\frac{1^2}{9.808(0.333)} + \frac{0.333^2}{2} = \frac{1^2}{9.808y_2} + \frac{y_2^2}{2}$$

$$y^2 = 0.633\text{m} \qquad \text{(by trial)}$$

3-19 Nonuniform Flow

The general equation describing nonuniform flow is

$$\frac{dE}{dx} = S_0 - S_f \tag{3-45}$$

where E = energy
x = distance
S_0 = channel slope
S_f = slope of the total energy line

Solution of nonuniform flow problems requires, in general, that Eq. (3-45) be integrated. The calculation must begin at a control (where the depth is known) and proceed in the direction in which the control is exercised. In most circumstances in sanitary engineering we deal with uniform channels and, very often, rectangular channels. This considerably simplifies the calculations. If we write Eq. (3-45) in finite difference form we obtain

$$\frac{\Delta E}{\Delta x} = \frac{\Delta(y + V^2/2g)}{\Delta x} = S_0 - S_f \tag{3-46}$$

S_f, from Eq. (3-26), is V^2/C^2R, whence

$$\frac{\Delta(y + V^2/2g)}{\Delta x} = S_0 - \frac{V^2}{C^2R} \tag{3-47}$$

for uniform channels S_0 is constant. Thus all terms in the equation are simple functions of y. We can therefore select appropriate values of y and calculate the intervals Δx between successive depths. This procedure is commonly called the *step method.*

Example 3-20 A rectangular channel 1.25 m wide and 30 m long with a flat bottom carries a flow of 0.5 m^3/s. The channel terminates in a free discharge. Plot the water profile in the channel. $n = 0.011$.

Solution
The depth at the free discharge will be critical.

$$y_c = \left(\frac{q^2}{g}\right)^{1/3} = \left(\frac{(0.5/1.25)^2}{9.808}\right)^{1/3} = 0.254 \text{ m}$$

In general, at any point along the channel,

$$R = \frac{A}{P} = \frac{1.25y}{1.25 + 2y}$$

$$S_0 = 0$$

$$S_f = \left(\frac{Vn}{R^{2/3}}\right)^2 = \left[\frac{(0.011)(0.5/1.25)}{y(1.25y/(1.25 + 2y))^{2/3}}\right]^2$$

Nominating values of y, we can calclate successively the values shown in the table below. Increments are chosen to be relatively small near the discharge, where the depth is changing rapidly. The depth at the upstream end is most readily determined graphically as shown in the plot of depth versus distance in Fig. 3-16.

y	V	E	ΔE	S_f	S_{favg}	Δx
0.254	1.575	0.380428		0.002939		
			0.000232		0.002842	0.082
0.260	1.538	0.380660		0.002744		
			0.001228		0.002600	0.472
0.270	1.481	0.381888		0.002456		
			0.002151		0.002332	0.922
0.280	1.429	0.384038		0.002208		
			0.002949		0.002101	1.404
0.290	1.379	0.386987		0.001994		
			0.003642		0.001900	1.917
0.300	1.333	0.390629		0.001807		
			0.009025		0.001653	5.460
0.320	1.250	0.399654		0.001499		
			0.010905		0.001379	7.908
0.340	1.176	0.410559		0.001259		
			0.012378		0.001165	10.625
0.380	1.053	0.436486		0.000918		

The step method is easily programmed and a number of procedures usable on programmable calculators[8] or digital computers[9] are readily available.

In the case of the flat-bottomed channel of Example 3-20, it is obvious that the depth of the water will increase without limit as the length of the channel increases. In channels for which $S_0 \neq 0$, the depth will increase until normal depth is attained and will remain at that level until some other factor causes it to change. It is of some interest to examine how the length of a channel affects its carrying capacity in circumstances in which the available energy is fixed. Example 3-21 considers such a case.

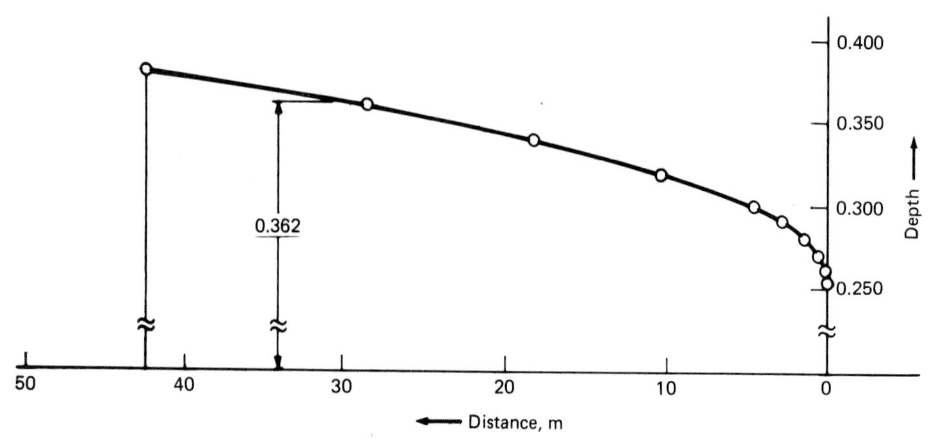

FIGURE 3-16
Water profile in flat-bottomed channel.

Example 3-21 A rectangular channel 1 m wide with $n = 0.012$ lies on a slope of 0.001. The channel connects a basin with a water surface 0.75 m above the channel bottom with a free discharge at the lower end. Determine the variation of flow with channel length.

Solution

The limiting cases are that in which the length is negligibly small and critical flow occurs at the entrance to the channel and that in which the channel is sufficiently long for normal flow to occur at the upper end. For critical flow,

$$E = y + \frac{q^2}{2gy^2} = 0.75 \text{ m}$$

$$y_c = \frac{2}{3}E = 0.50 \text{ m}$$

$$q = (gy_c^3)^{0.5} = 1.107 \text{ m}^2/\text{s}$$

$$Q = 1.107 \text{ m}^3/\text{s}$$

For flow at normal depth,

$$E = y + \frac{q^2}{2gy^2} = 0.75 \text{ m}$$

$$q = Q/1.0 = \frac{1}{n}AR^{2/3}S^{1/2} = \frac{1}{0.012}y\left(\frac{y}{1 + 2y}\right)^{2/3}(0.001)^{1/2}$$

$$= [(0.75 - y)2gy^2]^{1/2}$$

$$= 2.635y\left(\frac{y}{1 + 2y}\right)^{2/3}$$

Solving these simultaneously gives

$$y = 0.683 \text{ m}$$

$$q = Q = 0.785 \text{ m}^3/\text{s}$$

Finding the flow carried by channels of lengths intermediate between 0 and ∞ requires an iterative solution if the length is specified. It is generally easier to assume a depth at the downstream end intermediate between y_c and y_{normal} and calculate the backwater curve until $E = E_0$ (0.75 m in this case). The sum of the Δx values will give the length of the channel which corresponds to this discharge. The way in which discharge varies with length for the channel of Example 3-21 is illustrated in Fig. 3-17.

3-20 Open-Channel Flow Measurement and Control Structures

The most important controls in sanitary engineering structures are *sharp-crested weirs*. These devices are used to control the flow distribution in clarifiers, flow splitters and other basins and, at times, to measure flow.

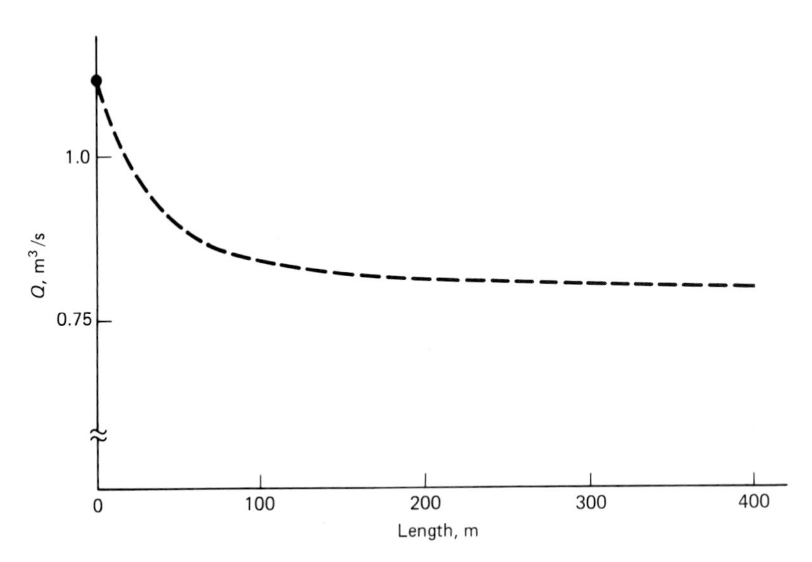

FIGURE 3-17
Variation of discharge with length in a channel of fixed slope and fixed available specific energy.

The simplest weir of this sort is the suppressed rectangular weir which extends across the entire width of the approach channel. The relation between discharge and head on this structure is

$$Q = \frac{2}{3} C_d L (2g)^{0.5} h^{1.5} \tag{3-48}$$

where L = weir length
h = vertical distance from the crest of the weir to the horizontal surface of the water some distance upstream
C_d = discharge coefficient (includes effect of approach velocity and contraction of flow in immediate vicinity of the weir)

C_d varies with the ratio of h to w, where w is the height of the weir above the channel bottom. When $h/w < 0.2$ (which is normally the case), C_d is given by

$$C_d = 0.611 + 0.08 \times \frac{h}{w} \tag{3-49}$$

Example 3-22 Determine the rate of flow and velocity in the approach channel when a head of 0.15 m exists on a sharp-crested rectangular weir 1.2 m long and 0.9 m high.

Solution

$$C_d = 0.611 + 0.08\left(\frac{0.15}{0.9}\right) = 0.624$$

$$Q = \frac{2}{3}(0.624)(1.2)(2g)^{0.5}(0.15)^{1.5} = 0.128 \text{ m}^3/\text{s}$$

$$V = \frac{Q}{A} = \frac{0.128}{1.2(0.9 + 0.15)} = 0.102 \text{ m/s}$$

The calculation is usually somewhat more complicated, since ordinarily flow is known but depth on the weir is not.

Example 3-23 The channel and weir of Example 3-22 receive a flow of 0.3 m³/s. What will be the depth and velocity in the channel?

Solution
This problem is not directly soluble, since C_d depends on h and h is unknown. On the basis of the solution of Example 3-22 we might assume $C_d \approx 0.63$. Since the flow is greater, the depth must be greater, and C_d increases with depth.

$$0.3 = \frac{2}{3}(0.63)(1.2)(2g)^{0.5}h^{1.5}$$

$$h = 0.262 \text{ m} \qquad C_d = 0.611 + 0.08\left(\frac{0.262}{0.9}\right) = 0.635$$

$$0.3 = \frac{2}{3}(0.635)(1.2)(2g)^{0.5}h^{1.5}$$

$$h = 0.261 \text{ m} \qquad C_d = 0.634$$

$$y = 0.9 + 0.261 = 1.161 \text{ m}$$

$$V = \frac{0.3}{1.161(1.2)} = 0.215 \text{ m/s}$$

One could then proceed to determine the upstream water profile as described in Art. 3-19.

Weirs which do not extend across the entire channel width produce a horizontal as well as a vertical contraction of the flow. This complexity of the flow pattern makes such weirs less reliable as measurement devices, and calibration is desirable. Rectangular contracted weirs may be described by

$$Q = \frac{2}{3}(0.611)(L - 0.2h)(2g)^{0.5}h^{1.5} \tag{3-50}$$

provided $L > 3h$ and the side contraction is greater than $3h$. Triangular weirs are described by

$$Q = \frac{8}{15}C_c \tan\frac{\alpha}{2}(2g)^{0.5}h^{2.5} \tag{3-51}$$

C_c is a contraction coefficient which varies with α, and α is the central angle of the triangle. For 90° weirs, C_c is about 0.585, which permits reduction of Eq. (3-51) to

$$Q = 0.312(2g)^{0.5}h^{2.5} \tag{3-52}$$

Example 3-24 An effluent weir in a clarifier consists of 90° V-notch weirs 225 mm on centers and has a design weir overflow rate of 250 $m^3/m \cdot day$. Determine the height of the water surface above the bottom of the notches at design flow and at twice design flow.

Solution

$$Q_{notch} = \frac{250}{1/0.225} = 56.25 \text{ m}^3\text{/day}$$

$$0.000651 \text{ m}^3\text{/s}$$

$$0.000651 = 0.312(2g)^{0.5}h^{2.5}$$

$$h = 0.047 \text{ m} \qquad \text{at design flow}$$

$$2 \times 0.000651 = 0.312(2g)^{0.5}h^{2.5}$$

$$h = 0.062 \text{ m} \qquad \text{at twice design flow}$$

To assure uniform flow, the notches must have a depth of at least 62 mm. V-notch weirs permit less variation of flow than flat weirs when irregularities are present. A weir designed as detailed in Example 3-24 and a flat weir of the same total length, each installed with a variation in elevation from side to side of 2.5 mm, would have a variation in discharge from side to side of about 17 percent and 50 percent respectively.

It may be noted that, in a rectangular weir, discharge varies with $h^{1.5}$ and, in a triangular weir, with $h^{2.5}$. One may ask if there exists a shape for which discharge varies linearly with head, and it develops that there is such a shape. The *proportional flow* or Rettger weir has the discharge equation

$$Q = K\frac{\pi}{2}(2h)^{0.5}h \tag{3-53}$$

and the weir is shaped (Fig. 3-18) according to

$$2xy^{0.5} = K \tag{3-54}$$

Proportional flow weirs are used to regulate the velocity in grit chambers. The design of such a weir is presented in Chap. 21.

The treatment of weirs has thus far presumed that the discharge is free, that is, that the water falls over the crest without any downstream restriction. If a downstream control causes the downstream level to exceed the height of the weir, the discharge will be affected or the upstream depth will be increased. For

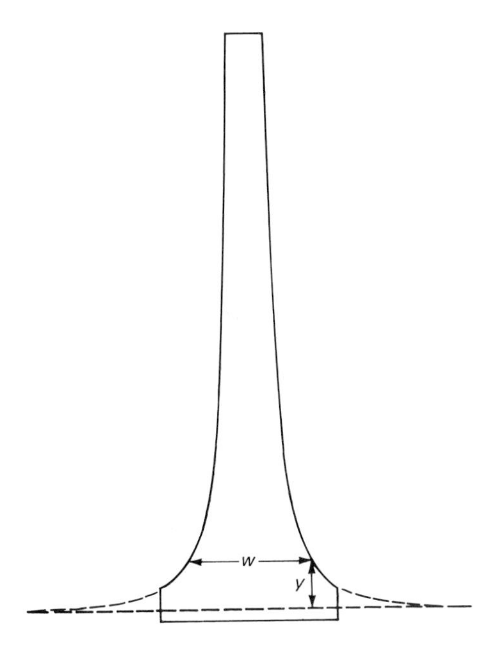

FIGURE 3-18
Proportional flow weir.

rectangular weirs the ratio of discharge of a submerged weir to that of a free weir is given by

$$\left(\frac{Q_s}{Q_f}\right)^2 = 1 - \left(\frac{h_2}{h_1}\right)^2 \tag{3-55}$$

where h_1 is the upstream depth above the weir crest and h_2 is the downstream depth above the weir crest.

Example 3-25 The weir of Example 3-23 ($Q = 0.3$ m^3/s, $L = 1.2$ m, $w = 0.9$ m) is flooded by a downstream control which causes the depth in the channel below the weir to be 1.05 m. What is the upstream depth?

Solution
A direct solution is not possible since Q_f (the free discharge at the new depth) is unknown. From Example 3-23, h_1 was 0.261. It is reasonable to expect that this will increase. If h_1 were 0.30,

$$\left(\frac{Q_s}{Q_f}\right)^2 = 1 - \left(\frac{0.15}{0.30}\right)^2$$

$$\frac{Q_s}{Q_f} = 0.866$$

At $h_1 = 0.3$, $C_d = 0.638$; therefore $Q_f = 0.371$ and $Q_s = 0.866 \times 0.371 = 0.322$ m^3/s. Since Q_s in fact is 0.3 m^3/s, h_1 must be less. At $h_1 = 0.295$,

$$\frac{Q_s}{Q_f} = 0.861 \qquad C_d = 0.637 \qquad Q_f = 0.362 \qquad Q_s = 0.311$$

At $h_1 = 0.290$,

$$\frac{Q_s}{Q_f} = 0.856 \qquad C_d = 0.637 \qquad Q_f = 0.352 \qquad Q_s = 0.302$$

The upstream depth is thus approximately 0.29 m rather than the 0.26 m for the free discharge.

Underflow gates such as sluice gates and shear gates are often used in water and wastewater plants. The discharge through such a structure can be calculated from the energy equation and is

$$Q = y_1 y_2 L \left(\frac{2g}{y_1 + y_2}\right)^{0.5} \tag{3-56}$$

where y_1 and y_2 are the water depths above and below the gate and L is its width. If the equation is written in terms of the gate opening w, $y_2 = C_c w$ and

$$q = C_d w (2gy_1)^{0.5} \tag{3-57}$$

where

$$C_d = \left[\frac{C_c^2}{1 + C_c(w/y_1)}\right]^{0.5} \times C_c$$

C_c is commonly taken as equal to 0.61.

Example 3-26 A flow of 0.5 m³/s passes under a sluice gate 1.5 m wide which is raised 0.15 m. What is the upstream depth?

Solution

Since y_1 is unknown, C_d is not calculable; however, Eq. (3-56) will permit a direct solution.

$$y_2 = C_c w = 0.61(0.15) = 0.0915$$

$$0.5 = 0.0915 y_1 (1.5) \left(\frac{2g}{0.0915 + y_1}\right)^{0.5}$$

$$y_1^2 - 0.677 y_1 - 0.0619 = 0$$

$$y_1 = 0.758 \text{ m}$$

Drowned sluice gates, in which a downstream control causes the depth below the gate to exceed $C_c w$, can be analyzed by the use of the specific energy and momentum equations. Figure 3-19 illustrates the condition at such a structure. The flow at section 2 is considered to pass through the area $C_c wL$. The turbulent volume above the constricted section has no net velocity.

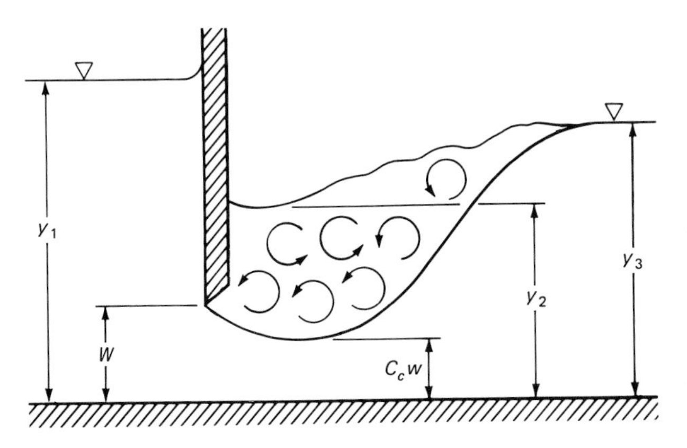

FIGURE 3-19
Drowned sluice gate.

Example 3-27 The gate of Example 3-26 has its discharge restricted by a downstream control which causes the depth a short distance below the gate to be 0.9 m. What is the upstream depth?

Solution
With reference to Fig. 3-19, $y_3 = 0.9$ m, $C_c w = 0.0915$ m. From sections 2 to 3, the momentum equation is used, since the energy loss between these points is unknown.

$$\frac{q^2}{gwC_c} + \frac{y_2^2}{2} = \frac{q^2}{gy_3} + \frac{y_3^2}{2}$$

$$y_2 = \left\{\left[\frac{(0.50/1.5)^2}{0.9g} + \frac{0.9^2}{2} - \frac{(0.5/1.5)^2}{0.0915g}\right] 2\right\}^{0.5}$$

$$y_2 = 0.767$$

From sections 2 to 1, the specific energy equation is used, since the energy loss through the gate will be very small.

$$y_1 + \frac{q^2}{2gy_1^2} = y_2 + \frac{q^2}{2g(wC_c)^2}$$

$$y_1 + \frac{(0.5/1.5)^2}{2gy_1^2} = 0.767 + \frac{(0.5/1.5)^2}{2g(0.0915)^2}$$

$$y_1 + \frac{0.005664}{y_1^2} = 1.444$$

$$y_1 = 1.442 \text{ m}$$

The *Parshall flume* is frequently used to measure open-channel flow. Typical details of such flumes are presented in Fig. 3-20 and Table 3-6. It is often necessary to raise the channel bottom at the flume in order to ensure that

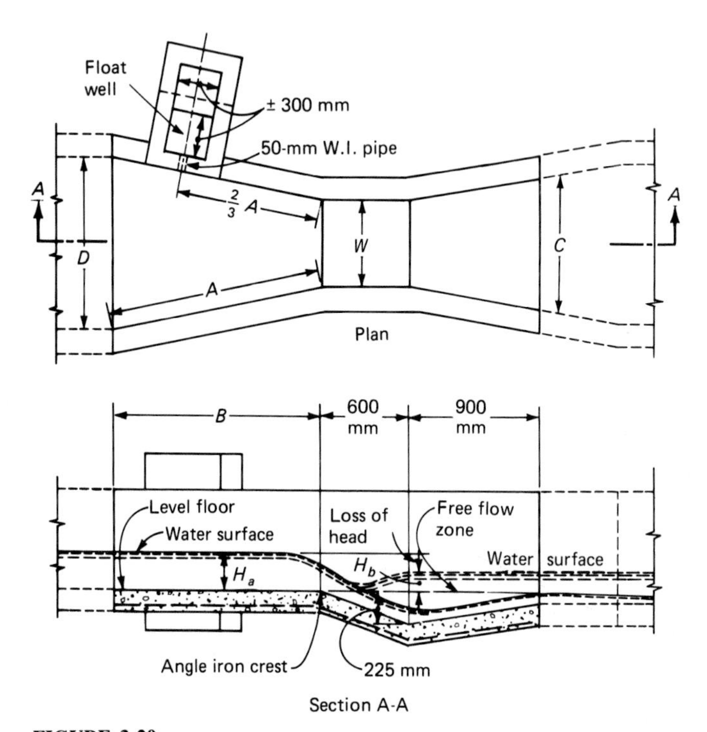

FIGURE 3-20
The Parshall flume.

TABLE 3-6
Standard dimensions and capacities of Parshall flumes (see Fig. 3-20)

Dimensions, mm*					Capacity, m³/min*	
W	*A*	*B*	*C*	*D*	Max.	Min.
300	1370	1345	600	845	27.4	0.6
600	1525	1495	900	1205	56.3	1.1
900	1675	1645	1200	1570	85.7	1.7
1200	1830	1795	1500	1935	115.4	2.1
1500	1980	1945	1800	2300	145.5	3.8
1800	2130	2090	2100	2665	176.0	4.5
2100	2290	2240	2400	3030	206.4	6.9
2400	2440	2390	2700	3400	237.2	7.9

* mm × 0.04 = in, m^3/min × 264.2 = gal/min

the upstream depth will be adequate. The required depth of the step may be determined from the energy and momentum equations. The flow through a Parshall flume is given by

$$Q = 2.23 \times 10^{-5} W \left(\frac{H_a}{0.305} \right)^{1.57 W^{0.026}} \tag{3-58}$$

provided $H_b < 0.7H_a$.

3-21 Transitions

Transitions in open channels include changes in direction, slope, or cross section which produce a change in the state of flow. Transitions, by definition, include controls.

Abrupt *expansions* produce energy losses analogous to those in pipe expansions (Art. 3-7). For practical purposes with subcritical flow and moderate expansions, it is satisfactory to calculate the energy loss using Eq. (3-13). It is important to note that in subcritical flow the water will rise in the downstream direction across an expansion.

Example 3-28 A channel 0.6 m wide expands abruptly to 1.2 m. The flow is 0.3 m³/s. Backwater calculations from a downstream control indicate a downstream depth of 0.76 m. What is the upstream depth?

Solution
The downstream specific energy is given by

$$E_2 = \frac{q_2}{2gy_2^2} + y_2 = \frac{(0.3/1.2)^2}{2g(0.76)^2} + 0.76 = 0.7656 \text{ m}$$

The energy loss across the expansion is

$$\Delta E = \frac{(V_1 - V_2)^2}{2g} = \frac{[0.3/0.6(y_1) - 0.3/(1.2)(0.76)]^2}{2g}$$

$$E_1 = E_2 + \Delta E$$

$$E_1 = \frac{(0.3/0.6)^2}{2gy_1^2} + y_1$$

Therefore

$$\frac{0.5^2}{2gy_1^2} + y_1 = \frac{(0.5/y_1 - 0.329)^2}{2g} + 0.7656$$

$$y_1 = 0.749 \text{ m} \qquad \text{(by trial)}$$

The surface, as predicted, rises across the expansion—in this case by about 10 mm (0.4 in). The calculated energy loss is 0.006 m. A satisfactory value for the loss could be obtained from the approximation

$$\Delta E = \frac{V_2^2}{2g}\left(\frac{b_2}{b_1} - 1\right)^2$$

$$\Delta E = \frac{[(0.3/(0.76)(1.2)]^2}{2g}(2 - 1)^2 = 0.0055$$

which is in error by less than 10 percent. Gradual expansions have lower energy losses. For an expansion with a 1-on-4 taper,

$$\Delta E = 0.3\frac{(V_1 - V_2)^2}{2g} \tag{3-59}$$

Contractions in open channels, as in pipes, produce lower losses than expansions. Losses in square-edge and round-edge contractions, respectively, are given by Eqs. (3-60) and (3-61).

$$\Delta E = 0.23\frac{V_2^2}{2g} \tag{3-60}$$

$$\Delta E = 0.11\frac{V_2^2}{2g} \tag{3-61}$$

where V_2 is the downstream velocity.

Changes in direction produce losses which may be approximated by

$$\Delta E = K\frac{V^2}{2g} \tag{3-62}$$

Values for K for various types of directional changes are presented in Table 3-7. Additional information may be found in Refs. 3, 10, and 11.

3-22 Lateral Inflow

A large number of sanitary engineering structures involve channels in which the flow varies in magnitude along the length. Filter backwash launders, clarifier effluent channels, and trickling filter underdrains are examples of channels in which such variations occur. Applications of the momentum equation to this situation in flat-bottomed channels yields

$$h = \left(y_L^2 + \frac{2Q^2}{gb^2y_L}\right)^{0.5} \tag{3-63}$$

where h = upstream depth
y_L = downstream depth
Q = flow at downstream end

The value of y_L is obtained from the downstream control. When y_L equals y_c (which is often the case),

TABLE 3-7
Loss coefficients in open channels

Type of change	K
Smooth 90° bend	$0.6b/r$*
Smooth 45° bend	$0.4b/r$
Smooth $22\frac{1}{2}$° bend	$0.3b/r$
Angular deflection	$(\theta/90)^{0.5}$*

* b = channel width; r = radius of curvature; θ = angular deflection

$$h = 1.73\left(\frac{q^2}{g}\right)^{1/3} \tag{3-64}$$

When h is known, the depth at any other point along the channel is obtainable from

$$y(h^2 - y^2) = \frac{2Q^2}{gb^2}\left(\frac{x}{L}\right)^2 \tag{3-65}$$

where y = depth
x = distance from upstream end
L = channel length

Equation (3-65) is based on a uniform addition of flow along the channel length. The depths obtained from these equations are only a first approximation, since energy losses are neglected.

Example 3-29 A clarifier has a circular effluent launder 30 m in diameter and 0.45 m wide. The total flow to the clarifier is 0.18 m³/s. One-half the flow goes around each side of the clarifier. Determine the required launder depth to provide a freeboard of 25 mm at the upstream end. The channel bottom is flat and the discharge is free.

Solution
Since the discharge is free,

$$y_L = y_c$$

$$y_L = \left(\frac{q^2}{g}\right)^{1/3} = \left\{\frac{[0.18/(2)(0.45)]^2}{g}\right\}^{1/3} = 0.160 \text{ m}$$

$$h = 1.73y_L = 0.277 \text{ m}$$

The approximate water surface at 1/10 points around the periphery and the resulting frictional losses are tabulated below. Energy losses are based on a value of $n = 0.014$. Because of the turbulence induced by lateral inflow and the curvature of the channel, n should be increased somewhat above normal values.

x/L	y	Q	S_f	L	HL
0	0.277	0	0		
				4.71	0.000039
0.1	0.276	0.009	0.0000166		
				4.71	0.000197
0.2	0.275	0.018	0.0000672		
				4.71	0.000525
0.3	0.272	0.027	0.0001557		
				4.71	0.001044
0.4	0.268	0.036	0.0002877		
				4.71	0.001789
0.5	0.263	0.045	0.0004721		
				4.71	0.002848
0.6	0.255	0.054	0.0007372		
				4.71	0.004333
0.7	0.246	0.063	0.0011028		
				4.71	0.006513
0.8	0.233	0.072	0.0016630		
				4.71	0.010136
0.9	0.214	0.081	0.0026412		
				4.71	0.023211
1.0	0.160	0.090	0.0072150		
				Total HL =	0.051 m

The upstream depth is therefore 0.277 + 0.051 = 0.328, and to provide a 25-mm freeboard the launder depth must be 353 mm.

In sloping channels receiving flow along their length, the upstream depth may be approximated by

$$h = \frac{1}{2}\left\{\left[L^2S^2 + 4\left(y_L^2 + \frac{2Q^2}{gb^2y_L} - y_LLS\right)\right]^{0.5} - LS\right\} \tag{3-66}$$

where L and S are the length and slope, respectively, and the other terms are as in Eq. (3-63). The depth at other points along the channel is calculable, for the case of uniform addition along the channel, from

$$-y^3 + y^2LS + y(h^2 + hLS) = \frac{2Q^2}{gb^2}\left(\frac{x}{L}\right)^2 \tag{3-67}$$

Example 3-30 Rework Example 3-29 using a channel with a slope of 0.001.

Solution

y_L is unchanged; $y_L = 0.16$.

$$h = \frac{1}{2}\left\{\left[47.1^2(0.001)^2 + 4\left(0.16^2 + \frac{2(0.09)^2}{g(0.45)^2(0.16)} - 0.16(47.1)(0.001)\right)\right]^{0.5} - 0.001(47.1)\right\} = 0.240 \text{ m}$$

It should be noted that the water surface from upstream to downstream will drop a distance of (0.240 − 0.160) + 47.1(0.001) = 0.127 m. The depths, friction slopes, and net change in the water surface are tabulated below.

x/L	y	Q	S_f	$S_f - S_0$	HL
0	0.240	0	0	—	
					0
0.1	0.244	0.009	0.001000	—	
					0
0.2	0.247	0.018	0.001000	—	
					0
0.3	0.248	0.027	0.001000	—	
					0
0.4	0.248	0.036	0.001000	—	
					0
0.5	0.246	0.045	0.001000	—	
					0
0.6	0.243	0.054	0.001000	—	
					0.0005
0.7	0.238	0.063	0.001204	0.000204	
					0.0022
0.8	0.229	0.072	0.001741	0.000741	
					0.0055
0.9	0.215	0.081	0.002608	0.001608	
					0.0184
1.0	0.160	0.090	0.007216	0.006216	
				Total HL =	0.027 m

The upstream water depth is thus 0.24 + 0.027 = 0.267 m. Adding the 25-mm freeboard gives an upstream depth of 0.292 m. The downstream channel depth is 0.292 + 47.1(0.001) = 339 mm.

Example 3-30, while lengthy, demonstrates several factors concerning channels carrying variable flows. First, one may note that little change in upstream depth occurs as a result of sloping the channel. Additionally, the downstream depth is greater, leading to both greater use of material and more difficult fabrication. It may further be noted that the water depth does not increase uniformly along the sloping channel, but reaches a maximum (in this case between 0.3 and 0.4 of the length). The water surface, of course, rises throughout the length of the channel.

3.23 Lateral Outflow

Distribution channels, entrance structures to clarifiers, trickling filter distributors, and similar devices are examples of lateral outflow. Improper design of such structures has led to circumstances in which the flow has been very poorly distributed.[12,13]

In cases in which the flow in the main distribution channel is subcritical (which is the only condition likely in sanitary engineering structures), a number of observations may be made. The energy losses will be small, since the velocity is low and such channels are short. As water is withdrawn from the channel over side weirs or through orifices, the flow and the velocity in the channel will decrease. Since the velocity head decreases and energy losses are negligible, the depth must increase in the downstream direction. The increase in depth, in itself, will cause the discharge per unit length of weir or per orifice to increase in the downstream direction. In addition, the orifice or weir coefficients are also affected by the channel velocity and increase with decreasing velocity. These factors can combine to provide substantially greater discharge at the downstream end of lateral outflow structures than that which occurs upstream.

Details of important design considerations are presented in Refs. 12 and 13. In general, the difference in discharge from upstream to downstream can be minimized by maximizing the differential head across individual orifices (that is, by making the openings small); by minimizing the velocity in the main channel; or by varying the channel cross section to maintain constant velocity.

Example 3-31 Water enters a clarifier through a rectangular channel 0.75 m wide at a rate of 1 m^3/s. The channel is 25 m long and has orifices spaced 0.5 m on centers (49 in all). Determine the variation in water elevation in the channel. The water depth at the upstream end of the channel is 1.0 m.

Solution

The water surface in the channel will vary according to

$$y + \frac{V^2}{2g} = \text{const}$$

$$y + \frac{0.09065Q^2}{y^2} = C$$

if friction losses are neglected. The depth at the upstream end is given as 1.0 m. The depth at the downstream end is thus

$$1 + \frac{0.0965}{1^2} = 1.097 \text{ m}$$

The differential head between the first and last orifices varies by 97 mm. This will produce a substantial difference in discharge in an orifice of reasonable size. It is even possible that the flow could reverse through the first few openings. Increasing the channel depth to 3 m would give a downstream depth of

$$3 + \frac{0.09065}{3^2} = 3.010 \text{ m}$$

and a variation of only 10 mm. If the channel were 1.5 m wide rather than 0.75 m, the downstream depth would be

$$3 + \frac{0.04533}{3^2} = 3.005 \text{ m}$$

Example 3-31 does not address the effect of varying orifice size or the effect of channel velocity on the orifice discharge coefficient. It is clear, however, that if the orifices are small the differential head across them will be large and variations in the differential will have a smaller effect on the flow.

PROBLEMS

3-1. The polymer of Example 3-1 is diluted to an absolute viscosity of 7.7×10^{-3} kg/(m · s) and a specific gravity of 1.01. The diluted polymer is pumped through 90 m of $\frac{1}{2}$-in schedule 40 pipe at a rate of 3.78 L/min. What is the lost head in the pipe? What would be the lost head if water were pumped? Is this difference significant?

3-2. An 8-in-diameter class 2 ductile iron pipe has an ID of 8.39 in (213 mm). What will be the friction loss in 300 m of this pipe carrying a flow of 0.1 m³/s? e for this pipe is 0.00015 in.

3-3. A 10-in-diameter schedule 40 steel pipe has an ID of 10.02 in (255 mm). How much flow can this pipe carry if the allowable head loss is 5 m in a length of 200 m? e for steel pipe is 0.00015 in.

3-4. A 36-in-diameter concrete pipe has an ID of 36 in (914.4 mm). Determine the flow this pipe will carry at a head loss of 1 m in 300 m. e is 0.003 in.

3-5. A flow of 0.22 m³/s is to be conveyed between two basins with a loss in the pipe (due to friction) of 1 m in 30 m. What size steel pipe is required? Schedule 40 steel pipe has actual diameters of 15.00, 13.124, 11.938, and 10.020 in for 16-in through 10-in sizes.

3-6. In Prob. 3-5, what pipe size will be required if the entrance and exit losses (1.5 velocity heads total) must also be accommodated at a total loss of 1 m?

3-7. Estimate the head loss coefficient K for a venturi meter with an entrance angle of 20° and an exit angle of 5° for contraction ratios β of 0.4, 0.5, 0.6, 0.7, 0.8, and 0.9. Express K in terms of the velocity head in the pipe. (*Hint:* see Example 3-6.)

3-8. A pipeline carries a pumped flow of 0.05 m³/s. The line contains three 90° bends, a gate valve, and a swing check valve. Determine the minimum check valve size and the total head loss if the pipe and all the fittings are the same size as the valve. The length of straight pipe is 250 m.

3-9. A flow nozzle in a 4-in schedule 40 pipe has a diameter of 3.00 in. What will be the differential head across the nozzle at a flow of 0.01 m³/s?

3-10. Develop a rating curve (Q versus $h^{0.5}$) for a 4.50-in orifice plate installed in a 6-in schedule 40 pipe. The curve should cover flows from 0 to 0.2 m³/s.

3-11. Uniform flow occurs in a rectangular channel 1.5 m wide at a depth of 1 m. If the Chézy coefficient is 100 $ft^{0.5}$/s, find f, n, and e.

3-12. At what depth will 0.6 m³/s flow uniformly in a rectangular channel 0.9 m wide which is made of steel-troweled concrete and lies on a slope of 0.00025?

3-13. What depth and width are necessary in Prob. 3-12 if the depth is to be held to that which gives an optimum hydraulic cross section?

3-14. 1.2 m³/s flows in a rectangular channel 1.5 m wide with $n = 0.012$. Accurately plot y versus E for depths of 0.25 to 2 m at 0.25-m increments on y. Use the same scale for y and E. From the diagram find:

(*a*) The critical depth

(*b*) The minimum specific energy

(*c*) E when $y = 1.2$ m
(*d*) y when $E = 1.5$ m
What type of flow occurs at depths of 0.25 m and 1.25 m? What channel slopes would be necessary to maintain these depths? What is the critical slope?

3-15. A rectangular channel 0.9 m wide has a flat bottom and carries a flow of 0.4 m^3/s. If the channel is 25 m long and terminates in a free discharge, how deep will the water be at the upstream end? $n = 0.012$.

3-16. A channel similar to that of Prob. 3-15, but on a slope of 0.0005, connects a free discharge with a basin which has a water surface 1 m above the channel bottom at the upstream end. What is the discharge?

REFERENCES

1. Victor L. Streeter and E. Benjamin Wylie, *Fluid Mechanics*, 6th ed., McGraw-Hill, New York, 1975.
2. John K. Vennard and Robert L. Street, *Elementary Fluid Mechanics*, 5th ed., John Wiley and Sons, New York, 1975.
3. F. M. Henderson, *Open Channel Flow*, Macmillan, New York, 1966.
4. L. F. Moody, "Friction Factors for Pipe Flow," *Transactions of American Society of Mechanical Engineers,* **66**:8, 1944.
5. "Flow of Fluids through Valves, Fittings, and Pipe," Technical Paper No. 410, Crane Co., New York, 1982.
6. Raymond E. Davis and Harvey H. Jordan, "The Orifice as a Means of Measuring Flow of Water through a Pipe," Bulletin No. 109, University of Illinois Experiment Station, 1918.
7. C. F. Cusick, *Flow Meter Engineering Handbook*, Minneapolis-Honeywell, Philadelphia, 1961.
8. Thomas E. Croley, *Hydrologic and Hydraulic Computations on Small Programmable Calculators*, Iowa Institute of Hydraulic Research, Iowa City, 1977.
9. *HEC-2 Water Surface Profiles—Users Manual*, U.S. Army Corps of Engineers, 1982.
10. A. Shukry, "Flow round Bends in an Open Flume," *Transactions of American Society of Civil Engineers*, **115**:75, 1950.
11. J. Allen and Sek Por Chee, "The Resistance to the Flow of Water round a Smooth Circular Bend in an Open Channel," *Proceedings of Institute of Civil Engineers* (London), **23**:423, 1962.
12. H. E. Hudson, R. B. Uhler, and R. W. Bailey, "Dividing Flow Manifolds with Square-Edged Laterals," *Journal Environmental Engineering Division, American Society of Civil Engineers*, **105**:EE4:754, 1979.
13. Junn-Ling Chao and R. Rhodes Trussel, "Hydraulic Design of Flow Distribution Channels," *Journal of Environmental Engineering Division, American Society of Civil Engineers*, **106**:EE2:321, 1980.

CHAPTER 4

RAINFALL AND RUNOFF

4-1 Hydrologic Data

Hydrology involves the study of the water of the earth—its precipitation; its movement over the surface and below the surface of the ground; its evaporation and transpiration from land, water, and plants; and its subsequent condensation and reprecipitation. In recent years, systems analysis has led to a better understanding of hydrologic processes and to extensive use of mathematical and computer simulation models.

Engineers are concerned with hydrology, since water supplies are taken from streams, reservoirs, and wells which are fed directly or indirectly by precipitation. They are also concerned with maximum and minimum rates of runoff; total volume of flow for flood, drought, and average conditions; and similar data which are necessary for design of reservoirs, spillways, storm sewers, and other hydraulic structures. Only a descriptive treatment of a few important principles of the science of hydrology is presented in this text. Additional information may be found in the publications listed in the references.

Hydrologic data are collected from many sources. Information dealing with surface and groundwater quality and quantity has been compiled by the U.S. Geological Survey (U.S.G.S.) and may be readily recovered from the STORET system.[1] The U.S. Environmental Data Service has long-term records of precipitation and other climatological information,[2–6] while the U.S.G.S. and the U.S. Army Corps of Engineers serve as sources of storm runoff and flood flow data. Individual states may also compile data, either selectively from federal sources, or from a combination of these and their own agencies.[7]

4-2 Measurement of Precipitation

Precipitation, including rain, snow, hail, and sleet is the primary source of water in streams, lakes, springs, and wells. In the absence of stream-flow records, precipitation data are the basis for estimates of flood magnitude, low flow, and basin yield.

The U.S. National Weather Service maintains observation stations throughout the country and reports daily, monthly, and annual precipitation in publications of the Environmental Data Service. The data are expressed in millimeters or inches of rainfall per hour, day, month, or year. Time averages are arithmetic means for 30 years. Space averages are usually statewide averages of little use to engineers. Details of rain gauges and rainfall measurement techniques may be found elsewhere.[8]

4-3 Precipitation Data and Analysis

As noted above, precipitation is commonly expressed in terms of average intensity during some time period. Mean annual precipitation is perhaps the commonest form of such data. Figure 4-1 presents the distribution of annual rainfall in the United States. Lines connecting points of equal precipitation are called isohyets and illustrations such as Fig. 4-1 are called isohyetal maps.

While mean annual precipitation is useful, the variation during the year and variations from point to point are also of major importance to the design engineer. A compilation of smoothed point rainfall data for the United States may be found in Refs. 5 and 6. This information is based on data from instantaneous, continuous, hourly, and daily measurements which have been adjusted to a 1440-min and 60-min basis rather than a calendar day and clock hour basis. The data are presented in the form of isohyetal maps for return periods ranging from 1 to 100 years and durations of 5 min to 24 h.

The curves are based on a partial series analysis (which permits more than one high value per time period) with empirical and theoretical curve fitting. Some internal inconsistencies exist among the maps, since the isohyetal lines do not reveal nonlinear interpolation used in their construction. It is possible that a 12-h value may be found to be larger than the 24-h value at the same return period or that the 50-year value may be found to be larger than the 100-year value at the same duration. These inconsistencies are within the margin of error of the curves. Overall, the standard error of estimate ranges from 10 percent for a 2-year return period on flat terrain to 50 percent for a 100-year value in rough terrain. For durations less than 60 min, Ref. 6 should be used. Reference 5 contains 30-min maps and presents a procedure for calculating 5-, 10-, and 15-min rainfalls; however, the data base is inferior to that of Ref. 6 and will not yield the same results.

The maps of Refs. 5 and 6 may be considered to present the probability of a rainfall of particular magnitude at a particular location. It should be noted, however, that the probability of an n-year rainfall in n years is not 1. The

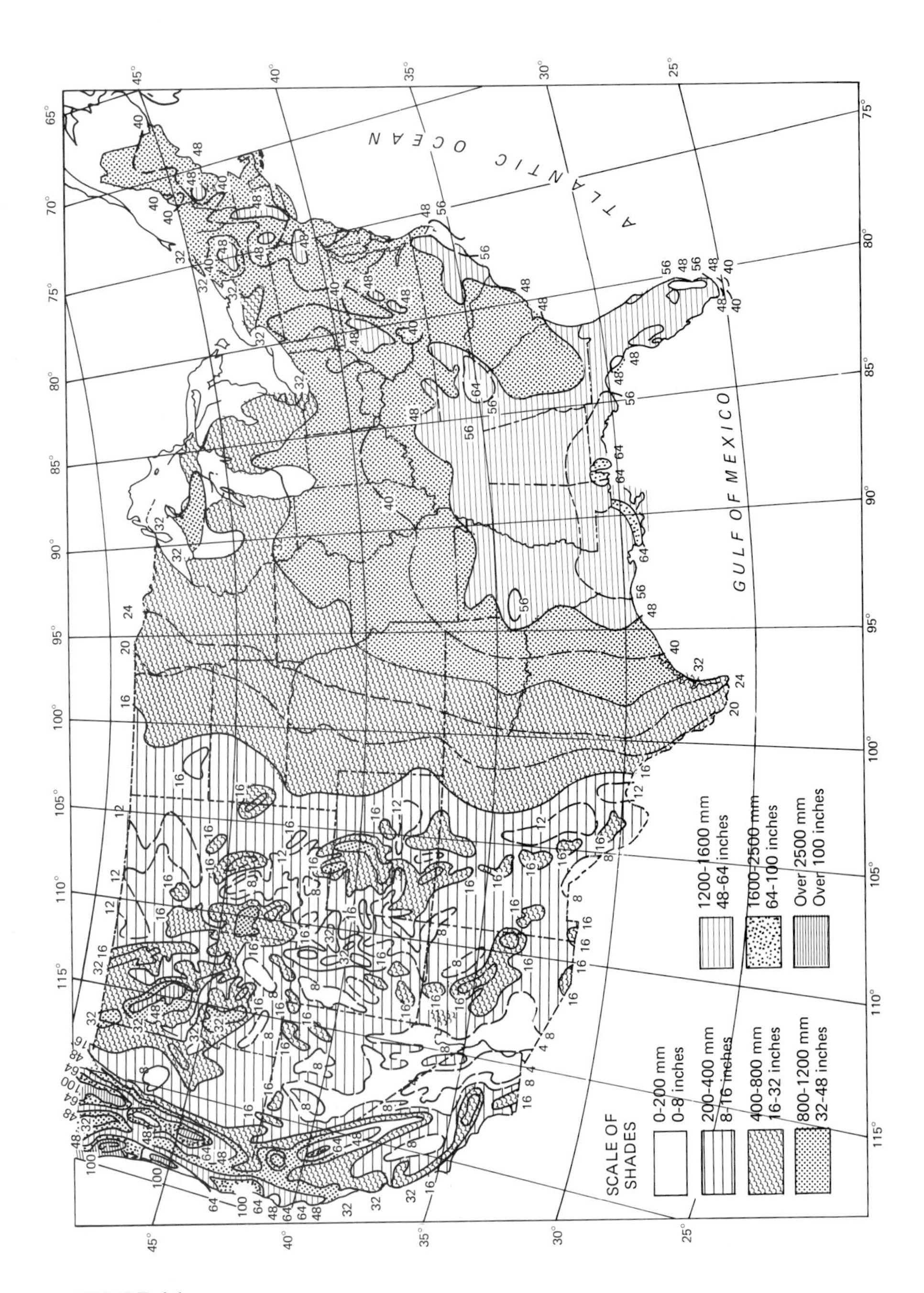

FIGURE 4-1
Normal annual precipitation in the United States.[3]

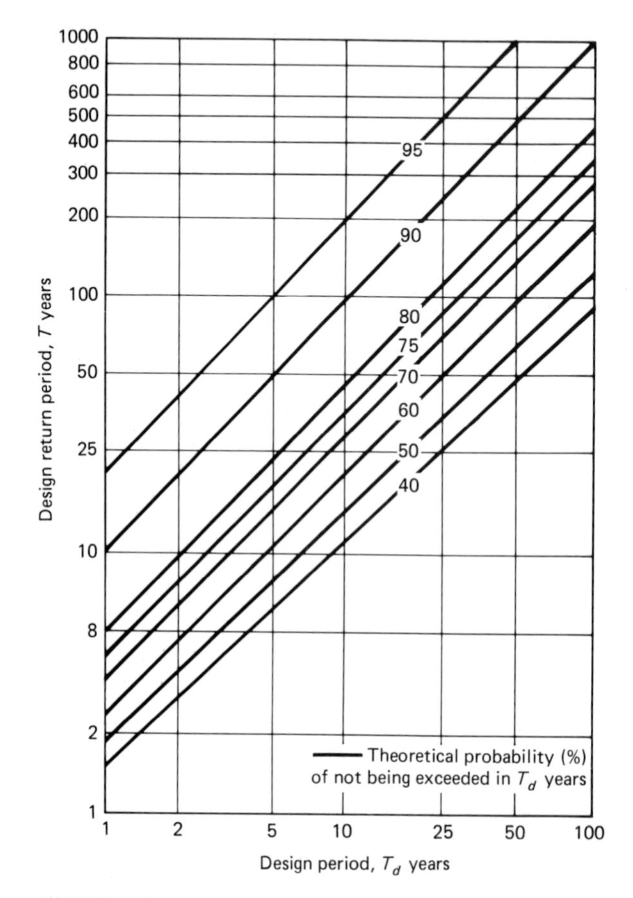

FIGURE 4-2
Relation among design return period, design period, and probability of not being exceeded in design period.[5]

probability that an n-year rain will occur at least once in the next n years is given by

$$P = 1 - \left(1 - \frac{1}{n}\right)^n \tag{4-1}$$

The relationship among design return period T, design period T_d, and probability of not being exceeded in T_d years is presented in Fig. 4-2.

Example 4-1 What design return period should be used to be approximately 90 percent certain the design will not be exceeded in the next 5 years?

Solution

From Fig. 4-2 with $T_d = 5$ years and $P = 90$ percent, $T = 50$ years. The design period is thus 50 years. In terms of rainfall magnitude, a 50-year value is approximately 50 percent greater than the 5-year value.

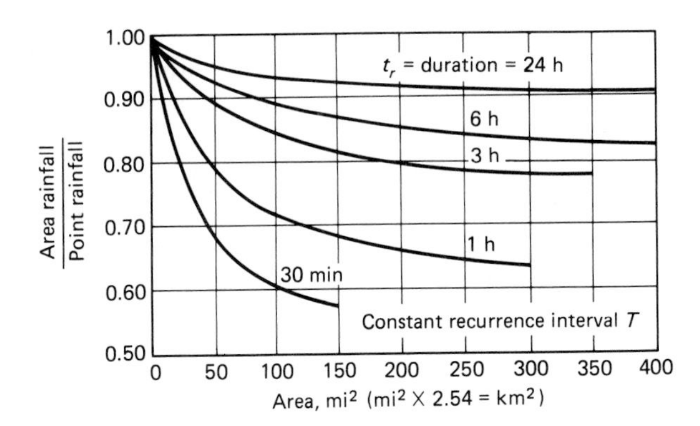

FIGURE 4-3
Areal variability of rainfall.[5]

In some circumstances it may be desirable to know what value of rainfall will never be exceeded. Reference 9 presents maps and graphs based upon a physical model and estimated meteorological parameters. The estimated probable maximum precipitation obtained from this source ranges up to 9 times the 100-year value.

The data on the isohyetal maps of Refs. 5 and 6 are smoothed point rainfall data which are applicable for areas of a few square kilometers at most. For larger areas the average depth may be expressed as a percentage of the point value through the use of Fig. 4-3.

Example 4-2 What is the average 2-year 3-h rainfall on an area of 200 mi^2 if the point value is 50 mm?

Solution
From Fig. 4-3, the ratio of area to point rainfall is 0.80.

$$\text{Rainfall} = 0.8\ (50) = 40 \text{ mm}$$

Individual station data is needed only for high-frequency storms. These can be very important in control of storm water pollution. For particular cities, it may be possible to obtain raw data on computer cards or tape. The data may then be analyzed statistically to develop localized intensity-duration curves. Chow and Yen[10] have presented a frequency analysis program utilizing either an exponential or gamma distribution. Use of such data is reasonable only when one is interested in short-duration and/or high-frequency storms, since only these can be reasonably assumed to vary from point to point in a real sense. For most purposes, the curves of Refs. 5 and 6 are superior to local data.

4-4 Runoff

Only part of the precipitation upon a catchment area will appear in the form of direct runoff. A portion is intercepted by vegetation (from which it later evaporates), a portion is held in depressions, and a portion infiltrates into the ground. A part of the infiltrated water is taken up by plant life and returned to the atmosphere through transpiration, while the remainder either moves through the ground (possibly emerging at springs or other surface waters) or is held by capillary action.

Since runoff appears as stream flow, it may be measured directly by determining the mean velocity and calculating the discharge. This should be done at a number of flows covering the range from minimum to maximum. From these data a rating curve may be constructed which relates discharge to stage, that is, stream depth. The depth may then be measured continuously by a float in a stilling well or by a pressure-sensing device to obtain a continuous record of flow versus time. Since stream beds change in shape, periodic recalibration of the gauge is necessary.

The U.S.G.S. operates an extensive network of stream-flow gauging stations throughout the nation. Use of the records of these stations[11,12] may make analysis of precipitation records unnecessary, since it is flow that is of concern. Smaller streams are usually not gauged, although short-term records for specific streams may be available from state agencies, the U.S. Army Corps of Engineers, the Bureau of Reclamation, or the Soil Conservation Service. Where no stream-flow records are available, precipitation data must be obtained and the watershed must be analyzed in order to predict the stream discharge. This procedure is less desirable than using actual runoff data. Even a few years of stream-flow records are of great value in relating precipitation and runoff.

The value of stream flow data as a predictor of future flows increases with the length of the period of record. The U.S.G.S. has concluded that a 1-year record is likely to vary so far from the mean that its use may involve very serious error. A 5-year period, while often giving results within 10 percent of the actual mean, may be in error by 30 percent or more. A 10-year record will give better results, but is unlikely to include extreme values. Since runoff events are assumed to be random occurrences, an adequate data base requires at least 30 years of records. Such records are available for most stations in the United States.

Long-term records not only give a better estimate of mean flow, but also permit evaluation of the probability of successive dry years or flood flows. Two or more successive dry years may deplete reservoirs if such events have not been considered in their design.

When only limited runoff records are available, it may still be possible to establish some relation between rainfall and runoff. Flow variations may then be calculated over a period of time as a function of recorded rainfall. This procedure is much less trustworthy than using actual runoff records and, if the procedure is used, careful consideration should be given to the difference between point and areal rainfall. In cases where no flow records exist, runoff records of other drainage areas of similar characteristics may be used to estimate flow from the ungauged basin. The characteristics which are important in making such comparisons include climate, vegetal cover, topography, soils, and surface geology.

4-5 Factors Which Affect Runoff

The discharge from a drainage area includes both surface and groundwater flow. Water which has entered the ground by infiltration may emerge at springs or, less visibly, at streams or other surface waters. Dry weather flow consists of drainage from surface impoundments and flow from groundwater, while during wet weather these sources are augmented by direct runoff from precipitation.

Precipitation is the most important single factor affecting the discharge from a basin. It is self-evident that the quantity of rainfall is important, but the distribution in both time and space may be equally significant. Rains which occur during the growing season may contribute very little to runoff and rains of low intensity may infiltrate with production of very little surface flow.

Solar radiation affects evaporation through its effect on temperature. Low temperatures permit accumulations of ice and snow which may produce rapid runoff during warmer weather. Additionally, frozen ground thaws rather slowly and very high runoff rates may occur as a result of rainfall following a period of low temperatures.

Local *topography and geology* influence both the timing and quantity of runoff. Steep slopes and impervious strata enhance the rate and quantity of discharge, while flat pervious deposits offer substantial opportunity for infiltration.

Evaporation is a function of temperature, wind velocity, and relative humidity. Evaporation rates are measured with a standardized pan[8] which is exposed to the atmosphere. Rates are expressed in terms of depth lost per unit time and the results are corrected to equivalent reservoir evaporation by multiplying them by a factor which varies from time to time and place to place but which is typically about 0.7. Figure 4-4 presents annual average evaporation from standard pans in the continental United States.

Evaporation from land surfaces is substantially less than that from open water, being reduced by shading by plants and limited availability of water in the soil. *Transpiration*, on the other hand, can be very significant at some seasons of the year. Most engineers lump evaporation from land surfaces and transpiration together in a single quantity, *evapotranspiration*, since the two effects are difficult to measure separately. Table 4-1[13] presents expected evapotranspiration from soils in which adequate water is available. A conservative approach to estimation of potential evapotranspiration, at least on an annual basis, is to take it as being equal to evaporation from a free-water surface with negligible heat-storage capacity.[14]

Interception includes that precipitation which is retained on leaves and other surfaces and never reaches the ground. Its magnitude can be substantial

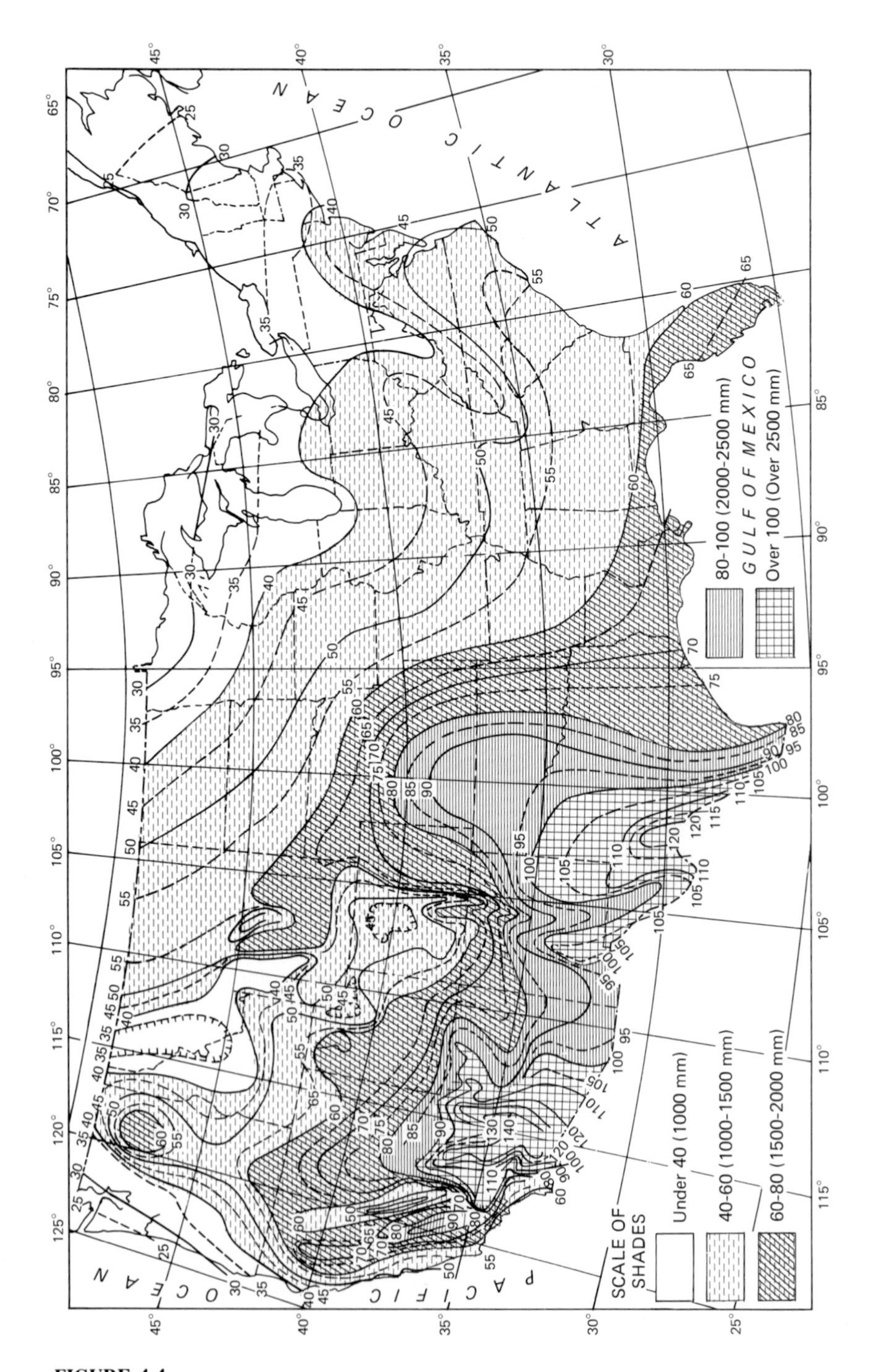

FIGURE 4-4
Mean annual pan evaporation in the United States.[3]

TABLE 4.1
Annual evapotranspiration on areas with different crops and vegetal cover

Vegetation	Location	Annual evapotranspiration, mm	Remarks
Alfalfa	New Mexico	1220	
Alfalfa and clover	Various	760–1220	
Field crops	New Mexico	760	
Field crops and native grasses	Colorado	533	During crop season
Field crops and native grasses	Oregon	305–460	During crop season
Garden truck and small grains	Idaho	305–460	During crop season
Meadows	Wyoming	380	
Citrus trees	California	660	
Peach trees	California	860	
Brush and grass	Texas and New Mexico	690–760	
Brush and grass	California	255–510	Located in a dry section
Alder, maple, and sycamore trees	California	1190	Ample water available

Source: Ref. 13.

on an annual basis in areas with heavy ground cover, but, since the intercepted water later evaporates, is included in evapotranspiration. On a short-term basis, interception may substantially reduce runoff peaks, since most interception occurs at the beginning of storm events as shown in Fig. 4-5. In urban hydrology, interception is sometimes considered as an initial abstraction from the rainfall or lumped with either depression storage or infiltration.

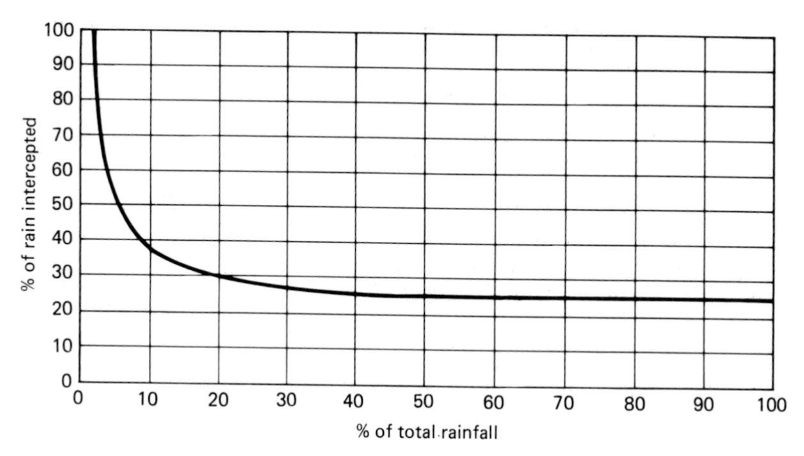

FIGURE 4-5
Percent of net rainfall lost to interception.

Depression storage is that water which is detained in low-lying areas during runoff events. This water will subsequently evaporate or infiltrate into the soil. Depression storage, like interception, has the effect of reducing runoff at the beginning of a rainfall event. Its effect upon catchment yield and peak flow is relatively small. Linsley[14] has suggested a rational model for depression storage which relates the volume detained to maximum storage capacity and rainfall excess (precipitation minus evaporation, interception, and infiltration):

$$V = S_d (1 - e^{-P_e/S_d}) \tag{4-2}$$

where V = volume stored at some time
S_d = total storage capacity
P_e = rainfall excess

All quantities are expressed in terms of depth. The relation of depression storage to time is not definable in general. In some cases it is considered as an initial abstraction which must be satisfied before runoff occurs, but it is more reasonable to recognize that the abstraction will be distributed nonlinearly. Three suggested relations between depression storage and rainfall are illustrated in Fig. 4-6. The magnitude of depression storage is related to topography, ground cover, and other factors. Typical values are on the order of 5 mm or less for soil and 1.5 mm or less for pavement.

Infiltration is affected by soil type, rainfall intensity, surface condition, and vegetation (which can alter soil porosity). Its measurement is difficult and subject to a variety of errors. A widely accepted model of the infiltration process is that suggested by Horton:

$$f = f_c + (f_0 - f_c)e^{-kt} \tag{4-3}$$

where f = infiltration capacity
f_c = equilibrium infiltration capacity
f_0 = initial infiltration capacity
k = experimental constant

Typical infiltration curves showing representative values of f_0 and f_c are shown in Fig. 4-7. Horton's equation presumes that the rainfall intensity i equals or exceeds f. If i is less than f, as given by the equation, then $f = i$. There are many other formulas for infiltration, but they are generally more complex and include additional variables which are difficult to evaluate. Some of the simulation models discussed in Chap. 13 employ other procedures for estimating infiltration.

4-6 Yield

The portion of the precipitation on a watershed which can be collected for use is called the *yield*. Yield includes both direct runoff and water which enters the ground before emerging as surface flow. *Safe yield* is the minimum yield recorded in the past. Draft is the quantity of water planned for use or actually

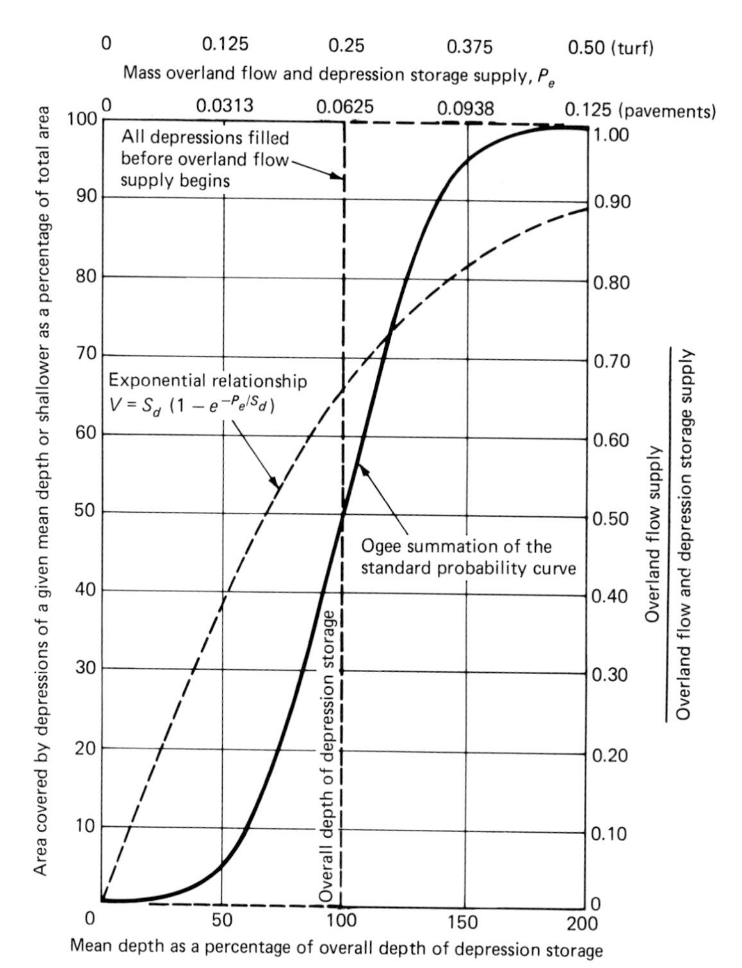

FIGURE 4-6
Hypothetical distributions of depression storage.

used. Unless the minimum daily flow of a stream is well above the maximum daily draft, it may be necessary to construct a reservoir to impound water during periods of high flow for use when flow is low.

Reservoirs may be built either to impound water for beneficial use or to retard flood flows. The two functions may be combined to some extent by careful operation.

A reservoir presents a water surface from which evaporation will occur, and this loss must be considered when yield is estimated. In addition, large quantities of water may escape through the reservoir bottom and be effectively lost from the surface supply. Site selection is extremely important if these losses are to be minimized.

Downstream water users may be entitled to certain minimum quantities of flow either by common law or prior appropriation. Such water rights must

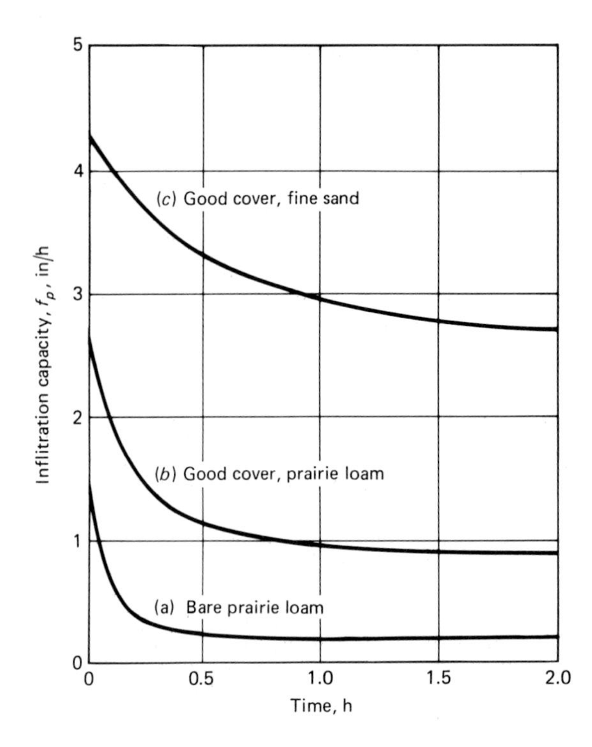

FIGURE 4-7
Typical infiltration curves.[13] Multiply values on vertical axis by 25.4 for mm/h.

either be purchased or satisfied by maintaining preexisting minimum flows. The latter requires that the draft be increased to include the flow which must be maintained.

4-7 Reservoir Storage

A *hydrograph* is a graphical record of flow versus time. An integrated or summation hydrograph, also known as a *mass diagram,* presents the accumulated total discharge as a function of time. Such mass diagrams, if they are to be useful, must be based on a substantial period of record—normally at least 30 years. Such a record, as noted above, may be expected to include periods of extreme low flow which can serve as a basis for assessing safe yield.

Figure 4-8 is an integrated hydrograph extending over a period of 3 years. The curved line *OA* represents the accumulated stream flow during this period, while line *OC* represents the accumulated draft. Strictly speaking, draft is not constant, but daily and other short-term variations are normally not of importance. Losses by evaporation and required minimum downstream flows may either be subtracted from the stream flow (which may result in negative flow during the summer) or be added to the draft.

The usual problem is the determination of the volume required to maintain a draft equal to that represented by *OC*. Tangents parallel to *OC* are constructed as shown. When the slope of *OA* is less than that of the tangents, the reservoir volume will be decreasing. When the slope is greater, the volume

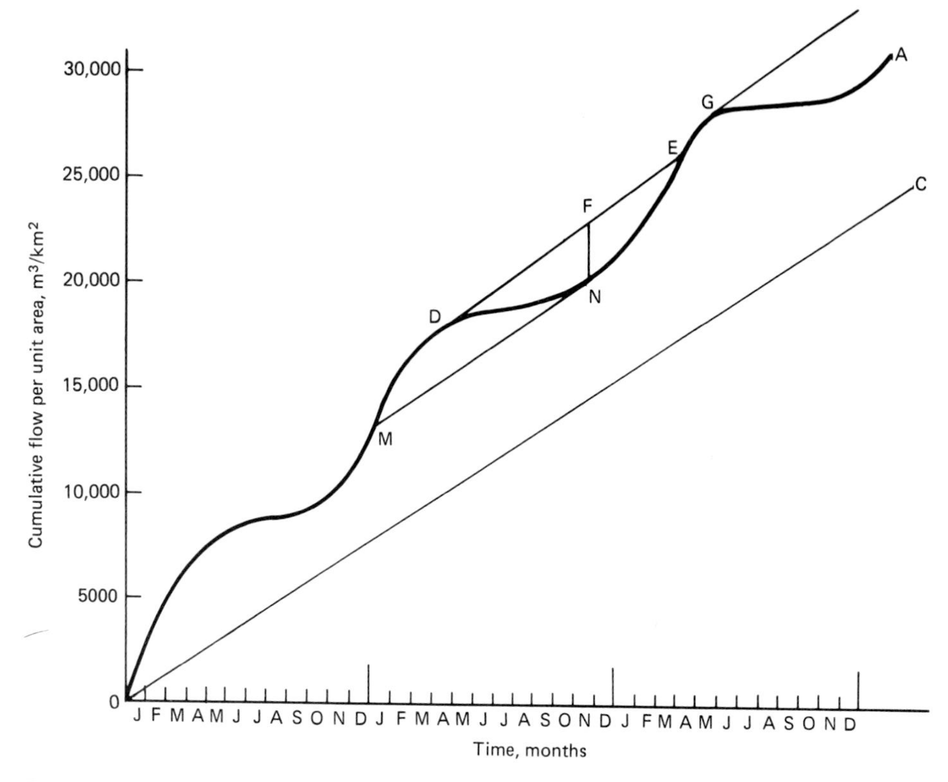

FIGURE 4-8
Integrated hydrograph.

will be increasing. The vertical distance between adjacent tangents represents the volume added or withdrawn during the intervening period. The largest such volume withdrawn (in this case *FN*) is the required reservoir volume. At point *D* the reservoir must be full, while at point *N* it will be empty. Thereafter it will fill (*N* to *E*) and then discharge excess flow (*E* to *G*). If the line *DE* does not intersect the curve, this indicates that the required draft cannot be maintained. The intersection may not occur for a period of several years—which indicates a prolonged period of low water in the reservoir.

The mass diagram selected for such an analysis must be based upon a detailed analysis of runoff duration data. The governing period is the driest period of record or that which is judged to have a sufficiently long return period (sufficiently low probability of occurrence).

One may also use a mass diagram to determine the maximum draft which can be maintained from a reservoir of fixed size. This is done by constructing parallel lines separated by a vertical distance equal to the storage available in such a manner that the lines are tangent to the curve as in Fig. 4-7. The maximum slope which permits the upper tangent to intersect the supply curve represents the maximum draft which can be maintained.

4-8 Flood Flows

Flood control for major rivers, prediction of flood stages, and prevention or mitigation of damage is largely under the control of the federal government in the United States. In this text floods and high runoff rates in general are considered primarily in the context of providing impounding reservoirs for relatively small watersheds and of managing urban drainage.

Long-term flow records of a stream are the best source of information concerning potential flood flows. For small streams, such records may not be available or may be of such short duration that they have little value other than for calibration of prediction models based on rainfall. Extreme events are important even for small areas, since hydraulic overload of spillways can cause failure of dams with resulting property damage and loss of life. Discharges from catchments of various sizes may be estimated by the techniques of Table 4-2. Of the three area classifications, only the two smaller are treated in this book.

It is often desirable to know the frequency of floods of particular magnitude. This information can be used to predict the probability of their occurrence in any particular period and to estimate the expected value of damages which might result from such events. Flood frequency data can generally be fitted to either the Gumbel or log-Pearson type 3 probability distribution.[14] A special type of the latter, the log-normal distribution, will plot as a straight line on log-normal probability paper. Data which fit the Gumbel distribution will plot as a straight line on extreme value paper.

Straight-line plots such as those described above may be extended to allow prediction of extreme events which might be expected once in 100 or 200 years. One should recognize that such a flood will not occur once in 200 years but, rather, that there is a 1-in-200 chance that it will occur in any year.

When loss of life is not likely to result from failure of a hydraulic structure, it may be more economical to design it for an event with a shorter return period (greater probability of occurrence) since the expected value of the repairs may be less than the cost of a larger structure. In the converse case, where great loss of life would result from failure, the design may be based on the probable maximum flood—that is, on the runoff resulting from the maximum probable rainfall (Art. 4-3). Floods exceeding previously estimated "maxima" have occurred. Extrapolation of data based on 30 to 60 years of record to long recurrence intervals may be misleading, since hydrological data do not always follow a known probability distribution.

TABLE 4-2
Flood analysis techniques for areas of different size

Catchment area, km^2*	Present practice
Less than 3	Overland flow hydrograph; simplified methods
3–5000	Unit hydrograph; flood frequencies; simulation techniques
Over 5000	Flood routing,[14,15] flood frequencies

* $km^2 \times 0.386 = mi^2$

4-9 Urban Hydrology

The problem of control of storm-water runoff in developed and developing areas has received substantial attention in recent years. Certain simple techniques which were widely used in the past may still be suitable for very small areas, but should not be considered as an acceptable basis for design or analysis of areas other than small suburban subdivisions. These methods include the use of empirical formulas which calculate peak flow based on watershed characteristics, the so-called *rational method*, and simplified applications of the more general SCS technique.[15] These procedures, and more recently developed computer simulation models, are discussed in Chap. 13.

PROBLEMS

4-1. What is the probability that a storm event with a return period of 2 years will occur at least once within the next 2 years?

4-2. What design return period should be used to be approximately 95 percent certain the design event will not be exceeded in the next 2 years?

4-3. The point rainfall data of Ref. 5 indicates the 10-year, 60-min rainfall for New Orleans is 3.3 in (83 mm). What value would you use for the metropolitan area of approximately 50 mi^2 (127 km^2) ?

4-4. A 24-h rainstorm has the pattern below. The infiltration capacity of the soil can be described by the equation $f = 0.75 + 4.25\,e^{-0.29t}$, where t is in hours. If depression storage and other initial abstrations amount to 3.5 mm/h for each of the first 3 h, determine the net effective rain (rainfall minus all losses).

Hour	Average intensity, mm/h	Hour	Average intensity, mm/h
1	10.3	13	3.8
2	12.5	14	5.8
3	8.2	15	7.1
4	8.1	16	6.6
5	5.6	17	6.3
6	2.0	18	2.3
7	2.0	19	1.8
8	2.3	20	1.5
9	2.0	21	0.7
10	1.5	22	0.5
11	2.8	23	0.3
12	3.0	24	0.3

4-5. The infiltration rate at the beginning of a storm is 100 mm/h and decreases to a constant value of 10 mm/h after 10 h. If the total infiltration during the 10 h amounts to 300 mm and the rainfall is always in excess of the infiltration capacity, what is the value of k in Horton's equation?

4-6. From the following record of average monthly stream flows during an extended period of drought, determine the required reservoir size to provide a uniform draft of 10,000 m^3/day.

Month	Monthly flow, $10^6 m^3$	Month	Monthly flow, $10^6 m^3$
January	0.18	August	0.08
February	1.02	September	0.07
March	1.32	October	0.04
April	0.51	November	0.10
May	0.87	December	0.26
June	0.67	January	0.20
July	0.19	February	1.10

REFERENCES

1. "Water Quality Control Information System (STORET)," U.S. Environmental Protection Agency.
2. "Climatological Data, National Summary," U.S. Environmental Data Service (monthly).
3. John L. Baldwin, *Climates of the United States,* U.S. Environmental Data Service, 1973.
4. "Guide to Standard Weather Summaries and Climatic Services," U.S. Environmental Data Service, 1973.
5. "Rainfall Frequency Atlas of the United States," Technical Paper 40, U.S. Department of Commerce, 1961.
6. "Five to 60-Minute Precipitation Frequency for the Eastern and Central United States," Technical Memorandum NWS Hydro-35, National Oceanic and Atmospheric Administration 1977.
7. Vernon B. Sauer, *Floods in Louisiana—Magnitude and Frequency*, 2d ed., U.S. Department of the Interior, Geological Survey, Baton Rouge, 1964.
8. Philip B. Bedient and Wayne C. Huber, *Hydrology and Floodplain Analysis*, Addison-Wesley, Reading, Mass., 1988.
9. "Seasonal Variation of the Probable Maximum Precipitation East of the 105th Meridian for Areas of 10 to 1000 Square Miles and Durations of 6, 12, 24, and 48 Hours" Report No. 33, U.S. Department of Commerce, 1956.
10. "Urban Stormwater Runoff: Determination of Volumes and Flowrates," U.S. Environmental Protection Agency, 1976.
11. "Water Resources Data," U.S. Geological Survey.
12. "Daily River Stages," U.S. National Weather Service.
13. "Hydrology Handbook," *Manual of Engineering Practice 28*, American Society of Civil Engineers, New York, 1949.
14. R. K. Linsley, Jr., M. A. Kohler, and J. L. H. Paulhus, *Hydrology for Engineers*, 2d ed., McGraw-Hill, New York, 1975.
15. Ven T. Chow, David R. Maidment, and Larry W. Mays, *Applied Hydrology*, McGraw-Hill, New York, 1988.
16. "Urban Hydrology for Small Watersheds," Technical Release No. 55, U.S. Department of Agriculture, 1975.

CHAPTER 5

GROUNDWATER

5-1 Sources and Quality

In Chap. 4 it was noted that some of the water which falls upon the earth infiltrates the soil. In addition, water in streams, lakes, and artificial impoundments and water spread over the surface of the land for either irrigation or disposal will percolate, in part, into the soil. A portion of the subsurface water is returned directly to the atmosphere by evaporation and transpiration, some is held by capillary forces, and the remainder moves downward until it encounters a more or less impermeable stratum. Water which is contained by an impermeable layer, or *aquiclude*, and which will flow to wells, springs, or other points of recovery is called groundwater.

Groundwaters are often superior in quality to surface waters (Chap. 8), are generally less expensive to develop for use, and usually provide a more certain supply. For these reasons, groundwater is generally preferred as a source for municipal and industrial water supplies. Against these common advantages it must be noted that groundwaters may be contaminated by toxic or hazardous materials leaking from landfills, waste treatment sites, or other sources (some natural) which may not be known to either the public or regulatory agencies.

Groundwater-bearing formations which are sufficiently permeable to yield usable quantities of water are called *aquifers*. When an aquifer is not overlain by an aquiclude it is said to be *unconfined*. The top of the saturated zone in such an aquifer is called the *water table* or *phreatic surface*. *Confined aquifers* consist of a water-bearing layer contained between two less permeable layers. The latter may be either aquicludes, if essentially impermeable, or *aquitards* if sufficiently permeable to let water enter and/or leave vertically, but not sufficiently permeable to carry a significant flow horizontally.

Water flow in an unconfined aquifer is analogous to that in an open channel, while flow in confined aquifers is analogous to that in pipes. The height to which water will rise in a tube penetrating a confined aquifer is called the *piezometric surface*. The height of this surface is a measure of the pressure in the aquifer—which may be sufficient to raise the water above the surface of the earth. *Artesian* wells are those which penetrate confined aquifers. *Flowing* wells are artesian wells in which the piezometric surface rises above the level of the ground.

5-2 Occurrence of Aquifers

The value of aquifers depends upon their ability to provide usable quantities of water. To a large extent this is a function of *porosity* and particle size. Porosity is the percentage of the total volume of a soil body that is occupied by voids. Some very porous materials, such as clays, may be very poor conductors of water because their particle size and interparticle openings are so small. The structure of clay can change from flocculated to dispersed as a result of exchange of monovalent and divalent ions on the individual particle surfaces. If the water contains primarily monovalent ions, the clay will be dispersed and relatively impermeable. In the presence of divalent ions, the clay will tend to flocculate into larger aggregates similar in size to sand. In this condition it will behave like sand insofar as transmission of water is concerned.

Sand, gravel, and sandstones are the most important sources of groundwater, but fissured limestones and shales, chalks, porous lava deposits, and even fragmented granite are usable in some areas. Various classifications of aquifers have been made. The simplest, based primarily upon extent, contains three groups.

The most important aquifers are extensive and thick formations of porous material deposited by water and wind. These are fairly uniform, and reasonably reliable information concerning their characteristics either is available or can be obtained with relatively few borings or from records of existing wells. Important examples of this group are:

1. The Tertiary deposits of sand and gravel which underlie the western plains
2. The sand and gravel of the eastern coastal plain, a strip bordering the Atlantic and Gulf coasts 160 to 320 km (100 to 200 mi) wide and extending from Long Island into Texas and up the Mississippi River to the Ohio River
3. The sandstones of the eastern part of the Dakotas and parts of Nebraska and Kansas
4. The sandstones of southern Wisconsin, northern Illinois, and eastern Iowa

Old lakes and river beds often contain deposits of sand and gravel. These deposits collect water from surface runoff and groundwater from higher land and conduct it underground in the general direction of surface flow. The old stream bed is often approximately parallel to an existing stream although there may be no surface evidence of the existence of the aquifer. The sand and gravel may be uniform or may alternate with layers of silt or clay. Such deposits are far less extensive than those of the first class but may be in hydraulic contact with surface waters which provide direct recharge. Cities situated beside rivers may thus be able to obtain very dependable supplies from shallow wells which penetrate such aquifers.

In northern states there are numerous deposits of glacial drift or till which were left by glaciers as they receded. These sand and gravel deposits are very irregular and limited in extent. They occur in old river beds, in thin strata in valleys, and in and along moraines. They are likely to be interspersed with and covered by layers of clay.

Drilling a well in areas in which there are no large producing wells is always a gamble. Despite the information obtainable from surface and subsurface exploration, a suitable aquifer may still not be encountered. Reconnaissance can begin most effectively with maps and aerial photographs from which areas of recharge such as fans, beach or dune deposits, and abandoned river beds can be identified. The presence of recharge areas can be a semiquantitative indication of potential yield.

Unconsolidated deposits are generally the first place one searches for groundwater since they are easily drilled, generally contain water, are more permeable than other formations, and usually offer a high yield. Alluvial valleys such as the flood plain of the Mississippi River offer good groundwater potential. Yields are not identical in all locations, but depend on proximity to recharge and the coarseness of the material. Surface appearances can be deceiving, and knowledge of subsurface geology is necessary before a construction program is started. Glacial deposits such as outwash plains contain well-sorted fine materials which have high permeability and storage capacity. Other glacial deposits such as kames, moraines, and eskers, while permeable, may be too well-drained and too small to be useful sources of water.

Coarse-grained *sedimentary rocks* such as sandstones have porosities ranging from 5 to 30 percent and permeabilities which are several orders of magnitude less than corresponding unconsolidated deposits. The most productive areas are along fault zones and thoroughly jointed zones. Fine-grained sedimentary rocks such as shale and claystone have high porosity but very low permeability. Fractured and jointed areas may yield small quantities of water, but most such materials are aquicludes.

Carbonate rocks yield water as a result of solution openings, jointing, and faults. Horizontal openings are more important than vertical, since the latter are likely to be sealed by sediments carried by surface runoff and wells are much more likely to strike horizontal than vertical openings. Limestone caverns offer very large quantities of water but are unlikely to be found except by chance.

Igneous and *metamorphic* rocks can also yield water as a result of fracture by faulting or weathering. Volcanic rocks are sometimes very permeable—large flows can be obtained from basalts, particularly when they are contained by dikes, faults, or ash deposits.

In addition to information from surface features, location of groundwater can be assisted by various geophysical methods. *Gravity surveys* can indicate the location of faults, folds, and intrusions and the depth to alluvial deposits. *Resistivity surveys* of soils can aid in locating water, since resistivity increases with increasing porosity, decreasing water content, and decreasing salt content. *Seismic surveys* are more accurate and generally more useful than other methods, but are also more expensive. The technique consists of creating a shock wave and measuring its time of arrival at a series of detectors or geophones. Properly designed seismic surveys can not only locate permeable strata but can also measure their depth. All the geophysical techniques (including interpretation of test well logs) require specialized expertise and substantial experience.

5-3 Groundwater Flow

As noted above, groundwater flow may be considered as analogous to open-channel and pipeline flow in unconfined and confined aquifers, respectively. As in other flow regimes (Chap. 3), there must be a hydraulic gradient in order for flow to occur. The velocity of flow is directly related to the hydraulic gradient and inversely related to the permeability of the aquifer. Since permeability of natural materials is seldom uniform, the hydraulic gradient is unlikely to be a straight line. Natural hydraulic gradients seldom exceed 0.2 to 0.4 percent. Velocities at gradients of this magnitude range from less than 100 m/year in sandstones to more than 100 m/day in coarse gravels.

Darcy's studies of groundwater flow in the 1850s led him to the relation

$$V = -k\frac{\partial h}{\partial l} = -ks \tag{5-1}$$

where V = superficial, or Darcy, velocity (not the actual velocity through the interstices of the soil)

h = drop in phreatic surface or groundwater table between two points

l = horizontal distance between those points

s = hydraulic gradient

k = constant dependent on characteristics of both fluid and solid medium

Reynolds demonstrated that Darcy's law is applicable provided the flow is laminar. This may be readily shown to be the case for all aquifers of practical importance. The only major exception occurs in the immediate vicinity of wells.

Darcy's coefficient may be evaluated by a number of laboratory techniques.[1] In general, these methods do not give results which are as useful as those obtained from field measurements. It is extremely difficult to obtain soil samples which are undisturbed, and even undisturbed samples are unlikely to be representative of aquifers—which are seldom uniform for any distance.

Recognizing that, in general, Eq. (5-1) is a vector equation, one may rewrite it as

$$\mathbf{V}_i = -k_i \frac{\partial h}{\partial \mathbf{i}} \tag{5-1a}$$

Defining a velocity potential Φ such that

$$\Phi = -kh + C \tag{5-2}$$

one can then write the generalized Darcy's law as

$$\mathbf{V}_{\mathrm{i}} = \frac{\partial \Phi}{\partial \mathbf{i}} \tag{5-3}$$

Considering the continuity equation for an incompressible element of the aquifer, one obtains

$$\nabla \mathbf{V} = 0 \tag{5-4}$$

which, upon substitution for **V** from Eq. (5-3), yields

$$\nabla^2 \Phi = 0 \tag{5-5}$$

Equation (5-5) is Laplace's equation for steady flow in homogenous and isotropic media. For the two-dimensional case (which is often adequate) this may be written as

$$\nabla^2 \Phi = \frac{\partial^2 \Phi}{\partial \mathbf{x}^2} + \frac{\partial^2 \Phi}{\partial \mathbf{y}^2} = 0 \tag{5-6}$$

In cylindrical coordinates this becomes:

$$\nabla^2 \Phi = \frac{1}{\mathbf{r}} \frac{\partial}{\partial \mathbf{r}} \left(r \frac{\partial \Phi}{\partial \mathbf{r}} \right) + \frac{1}{\mathbf{r}^2} \frac{\partial^2 \Phi}{\partial \boldsymbol{\theta}^2} = 0 \tag{5-7}$$

where

$$r = (x^2 + y^2)^{0.5}$$

and

$$\theta = \tan^{-1}(y/x)$$

For purely radial flow in the horizontal plane there is no variation of Φ with θ, thus

$$\frac{1}{r} \frac{\partial}{\partial r} \left(r \frac{\partial \Phi}{\partial r} \right) = 0 \tag{5-8}$$

from which,

$$\Phi = C_1 \ln r + C_2 \tag{5-9}$$

Defining the boundary conditions $\Phi = 0$ at $r = r_w$ and $\Phi = kh$ at $r = R$, where h is the head loss from $r = R$ to $r = r_w$, gives

$$\Phi = \frac{kh}{\ln (R/r_w)} \ln \frac{r}{r_w} \tag{5-10}$$

and

$$V = \frac{\partial \Phi}{\partial r} = \frac{kh}{r \ln (R/r_w)} \tag{5-11}$$

At any radius from the well,

$$Q = \int_0^{2\pi} DrV \, d\theta \tag{5-12}$$

where D is the thickness of the aquifer. Substituting for V from Eq. (5-11) and integrating, for the case in which D is constant, gives

$$Q = \frac{2\pi kD(h_2 - h_1)}{\ln (R/r_w)} \tag{5-13}$$

in which $(h_2 - h_1)$ is the head loss between $r = R$ and $r = r_w$. Equation (5-13) applies to confined aquifers in which D is constant and h_2 and h_1 are the elevations of the piezometric surface above an arbitrary datum. In this equation the product kD is often set equal to T, the transmissivity of the aquifer. Where D can vary, as in unconfined aquifers, if h is measured from the underlying aquiclude, then $D = h$ and Eq. (5-13) becomes

$$Q = \frac{\pi k(h_2^2 - h_1^2)}{\ln (R/r_w)} \tag{5-14}$$

The two equations above are called, respectively, the Theim and Dupuit equations. They are based on the assumptions that the flow is horizontal and occurs in an aquifer of infinite extent. While these assumptions can never be said to be valid, the equations are nevertheless useful—particularly in field evaluations of the hydraulic characteristics of aquifer materials.

In the case of unsteady flow (in which $\nabla^2 \neq 0$), it is necessary to include the flow produced by release from storage. If we define S as the volume released per unit volume of aquifer material per unit drop in phreatic surface, the Laplace equation becomes

$$\nabla^2 h = \frac{S}{kD}\frac{\partial h}{\partial t} = \frac{S}{T}\frac{\partial h}{\partial t} \tag{5-15}$$

which, in cylindrical coordinates, is

$$\frac{1}{r}\frac{\partial}{\partial r}\left(r\frac{\partial h}{\partial r}\right) + \frac{1}{r^2}\frac{\partial^2 h}{\partial \theta^2} = \frac{S}{T}\frac{\partial h}{\partial t} \tag{5-16}$$

For uniform radial flow, as before,

$$\frac{\partial h}{\partial \theta} = 0$$

and

$$\frac{\partial^2 h}{\partial r^2} + \frac{1}{r}\frac{\partial h}{\partial r} = \frac{S}{T}\frac{\partial h}{\partial t} \tag{5-17}$$

This equation has a solution of the form

$$s = \frac{Q}{4\pi T} \int_0^\infty \frac{e^{-u}}{u} du \tag{5-18}$$

where s is the drawdown in the phreatic surface at any point and

$$u = \frac{r^2 S}{4Tt}$$

The integral

$$\int_0^\infty \frac{e^{-u}}{u} du$$

is called the *well function* of u, $w(u)$, and is equal to

$$w(u) = -0.5772 - \ln u + u - \frac{u^2}{2 \cdot 2!} + \frac{u^3}{3 \cdot 3!} - \cdots \tag{5-19}$$

which is convergent for all u.

The closed-form solutions presented above are useful for many purposes, but modern analysis of groundwater flow is more commonly based upon finite-element models which permit variation of aquifer characteristics within the flow field and which often include the ability to predict transport of contaminants.[2,3,4]

5-4 Equilibrium Analysis of Wells

Equations (5-13) and (5-14), as noted above, are based on the assumptions that the flow rate is constant, that the well is screened through the entire thickness of the aquifer, that the aquifer is uniform and of infinite extent, and that water is released immediately in response to a drop in the piezometric surface. A number of corrections can be applied to these equations to correct for recharge, seepage through aquitards, nonuniformity of the aquifer, partial penetration of the well, and losses in the immediate vicinity of the well. These matters are treated in detail in other texts.[1,5]

The equilibrium equations are particularly useful in field evaluations of aquifer characteristics. By measuring the level of the piezometric surface in small observation wells and the flow in the main well, one can calculate k or T and obtain a value which represents an effective average for the aquifer.

Example 5-1 A well in an unconfined aquifer is pumped at a rate of 25 L/s (400 gal/min). The thickness of the aquifer is 15 m and the elevation of the phreatic surface is 12.5 m above the underlying aquiclude at an observation well 20 m away from the well and 14.6 m above at a well 50 m away. What is the value of k for this aquifer?

Solution
Applying Eq. (5-14) gives

$$0.025 = \frac{\pi k(14.6^2 - 12.5^2)}{\ln (50/20)}$$

$$0.025 = k(195)$$

$$k = 0.0001 \text{ m/s} = 11.1 \text{ m/day}$$

Equation (5-14) should not be used to calculate the drawdown at the pumped well since it does not consider the vertical component of the velocity, which becomes significant near the well screen. Equation (5-13), on the other hand, may be applied at the well since in confined aquifers the flow is, in fact, horizontal.

Example 5-2 A well in a confined aquifer with a thickness of 15 m produces a flow of 25 L/s (400 gal/min). The height of the phreatic surface is at an elevation of 114.6 m at an observation well 50 m away and at 112.5 m at an observation well 20 m away. Find k and T for the aquifer and estimate the height of the phreatic surface at the 0.5-m-diameter well.

Solution
Applying Eq. (5-13) gives

$$0.025 = \frac{2\pi k(15)(114.6 - 112.5)}{\ln (50/20)}$$

$$k = 0.0001 \text{ m/s} = 10.0 \text{ m/day}$$

$$T = kD = 150 \text{ m}^2\text{/day}$$

At the well,

$$0.025(86{,}400) = \frac{2\pi(150)(112.5 - h)}{\ln (20/0.25)}$$

$$h = 102.5 \text{ m}$$

The equation could equally well be written for the observation well located 50 m away. The same answer is obtained regardless of which pair of points is used.

5-5 Nonequilibrium Analysis of Wells

Equation (5-18) was developed for confined aquifers but may be used in analysis of unconfined aquifers as well—provided the drawdown is small compared to the aquifer thickness. With known values of S and T the drawdown at any point in the well field may be readily calculated as a function of time. In general, as t increases, s (the drawdown) increases at a decreasing rate, but never becomes constant. If one considers the relation

$$u = \frac{r^2 S}{4Tt}$$

and that r may range from a few to several hundred meters, S from 10^{-3} to 10^{-5} or less, and T from 1 to several thousand, it is clear that, for reasonable values of t,

$$u < 1$$

Equation (5-19) may thus be simplified by neglecting the higher-order terms to yield

$$w(u) \approx -0.5772 - \ln u \tag{5-20}$$

Thus,

$$s = \frac{Q}{4\pi T}(-0.5772 - \ln u) = \frac{Q}{4\pi T} \ln \left(\frac{2.25Tt}{r^2 S}\right) \tag{5-21}$$

Equation (5-21) is called Jacob's approximation and may be used in nonequilibrium analyses provided that r is relatively small and t is relatively large.

Example 5-3 An aquifer has $T = 150$ m²/day, $S = 10^{-4}$, and provides a flow of 25 L/s (400 gal/min) to a well. Find the drawdown at two observation wells located 20 and 50 m from the producing well after 1 day, 30 days, 1 year, and 5 years.

Solution

Applying Jacob's approximation gives

$$s = \frac{0.025(86{,}400)}{4\pi(150)} \ln \left[\frac{2.25(150)t}{r^2(10^{-4})}\right]$$

For this equation, calculations can be tabulated as follows.

Time t, days	**r, m**	**s, m**	**r, m**	**s, m**	**Δs, m**
1	20	10.36	50	8.26	2.10
30	20	14.26	50	12.16	2.10
365	20	17.12	50	15.02	2.10
1825	20	18.96	50	16.86	2.10

One may observe that, as predicted, the drawdown increases with time and, although the rate of increase decreases, the latter never approaches zero. One may further observe that the difference between the drawdowns at the two observation wells very quickly becomes constant and that Eq. (5-13) is therefore valid even though the assumption of steady-state conditions is not.

5-6 Interference in Aquifers

Laplace's equation [Eq. (5-5)] and the subsequent derivations were all based on consideration of a single well in an infinite aquifer. Noting, however, that Laplace's equation is linear, one may conclude that the combined effect of multiple wells on the piezometric surface may be determined by simple superposition. In general one can conclude that the head h_i required at any point to sustain a flow Q_i of the ith well in a confined aquifer is given by

$$h_i = \frac{Q_i}{4\pi T} \ln [(x - x_i)^2 + (y - y_i)^2] + C_i \tag{5-22}$$

To obtain simultaneous discharges, $\sum_1^n Q_i$, the head required will be $\sum_1^n H_i$.

$$h(x, y) = \frac{1}{4\pi T} \sum_1^n Q_i \ln [(x - x_i)^2 + (y - y_i)^2] + C \tag{5-23}$$

For unconfined flow, the equation is

$$h^2(x, y) = \frac{1}{2\pi k} \sum_1^n Q_i \ln [(x - x_i)^2 + (y - y_i)^2] + C \tag{5-24}$$

In either case, C is a constant which must be evaluated for each problem.

Example 5-4 Two wells are drilled 50 m apart in an aquifer which has $k = 0.05$ cm/s. The original height of the phreatic surface above the aquiclude is 12.2 m. The wells are pumped at a rate of 10 L/s. It is estimated that the effect of the pumping extends a distance of 600 m from the center of the well group. Determine the shape of the drawdown curve. The well diameters are 0.5 m.

Solution
Choosing the origin midway between the two wells,

$$h^2 = \frac{1}{2\pi k} [Q_1 \ln (x + 25)^2 + Q_2 \ln (x - 25)^2] + C$$

Substituting $k = 43.2$ m/day and $Q_1 = Q_2 = 864$ m³/day gives

$$h^2 = 3.18[\ln (x + 25)^2 + \ln (x - 25)^2] + C$$

C is found from the condition $h = 12.2$ at $x = 600$.

$$12.2^2 = 3.18[\ln (625)^2 + \ln (575)^2] + C$$

$$C = 67.48 \text{ m}^2$$

Calculating the drawdown, we compile the following table.

x, m	h, m
0	10.41
15	10.28
24.75	9.14
25.25	9.14
50	10.74
100	11.21
300	11.83
600	12.20

If only one well were pumping, the calculated drawdown at the well would be about 2.24 m rather than 3.06 m.

5-7 Groundwater Contamination

Groundwaters may be contaminated by both naturally occurring and artificial materials. To a certain extent, anything with which water comes in contact will be dissolved in or mixed with the flow. Pollutants, once introduced into groundwater, may be carried for very long distances and will be very difficult to remove—although natural processes such as adsorption, biodegradation, radioactive decay, ion exchange, and dispersion may reduce concentrations to some extent.

Advection-dispersion theory helps in understanding the various phenomena which affect the transport of contaminants in groundwater and may be useful in the design of remediation techniques. It is extremely difficult, however, to match the mathematical theory and the physical reality and one must not rely blindly on the predictions of sophisticated groundwater contaminant models.

Diffusion and dispersion distribute solutes relative to the bulk flow, resulting in lower concentrations but a larger volume of contaminated water. At low velocities in fine media, diffusion is most important, while in coarser material or at higher velocities, dispersion dominates. In most circumstances of practical interest, both are important. Measurement of dispersion and diffusion is extremely difficult since differences in advection in different layers of the aquifer, retardation of contaminants by chemical interaction with the soil, differences in density of contaminants and water, and similar phenomena may cause variations in concentration which are greater than those resulting from the characteristics which are being evaluated.[6]

Nonaqueous-phase liquids (NAPLs) are a source of particular concern with respect to groundwater. *Light* NAPLs (LNAPLs) include such materials as gasoline, heating oil, and kerosene. These are very widespread in soils because of leakage from underground storage tanks. LNAPLs tend to float on the groundwater, penetrating the capillary fringe and depressing the water surface. When the source of leakage is controlled, the soil will remain contaminated and the floating layer will serve as a long-term source of contamination. *Dense* NAPLs (DNAPLs) are a much more significant problem from a health standpoint. These include such compounds as trichloroethane, carbon tetrachloride, pentachlorophenols, dichlorobenzenes, tetrachloroethylene, and creosote. Toxic concentrations of such materials are extremely low—10 kg of tetrachloroethylene, for example, can contaminate 10^8 L of water. DNAPLs can be quite mobile in groundwater. Their viscosity is low, their density great, and their solubility low. Upon entering the soil, they will plunge in the aquifer until they reach an aquiclude, where their depth will increase until flow occurs. Flow may occur at right angles to or even against the general direction of water motion. NAPLs have a residual concentration below which they are not mobile in soil; nonetheless, they can be slowly leached from the soil by water, which greatly extends the time of contamination of the aquifer.

Adsorption of chemical species on soil particles may be effective in removing them from the flow, but the processes involved are quite complex and difficult to predict without detailed knowledge of the chemical environment. *Cosolvents* (materials which are miscible with water and solvents for hydrophobic organics) may greatly increase the mobility of otherwise adsorbed compounds. Changes in pH can also affect the partition between soil and water, as can chemically or biologically mediated replacement of functional groups. An example of the latter is the sequential conversion of tetrachloroethylene to vinyl chloride. The former is relatively strongly bound to soil, while the latter is quite mobile and very toxic.

Biodegradation within the soil-water system may reduce concentrations of contaminants. Relatively large numbers of microorganisms are found in subsurface strata, and a large number of organic materials have been observed to be metabolized in such environments both in nature and in laboratory experiments. Not all organics can be expected to be degraded under all circumstances. Factors which can limit biological activity include pH, salinity, synthetic chemicals, heavy metals, osmotic pressure, hydrostatic pressure, lack of trace nutrients or free water, toxicity of substrate or products, and the absence of acclimated organisms.

Historically, the problems associated with contaminated aquifers have been handled by abandoning the affected wells and seeking an alternative source of water. This may still be the only economic choice in circumstances where large volumes of the aquifer are affected and/or the source of contamination is unknown. In cases where the source and extent of the problem can be well defined, the techniques discussed below may be applicable.

5-8 Renovation of Contaminated Groundwater

Contaminated aquifers are presently restored or the contamination is limited by techniques which may be classified as containment, extraction, or in situ degradation.

Containment is applied only to materials of limited mobility which cannot be economically handled by techniques of extraction or in situ treatment. Containment does not really solve the problem, is costly, and is uncertain in the long term. Contaminated zones of the aquifer may be isolated by a combination of slurry walls, clay caps, interceptor trenches, or other hydraulic barriers. These methods prevent flow from entering or leaving the contaminated area. The barriers must be designed to last forever—a very restrictive design criterion which can probably not be met in a real sense.

Extraction methods include excavation and pump-and-treat systems. Excavation is applied to relatively immobile materials which cannot be economically removed by pumping. The process is costly, may not be entirely effective, and transfers the problem to another location since the excavated materials must be disposed of in some manner. Pump-and-treat systems are applicable only to easily transported materials, but even with these the method requires a long-term commitment to the process until all the residual material is extracted. Nonsoluble contaminants are not removed by such systems. The pumping system involves multiple wells located so as to intercept all of the flow from the contaminated zone. The treated water may be injected up-gradient to increase the flow rate locally and shorten the time required to purge the aquifer. Treatment of the extracted water depends upon the specific contaminants and may include stripping, chemical oxidation or precipitation, adsorption, or biological processes.

In situ degradation, when it is applicable, is a relatively short-term process which treats both mobile and immobile contaminants and which has a fairly well-defined end point. In situ processes may be chemical or biological. The chemical processes include oxidation, reduction, hydrolysis, polymerization, etc., depending on the particular contaminants. Treatment chemicals are injected up-gradient and their concentration is monitored in down-gradient wells. When the process is complete the chemicals will pass through the treatment zone unchanged in concentration. Biological processes may require addition of oxygen and trace nutrients such as nitrogen and phosphorus. The progress of treatment can be monitored, as with the chemical processes, by observing the residual concentration of the added materials down-gradient of the treatment zone.

The treatment processes described above require very large investments of time and money and may not be completely effective. The public danger presented by groundwater contamination and the enormous cost of correction or containment justify very stringent measures to prevent the creation of such problems.

5-9 Recharge of Aquifers

The recharge of aquifers afforded by natural hydrologic processes is sometimes deliberately augmented either to create barriers to other flow or to restore water to the aquifer for other use.[7–10]

Saltwater intrusion into coastal aquifers can be prevented by injecting water through recharge wells. Such injection wells raise the phreatic or piezometric surface and create freshwater flow toward the sea. Figure 5-1 illustrates a recharge system used near Los Angeles where heavy pumping of inland wells had caused depression of the piezometric surface and intrusion of saltwater. The increase in salt content had made it necessary to abandon some wells, which were recovered for productive use as a result of the recharge system.[7]

Some states require that well water which has been used only for cooling be returned to the aquifer. The recharge wells must be properly located so that their area of influence does not interfere with that of the producing wells, or the warm water will be drawn to the latter. The recharge wells must be as carefully

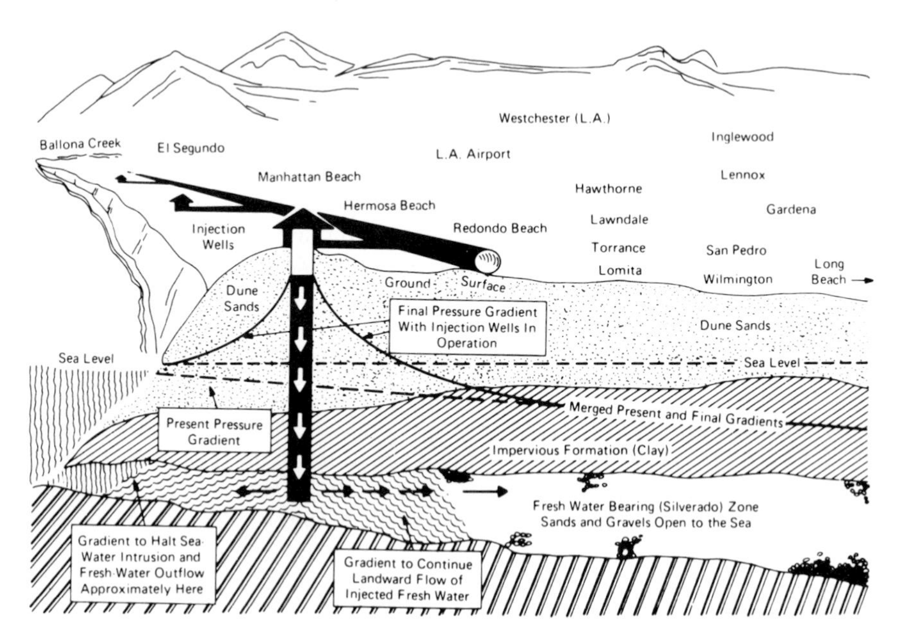

FIGURE 5-1
Recharge well operated to establish a freshwater barrier to saltwater encroachment.[7] (*Reprinted from Journal of American Water Works Association,* ***66****, by permission of the Association. Copyright 1974, American Water Works Association, Inc., 6666 W. Quincy Ave., Denver, CO, 80235.*)

developed as the producing wells, and it may be necessary to treat the water before it is returned to the ground. The character of the water may be changed substantially simply as a result of withdrawing it from the aquifer. Metals may be precipitated from solution, gas solubilities are reduced by the increased temperature and reduced pressure, and bacterial or algal growth may be stimulated. It is not unusual for filtration and chlorination to be required before recharge of cooling water.

Groundwater recharge through land disposal of treated sewage is an alternative which is routinely considered in wastewater facility planning.[9,10] A variety of industrial wastes can be handled in a similar fashion, provided they contain no materials which could pass through the soil mantle and contaminate the aquifer. In some cases the recharge of groundwater may simply be incidental to disposal of wastewater.

5-10 Well Construction

Wells are usually constructed by organizations which specialize in this work and which contract their services to the municipality or other owner. In order to ensure that the result of a groundwater project is satisfactory, the resource to be tapped must be studied carefully and the specifications and other contract documents must be carefully drawn. Analyses of aquifers and aquifer characteristics should be made by persons experienced in groundwater hydrology. The advice and data of the U.S.G.S. and state agencies should always be obtained during planning of new well development.

Contracts for well construction should address the following issues:

Output and the conditions under which it is desired should be specified. If a guaranteed flow is required a bonus may be provided or a penalty assessed for each unit of output above or below the specified value.

Out-of-plumbness should be limited to not more than one well diameter per 30 m (100 ft).

Out-of-straightness should be limited to that which will permit free passage of a 10 m (33 ft) blank with an outside diameter 12 mm (0.5 in) less than that of the well casing.

Screen characteristics (material, slot opening, and perhaps the manufacturer) should be specified. The screen is selected so as to prevent entry of the aquifer material into the well while still permitting free flow of water. The size of the opening depends on the material of the aquifer, typically ranging upward from 0.01 mm (0.004 in). The net open area should provide velocities less than 50 to 100 mm/s (2 to 4 in/s). The contract should provide a cost per unit length of screen so that adjustments in price can be made if initial estimates of required screen length are incorrect.

Casing type and weight should be specified, including overlap and packing required at changes in size. The casing should be cemented from the ground surface to the first aquiclude.

Cleaning and disinfection of the finished well should be included in the contract. The requirement should include bacteriological testing of the well output.

Development and testing are a part of the project and the conditions should be carefully specified, particularly if a guaranteed flow has been included. Development is a process which increases the permeability of the aquifer material in the immediate vicinity of the well by removal of fines. The process involves reversal of flow through the screen coupled with relatively high velocities. The flow reversal may be provided by surging the well with a piston, applying air pressure, or turning off a pump which has no check valve. This will permit the column of water to fall and flow outward through the screen. If the aquifer material is quite uniform, flow reversal may not be effective. In such cases an artificial gravel pack (Fig. 5-2) may be placed around the screen to reduce losses in the immediate vicinity of the well. The gravel pack thickness is typically 75 to 225 mm (3 to 9 in).[11]

Water supply wells may be characterized as either shallow or deep. *Shallow wells* are those which have depths less than 30 m (100 ft). Such wells are not particularly desirable for municipal supplies since the aquifers which

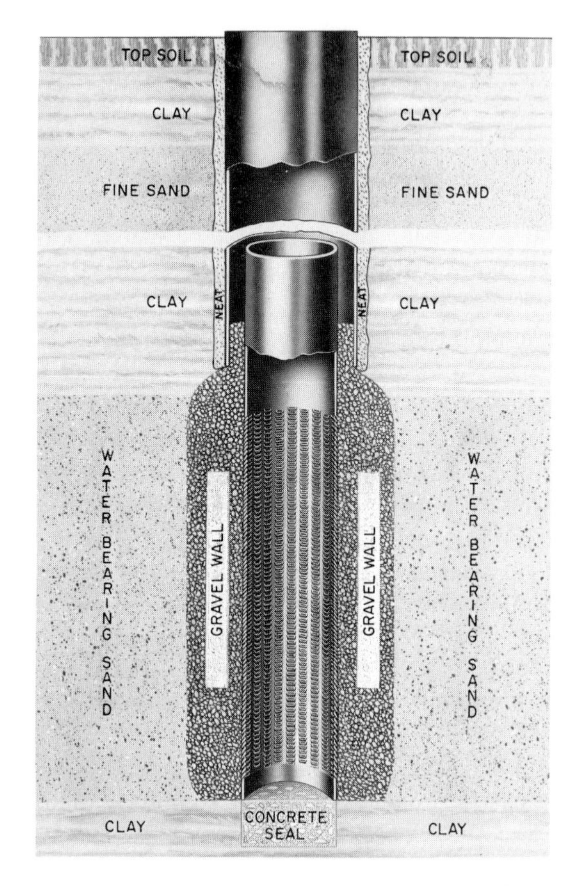

FIGURE 5-2
Gravel-packed well. (*Courtesy Layne and Bowler, Inc.*)

they tap are likely to fluctuate considerably in depth, making the yield somewhat uncertain. Municipal wells in such aquifers will cause a reduction in the phreatic surface that will affect nearby private wells, which are more likely to utilize shallow strata. Such interference with private wells may result in damage suits against the community.

Shallow wells may be dug, driven, or bored. *Dug* wells are excavated by hand or by a variety of unspecialized excavation equipment. Their diameter is generally 0.5 m (1.5 ft) or more and their depth less than 15 m (50 ft). Such wells are usually lined or cased with concrete or brick. *Driven* wells consist of a pipe casing terminating in a point slightly greater in diameter than the casing. The well is driven in the same manner as a pile—usually with a drop hammer. Since the screen may be damaged during driving or be clogged by fines through which it passes, the screen, casing, and driving point may be arranged so that the casing can be raised after driving to clear the screen. Driven wells are usually 25 to 75 mm (2 to 3 in) in diameter and are used only in unconsolidated materials. *Bored* wells range from 25 to 900 mm (1 to 36 in) in diameter and are constructed in unconsolidated materials. The boring is accomplished with augers which are filled with soil and then drawn to the surface to be emptied. The casing may be placed after the well is completed in relatively cohesive materials, but must advance with the well in noncohesive strata.

Deep wells are the usual source of groundwater for cities. Deep wells tap thick and extensive aquifers which are not subject to rapid fluctuations in piezometric level and which provide a large and uniform yield. Deep wells typically yield water of more constant quality than shallow wells, although the quality is not necessarily better. Deep wells are constructed by a variety of techniques, the more important of which are discussed below.

The *standard method* is used in both consolidated and unconsolidated materials. A truck-mounted derrick is set up over the well to permit placing casing and tool strings. Drilling is effected by a cable-suspended bit which is alternately raised and dropped by a walking beam. The flexibility of the cable and the tool string permits the bit to rebound and thereby prevents it from jamming in the hole. As drilling proceeds, the cable is lengthened. The tool string is periodically withdrawn and the loose material in the hole is bailed out by a bucket with a flap valve at the bottom. A modification of this technique uses steel tubes in the tool string through which the loose material can be carried to the surface. As the string falls, water and solid material rise in the tubes, while check valves prevent reversal of flow when the string is raised.

The *California* or *stovepipe* method is used in unconsolidated alluvial deposits and consists of pushing short lengths of steel casing into the earth with hydraulic jacks. Two concentric tubes are driven, one slightly smaller than the other, with the joints on the concentric tubes staggered. The well casing is driven even with or ahead of the excavation, which is done with a sand bucket or a small clamshell. When the desired depth has been reached, a special tool is used to perforate the casing. Wells up to 1 m (40 in) in diameter have been constructed with this method.

Rotary drilling involves rotating a pipe string which has a hardened cutting tool at the lower end. Water with a mixture of additives (called drilling mud) is pumped down the well either through the pipe or between the pipe and the side of the well. The return flow carries the cuttings to the top of the well, where they are separated from the flow, which is then returned to the well. The casing may be advanced as drilling proceeds or may be placed after drilling is completed. The drilling mud, if properly proportioned, will prevent the sides of the hole from collapsing.

Jetting is useful for small wells in unconsolidated deposits. A relatively high velocity stream of water is directed downward through a nozzle at the bottom of a pipe string as the string is raised, turned, and lowered. The high-velocity flow washes out the material, and the casing either drops by its own weight or is driven as the hole advances.

Core drilling employs a ring fitted with diamond or hardened steel teeth. The ring is rotated while a stream of water washes cuttings from the working face. The core rises within the ring as drilling advances and must be periodically broken off and withdrawn from the well. This technique is used only in consolidated or cemented materials.

5-11 Well Maintenance

Wells do not have an infinite life, and their output is likely to reduce with time as a result of hydrological and/or mechanical factors. Reduction in flow may result from depression of the phreatic surface in the well vicinity or from increased losses through the screen. Either of these phenomena will decrease the head available at the pump and thus decrease the flow.

Corroded casings or screens are sometimes withdrawn and replaced, but this is a difficult undertaking which is not always successful. It may be cheaper to simply construct a new well.

Incrustations on screens and adjacent aquifer material result from chemical or biological reactions at the air-water interface in the well. Reduced pressure resulting from pumping will cause the pH of the water to increase because of loss of carbon dioxide. This, in turn, may permit precipitation of calcium carbonate. Oxygen introduced by the well may react with iron or manganese, rendering them insoluble, or may be utilized by a species of iron-oxidizing bacteria called *Crenothrix*. Incrustation, from whatever source, may normally be remedied by acid treatment with hydrochloric acid inhibited by addition of gelatin.[12] The acid is placed in the well to a level slightly above the screen and is agitated with a plunger every 2 h for about 12 h. The well is then pumped, with surging, to remove the acid and the loosened deposits. Incrustation may also be removed, depending upon its cause, by pressurizing the well with dry ice, by chlorination, or by adding polyphosphates in concentrations of 30 to 40 g/L.[12]

PROBLEMS

5-1. An artesian well is pumped at a rate of 1.6 m^3/min. At observation wells 150 and 300 m away, the drawdowns are 0.75 and 0.6 m respectively. The average thickness of the aquifer is 6 m. Compute the transmissibility of the aquifer.

5-2. An aquifer has a cross-sectional area in the direction of flow of 100,000 m^2, k = 20 m/day, and $s = 5 \times 10^{-4}$. What is the net flow in the aquifer?

5-3. A confined aquifer has a transmissibility of 50 m^2/day. If $s = 3 \times 10^{-4}$, what is the flow per square meter of aquifer material?

5-4. Water flows through sand with a particle size of 0.5 mm. What is the maximum velocity of laminar flow through this material (see Chap. 3)? What gradient is required to produce such a velocity? Could such a gradient occur under natural conditions? What does this tell us about the general applicability of Darcy's law?

5-5. A well with a radius of 0.4 m penetrates a confined aquifer with k = 20 m/day and thickness = 35 m. The well is pumped so that its drawdown is maintained at 5 m. Assuming that the drawdown is essentially zero 600 m from the pumped well, what is its discharge?

5-6. A well is pumped at a rate of 0.75 m^3/min. At an observation well 30 m away, the following drawdowns were noted as a function of time:

Days	m	Days	m	Days	m
1	0.75	6	3.45	30	7.47
2	1.30	8	4.02	40	8.24
3	1.90	10	4.57	60	9.34
4	2.45	15	5.60	80	10.10
5	3.00	20	6.37	100	10.66

Determine the values of T and S using Jacob's approximation. Use these values to predict the drawdown 300 m away from the well and at the well itself (diameter = 0.2 m) after pumping for 2 years at a rate of 1.0 m^3/min.

5-7. Rework Prob. 5-5 assuming that the drawdown at the well (5 m) results from pumping that well and another identical well located 50 m away at identical rates.

5-8. Using the principle of superposition, one can consider a recharge well to be a well with negative flow. Direct application of the Dupuit or Theim equations will then permit calculation of the amount the groundwater table will be raised. Using this principle, determine what flow would be required in a recharge well to raise the phreatic surface by 5 m a distance of 50 m from the well. Assume that $T = 25$ m^2/day and that the recharge well has negligible effect at a distance of 500 m.

5-9. Rework Prob. 5-8 assuming the increase in depth results from two identical wells located 100 m apart. What is the flow in each well? Is the total flow greater or less if two wells are used?

REFERENCES

1. Herman Bouwer, *Groundwater Hydrology*, McGraw-Hill, New York, 1978.
2. P. K. M. van der Heijde et al., "Groundwater Management: The Use of Numerical Models," *Water Resources Monograph 5*, 2d ed., American Geophysical Union, Washington, D.C., 1985.
3. S. Gorelick "A Review of Distributed Parameter Groundwater Management Modeling Methods," *Water Resources Research*, **19**:2, 1983.
4. I. Javendel et al., "Groundwater Transport: Handbook of Mathematical Models," *Water Resources Monograph 10*, American Geophysical Union, Washington, D.C., 1984.
5. C. W. Fetter, *Applied Hydrogeology*, 2d ed., Merrill Publishing, Columbus, Ohio, 1988.
6. R. W. Gillham et al., "An Advection-Diffusion Concept for Solute Transport in Heterogeneous Unconsolidated Geological Deposits," *Water Resources Research*, **20**:3, 1984.
7. John M. Toups, "Water Quality and Other Aspects of Ground-Water Recharge in Southern California," *Journal of American Water Works Association*, **66**:149, 1974.
8. David K. Todd, "Salt Water Intrusion and Its Control," *Journal of American Water Works Association*, **66**:180, 1974.
9. S. Reed, "Wastewater Management by Disposal on the Land," U.S. Army Corps of Engineers Cold Regions Research and Engineering Laboratory, Special Report No. 171, 1972.
10. Herman Bouwer, "Renovating Municipal Wastewater by High Rate Infiltration for Ground Water Recharge," *Journal of American Water Works Association*, **66**:159, 1974.
11. Fletcher G. Driscoll, *Groundwater and Wells*, 2d ed., Johnson Division, St. Paul, 1986.
12. W. Arceneaux, "Operation and Maintenance of Wells," *Journal of American Water Works Association*, **66**:199, 1974.

CHAPTER 6

AQUEDUCTS AND WATER PIPES

6-1 Conveyance and Distribution

Water, whether it be drawn from surface or ground supplies, must be conveyed to the community and distributed to the users. Conveyance from the source to the point of treatment may be provided by aqueducts, pipelines, or open channels, but once the water has been treated it is distributed in pressurized closed conduits. *Pumping* may be necessary to bring the water to the point of treatment and is nearly always a part of the distribution system. This chapter deals with the types of conduits and the materials used in conveying and distributing water.

6-2 Aqueducts

The term *aqueduct* usually refers to conduits constructed of masonry and built at the hydraulic gradient. Such structures are operated at atmospheric pressure and, unless the available hydraulic gradient is very large, tend to be larger and more expensive than pipelines operated under pressure. The advantages of aqueducts include the possibility of construction with locally available materials, longer life than metal conduits, and lower loss of hydraulic capacity with age. Their disadvantages include the need to provide the ultimate capacity initially and the likelihood of interference with local drainage.

6-3 Stresses in Pipes

Pipe used in water conveyance and distribution is always of circular cross section. The stresses which the pipe must resist are produced by the static pressure of the water, centrifugal forces caused by changes in direction of flow, external loads, changes in temperature, and sudden changes in velocity. The later phenomenon is called *water hammer*. The magnitude of the stresses resulting from these causes may be calculated by the methods of applied mechanics.

Internal pressure of any kind produces a hoop stress given by

$$\sigma_h = \frac{rP}{t} \tag{6-1}$$

and a longitudinal stress given by

$$\sigma_l = \frac{rP}{2t} \tag{6-2}$$

where r = radius
P = internal pressure
t = pipe wall thickness

Water hammer results from the sudden stopping or slowing of flow in a conduit. The kinetic energy of the water moving through the pipe is converted into potential energy stored in the water and the walls of the conduit through the elastic deformation of both. The water is compressed and the pipe material is stretched. Under the worst conditions, this produces a pressure which will not exceed

$$P_h = \rho V \left[\frac{1}{\rho(1/K + d/Et)} \right]^{0.5} \tag{6-3}$$

where V = velocity of flow
d = pipe diameter
t = pipe wall thickness
K = bulk modulus of elasticity of water
E = modulus of elasticity of pipe

The pressure resulting from water hammer must be added to the normal internal pressure, but may be less than the value calculated from Eq. (6-3), depending on the length of the pipeline, the rate at which the velocity is reduced, and the degree of relief produced by special valves, surge tanks, air chambers, or similar devices.[1,2]

At *changes in direction*, the principle of conservation of momentum requires that a force be applied to the moving water. This force may be reacted by friction of the pipe against the soil, by suitable buttressing, or by developing additional tension in the pipe. In unrestrained pipes, an additional longitudinal tension can develop, equal to

$$\sigma_l = \frac{d}{2t} \rho V^2 \sin \frac{\alpha}{2} \tag{6-4}$$

in which ρ is the density of water and α is the angle of deflection of the pipe. Displacements may be produced at changes in pipe direction as a result of changes in pipe length resulting from this stress. Additionally, joints which do not provide a positive mechanical connection (Art. 6-4) may separate because of their inability to transfer the load from one section to the next. Such displacements or separations can be prevented by suitable buttressing. The design force for such a buttress is given by

$$F = \left(\frac{\pi d^2}{2}\right)(V^2\rho - P)\sin\frac{\alpha}{2} \tag{6-5}$$

Thermal stresses are calculated in the usual fashion from

$$\sigma_T = C\Theta E \tag{6-6}$$

where C = coefficient of thermal expansion
Θ = change in temperature
E = modulus of elasticity of pipe material

This stress is produced, of course, only if the pipe is restrained so that it cannot change in length. Such restraint is normally provided by buttresses and soil friction.

The vertical load on a buried pipe depends on the stiffness of the pipe, the dimensions of the trench, the characteristics of the backfill material, the care of bedding, and the imposed surface loads. Water pipe is manufactured in a variety of thicknesses and strengths. For specific bedding conditions (Fig. 6-1) and known surface loads and working pressures, the pipe class (which specifies its strength) may be selected from prepared tables in standard design manuals.[3,4]

6-4 Pipelines

Pipelines are commonly constructed of reinforced concrete, asbestos cement, ductile iron, steel, or plastic and are located below the ground surface only so far as is necessary to protect them against freezing and surface loads and to avoid other subsurface structures. In locations in which the ground (and pipe) elevations vary by very large amounts, high pressures at low points may be avoided by breaking the hydraulic gradient with overflows or auxiliary reservoirs or by installing special pressure-reducing valves.

At low points in the system, valved blowoff branches or hydrants are provided to drain the line and permit removal of sediment. High points in the line should be kept below the hydraulic grade line, since negative pressure at such locations will lead to accumulation of gases which eventually may block the flow. High points should be provided with vacuum- and air-relief valves (Art. 6-10) to admit air when the line is being emptied and to release air which is initially in the line or which accumulates during use. Admitting air is particularly important with thin-walled pipe (such as steel pipe), which may buckle under compressive loads.

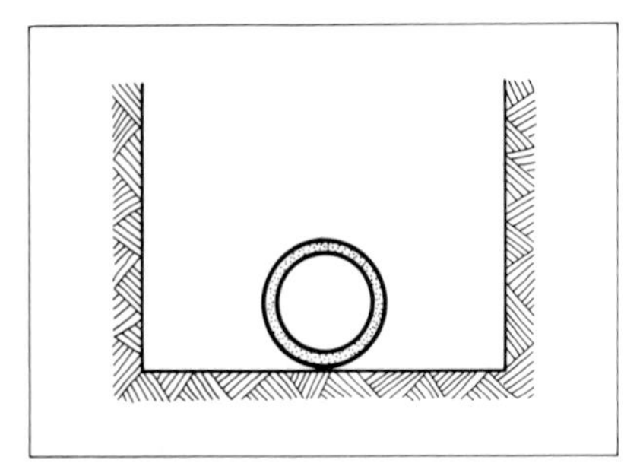

Type 1 Flat-bottom trench. Loose backfill.

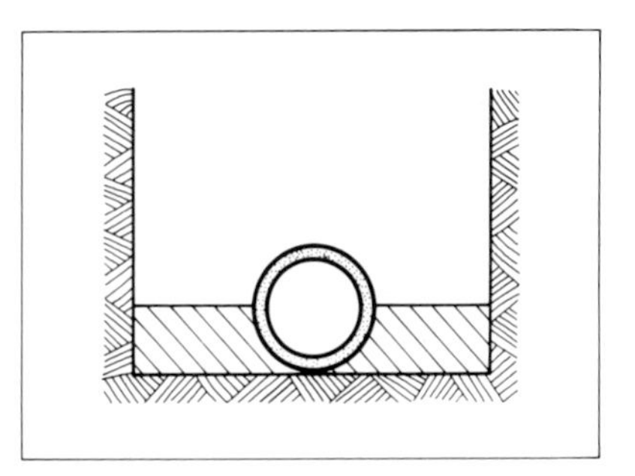

Type 2 Flat-bottom trench. Backfill lightly consolidated to centerline of pipe.

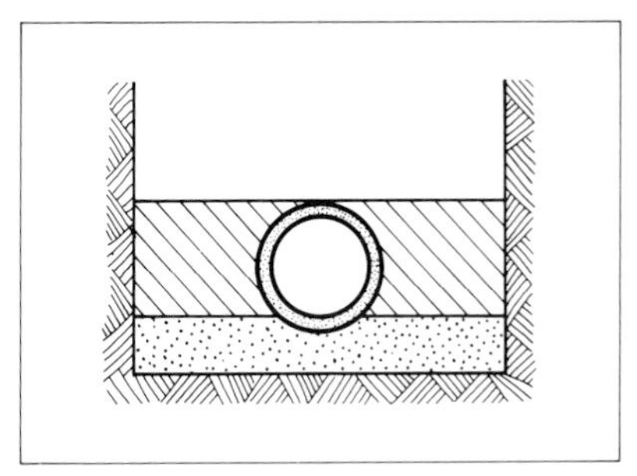

Type 3 Pipe bedded in 100-mm minimum loose soil. Backfill lightly consolidated to top of pipe.

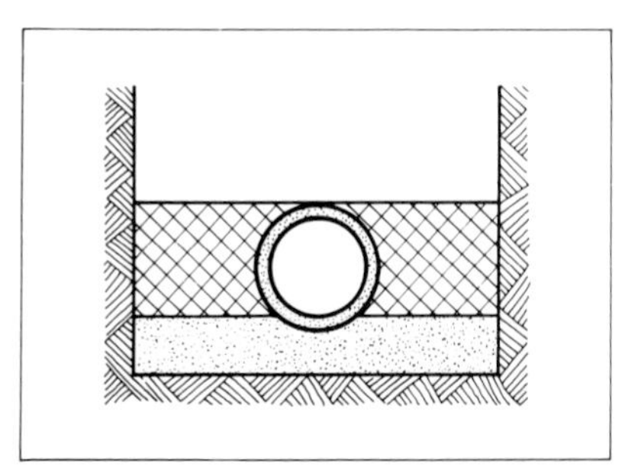

Type 4 Pipe bedded in sand, gravel, or crushed stone to depth of $\frac{1}{8}$ pipe diameter. 100-mm minimum. Backfill compacted to top of pipe. (Approx. 80 percent Standard Proctor, AASHTO T-99)

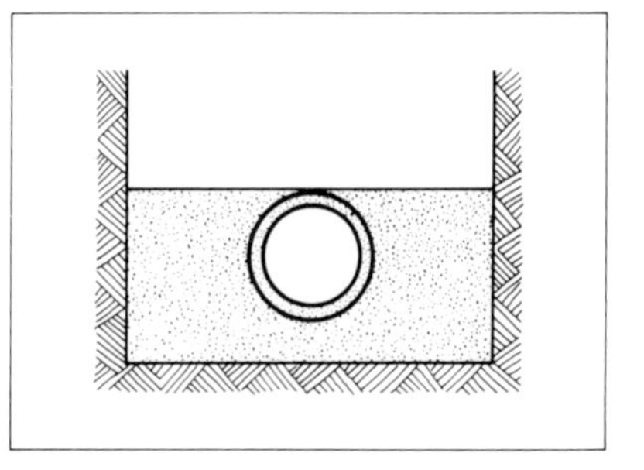

Type 5 Pipe bedded to its centerline in compacted granular material. 100-mm minimum under pipe. Compacted granular or select material to top of pipe. (Approx. 90 percent Standard Proctor, AASHTO T-99)

FIGURE 6-1
Standard pipe bedding conditions.

In selecting the type of material and pipe size to be used, one should consider carrying capacity, durability, maintenance cost, and first cost. The character of the water and its potential effect upon pipe of different materials is an important consideration as well.

6-5 Iron Pipe

Iron pipe has been used for water conveyance for over 300 years—in fact, cast iron pipe installed in 1664 in Versailles in France is still in use. Ductile iron has largely replaced cast iron in new construction, since for a given strength it is lighter and is less brittle. Although its wall thickness is less, ductile iron pipe has the same outer diameter as cast iron of the same nominal size. Despite its greater hydraulic capacity, it may thus be used interchangeably with cast iron and can be used with cast iron fittings. Ductile iron is produced by adding a magnesium alloy to an iron of very low phosphorus and sulfur content. The resulting crystalline microstructure is quite different from that in ordinary gray cast iron and imparts additional strength, toughness, and ductility.

Iron pipe is manufactured by casting in molds, with the mold being rotated about its longitudinal axis. Such centrifugally cast pipe is normally manufactured in 5.5-m (18-ft) lengths.

Fittings are manufactured by casting in sand molds with solid patterns and core boxes. Standard fittings (Fig. 6-2) include bends of various angles, tees, crosses, concentric and eccentric reducers, offsets, wyes, reducing fittings, and a large assortment of specialties such as wall sleeves, cutting-in fittings, flares, and valve boxes.

Iron pipe is extremely durable and may be expected to have a service life in excess of 100 years. It is, however, subject to corrosion, which may produce a phenomenon called tuberculation, in which scales of rust coat the inside of the pipe, reducing its diameter and increasing its relative roughness. The combination of these effects may produce a reduction in hydraulic capacity of 70 percent or more. For this reason, it is not uncommon to line iron pipe with cement or bituminous material.

The cement lining consists of a 1:2 portland cement mortar which is applied centrifugally and is tapered slightly at the ends. The lining may be cured by coating it with a bituminous seal or by storing the lined pipe in a moist environment. Cement-lined pipe retains good hydraulic properties and does not deteriorate with age as long as the lining is intact. No damage is done by normal cutting or tapping procedures and minor cracks have been shown to be self-healing when placed in contact with water.[5]

External corrosion of iron pipe is seldom a major problem because of its relatively great wall thickness. In particularly unfavorable soil conditions, it may be encased in polyethylene tubes as construction proceeds.[6] This technique has been demonstrated to be effective in protecting the pipe against external corrosion.

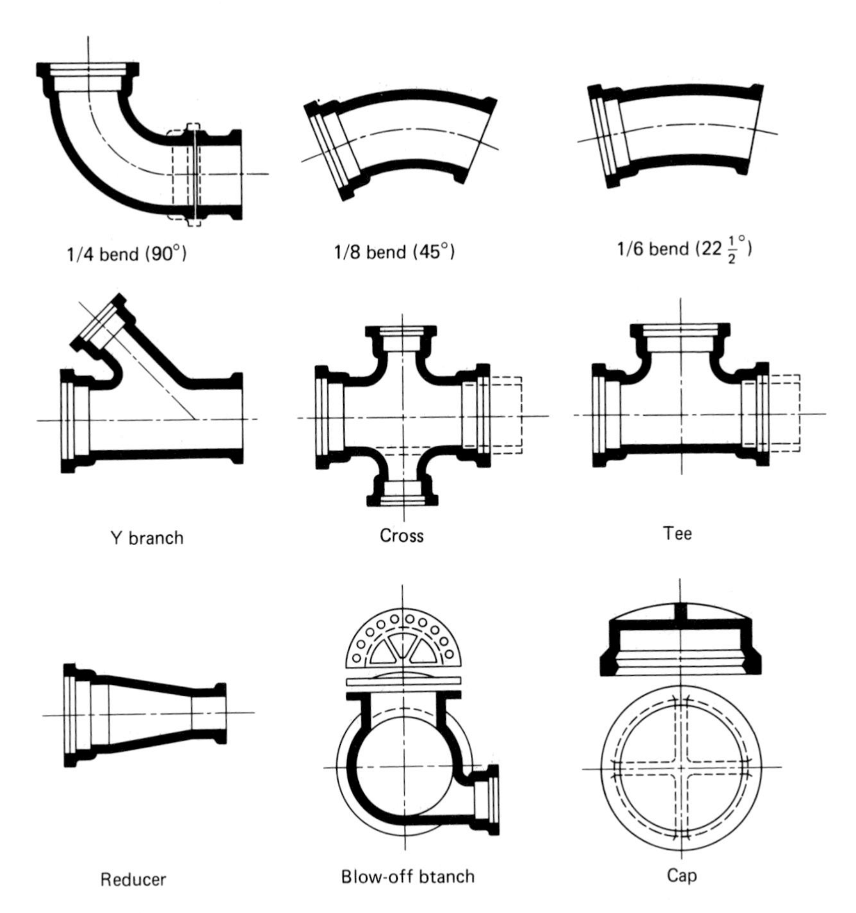

FIGURE 6-2
Some standard bell-and-spigot fittings.

Iron pipe is assembled by a variety of techniques depending, in part, on the circumstances in which it is being used. Typical joints are illustrated in Fig. 6-3. Bell-and-spigot joints, in which lead is used as the sealing material after the joint is packed with oakum yarn, have largely been replaced by push-on joints with rubber gaskets. The latter are much more readily assembled and are less likely to develop leaks as a result of displacements occurring after construction is complete. Neither bell-and-spigot nor push-on joints are able to resist longitudinal forces such as those developed at changes in direction. For this reason, such joints must frequently be buttressed. Minor changes in direction can be accommodated in either type by deflecting adjoining sections by up to 5°, depending upon the diameter.

Mechanical joints are available both with and without a locking ring. The unlocked joint cannot resist much thrust and is sometimes used with threaded tie rods which transfer longitudinal loads to adjoining sections in order to

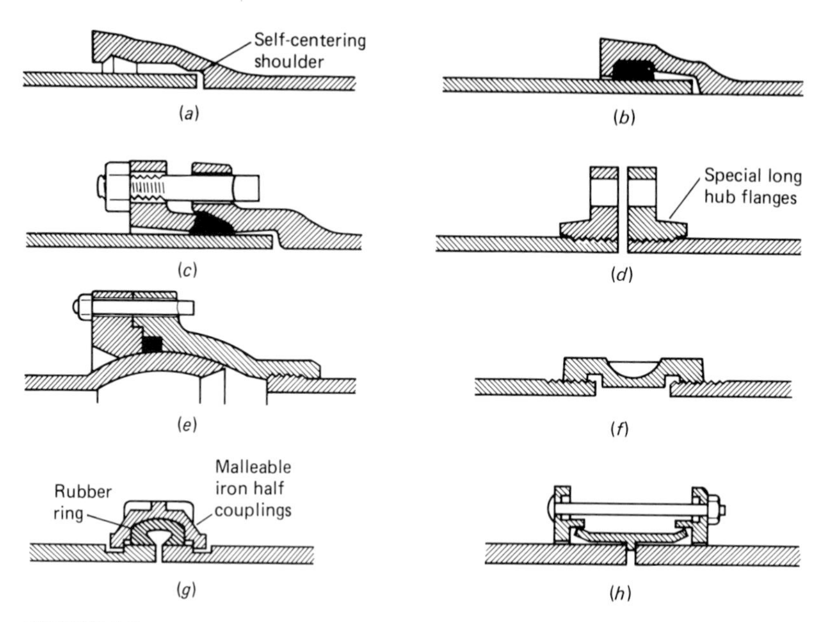

FIGURE 6-3
Iron pipe joints. (*a*) Bell-and-spigot; (*b*) push-on; (*c*) mechanical; (*d*) flanged; (*e*) ball; (*f*) threaded; (*g*) victaulic coupling; (*h*) Dresser coupling.

develop more soil resistance. Such joints can be deflected by up to 8°, depending upon the diameter.

Flanged pipe is manufactured by threading the pipe ends and screwing the flanges on to the ends. The flanges are machined with smooth parallel faces and a gasket is inserted to ensure a watertight fit. Such joints are able to resist both thrust and moment. They do not permit any deflection of the pipe sections relative to one another. Couplings which will permit accommodation of minor errors in alignment must be incorporated in pipe assemblies with flanged joints. Flanged joints are never buried, since corrosion could make their later disassembly very difficult and settlement of the soil could cause failure of the pipe itself.

Flexible ball joints permit deflection of adjoining sections by up to 15°. A rubber gasket ensures a watertight seal in an otherwise all-metal joint. This pipe is used in circumstances in which large deformations are anticipated, particularly in river crossings, where adjoining sections are assembled on a barge and lowered over the side in a continuous process.

Threaded joints are seldom used in iron pipe larger than that found in household plumbing. Most threaded iron pipe used in large-scale engineering projects is provided with flanges.

Victaulic joints consist of two semicircular housings which are bolted together around the pipe. The housings engage grooves in the pipe ends and enclose a rubber ring which serves as a gasket. Joints of this type are used in applications similar to those where flanged pipe might be used.

Dresser couplings consist of a central ring and two external rings which are clamped together against the central ring, forcing gaskets beneath the central ring. Such couplings permit a degree of rotation of the union and some misalignment of the pipe centerlines and accommodate longitudinal motion. They are often used in flanged or victaulic pipe systems to permit assembly and disassembly of these rigid joints.

6-6 Steel Pipe

Steel may be used for water lines, particularly in circumstances where diameters are large and pressures are high. Steel has economic advantages in such circumstances, since it is stronger and thus lighter for a given strength. Steel pipe, weight for weight, is cheaper than iron pipe, is more easily transported and more easily assembled. In the circumstances in which it is commonly used, it may be subject to failure because of negative pressures developed during transient conditions, since its relatively thin walls buckle readily. Its relatively thin walls also make steel pipe more likely to be structurally damaged by corrosion than iron. Under favorable conditions, its life may exceed 50 years. Unfavorable conditions include corrosive water and corrosive soils. Modern steel pipe is assembled by welding, but riveted or lock-bar pipe is still found in use in older systems. Such pipe is likely to leak somewhat at the joints as a result of corrosion or deterioration of the caulking.

Steel pipe must be cleaned of any mill scale and is then normally coated with tar or a bituminous enamel.[7] Scars or other defects in the coating caused by handling and placement of the pipe must be repaired before it is put in service. All coatings tend to lose elasticity and adhesion with time, and some are pervious to water under high pressure. For these reasons, recoating may be necessary at regular intervals. Coating inside and outside with portland cement mortar[8] has been shown to provide good protection against external and internal corrosion, increased resistance to buckling, and improved hydraulic properties.

6-7 Concrete Pipe

Concrete cylinder pipe is frequently used in water conveyance. It is manufactured by wrapping high-tensile-strength wire about a steel cylinder which has been lined with centrifugally placed cement mortar. The wire is wound tightly to prestress the core and is covered with an outer coating of concrete. For lower pressures, a similar pipe is manufactured without the prestressing wire. Where leakage is not important, reinforced-concrete pressure pipe or unreinforced-concrete pipe may be used. Concrete pipe is manufactured in lengths ranging from 3.7 to 4.9 m (12 to 16 ft) and is rated for static pressures up to 2700 kPa (400 lb/in^2). Joints are illustrated in Fig. 6-4 and consist of concentric circular steel rings which are sealed with a rubber gasket. The steel rings can be welded to each other if necessary to develop thrust resistance in the joint.

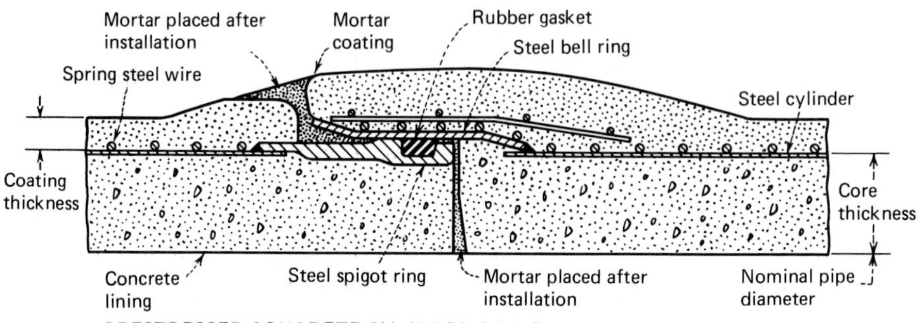

PRESTRESSED CONCRETE CYLINDER PIPE, RUBBER AND STEEL JOINT

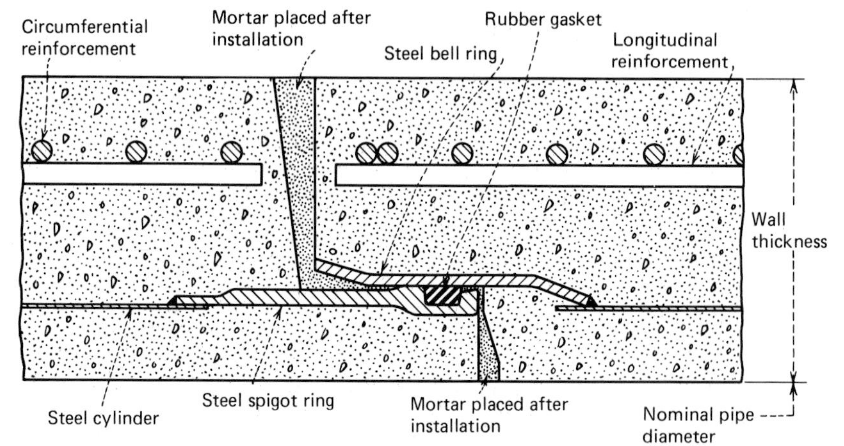

CONCRETE CYLINDER PIPE, NOT PRESTRESSED, RUBBER AND STEEL JOINT

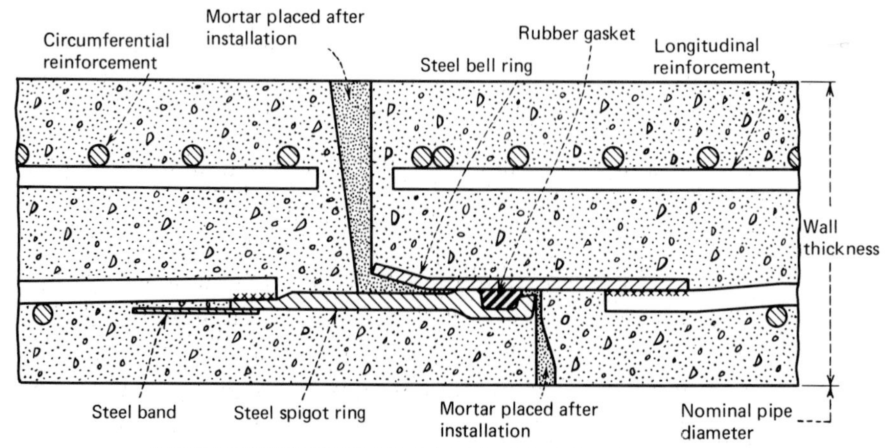

CONCRETE PRESSURE PIPE, RUBBER AND STEEL JOINT

FIGURE 6-4
Concrete pipe joints.

Adjoining sections can be deflected to permit gradual curvature, or specially made sections with beveled ends can be used. The latter permit deflection angles of up to 5°.

Fittings may be manufactured in the same manner as the pipe or may be constructed in place of reinforced concrete. Concrete cylinder pipe is commonly manufactured specifically for particular projects, hence special sections are not difficult to obtain.

Concrete pipe (except in the case of acid waters) is not subject to corrosion and suffers no loss in hydraulic capacity with age. A reasonable estimate of its service life is 75 years.

6-8 Asbestos Cement Pipe

Asbestos cement pipe is composed of a mixture of portland cement and asbestos fiber which is built up on a rotating steel mandrel and then compacted with steel pressure rollers. This pipe has been used for over 60 years in western Europe and the United States. Over 2.4 million km (1.5 million mi) is in service worldwide. Since it has a very smooth inner surface, it has excellent hydraulic characteristics.

Asbestos has been shown to be a carcinogen when the fibers are inhaled, and there is some evidence, although it is debatable, that asbestos fibers in water may cause intestinal cancers as well. No mandatory standard for asbestos in water had been established as of 1989. Asbestos fibers are found in some natural waters[9] and can be leached from asbestos cement pipes by very aggressive waters—those that dissolve the cement itself.[10] Some water utilities no longer use asbestos cement pipe in new construction.

Iron fittings are used with asbestos cement pipe. The outer diameter of the asbestos cement pipe is identical to that of ductile or gray cast iron pipe. Small connections are made by tapping, as with iron pipe. Joints consist of cylindrical sleeves that fit over the ends of adjoining sections. Both the pipe and the sleeve are grooved to retain rubber rings, which serve as gaskets. The joints can be deflected up to 12°.

6-9 Plastic Pipe

Plastic pipe is manufactured of both solid and fiber-reinforced materials. Such pipe is widely used both in domestic plumbing and in water distribution systems, since it is far easier to handle and install and generally cheaper than traditional materials such as iron and concrete. The long-term performance of these materials can be established only by the passage of time. Cold flow, age embrittlement, or installation stresses may affect the long-term serviceability of plastics. Some manufacturers offer a 25-year prorated warranty on both materials and labor. The American Water Works Association has established standards for polyvinyl chloride, polyethylene, polybutylene, and glass-fiber-reinforced thermosetting resin pipe.[11–14]

Small-diameter plastic pipes are joined by solvent welding in cylindrical sleeves. Larger-diameter lines have bell-and-spigot push-on connections and are compatible with cast iron fittings.

6-10 Valves and Appurtenances

A variety of valves and specialized appurtenances are used in water distribution systems. *Gate valves* (Fig. 6-5) are most commonly used for on-off service since they are relatively inexpensive and offer relatively positive shutoff. Gate valves are located at regular intervals throughout distribution systems so that breaks in the system can be readily isolated. The Insurance Services Office (which regulates fire insurance rates) requires that valves be so located that breaks will not shut down primary feeders or pipes longer than 150 m (500 ft) in high-value districts and 250 m (750 ft) in other districts. It is desirable to place all valves in manholes, although smaller valves may be buried, with access being afforded by a valve box made of iron or plastic (Fig. 6-6). Gate valves are manufactured with threaded, flanged, bell-and-spigot, or combination ends.

Valves which are operated frequently, such as those in treatment plants, must be designed to be resistant to wear and are often provided with hydraulic or electric operators. Most gate valves will operate properly only when in-

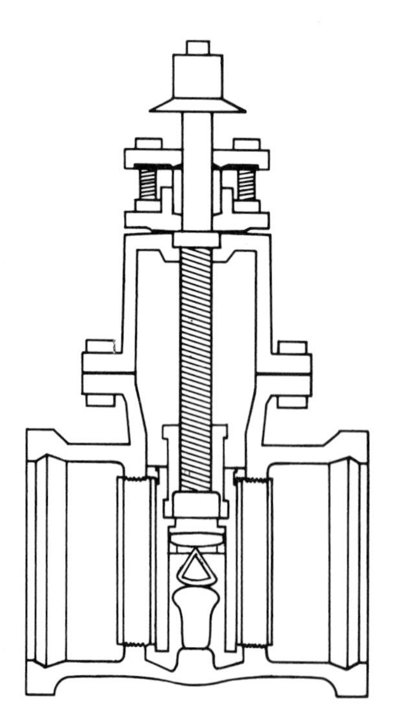

FIGURE 6-5
Iron-body double-disk gate valve.

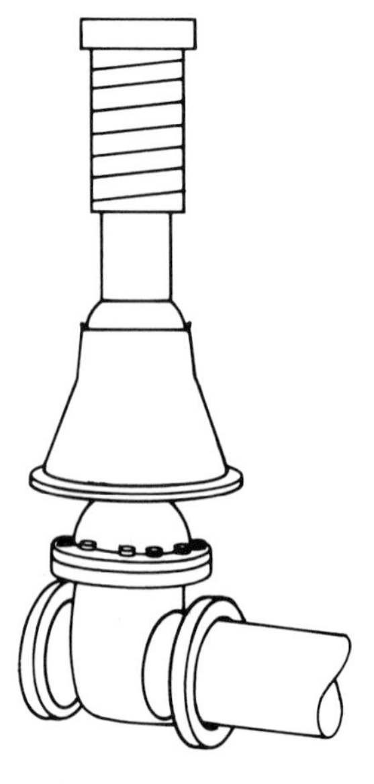

FIGURE 6-6
Valve box.

stalled in a vertical position. For other applications, special valves with disk tracks can be provided. Large valves, even at low pressure, are subjected to large forces when they are in the closed position. Geared operators and small bypass valves (Fig. 6-7) may be used in such cases. The bypass valve will equalize the pressure on the main valve and will reduce the potential for water hammer when the large valve is shut.

Check valves permit water to flow in only one direction and are commonly used to prevent reversal of flow when pumps are shut off. Check valves installed at the end of a suction line are called *foot valves*. These prevent draining of the suction line and loss of prime when the pump is shut down. Check valves are also installed on the discharge side of pumps to reduce hammer forces on the pump mechanism. Such valves may be simple swing-check or ball devices in small lines but are designed to close slowly in large lines—often with discharge of some water through a bypass.

Globe and *angle valves* are seldom used in water distribution systems.

FIGURE 6-7
Geared valve with bypass. (*Courtesy Clow Corporation.*)

Their primary application is in household plumbing where their low cost outweighs their poor hydraulic characteristics.

Plug valves consist of a tapered plug which turns in a tapered seat. A hole in the plug, when the valve is open, coincides with ports in the seat, and these, in turn, are extensions of the pipe in which the valve is placed. Such valves, when open, offer practically no resistance to flow.

Butterfly valves are widely used in both low- and high-pressure applications. In large sizes, they are substantially cheaper, more compact, easier to operate, and less subject to wear than gate valves. They are not suitable for liquids that contain solid material which might prevent their complete closure.

Air-vacuum and air relief valves are provided in long pipelines to permit release of air which accumulates at high points and to prevent negative pressures from building up when the lines are drained. These valves are automatic in operation, opening to release accumulated air and closing when the pipe is full of water.

Pressure-regulating valves automatically reduce the pressure on the downstream side to any desired level. They function by using the upstream pressure to throttle the flow through an opening similar to that in a globe valve. The throttling valve will close (or open) until the downstream pressure reaches the preset value.

Backflow preventers are automatic valves which are designed to prevent contamination of water supplies by transient unfavorable pressure gradients which might otherwise cause reversal of flow. They use either double check valves or reduced-positive-pressure valves (Fig. 6-8). The former close when flow reverses and the latter when the pressure drops, thus providing an additional margin of safety. The type used depends upon the application and the risk to the general public.[15]

Fire hydrants consist of a cast iron barrel with a bell or flange at the bottom which connects to a branch from the water main (Fig. 6-9). The American Water Works Association has developed specifications for hydrants which include the following major features:

1. The hydrant shall be sufficiently slow in closing that water hammer will not exceed the working pressure or 400 kPa (60 lb/in^2), whichever is greater.
2. The hydrant shall be fabricated so that the valve will remain closed if the upper portion of the barrel is broken off.
3. When the discharge is 0.95 m^3/min (250 gal/min) from each hose outlet, the energy loss in the hydrant will not exceed 10 kPa for two-way, 20 kPa for three-way, and 30 kPa for four-way designs.
4. To prevent freezing, a drip valve must be provided to drain the barrel when the main valve is closed.
5. Hydrant outlets should conform with the national standard in order to permit interchange of fire-fighting equipment between adjoining communities.

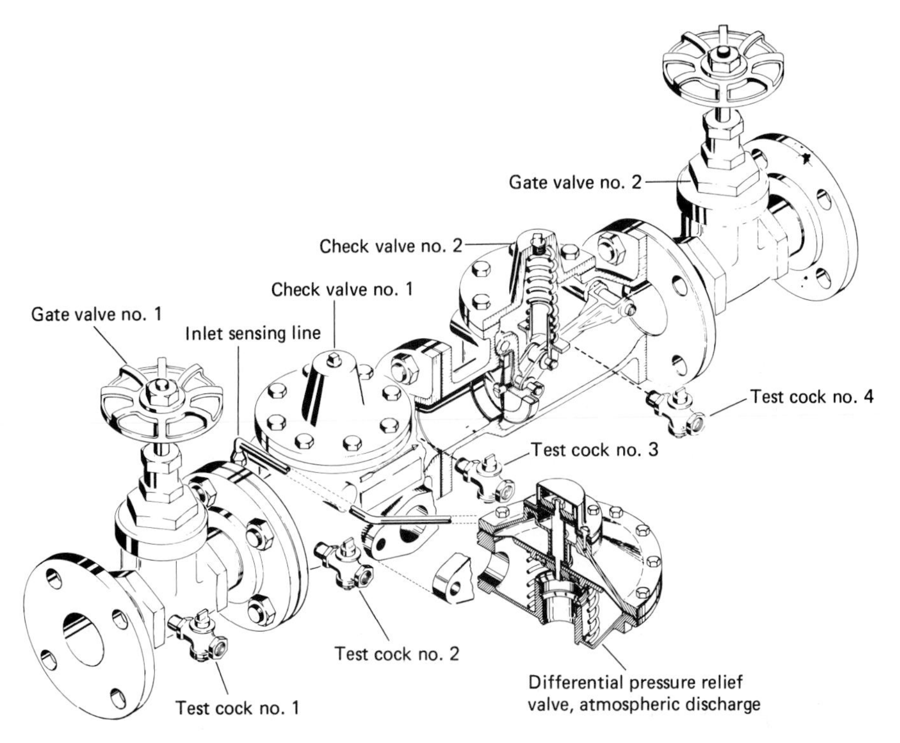

FIGURE 6-8
Reduced-positive-pressure backflow preventer. (*Courtesy Cal-Val Co.*)

6-11 Effects of Pipe Materials on Water Quality

The quality of water can be unfavorably affected by the pipe materials through which it is conveyed. Such effects depend, in part, on the initial character of the water as well as the pipe material. Water which is acidic and low in dissolved solids is particularly likely to attack cement or any metals with which it comes in contact, and some of the materials which are dissolved may be harmful to health, besides being esthetically undesirable.

Iron dissolved from iron or steel pipe can produce red color and contribute a metallic taste to water. Lead may be dissolved from lead pipelines still in use in some communities or from soldered joints in household copper lines.[16] Plastic pipe, under some circumstances, may permit organic materials to pass through the wall and contaminate the water being conveyed.[17,18] Such contamination can only occur, of course, if the outside of the pipe is exposed to organics to which it is permeable. As noted above, corrosive waters may dissolve the cement and release the asbestos fibers in asbestos cement pipe. Although the health impact of asbestos fibers in water is still uncertain at the time of this writing, it is possible that they may be harmful. Among other things, asbestos in water may be released into the air from showering, cooking, and cleaning.

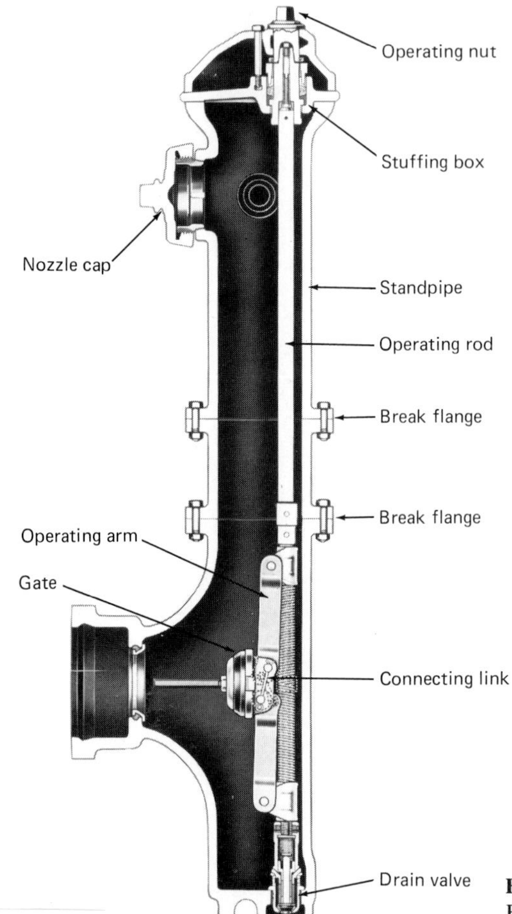

FIGURE 6-9
Fire hydrant. (*Courtesy Clow Corporation.*)

6-12 Corrosion and Its Prevention

Corrosion may be defined as the conversion of a metal to a salt or oxide with a loss of desirable properties such as mechanical strength.[19] Corrosion may occur over an entire exposed surface or may be localized at micro- or macroscopic discontinuities in the metal. In all types of corrosion an electron transfer must occur, and in most reactions of interest in environmental engineering this transfer occurs either between dissimilar metals or between different areas on a single material. The zone which releases electrons is called anodic, while that which accepts them is called cathodic, as in other electrical circuits.

At the anode, an oxidation reaction takes place which can be represented as

$$Fe \rightarrow Fe^{++} + 2e^{-} \tag{6-7}$$

in which it is seen that the iron enters into solution. For this reaction to proceed a simultaneous reduction reaction must occur, which commonly is the reduction of the dissolved oxygen present in the water or the reduction of hydrogen ion to hydrogen gas.

$$O_2 + 4e^- + 2H_2O \rightarrow 4OH^- \tag{6-8}$$

$$2H^+ + 2e^- \rightarrow H_2 \tag{6-9}$$

Different reactions may occasionally occur with other oxidizing agents if they are present.

Subsequent reactions between the oxidized iron and the hydroxide ion produced by oxygen reduction or liberated by hydrogen reduction may occur and lead to the precipitation of insoluble products such as Fe_2O_3, $Fe(OH)_2$, and $Fe(OH)_3$. Accumulation of these products may slow the rate of the reaction by interfering with oxygen diffusion to the metal surface.

A variety of complex reactions can occur in the oxidation of iron. These reactions are dependent on either pH or electrode potential, or both. A simplified Pourbaix diagram[19] is presented in Fig. 6-10, which indicates the zones in which corrosion will or will not occur in systems involving iron and pure water. Corrosion can occur in the region labeled "Passivity", but ordinarily does not. In real systems, the diagram is complicated by reactions with other dissolved substances. The most important observation that can be made from this diagram is that corrosion of iron is possible under all conditions occurring in water and waste treatment save, perhaps, softening and phosphorus precipitation with lime.

Since two reactions, an oxidation and a reduction, must occur for aqueous corrosion to proceed, the process can be slowed or halted by interfering with either. The techniques employed include cathodic protection, anodic protection, inhibition, and application of metallic or chemical coatings.

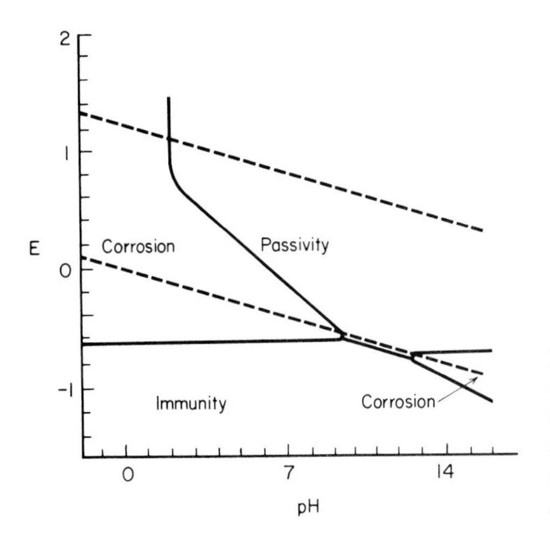

FIGURE 6-10
Simplified Pourbaix diagram for iron.[19] *(From The Fundamentals of Corrosion, 2d ed., by J. C. Scully. Copyright 1975. Used with permission of the author and publisher.)*

Galvanic protection forces the entire metal surface to act as a cathode. Since corrosion occurs only at anodic areas, such a procedure prevents loss of metal. Cathodic protection can be provided by either the impressed current or the galvanic technique. *Impressed-current* protection involves applying a dc voltage to the metal so that electrons will flow into it at a rate at least equal to that at which they left under conditions of corrosion. The anode may be any conducting material. If it is prone to corrosion it may be replaced at intervals. A number of commercial firms design and manufacture impressed-current cathodic protection systems. Design details are presented elsewhere.[20]

Galvanic protection utilizes a sacrificial anode—a material of higher corrosion potential than that which it is desired to protect. Since such a material will be anodic with respect to the protected metal, the latter will be entirely cathodic and will not corrode. Materials used to protect iron and steel include magnesium, zinc, and aluminum. Magnesium is most commonly used in water works practice, since it does not form dense oxide films or exhibit passivation with respect to iron. The anode will corrode and must eventually be replaced.

Anodic protection is a passivation technique in which impression of an external voltage greatly reduces the rate of corrosion of a metal. A relatively high current density may be necessary to achieve passivation, but far less is required for maintenance of protection. A breakdown of the polarized film on the metallic surface may produce extremely rapid corrosion since a small area may become anodic to the entire surface. Variations in applied voltage, once passivation has been achieved, may also produce rapid corrosion.[20,21]

Inhibition is effected by the deposition or adsorption of ions on metallic surfaces. Chemicals used for this purpose include chromates, nitrates, phosphates, molybdates, tungstates, silicates, benzoates, etc.[19] Such coatings are effective in reducing corrosion in neutral or alkaline (but not acid) solutions.

Metallic coatings can be applied by hot dipping, metal spraying, cladding, vapor deposition, electroplating, metalliding and mechanical plating. The film formed ranges from 2×10^{-4} to 5 mm and may serve as a final cover or as a base for another protective coating. Coatings may be either anodic or cathodic with respect to the base metal. If the coating is anodic, it will provide galvanic protection to the base metal in the event the coating is broken. Noble (cathodic) coatings may produce rapid pitting in the base metal when discontinuities occur.

Chemical coatings include paints, coal tar preparations, asphalt, epoxy materials and, as mentioned earlier, cement. These materials generally serve to isolate the metal from the aqueous environment. Paints, in addition to isolating the surface, may provide other protection. Zinc chromate paints, for example, include an element of cathodic (galvanic) protection and an element of inhibition provided by the chromate, and are mildly alkaline as well.

Inert materials are sometimes substituted for traditional materials which are subject to corrosion. Asbestos cement and plastic pipe are examples of such substitution, although in this case corrosion properties are not the only factor involved. Plastic materials reinforced with fiber glass and compounded to have superior resistance to ultraviolet light (which can cause rapid deterioration of some resins) are now used in many applications in which sheet metal was used in the past.[22] Weirs, launders, and other light structures are typical examples of applications in which reinforced plastics may be used.

PROBLEMS

6-1. A steel pipe line with a nominal diameter of 36 in has an ID of 34.50 in (899 mm) and a wall thickness of 0.75 in (19 mm). The line is 500 ft (152.5 m) long and is embedded at each end in a rigid concrete structure. The temperature of the line may vary from -20 to $+120°F$. What is the maximum thrust which can be developed by temperature changes? What stress does this produce in the pipe?

6-2. A steel pipeline with a nominal diameter of 24 in has an ID of 22.624 in (575 mm) and a wall thickness of 0.688 in (17.5 mm) carries a flow of 28 ft^3/s (794 L/s). The pipe makes a short-radius 90° turn. What is the force developed in a buttress at the bend? What stress would be developed in the pipe by this change in direction if no buttress were provided?

6-3. What is the maximum stress which could be produced in the pipe of Prob. 6-2 as a result of suddenly closing a valve in the line? (K for water is approximately 10 percent of the value of E for steel.)

REFERENCES

1. John Parmakian, *Water Hammer Analysis*, Prentice-Hall, New York, 1955.
2. E. Benjamin Wylie and Victor L. Streeter, *Fluid Transients*, FEB Press, Ann Arbor, Mich., 1983.
3. "American National Standard for Thickness Design of Ductile Iron Pipe," AWWA C150, American Water Works Association, Denver, 1986.
4. *Concrete Pipe Design Manual*, American Concrete Pipe Association, Vienna, Va., 1987.
5. Ernest F. Wagner, "Autogenous Healing of Cracks in Cement Mortar Linings for Gray-Iron and Ductile-Iron Water Pipe," *Journal of American Water Works Association*, **66**:358, 1974.
6. "American National Standard for Polyethylene Encasement for Gray and Ductile Cast-Iron Piping for Water Mains and Other Liquids," AWWA C150, American Water Works Association, Denver, 1982.
7. William R. Kinsey, "Steel Water Pipe Design, Lining, Coating, Joints and Installation," *Journal of American Water Works Association*, **65**:786, 1973.
8. "AWWA Standard for Cement-Mortar Protective Coating for Steel Water Pipe," AWWA C205, American Water Works Association, Denver, 1985.
9. Roger C. Bales, Dale D. Newkirk, and Steven B. Hayward, "Chrysolite Asbestos in California Surface Waters: From Upstream Rivers through Water Treatment," *Journal of American Water Works Association*, **76**:5:66, 1984.
10. James S. Webber, James R. Covey and Murray Vernon King, "Asbestos in Drinking Water Supplied through Grossly Deteriorated A-C Pipe," *Journal of American Water Works Association*, **81**:2:80, 1989.
11. "AWWA Standard for Polyvinyl Chloride (PVC) Pressure Pipe," AWWA C900, American Water Works Association, Denver, 1981.
12. "AWWA Standard for Polyethylene (PE) Pressure Pipe," AWWA C901, American Water Works Association, Denver, 1978.
13. "AWWA Standard for Polybutylene (PB) Pressure Pipe," AWWA C902, American Water Works Association, Denver, 1978.

14. "AWWA Standard for Glass-Fiber-Reinforced Thermosetting-Resin Pressure Pipe," AWWA C950 American Water Works Association, Denver, 1981.
15. Gustave J. Angele, Sr., *Cross Connections and Backflow Prevention*, 2d ed., American Water Works Association, Denver, 1974.
16. Hudson H. Birden, Jr., Edward J. Calabrese, and Ann Stoddard, "Lead Dissolution From Soldered Joints," *Journal of American Water Works Association*, **77**:11:66, 1985.
17. William G. Leseman, "Water Contamination Caused by Gasoline Permeating a Polybutylene Pipe," *Journal of American Water Works Association*, **78**:11:39, 1986.
18. Alan R. Berens, "Prediction of Organic Chemical Permeation through PVC Pipe," *Journal of American Water Works Association*, **77**:11:57, 1985.
19. J. C. Scully, *The Fundamentals of Corrosion*, 2d ed., Pergamon Press, Oxford, 1975.
20. Walter G. v. Baeckmann, "Cathodic Protection of Underground Pipelines with Special Reference to Urban Areas," *Journal of American Water Works Association*, **66**:466, 1974.
21. Michael Henthorne, "Cathodic and Anodic Protection for Corrosion Control," *Chemical Engineering*, **78**:27:73, 1971.
22. Walter A. Szymanski and Robert C. Taylor, "Fiber Glass Reinforced Polyesters and Furan Resins for Water and Waste Treatment Systems," *Journal of American Water Works Association*, **68**:228, 1976.

CHAPTER 7

COLLECTION AND DISTRIBUTION OF WATER

7-1 Intakes

Surface sources of water are subject to wide variations in flow, quality, and temperature, and intake structures must be designed so that the required flow can be withdrawn despite these natural fluctuations. The intake itself normally consists of an opening (frequently screened in some manner) and a conduit which conveys the flow to a sump from which it may be pumped to the treatment plant. In locating intakes, one must consider anticipated variations in water level, navigation requirements, local currents and patterns of sediment deposition and scour, spatial and temporal variations in water quality, and the quantity of floating debris.

Impounding reservoirs are subject to rather wide variations in depth and thus require intake structures which will permit withdrawal over a wide range of elevations. It is normally not satisfactory to locate a single inlet at the bottom, since the water quality in reservoirs varies with both time and depth. The quality is usually best close to the surface, although this may not be true for brief periods in spring and fall when overturns may occur.

The lower levels of deep impoundments are normally cool and change very little in temperature during the year. The surface, on the other hand, varies in temperature with the air and, during most of the year, is warmer than the lower levels. The water temperature decreases slowly with depth until, at some level where wind-driven mixing currents become ineffective, it decreases rapidly to the uniform bottom temperature in a short vertical distance. This region of rapid temperature change is called the *thermocline*. In the fall, as the air temperature decreases, the surface layers will cool and sink, displacing the lower layers and driving them to the surface. A similar phenomenon may occur in spring as the water from thawing ice reaches its maximum density and sinks toward the bottom. The water at the bottom of an impoundment is normally low in dissolved oxygen and high in organic matter. It is desirable to avoid drawing this water into the intake, hence the optimum elevation for withdrawal is likely to change during the year.

An intake designed for an impoundment with an earthen dam is illustrated in Fig. 7-1. In concrete or masonry dams the intake may be constructed in the dam itself.

Lake intakes should be located as far as possible from sources of pollution, and one should consider wind and current effects on the motion of contaminants. In particular, winds may stir up sediment from the bottom which may be carried to the intake if it is located in shallow water or too close to the bottom. Inlet velocities should be less than 0.15 m/s (0.5 ft/s) to avoid trapping excessive quantities of floating material, sediment, ice, or fish. A water depth of 6 to 9 m (20 to 30 ft) is necessary to prevent blocking of the intake by ice jams which may fill the lake to the bottom in shallower depths. *Anchor ice* may form beneath the surface on screens, gates, and valves which are cooled below the water temperature by conduction through connecting appurtenances exposed to air inside the intake structure. *Frazil ice* crystallizes within the water in the form of needlelike masses which may adhere to anchor ice already formed in the intake or which may plug screens. Accumulations of ice have been removed by forcing compressed air through the blocked openings, but this technique is

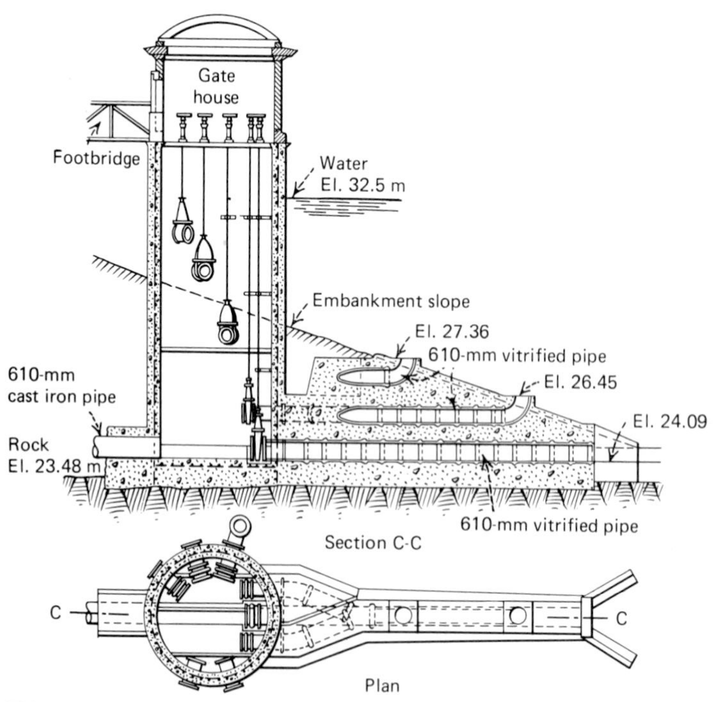

FIGURE 7-1
Reservoir intake, Lexington, Ky.

not always successful. Preventive measures generally involve heating the screens and the air within the structure to keep them at a temperature slightly above the freezing point of water. Figure 7-2 shows the Wilson Avenue intake of Chicago. This structure is located 3.2 km (2 mi) offshore in 11 m (35 ft) of water and has eight radial ports and movable bar screens.

Submerged cribs (Fig. 7-3) are used by smaller communities. Their depth is dictated in part by navigation requirements and in part by the need to locate the intake above high concentrations of sediment which may be suspended by wind action. If entrance velocities are low, no regular attendance is necessary.

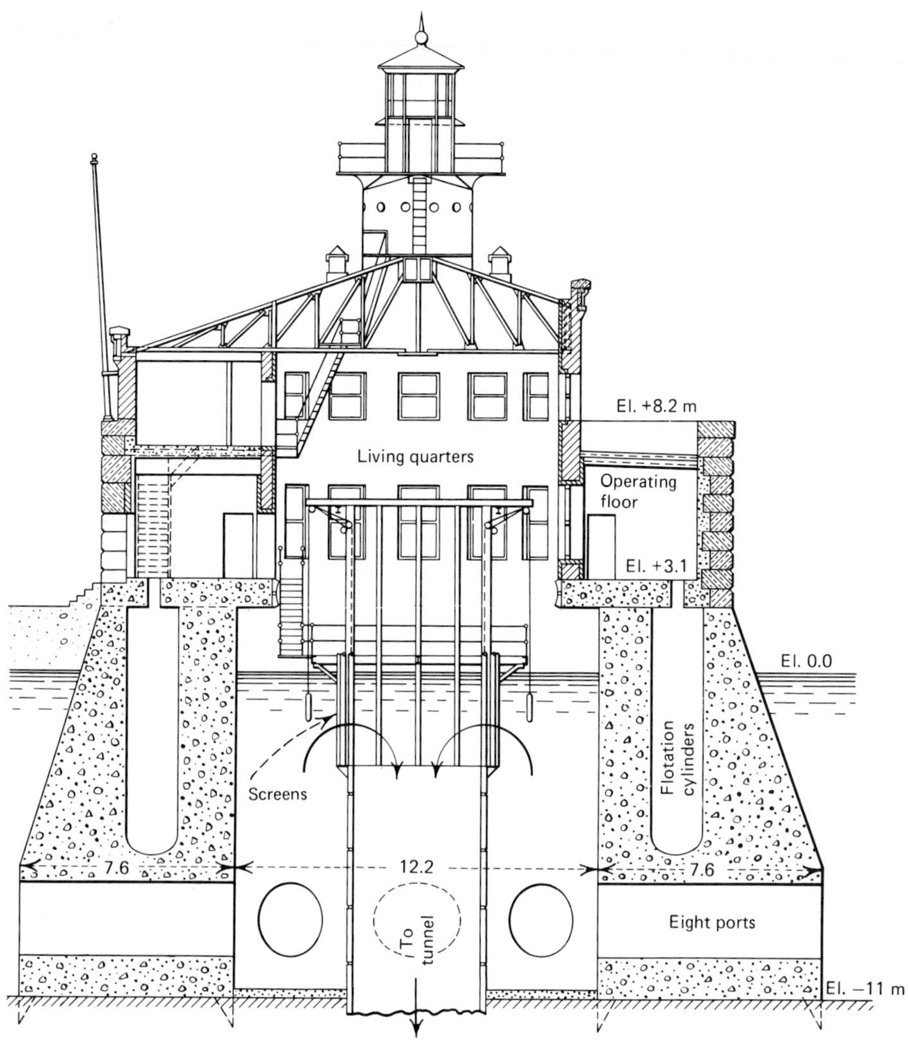

FIGURE 7-2
Lake intake, Chicago.

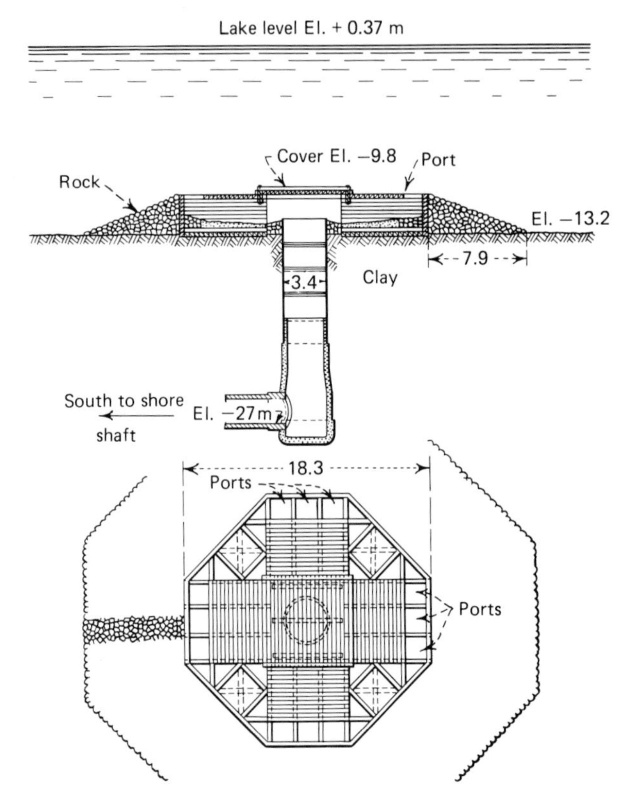

FIGURE 7-3
Submerged crib intake.

The example shown consists of a wooden crib protected by riprap. Pipe intakes have also been used by small cities. This type of structure may consist only of a pipe projecting into the water with a screened opening on the end. More elaborate pipe intakes may be supported above the bottom on piers or pile bents. Multiple screened inlets may by connected to a single pipe to reduce inlet velocities. An intake of this sort is shown in Fig. 7-4.

River intakes should be designed, when possible, to withdraw water from slightly below the surface in order to avoid both sediment in suspension at lower levels and floating debris. Some large cities, notably St. Louis and Cincinnati, have built elaborate river intakes resembling bridge piers with ports at various depths to accommodate variations in water level. Small cities may use simple pipe intakes located so that they are sufficiently below the low-water level that river traffic is not impeded. Such intakes must also be above the bottom so that materials being carried in traction will not cover them. These requirements often dictate that the intake opening be in the main channel, which may be quite far from the normal bank.

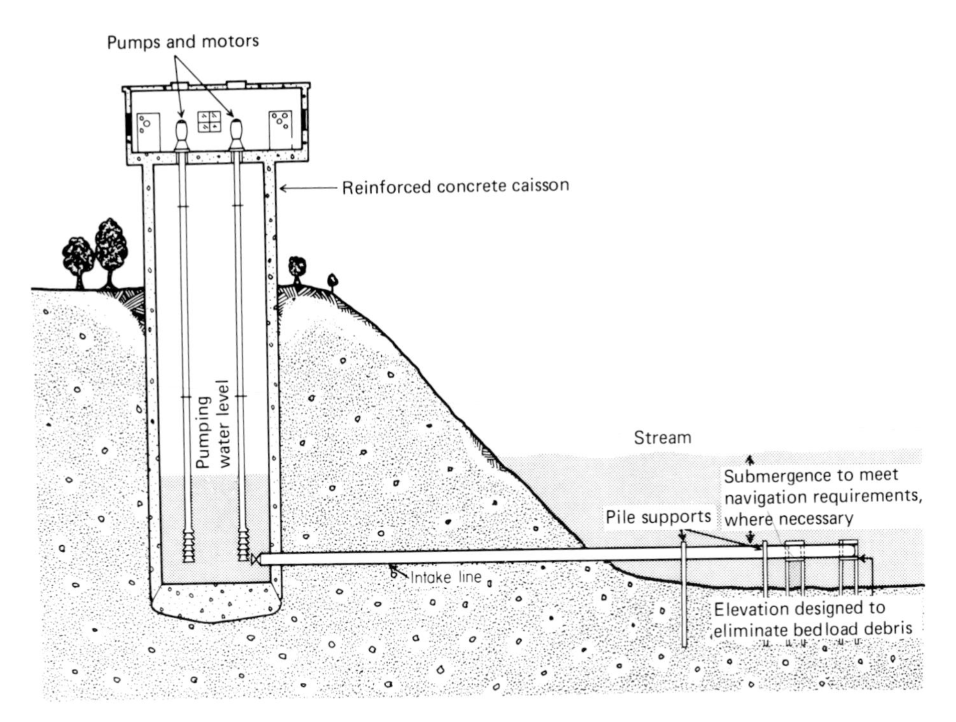

FIGURE 7-4
Screened pipe intake. (*Courtesy The Ranney Company.*)

Shore intakes may be constructed when the main channel of the river is at or near the bank. Low diversion dams may be built upstream to ensure that the total flow goes past the inlet during low river stages, thus minimizing silting of the channel.

Screens are particularly desirable in river intakes, since large quantities of suspended material might otherwise enter the structure. In some areas, waterlogged plant material may be carried below the surface and be drawn into the inlet—as may fish which venture too close. Automatically cleaned bar screens or movable fine screens are frequently necessary to prevent this material from clogging pumps.

Intakes which are located far from the shore of either a river or lake deliver the flow to a pipe conduit buried below the bottom. Velocities in the conduit must be sufficiently great to prevent deposition of sediment, normally in the range of 0.3 to 0.6 m/s (1 to 2 ft/s). The conduit terminates in a sump or well from which the flow is pumped to the treatment plant. Figure 7-5 shows a lake intake, conduit, and pumping station. Where rivers are contained by protective levees, the pumping station must be located within the flood plain. The pumped flow then passes over the top of the levee.

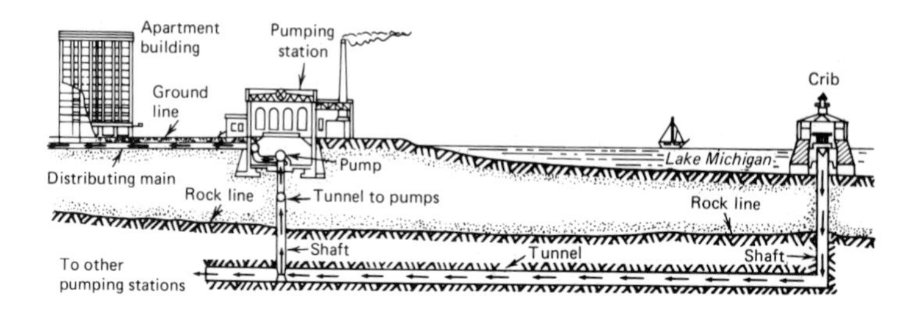

FIGURE 7-5
Intake, conduit, and pumping station, Chicago. (*Civil Engineering*, **20**:*11, November 1950*.)

7-2 Methods of Distribution

Water may be distributed by gravity, by pumps alone, or by pumps in conjunction with on-line storage. *Gravity distribution* is possible only when the source of supply is located substantially above the level of the city. This is the most dependable technique, provided there are multiple well-protected conduits carrying the flow to the community. High pressure for fire fighting may require use of motor pumping trucks, and low-lying areas may need to be isolated to prevent excessive pressure (Art. 6-4).

Pumping without storage is the least desirable method of distribution, since it provides no reserve flow in the event of power failure and pressures will fluctuate substantially with variations in flow. Since the flow must be constantly varied to match an unpredictable demand, sophisticated control systems are required. Peak water use and thus peak power consumption are likely to coincide with periods of already high power use, increasing power costs. Systems of this kind have the advantage of permitting increased pressure for fire fighting, although individual users must then be protected by pressure-reducing valves.

Pumping with storage is the most common method of distribution. Water is pumped at a more or less uniform rate, with flow in excess of consumption being stored in elevated storage tanks distributed throughout the system. During periods of high demand, the stored water augments the pumped flow, thus helping to equalize the pumping rate and to maintain more uniform pressure in the system. It may be economical, in some cases, to pump only during off-peak hours to minimize power costs.[1]

The stored water provides a reserve for fire flow and ensures a reliable general-purpose flow when power fails. Motor pumpers may be used to provide the pressure necessary for fire fighting or a fire pump may be operated at the pumping plant. The latter technique has the same drawbacks as in pumping without storage; in addition, it requires that the storage be isolated from the system when the fire pump is run.

7-3 Storage

Water is stored to equalize pumping rates in the short term, to equalize supply and demand in the long term, and to furnish water during emergencies such as fires and loss of pumping capacity.

Elevated storage may be provided by earthen, steel, or concrete reservoirs located on high ground or by standpipes or tanks raised above the ground surface. Standpipes are cylindrical structures, usually of relatively small diameter. The amount of water available for fire protection from a standpipe is that volume above the level which provides a residual pressure of 140 kPa (20 lb/in^2) at the fire pumps. Elevated tanks (Fig. 7-6) are designed and constructed of steel in capacities up to 15,000 m^3 (4×10^6 gal) by firms which specialize in this work. In large systems, a number of elevated tanks may be located at points selected to minimize pressure variations during periods of high consumption.

FIGURE 7-6
Elevated steel tank. (*Courtesy Chicago Bridge and Iron Company.*)

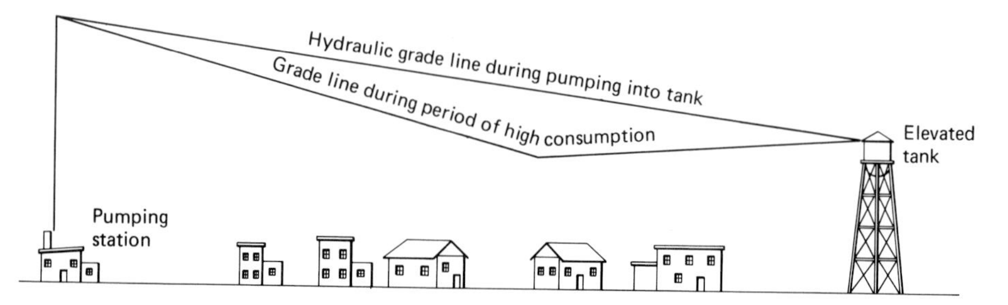

FIGURE 7-7
Effect of elevated storage on pressure.

Normally, elevated storage is located so that zones of high consumption lie between the pumping station and the tanks as shown in Fig. 7-7. During periods of high use, the district will be fed from both sides, which reduces the pressure drop to about one-quarter that which would exist if flow were only from one direction.

The capacity of the elevated storage tanks depends upon the flow variations expected in the system. Equalization of the pumping rate, that is, provision of sufficient capacity to permit pumping at a constant rate, normally requires storage equal to 15 to 30 percent of the maximum daily use. Figure 7-8 shows the pattern of water use on the maximum day for a community of 43,000 persons which has an average daily consumption of 22,750 m^3/day. The average rate of use on the peak day is 41,466 m^3/day or 964 L/day per capita. The maximum hourly rate, at 6:30 p.m., is 1525 L/day per capita. Clearly, if it is possible to pump at the average rather than the peak rate, the pumping station will be smaller and energy costs will be lower. In order to equalize the rate, the water pumped when use is less than average must be stored for use when the rate is higher than average. The cross-hatched area above the average rate is equal to that below and either represents the storage volume required. In this case the volume is 60,840 m^3—147 L per capita, or 28 percent of the average daily consumption of 529 L/day per capita. The calculated amount must be increased to provide for future growth in population.

Additional storage beyond that necessary to equalize pumping may be required to provide for fire protection. The Insurance Services Office, which bases insurance rates partly on the level of fire protection, considers a water supply adequate if it can furnish the required fire flow in addition to the average consumption on the maximum day. The required fire flow may be entirely pumped or may be provided by a combination of pumping and storage in excess of that required to equalize normal demand.

Storage may also be necessary to equalize demand over a lengthy period of high use, such as a cold period in winter or a dry period in summer. Storage of this type is particularly desirable when the source or treatment plant is limited in capacity. The required storage can be determined only from the study

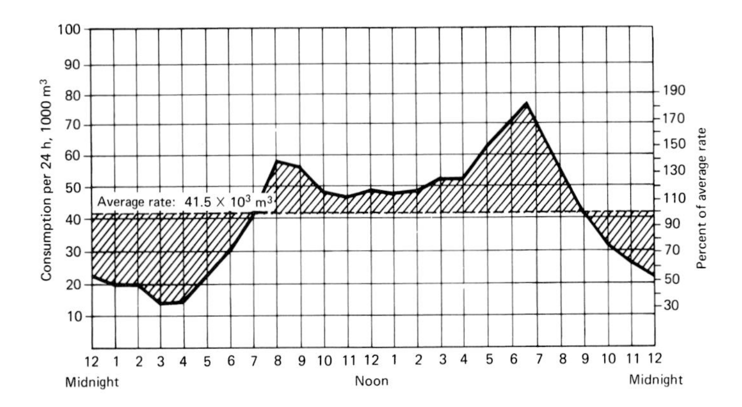

FIGURE 7-8
Diurnal variations in water consumption.

of extensive records of water consumption. Data from periods of high use are used to construct mass diagrams from which the required storage is obtained. The calculated amount must be increased to provide for future growth. When flow data are not available, the Goodrich formula (Chap. 2) may be used. This matter is also discussed in Art. 10-10.

Elevated storage tanks in areas of high consumption and low pressure will increase pressure during periods of peak use without increasing the size of the water mains. The tanks fill during the night when consumption is low and pressure is high. When use is high, the tanks provide water to the system and maintain the pressure in their vicinity. Elevated tanks are commonly provided with automatic valves which close when the tank is full and open when the pressure in the mains falls below that at the bottom of the tank.

7-4 Flow Estimation

Design of a water distribution system requires that the flows to each section of the community be estimated with reasonable accuracy. Chapter 2 deals with estimation of the total quantity of water required at some specific time in the relatively near future. Distribution systems are designed to provide for ultimate development, since the design life of pipe systems may well exceed 100 years.

The first step in design of a water distribution system thus involves prediction of the future development pattern. Many communities have master development plans which establish the allowable uses of various subareas—industrial, commercial, public, and residential. Such plans, when they exist, are the best point of beginning since water consumption can normally be related to land use. Multivariate models which relate water consumption to the customer class have been developed for individual communities.[2] Such models require that local use patterns be carefully analyzed in order to disaggregate the total consumption.[3]

Industrial use is industry-specific and is the most difficult to predict. Assuming that an industrial area will have an average water consumption equal to that of a high-density residential zone [20 L/($m^2 \cdot$ day) or 21,000 gal/(acre $\cdot$ day)] is a reasonably conservative approach for design of water mains. Some industries may require more than this amount, but most will use far less.

Commercial consumption is also specific to the use, being highest for hotels and hospitals [up to 330 L/($m^2 \cdot$ day) or 350,000 gal/(acre $\cdot$ day)]. Offices and retail sales facilities may range up to 90 L/($m^2 \cdot$ day) or 100,000 gal/(acre $\cdot$ day) for multistory construction. A reasonable average value for undefined commercial development is 40 L/($m^2 \cdot$ day) or 45,000 gal/(acre $\cdot$ day) applied to the land area actually covered by structures, not including parking lots or grassed areas.

It is easier to evaluate areas which are zoned residential since anticipated population densities may be established from the residential classification. Table 7-1 presents the range of population densities found in areas of different character. In areas already developed, of course, the population may be determined with considerable certainty. Once an estimate has been made of population density, the average and peak flows can be determined using the *domestic* consumption figures and the Goodrich formula of Chap. 2.

Water is actually removed from the distribution system of a city at a very large number of points. Each user normally has an individual connection so that small flows may be withdrawn at intervals of 15 m (50 ft) or less. It is not reasonable to attempt to analyze a system with this degree of detail. Rather, the individual flows are concentrated at a smaller number of points, commonly at the intersections of the streets. The distribution system can then be considered to consist of a network of nodes (corresponding to points of flow withdrawal) and links (pipes connecting the nodes). The estimated water consumption of the areas contained within the links is distributed to the appropriate nodes.

In addition to providing water for industrial, commercial, and residential use, the distribution system also serves the very important function of fire protection. The amount of water required for fire control is dependent on the character of construction in the area being considered (see Chap. 2). The maximum flow required for an individual fire is 45.4 m^3/min (12,000 gal/min),

TABLE 7-1
Typical population densities

	Population density	
Residential zone classification	**People per mi²**	**People per km²**
Single-family dwellings, large lots	3200–9600	1250–3700
Single-family dwellings, small lots	9600–22,500	3700–5500
Multiple-family dwellings	22,500–64,000	5500–25,000
Apartments or tenements	64,000–640,000	25,000–250,000

but in large communities the possibility of concurrent fires should also be considered. In residential districts the required flow ranges from a minimum of 1.9 m^3/min (500 gal/min) to a maximum of 9.5 m^3/min (2500 gal/min).

Hydrants must be placed throughout the community so that the necessary fire flows can be withdrawn wherever they are required. A single hose stream is considered to be 0.95 m^3/min (250 gal/min), hence maximum fire flow requires 48 hose connections. The minimum area served by a single hydrant is commonly taken as 3720 m^2 (40,000 ft^2), thus minimum spacing is approximately 60 m (200 ft). In no case should hydrants be more than 150 m (500 ft) apart. Hydrants are ordinarily located at street intersections so that hoses may be run in any direction. In high-value districts additional hydrants may be necessary in the middle of long blocks.

7-5 Pressure Required

The pressure in municipal distribution systems ranges from 150 to 300 kPa (20 to 40 lb/in^2) in residential districts with structures of four stories or less to 400 to 500 kPa (60 to 75 lb/in^2) in commercial districts. Pressures less than 350 kPa (42.5 lb/in^2) will not supply 150 kPa (20 lb/in^2) at the top of a six-story building, while pressures less than 200 kPa (30 lb/in^2) are inadequate for four-story buildings. During heavy fire demand when pumper trucks are used, a drop in pressure to not less than 150 kPa (20 lb/in^2) in the vicinity of the fire is permissible.

The American Water Works Association recommends a normal static pressure of 400 to 500 kPa (60 to 75 lb/in^2) since this will supply ordinary uses in buildings up to 10 stories in height, will supply sprinkler systems in buildings of four or five stories, will provide useful fire flow without pumper trucks, and will provide a relatively large margin of safety to offset sudden high demand or closure of part of the supply system.

Pressures in the range of 150 to 300 kPa (20 to 40 lb/in^2) are adequate for normal use and may be used for fire supply in small towns. Fire demand may be met in a variety of ways when normal pressures are low. Turning on special high-pressure fire pumps at the pumping station will increase the pressure throughout the system and may be able to provide the required flow. The high pressures required in this technique may cause failures in household plumbing or even in the mains themselves. Dual systems are used at times in high-value districts, with a separate fire system maintained at pressures of up to 2100 kPa (300 lb/in^2). This technique is quite expensive and is not applicable on a citywide basis. The most common method of providing the necessary pressure is to use motor pumper trucks which take flow from the hydrants in the vicinity and boost the pressure to the level necessary to reach the upper levels of tall buildings (up to 1000 kPa or 150 lb/in^2). Very tall buildings must provide their own fire protection through internal pumping systems and standpipes which are connected to external wall hydrants.

7-6 The Pipe System

The network of pipes which makes up the distribution system may be subdivided into primary or arterial lines, secondary lines and small distribution mains.

The *primary or arterial mains* form the basic structure of the system and carry flow from the pumping station to and from elevated storage tanks and to the various districts of the city. These lines are laid out in interlocking loops with the mains not more than 1 km (3000 ft) apart. Looping assures continuous service even if a portion of the system is shut down for repairs and provides flow from two directions for fire demand. The arterials should be valved at intervals of not more than 1.5 km (1 mi) and all smaller lines connecting to them should be valved so that failure in the smaller lines does not require shutting off the larger. Large primary mains should be provided with blowoff valves at low points and with air and vacuum relief valves at high points.

The *secondary lines* form smaller loops within the primary mains and run from one primary line to another. They are located at spacings of two to four blocks and thus serve to provide large amounts of water for fire fighting without excessive pressure loss.

The *small distribution mains* form a grid over the entire service area—supplying water to every user and to the fire hydrants. They are connected to primary, secondary, or other small mains at both ends and are valved so that the system can be shut down for repairs without depriving a large area of water. The size of the small mains is generally dictated by fire flow except in residential areas with very large lots.

Velocities at maximum flow, including fire flow, normally do not exceed 1 m/s (3 ft/s), with an upper limit of 2 m/s (6 ft/s), which may occur in the immediate vicinity of large fires. The size of the small distribution mains is seldom less than 150 mm (6 in) with cross mains located at intervals of not more than 180 m (600 ft). In high-value districts the minimum size is 200 mm (8 in), with cross-mains at the same maximum spacing. Major streets are provided with lines not less than 305 mm (12 in) in diameter.

Lines which provide only domestic flow may be as small as 100 mm (4 in) but should not exceed 400 m (1300 ft) in length if dead-ended or 600 m (2000 ft) if connected to the system at both ends. Lines as small as 50 and 75 mm (2 and 3 in) are sometimes used in small communities. The length of such lines should not exceed 100 m (300 ft) if dead ended and 200 m (600 ft) if connected at both ends. Dead ends should be avoided whenever possible, since the supply is less certain and the lack of flow in such lines may contribute to water quality problems.

7-7 Design of Water Distribution Systems

The detailed design of a water distribution system is affected by local topography, existing and expected population densities, and commercial and industrial demand. First, the flow must be disaggregated to individual subareas of the system as described in Art. 7-4. Next, a system of interlocking loops must be laid out as described in Art. 7-6. The disaggregated flows are then assigned to the various nodes of the system. The design then involves determination of the sizes of the arterials, secondary lines, and small distribution mains required to ensure that the pressures and velocities desired in the system are maintained under a variety of design flow conditions. These design conditions are based on the maximum daily flow rate plus one or more fires, depending on the size of the community. The fire flow rate depends upon the character of the individual subarea as discussed in Arts. 2-6 and 7-4. In general, those fire locations which are most distant either vertically or horizontally from the pumping plant will be critical for design; however, it is usually necessary to assume various fire locations in order to ensure that all areas are adequately protected.

Consideration of the design problem described above leads to the obvious conclusion that, in general, there are many possible solutions which will satisfy the design constraints. The task then becomes determining the "best" solution. Such an optimization problem for a looped pipe network is very complicated since the distribution of flows in the pipes is a function of the design, hence simplified techniques are often used.[4] Even so, the optima for the various design conditions will not be the same, so that the final design which satisfies all required conditions may not be an optimum for any particular condition.

The usual engineering approach to design of looped pipe systems involves layout of the network as described in Art. 7-6, assignment of estimated pipe sizes (perhaps the minima of Art. 7-6), and calculation of resulting flows and head losses. The pipe sizes are then adjusted as necessary to ensure that the pressures at the various nodes and the velocities in the various pipes meet the criteria established for the community. For a given set of pipe sizes, the calculation of flows and pressures is normally a reasonably straightforward task which can be performed in a variety of ways.

The *Hardy Cross* method[5] and its modifications have been used in design and analysis of water distribution systems for many years. The method is based upon the hydraulic formulas of Chap. 3, which are used to calculate the energy losses in the elements of the system. It is not unusual to neglect the losses in fittings, since these will be small with respect to those in long pipes. The energy loss in any element of the system may be expressed as

$$h_i = k_i Q_i^x \tag{7-1}$$

where h_i = energy loss in element i
Q_i = flow in that element
k_i = constant depending on pipe diameter, length, type, and condition
x = 1.85 to 2 normally, depending on equation used

For any pipe in a loop of the system, the actual flow will differ from an assumed flow by an amount Δ:

$$Q_i = Q_{i0} + \Delta \tag{7-2}$$

where Q_i = actual flow in pipe
Q_{i0} = assumed flow
Δ = required correction

Substituting Eq. (7-2) in (7-1) gives

$$k_i Q_i^x = k_i[Q_{i0}^x + xQ_{i0}^{(x-1)}\Delta + \cdots] \tag{7-3}$$

The remaining terms in the expansion may be neglected if Δ is small compared to Q_i. For any loop, the sum of the head losses about the loop must be equal to zero. This statement is mathematically equivalent to saying there is only one pressure at any point. Thus, for any loop,

$$\sum_1^n k_i Q_i^x = 0 \tag{7-4}$$

where n is the number of pipes in the loop. Then, from Eq. (7-3),

$$\sum_1^n k_i Q_i^x = \sum_1^n k_i Q_{i0}^x + \sum_1^n x k_i Q_{i0}^{(x-1)}\Delta = 0 \tag{7-5}$$

Equation (7-5) may then be solved for the correction:

$$\Delta = -\frac{\sum_1^n k_i Q_{i0}^x}{\sum_1^n x k_i Q_{i0}^{(x-1)}} = -\frac{\sum_1^n h_i}{x\sum_1^n h_i/Q_{i0}} \tag{7-6}$$

The procedure may be outlined as follows:

1. Disaggregate the flow to the various blocks or other subareas of the community.
2. Concentrate the disaggregated flows at the nodes of the system.
3. Add the required fire flow at appropriate nodes.
4. Select initial pipe sizes using the criteria of Art. 7-6.
5. Assume any internally consistent distribution of flow. The sum of the flows entering and leaving each node must be equal to zero.
6. Compute the head loss in each element of the system. Conventionally, clockwise flows are positive and produce positive head loss.
7. With due attention to sign, compute the total head loss around each loop:

$$\sum_i^n h_i = \sum_1^n k_i Q_{i0}^x$$

8. Compute, without regard to sign, the sum

$$\sum_1^n k_i Q_{i0}^{x-1}$$

9. Calculate the correction for each loop from Eq. (7-6) and apply the correction to each line in the loop. Lines common to two loops receive two corrections with due attention to sign.
10. Repeat the procedure until the corrections calculated in step 9 are less than some stipulated maximum. The flows and pressures in the initial network are then known.
11. Compare the pressures and velocities in the balanced network to the criteria of Arts. 7-5 and 7-6. Adjust the pipe sizes to reduce or increase velocities and pressures and repeat the procedure until a satisfactory solution is obtained.
12. Apply any other fire flow conditions which may be critical and reevaluate the velocities and pressure distribution. Adjust the pipe sizes as necessary.

The procedure outlined above is almost always performed on a computer. A simple example using tabular calculations is presented here in order to aid in understanding the technique.

Example 7-1 Figure 7-9 represents a simplified pipe network. Flows for the area have been disaggregated to the nodes, and a major fire flow has been added at node *G*. The water enters the system at node *A*. Pipe diameters are based on the flows and the criteria discussed above. The calculations are tabulated in Tables 7-2 through 7-4, and the corrected flows after each iteration are shown on Fig. 7-10. The calculations are continued in this example until the corrections are less than 0.2 m^3/min (50 gal/min).

The network is divided into the loops *ABHI*, *BEFGH*, and *BCDE*. Any other system might be used (*ABCDEFGHI*, *ABHI*, and *BCDE*, for example), provided all lines are included in at least one loop. In Table 7-2 the pipe identification, assumed flows, length, and diameter are listed in the first four columns. The slope of the hydraulic grade line (in this case calculated from the Hazen-Williams equation with $C = 100$) is tabulated in the fifth column. The head loss in each line is the product of s and the length and is positive or negative depending on the direction of the flow in each line.

Columns 6 and 7 are summed and the correction calculated as shown. Note that the flows in lines common to two loops are positive in one loop and negative in the other. The calculated corrections are applied, with attention to sign, to the flows in each loop. Lines common to two loops receive both corrections. The corrected flows entered in Table 7-3 are then reanalyzed in the same fashion to yield a second set of corrections. The iteration in Table 7-4 gives corrections which are equal to or less than the stipulated maximum. The last corrections are applied to yield the final flows of Fig. 7-10.

The balanced network must then be reviewed to assure that the velocity and pressure criteria are satisfied. The velocities vary from 2.13 m/s (7 ft/s) in line *AB* to 0.22 m/s (0.72 ft/s) in line *ED*. The velocities in lines *AI*, *IH*, *AB*, *BE*, and *EF* exceed the criteria suggested above, and these lines might be increased in diameter. The pressure drop from node *A* to node *G* is 49 m of water or 480 kPa. If the pressure at node *A* were 500 kPa (Art. 7-5), the pressure at node *G* would be only

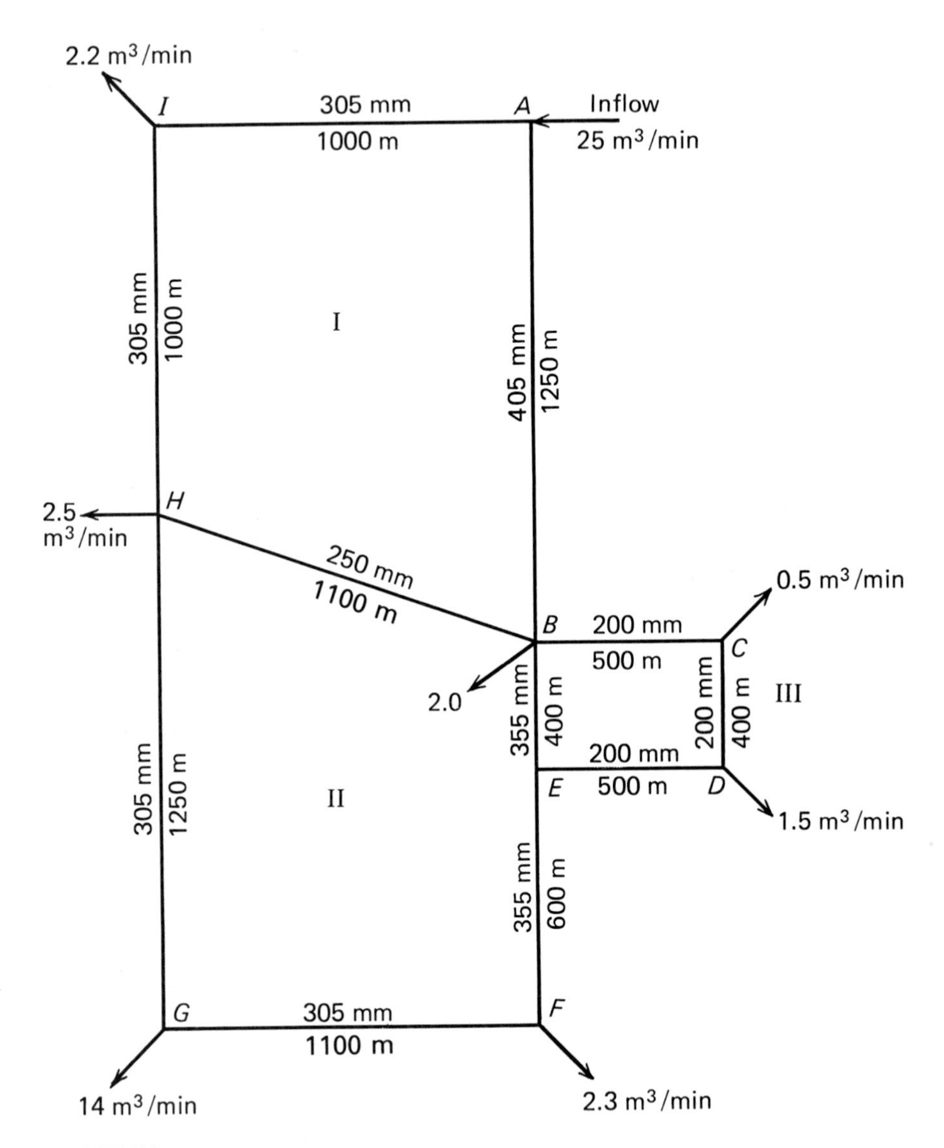

FIGURE 7-9
Simplified distribution system.

20 kPa—well below the normal minimum of 150 kPa. The pressure calculation here assumes points *A* and *G* are at equal elevations. If they are not, the static difference must be included in the calculation. The pipe sizes are evidently somewhat inadequate with regard to both velocity and pressure. The deficient lines should be increased in size and the procedure repeated until adequate pressure and velocity are obtained.

TABLE 7-2
Hardy Cross analysis—first correction

Loop I

Line	Flow, m^3/min	Dia., m	Length, m	s	h, m	h/Q, m/(m^3 · min)
AB	13	0.40	1250	0.0110	13.75	1.058
BH	3	0.25	1100	0.0033	3.63	1.815
HI	−9.8	0.30	1000	−0.0260	−26.00	2.653
IA	−12	0.30	1000	−0.0380	−37.80	3.150
					−46.42	8.676

$$\Delta_I = -\frac{-46.42}{1.85(8.676)} = 2.9$$

Loop II

Line	Flow, m^3/min	Dia., m	Length, m	s	h, m	h/Q, m/(m^3 · min)
BE	7.5	0.35	400	0.0075	3.00	0.400
EF	7.0	0.35	600	0.0066	3.96	0.566
FG	4.7	0.30	1000	0.0067	6.68	1.423
GH	−9.3	0.30	1250	−0.0236	−29.54	3.177
HB	−2.0	0.25	1100	−0.0033	−3.63	1.815
					−19.53	7.381

$$\Delta_{II} = -\frac{-19.53}{1.85(7.381)} = 1.4$$

Loop III

Line	Flow, m^3/min	Dia., m	Length, m	s	h, m	h/Q, m/(m^3 · min)
BC	1.5	0.20	500	0.0058	2.91	1.937
CD	1.0	0.20	400	0.0028	1.10	1.110
DE	−0.5	0.20	500	−0.0008	−0.38	0.762
EB	−7.5	0.35	400	−0.0075	−3.00	0.400
					0.63	4.209

$$\Delta_{III} = -\frac{0.63}{1.85(4.209)} = 0.1$$

As noted above, the procedure outlined in the example is carried out by digital computer techniques. The basic Hardy Cross technique is readily programed[6] and is often used as an exercise in introductory computer courses. The capability of commercially available models includes simulation of in-line pumping and storage.[7] Recently developed models use linear theory or Newton-Raphson techniques to solve somewhat differently formulated energy equations.[8,9] These solution methods converge more certainly than the Hardy

TABLE 7-3
Hardy Cross analysis—second correction

Loop I

Line	Flow, m³/min	Dia., m	Length, m	s	h, m	h/Q, m/(m³ · min)
AB	15.9	0.40	1250	0.0157	19.65	1.236
BH	3.5	0.25	1000	0.0094	10.34	2.954
HI	−6.9	0.30	1000	−0.0136	−13.60	1.971
IA	−9.1	0.30	1000	−0.0227	−22.70	2.495
					−6.31	8.656

$\Delta_{I} = 0.4$

Loop II

Line	Flow, m³/min	Dia., m	Length, m	s	h, m	h/Q, m/(m³ · min)
BE	9.0	0.35	400	0.0105	4.20	0.467
EF	8.4	0.35	600	0.0093	5.58	0.664
FG	6.1	0.30	1000	0.0108	10.80	1.770
GH	−7.9	0.30	1250	−0.0175	−21.88	2.769
HB	−3.5	0.25	1000	−0.0094	−10.34	2.954
					−11.64	8.624

$\Delta_{II} = 0.7$

Loop III

Line	Flow, m³/min	Dia., m	Length, m	s	h, m	h/Q, m/(m³ · min)
BC	1.4	0.20	500	0.0051	2.55	1.821
CD	0.9	0.20	400	0.0023	0.92	1.022
DE	−0.6	0.20	500	−0.0011	−0.55	0.917
EB	−9.0	0.35	400	−0.0105	−4.20	0.467
					−1.28	4.227

$\Delta_{III} = 0.2$

Cross procedure and, in some cases, do not require that the continuity equations be satisfied by an initial set of flow assumptions.

As with any solution to an engineering problem, the predictions of pressure and flow rates are only as accurate as the assumptions or measurements used to formulate the equations.

Appropriate values for friction losses, actual pump performance, and similar factors must be carefully defined.[10] When the model has been properly calibrated, predicted pressures in actual systems have been found to be within 35 to 70 kPa (5 to 10 lb/in²) of measured values.[11]

7-8 Construction of Water Distribution Systems

Water lines are normally installed within the rights-of-way of the streets. Cover provides protection against traffic loads and freezing and varies from as little as

TABLE 7-4
Hardy Cross analysis—third correction

Loop I

Line	Flow, m³/min	Dia., m	Length, m	s	h, m	h/Q, m/(m³ · min)
AB	16.9	0.40	1250	0.0165	20.63	1.265
BH	3.2	0.25	1000	0.0080	8.80	2.750
HI	−6.5	0.30	1000	−0.0122	−12.20	1.877
IA	−8.7	0.30	1000	−0.0209	−20.90	2.402
					−3.67	8.294

$\Delta_I = 0.2$

Loop II

Line	Flow, m³/min	Dia., m	Length, m	s	h, m	h/Q, m/(m³ · min)
BE	9.5	0.35	400	0.0116	4.64	0.488
EF	9.1	0.35	600	0.0107	6.42	0.705
FG	6.8	0.30	1000	0.0132	13.20	1.941
GH	−7.2	0.30	1250	−0.0147	−18.38	2.552
HB	−3.2	0.25	1000	−0.0080	−8.80	2.750
					−2.92	8.436

$\Delta_{II} = 0.2$

Loop III

Line	Flow, m³/min	Dia., m	Length, m	s	h, m	h/Q, m/(m³ · min)
BC	1.6	0.20	500	0.0066	3.30	2.063
CD	1.1	0.20	400	0.0033	1.32	1.200
DE	−0.4	0.20	500	−0.0005	−0.25	0.625
EB	−9.5	0.35	400	−0.0116	−4.64	0.488
					−0.27	4.376

$\Delta_{III} = 0.03$

0.75 m (2.5 ft) in the south to as much as 2.4 m (8 ft) in the north. Trench width must be great enough to provide room to join the pipe sections and install required fittings. Clearance of about 150 mm (6 in) on either side is normally adequate. This requires a trench width of about 1760 mm (68 in) for a 1220-mm (48-in) pipe. The trench width must be increased at joints and fittings. An extra depth of 150 mm (6 in) and an extra width of 250 mm (10 in) on either side should be provided for a distance of 900 mm (3 ft) at the joints.

Since water line trenches are relatively shallow, they are unlikely to require bracing except in unstable soils. In rock excavation, the trench should be cut to a level at least 150 mm (6 in) below the final grade of the pipe and a cushion of sand or clean fill should be placed between the rock and the pipe.

Backfill material should be free of cinders, refuse, and large stones. Careful backfilling decreases the load on the pipe and will decrease the proba-

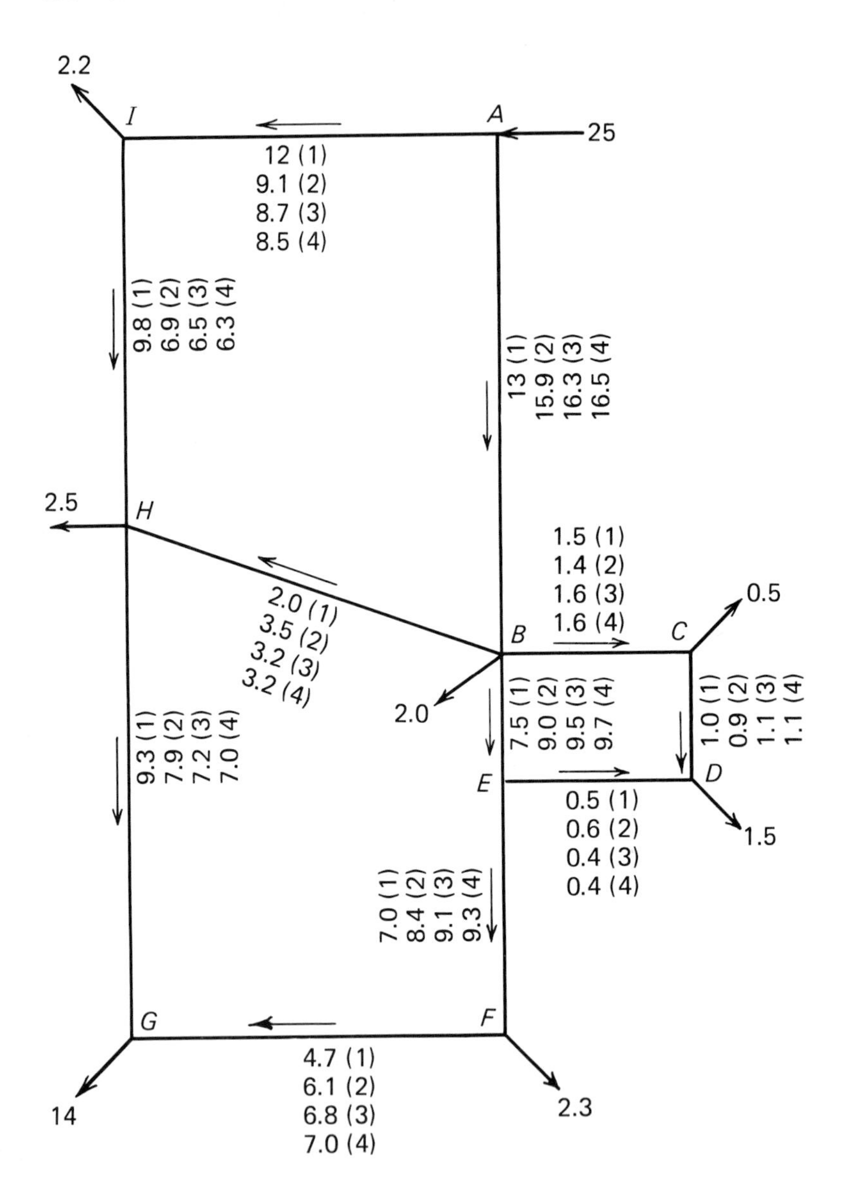

FIGURE 7-10
Corrected flows in distribution system.

bility of failure. Standard pipe bedding conditions are illustrated in Fig. 6-1. Type 2 bedding is commonly used and requires that the fill material be placed by hand up to the centerline of the pipe in carefully tamped layers of not more than 75 mm (3 in). From the centerline to 300 mm (12 in) above the top of the pipe, the fill should be placed by hand. From 300 mm (12 in) above the top of the pipe to the top of the trench, stones no larger than 200 mm (8 in) may be used as part of ordinary fill. Filling should not be done with frozen material or in freezing weather.

Before pipe sections or fittings are placed in the trench, they should be carefully inspected to ensure they have not been damaged in transit. Although it is more common to use machinery to raise the pipe from the truck and to lower it into the trench, small iron and concrete pipe and plastic pipe can be lowered into the trench by hand with rope and a rolling hitch. Iron pipe may be tested for cracks by striking it lightly with a hammer. Sound pipe will ring like a bell. Pipe with internal rubber gaskets or cement lining should not be lifted with hooks, since these may damage the interior.

Hydrants are intended to provide fire protection but may also be used to release air at high points and blow off sediment at low points in the system. The hose pumper cap should be not more than 300 mm (12 in) nor less than 150 mm (6 in) from the gutter face of the curb nor closer than 150 mm (6 in) to the sidewalk. The hydrant is connected to the main by a side branch not less than 150 mm (6 in) in diameter and must be butressed or otherwise restrained. Hydrants are sometimes placed on a concrete base, but must be surrounded by about 0.1 m^3 (3 ft^3) of gravel or crushed stone in order to permit draining of the hydrant after it has been used. Hydrants which remain closed if the barrel is broken should be used in new construction. A valve in the side branch is also desirable to permit repairs to be made with minimum water loss and minimum interruption of service.

Valves are located at distances of not more than 150 m (500 ft) in high-value districts and 250 m (750 ft) in other areas in order to minimize loss of service when repairs are required. Each valve should be placed in a manhole or in a valve box (Fig. 6-6) and should be set upon a concrete base. At intersections of large mains, the valves may be placed together in a single vault.

Contamination of new water mains is unavoidable because the pipes are stored in the open and, when placed in trenches, they may be flooded during construction. Before disinfection, the pipe should be flushed with a foam or rigid "pig" which is driven through by water pressure or pulled through by a cable. This technique will remove dirt, tools, and other solid material which may have been left in the lines. Flushing at a velocity of at least 0.75 m/s (2.5 ft/s) will remove fine sediments. After pigging and flushing, the lines are filled with water containing a free chlorine residual of at least 1 mg/L. A free residual of at least 0.5 mg/L must remain after 24 h of contact. Bacteriological analyses of the water should then be conducted to ensure that the total bacterial count does not exceed 500/mL and that no coliform bacteria are present. If either standard is not met, the line should be filled with water containing at least 50 mg/L available chlorine, which should not fall below 25 mg/L after 24 h.[12]

Testing is a normal requirement for new construction. The specifications of the American Water Works Association require that no installation be accepted unlesss the leakage is less than

$$L = \frac{ND(P)^{0.5}}{C} \tag{7-7}$$

where L = allowable leakage
N = number of joints in the length tested
D = nominal pipe diameter
P = average pressure during test
C = constant equal to 32.6 (m^3/h, mm, kPa) or 1850 (gal/h, in, lb/in^2)

The lengths tested are generally less than 300 m (1000 ft). The pipe is carefully filled with water to ensure that no air is trapped in the section. A pressure of 50 percent above the normal operating pressure is applied and maintained by a hand pump for at least 30 min.

Service pipes are those lines extending from the mains to the user's meter. Service pipes are connected to corporation cocks which, with special tools, can be installed in the mains while the mains are pressurized. The line from the corporation cock to the meter is from 20 mm ($\frac{3}{4}$ in) to 50 mm (2 in) in diameter, depending on the distance involved and the rate of consumption. Service lines may be of plastic, copper, or iron and usually incorporate a "gooseneck" to provide for some relative displacement between the two lines. Service lines larger than 50 mm (2 in) are ordinarily not tapped into the mains. Such connections may be made with standard fittings, or a number of small lines tapped into the main may be joined to supply the service.

7-9 Distribution System Maintenance

Maintenance of distribution systems includes occasional cleaning, servicing of valves and hydrants, leak surveys, repairs, disinfection of repaired sections, and, in some areas, thawing of frozen lines.

As pipelines age, accumulations of sediment, rust, and bacterial growths may substantially reduce their hydraulic capacity. *Cleaning* with a variety of tools, including pigs with wire brushes on their perimeter, will restore a large part of the lost capacity. The lower end of the pipe is broken and a 45° branch is installed to bring the line to the street surface. A special sleeve is installed at the upper end and a small float attached to a cable is inserted. The upper valve is then opened and the flow carries the float and attached cable through the line. When the cable has reached the lower end, a larger cable is attached and drawn through. A scraper is then attached to the cable and pulled through the pipe, dislodging the accumulated material. The improvement is not permanent and recleaning may be necessary at regular intervals. Plastic, cement, asbestos cement, and cement-lined pipes are less likely to require cleaning than ordinary iron pipe.

Hydrants may leak at the hose outlets, indicating damage to the valve or valve seat. Leaking hydrants should be removed and replaced and be reconditioned in a shop. Field repairs are limited to replacement of nozzle caps and their chains and replacement of worn operating nuts.

Valves are often found to be defective on inspection, hence a regular program of testing is advisable. Valves may be found to be inaccessible, inoperable, or closed. Inaccessibilty results from covering the valve with earth or pavement or from filling the valve box with dirt. Since rapid closing of valves is important in the event of breaks, accessibilty must be maintained.

Valves which are not operated for many years may become frozen as a result of corrosion and the stem may break when the valve is forced. Sediment may accumulate in the bottom of the valve and prevent it from closing completely. Both conditions can be prevented by a regular program of valve inspection in which the valves are operated periodically. Slowly closing the valve will increase the velocity and aid in sweeping out accumulated debris. Regular operation will prevent freezing due to corrosion. Valves may leak around the stem when they are operated, requiring uncovering and repacking of the gland.

Valves are frequently left closed after repairs have been made to the system. The loss in capacity is not evident until a high flow is required, at which time it may not be possible to locate and correct the problem expeditiously. Records of repairs should require the supervisor to note the time at which valves are closed and reopened. This will help in preventing the problem.

Leak surveys are conducted when it appears from comparison of pumping records and users' meter readings that excessive leakage is occurring. Individual sections of mains may be tested by isolating them by closing valves at either end and supplying the flow through a fire hydrant. A meter in the line to the hydrant measures the flow, and a high flow at night, when normal consumption is low, indicates the presence of a leak.

The exact location of a leak may be difficult to establish, although melted ice or snow in the winter or green grass during a drought may be helpful indications. Measuring the pressure at household hose cocks along the line can be used to plot the hydraulic gradient. A sudden change in the slope of the hydraulic gradient may indicate the approximate location of the leak.

Disinfection of existing pipes is necessary after repairs or modifications which involve cutting the pipe. Flushing of the repaired section through adjacent hydrants and application of the procedure used for new mains is usually successful. Bacterial growths in old pipes may produce taste and odor problems which can often be corrected by chlorination of the sections affected. Chlorine is injected into the mains and flow from the fire hydrants is tested until it shows a substantial chlorine residual.

Thawing of frozen pipes, particularly household services, may be required from time to time in the north. Electrical thawing is the most common technique and is usually provided by a truck-mounted generator which provides a current of 500 A at 50 to 100 V. When household services are thawed, the pipe must be separated from the internal house piping before the current is applied. Connections are made adjacent to the meter and at the nearest hydrant. The amperage required for quick thawing of household services is

TABLE 7-5
Amperage required for thawing service pipes

Size (mm)	Size (in)	Current, A: Wrought iron, steel, lead	Current, A: Copper
12	$\frac{1}{2}$	200	500
19	$\frac{3}{4}$	250	625
25	1	300	750
31	$1\frac{1}{4}$	450	1000
38	$1\frac{1}{2}$	600	1500
50	2	800	2000

TABLE 7-6
Amperage required for thawing iron mains

Cast-iron pipe size (in)	Cast-iron pipe size (mm)	Current, A
4	100	350–400
6	150	500–600
8	200	700–900
10	250	1000–1300

presented in Table 7-5. Mains may be thawed by making the connections to adjacent hydrants. The length thawed at one time normally does not exceed 200 m (600 ft). Table 7-6 presents amperages required for thawing cast iron mains. The time required to thaw frozen pipes at the amperages listed is on the order of a few minutes for household services and from 30 min to 2 h for iron mains. Reducing the amperage greatly increases the time required.

Nonmetallic pipe, of course, cannot be thawed by resistance heating; however, its lower thermal conductivity makes freezing less likely. Thawing of nonmetallic pipe is possible by forcing a small steam line into the conduit. Quick action to thaw frozen lines is desirable since freezing, if it is extensive, can develop sufficient force to rupture the pipe.

7-10 Protection of Water Quality in Distribution Systems

Although the water produced by public supply systems generally meets stringent quality standards as it leaves the pumping station, it may deteriorate as it passes through the distribution system. As has been noted above, some organic materials may pass through the walls of plastic pipe, metals may be dissolved from pipe or solder, and asbestos fibers may be released from asbestos cement pipe. Additionally, autotrophic bacteria can grow within the pipes using the carbonate and bicarbonate ion alkalinity of the water as the sole carbon source. The organic matter produced by this growth can then support other microbial life and result in taste, odor, and color problems.

The greatest potential hazard in distribution systems is associated with cross-connections to nonpotable waters. There are many connections between potable and nonpotable systems—every toilet constitutes such a connection—but cross-connections are those through which backflow can occur. Facilities which are particularly likely to have cross-connections through which dangerous materials can enter the distribution system include hospitals, metal plating and chemical plants, car washes, laundries, and dye works.[13]

When connections to potentially contaminated systems are required, the public system should be protected by backflow preventers (Art. 6-8). Regular inspection and testing of these devices is essential to ensure that they remain in proper working order. Testing procedures and schedules may be found in Ref. 13. In relatively low-risk situations, such as may exist in individual residences, adequate protection may be provided by air gaps and vacuum breakers. The latter are not reliable and should not be used for industrial or commercial connections.

City regulation of plumbing fixture types, connections to nonpotable systems, careful inspection of all new connections, and regular reinspection of industrial and commercial users will reduce the risk of contamination from cross-connections. Control of water stability through treatment processes will minimize the risk of dissolution of materials in the pipes. Maintenance of an adequate chlorine residual in the distribution system will minimize biological growth.

PROBLEMS

7-1. The pipe system of Fig. 7-9 is modified as a result of the analysis presented in the text. All 305-mm (12-in) lines are increased to 355 mm (14 in), the 355-mm (14-in) lines are increased to 405 mm (16 in), and the 405-mm (16-in) line is increased to 450 mm (18 in). Determine the flow in each line, the velocity in each line, and the pressure at point G assuming points A and G are at the same elevation and the pressure at A is 500 kPa (75 lb/in^2). Is the system now satisfactory?

7-2. From the following hourly demand figures, find the uniform pumping rate which will just meet the demand and the volume of storage required to meet maximum hourly demand.

Time, a.m.	**Rate, m^3/min**	**Time p.m.**	**Rate, m^3/min**
12 (midnight)	7.2	12 (noon)	23.9
1	7.1	1	24.5
2	7.0	2	24.4
3	7.1	3	24.2
4	7.0	4	24.3
5	7.3	5	25.6
6	7.5	6	26.9
7	12.2	7	34.0
8	18.9	8	33.0
9	21.0	9	20.0
10	22.8	10	8.4
11	23.5	11	7.6

7-3. Rework Prob. 7-2 assuming water is pumped only from 6 a.m. to 6 p.m.

7-4. Rework Prob. 7-2 to include a fire flow requirement of 10 m^3/min for 10 h.

7-5. In the pipe systems (*a*), (*b*), and (*c*) in the figure, find the flow and velocity in each pipe and the pressure at points *A*. Assume all nodes are at identical elevations.

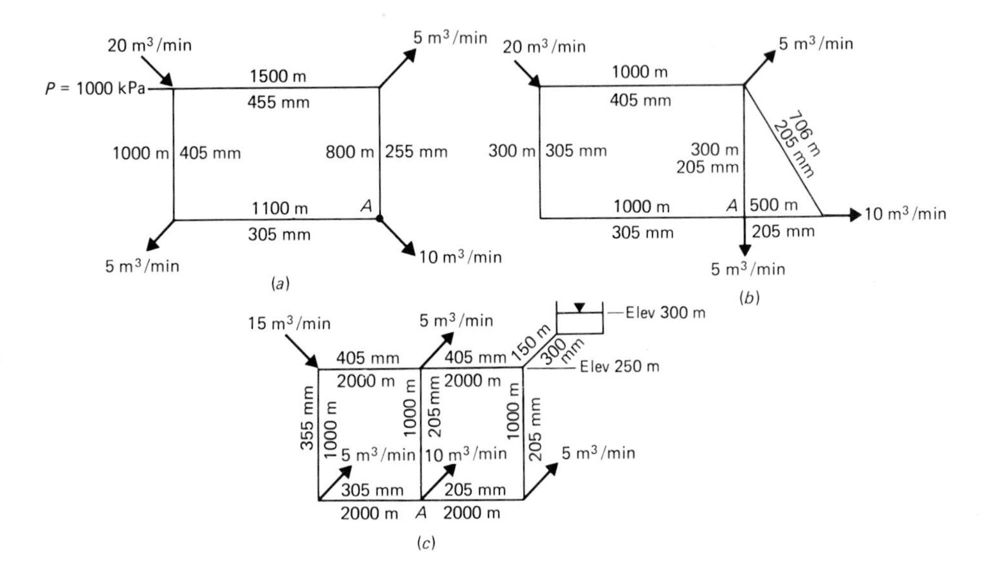

7-6. A 305-mm (12-in) pipeline contains 55 sections of pipe, 10 tee sections, and 2 gate valves. Each section or fitting contributes one joint. The pipe is tested at a pressure of 750 kPa (112.5 lb/in²), and 0.6 L (0.16 gal) is added to the line during a 30-min period. Does this satisfy AWWA requirements?

REFERENCES

1. Bruce S. Aptowicz, Norman C. Weintraub, and Charles Zitomer, "Using Elevated Storage and Off-Peak Pumping to Control Energy Costs," *Journal of American Water Works Association*, **79**:11:46, 1987.
2. Rudolf C. Metzner, "Demand Forecasting: A Model for San Francisco," *Journal of American Water Works Association*, **81**:2:56, 1989.
3. Timothy D. Hirrel, "Automating Water Consumption Data Manipulation," *Journal of American Water Works Association*, **78**:8:49, 1986.
4. Hermann M. Orth, *Model-based Design of Water Distribution and Sewage Systems*, John Wiley and Sons, New York, 1986.
5. Hardy Cross, "Analysis of Flow in Networks of Conduits or Conductors," Bulletin 286, University of Illinois Experiment Station, 1936.
6. Theodore V. Hromadka II, Timothy J. Durbin, and Johannes J. Devries, *Computer Methods in Water Resources*, Lighthouse Publications, Mission Viejo, Calif., 1985.
7. Stuart M. Alexander, Norman L. Glenn, and Donald W. Bird, "Advanced Techniques in the Mathematical Modeling of Water Distribution Systems," *Journal of American Water Works Association*, **67**:343, 1975.
8. D. J. Wood and A. G. Rayes, "Reliability of Algorithms for Pipe Network Analysis," *Journal of Hydraulics Division, American Society of Civil Engineers*, **107**:HY10:1145, 1981.
9. Ronald U. Harris, "Computer Modeling in Water System Planning and Design," *Journal of American Water Works Association*, **76**:7:78, 1984.

10. Thomas M. Walski, "Assuring Accurate Water Distribution System Model Calibration," *Journal of American Water Works Association*, **77**:12:38, 1985.
11. A. Lee Cesario and J. O. Davis, "Calibrating Water System Models," *Journal of American Water Works Association*, **76**:7:78, 1984.
12. Ralph W. Buelow, "Disinfection of New Water Mains," *Journal of American Water Works Association*, **68**:283, 1976.
13. Gustave J. Angele, Sr., *Cross Connections and Backflow Protection*, 2d ed., American Water Works Association, Denver, 1974.

CHAPTER 8

QUALITY OF WATER SUPPLIES

8-1 Water and Its Impurities

Water, perhaps because it is ubiquitous in nature, seems a simple material. It is, however, quite complex in a chemical sense. The molecule itself is not symmetric, and its unbalanced nature imparts a polar character which permits individual molecules to join by hydrogen bonding, forming arrays which are in a constant state of flux. A simplified representation of this circumstance is:

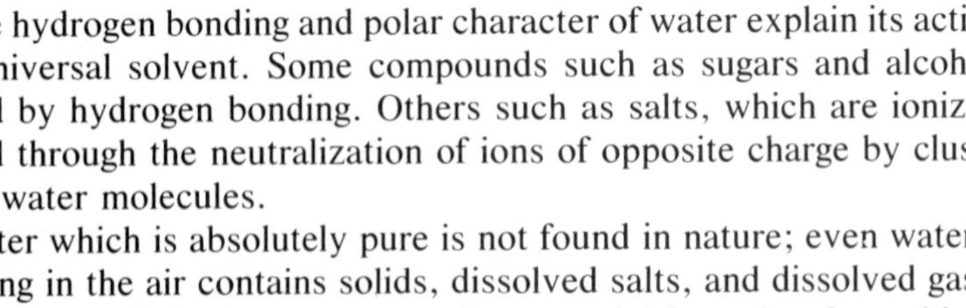

$$\left[H^{+} \diagdown O^{-} \diagup H^{+}\right]_n = -O^{-} \begin{matrix} H^{+} \\ H^{+} \end{matrix} \quad \cdots \quad O^{-} \quad \cdots \quad H^{+}$$

$n \approx 100$ at 20°C

The hydrogen bonding and polar character of water explain its action as a nearly universal solvent. Some compounds such as sugars and alcohols are dissolved by hydrogen bonding. Others such as salts, which are ionized, are dissolved through the neutralization of ions of opposite charge by clusters of oriented water molecules.

Water which is absolutely pure is not found in nature; even water vapor condensing in the air contains solids, dissolved salts, and dissolved gases. As condensed water falls, it sweeps up other material from the air, and becomes still more contaminated on reaching the ground, running over the surface and percolating through the various strata of the soil. Some contaminants may be removed by passage through the soil as a result of filtration and exchange and adsorption reactions; some may be removed in surface waters by sedimentation and biological activity; some may be removed by specific engineered processes in water treatment plants. At the same time, new impurities are introduced by dissolution and exchange reactions in the soil, others by accumulation of decomposition products and discharge of wastes in surface waters, still others by treatment processes and reactions with the materials of the distribution system.

The materials found in water may be divided into living organisms and solid or dissolved organics and inorganics. Not all of these are harmful, and some may even be desirable for health, esthetic, or technical reasons. A *potable* water is one that is safe to drink, pleasant in taste, and suitable for domestic purposes. A *contaminated* or *polluted* water is one that contains suspended or dissolved material which makes it unsuitable for its intended use. A potable water may be considered contaminated with respect to some industrial uses.

8-2 Waterborne Diseases

Communicable diseases which may be transmitted by water include bacterial, viral, and protozoal infections. The bacterial diseases include typhoid, paratyphoid, salmonellosis, shigellosis, bacillary dysentery, asiatic cholera, Legionnaires' disease, and Pontiac fever.[1] Viral diseases associated with water include hepatitis, poliomyelitis, and those cases of gastroenteritis attributable to Norwalk virus and rotavirus.[2] Protozoans such as *Giardia* and *Cryptosporidium* can produce gastroenteritis and are very resistant to disinfectants.[3] Certain fungi, notably *Aspergillus*, are human pathogens[4] while schistosomiasis is caused by a worm which may be transmitted through water via a snail carrier. Other infectious diseases are not generally transmitted by water.

Organisms which cause infectious diseases are normally spread through the fecal and urinary discharges of sick persons and carriers, although there are some animal and soil reservoirs of protozoa and bacteria associated with gastroenteritis. Protection of water supplies against these agents is thus normally a matter of preventing discharges of inadequately treated wastewater into the source and provision of the treatment techniques described in Chaps. 9 through 11.

Not all outbreaks of the diseases cited above can be attributed to public water supplies. Flies or human carriers of disease may contaminate food or beverages; oysters and other shellfish may be contaminated by sewage discharges; Legionnaires' disease is most commonly spread by contaminated air-conditioning systems; and viral diseases are also spread by personal contact, food, and beverages. The characteristics of an outbreak of waterborne disease are described in Art. 8-9.

A great many microorganisms are found in water, most being of no health significance. It is difficult to test for the disease-causing species, whether viral, bacterial, fungal, or protozoal, since most grow rapidly only in their host. As noted above, these organisms are transmitted primarily through the feces and urine of infected persons. Water which shows evidence of such contamination is thus considered to be unfit for consumption. The possibility of such contamination is usually assessed by determining the number of coliform bacteria. *Escherichia coli* is excreted in enormous numbers—up to 4×10^{10} organisms per person per day. Pathogens are found in far smaller numbers and tend to die off more rapidly than the coliforms under the conditions found in natural waters and in water and wastewater treatment plants. Thus, while the presence of coliforms is not proof that the water is dangerous, the absence of this group is taken as evidence that it is free of pathogens.

Bacterial counts are performed by filtering a measured volume of water through a cellulose acetate or glass filter with openings less than 0.5 μm. The bacteria retained upon the filter are then incubated in a special nutrient medium which will produce readily distinguishable colonies of the species being evaluated. The incubation temperature and the medium used permit enumeration of coliforms, fecal coliforms, and fecal streptococci.

8-3 Inorganic Contaminants

Inorganic contaminants include both suspended and dissolved materials. Suspended materials are undesirable for esthetic reasons, but their primary effect on quality lies in their ability to shield microorganisms from disinfectants. Dissolved inorganics which have health effects include aluminum, arsenic, barium, cadmium, chromium, fluoride, lead, mercury, nitrate, selenium, and silver. The Environmental Protection Agency (EPA) has established a maximum contaminant level (MCL) for all of these save aluminum (see Table 8-1). Elevated concentrations of aluminum in water have recently been associated with neuropathological disorders such as Alzheimer's disease and presenile dementia as well as physiological problems among dialysis patients.[5] Maximum contaminant levels have also been established for gross alpha and gross beta activity and the specific radionuclides radium-226 plus radium-228, strontium-90, and hydrogen-3. Radon may also be found in water supplies, and while it is not known to be harmful when ingested in small concentrations, its presence in the water may lead to contamination of the air, where it may contribute to lung cancer.[6,7]

The inorganic materials for which MCLs have been established are generally toxic in one manner or another. *Arsenic* is a well-known poison which can be fatal in high doses. *Barium*, in soluble form, is very toxic, while *cadmium* and *chromium* can exhibit either acute or chronic toxicity, depending on the concentration. *Fluoride*, sometimes added to water as a means of making teeth more resistant to decay, in higher concentration can cause permanent discoloration and loss of teeth and embrittlement of bones. *Lead* and *mercury* are usually associated with chronic effects on the nervous system. *Nitrate* can interfere with oxygen transfer in the blood of infants, since it can be reduced to nitrite in immature digestive systems and, in that form, complex with hemoglobin. *Selenium* can produce gastrointestinal and dental problems, while *silver* is toxic in large doses.

TABLE 8-1
Drinking Water Standards of EPA

Contaminant	Limit
Primary standards (maximum contaminant level, MCL)	
Total coliforms (av. number/100 mL)	1
Total coliforms (max. number/100 mL)	5
Turbidity (ntu)	1–5
Inorganic chemicals (mg/L)	
Arsenic	0.05
Barium	1.0
Cadmium	0.01
Chromium	0.05
Fluoride	0.7–2.4
Lead	0.05
Mercury	0.002
Nitrate (as N)	10.0
Selenium	0.01
Silver	0.05
Radionuclides (pCi/L)	
Gross alpha	15
Ra-226 + Ra-228	5
Gross beta	50
H-3	20,000
Sr-90	8
Organic chemicals (μg/L)	
Endrin	0.2
Lindane	40
Methoxychlor	100
Toxaphene	5
2,4-D	100
2,4,5-TP	10
Trihalomethanes	100
Benzene	0.05
Carbon tetrachloride	0.05
1,2 Dichloroethane	0.05
Trichloroethylene	0.05
Para-dichlorobenzene	0.75
1,1 Dichloroethylene	0.07
1,1,1 Trichloroethane	2.0
Vinyl chloride	0.02
Secondary standards (recommended contaminant level, RCL)	
Chloride	250 mg/L
Color	15 units
Copper	1 mg/L
Iron	0.3 mg/L
Manganese	0.05 mg/L
Odor	3 TON
pH	6.5–8.5
Sulfate	250 mg/L
Total dissolved solids	500 mg/L
Zinc	5 mg/L

Radionuclides are widely distributed in nature and some can be biologically concentrated. MCLs range from 20,000 pCi/L for hydrogen-3 (which passes through the body) to 5 pCi/L for radium (which accumulates in the bones). Uranium-238, the commonest isotope of uranium, is soluble in the milligram per liter range, but, while it is present in all ground water, is seldom found in such concentrations. It is more dangerous as a chemical toxin to the liver than as a source of radioactivity. Radium is relatively immobile in most environments, but exceeds the MCL in many groundwaters. Strontium-90 and cesium-137 are found in high concentration in ore processing and reactor wastes. Total production from these sources amounts to millions of curies per month, which, when compared to the MCL of 8 pCi/L for strontium, indicates the magnitude of the potential contamination problem.

Turbidity is a measure of the presence of suspended solid material. Aside from the esthetic undesirability of turbidity, suspended solids can shelter microorganisms from the action of disinfectants and are an indication of inadequate treatment of water. Suspended solids are readily removed by coagulation, sedimentation, and filtration; thus, if high concentrations are found, something is likely to be malfunctioning in the treatment plant and the water may be otherwise unsuitable.

8-4 Organic Contaminants

Both naturally occurring and artificial organics are found in water. The former may be associated with color, taste, or odor while the latter, in some cases, may be toxic or carcinogenic. The organic materials in water may be altered by treatment processes, which sometimes make them more dangerous or unpleasant than they were before treatment.

Chlorinated hydrocarbons are used as pesticides and herbicides. These materials are relatively persistent both in nature and within the human body. Many have been shown to produce carcinogenic effects in laboratory animals. EPA has established MCL values for the insecticides endrin, lindane, methoxychlor, and toxaphene and for the herbicides 2,4-D and 2,4,5-TP (Table 8-1).

Trihalomethanes may enter water from industrial processes, but most commonly are formed during chlorination of water containing naturally occurring organics such as humic acid. The trihalomethanes are single-carbon organics with three of the carbon bonds being occupied by halogens such as chlorine, bromine, or iodine. Chloroform is the most commonly occurring trihalomethane, but brominated forms are often encountered as well. The trihalomethanes are carcinogens, hence their presence in public water supplies is undesirable. The MCL established by EPA (Table 8-1) is an annual running average, not a single value, hence occasional higher values do not necessarily violate the standard.

Volatile organic chemicals (VOCs) are industrial chemicals which have been found to be widely distributed in both surface waters and groundwaters.[8]

Many of these substances are either known or suspected to be carcinogens. As of 1989, EPA had established MCLs for benzene, carbon tetrachloride, 1,2-dichloroethane, 1,1-dichloroethylene, *para*-dichlorobenzene, 1,1,1-trichloroethane, trichloroethylene, and vinyl chloride. Thirty-three other organic chemicals are scheduled to be regulated by 1991, and it is likely that the list will be expanded as further health information and improved analytical techniques become available.

8-5 Common Constituents of Natural Waters

All natural waters contain some dissolved mineral matter. In fact, we are so accustomed to these impurities that distilled water tastes flat and unpleasant. The commonly encountered cations are sodium, potassium, calcium, magnesium, iron, and manganese, which are associated with the anionic species bicarbonate, carbonate, sulfate, and chloride. EPA has established a secondary standard called a recommended contaminant level (RCL) for some of these and a few other contaminants which do not have chronic or toxic health impacts (Table 8-1).

The divalent cations contribute to *hardness*, which is sometimes defined as the ability to neutralize soap. Hardness can be precipitated by heating water, hence it can produce damaging deposits in boilers, hot-water heaters, and hot-water lines. Hardness is not harmful to health—in fact, there is some evidence that soft water may be a contributing factor in some ailments, particularly cardiovascular disease[9].

Iron and manganese contribute to hardness, but their more important effect results from their oxidation and subsequent precipitation. This causes metallic tastes and discolored water which stains clothes, cooking utensils, and plumbing fixtures. Both iron and manganese may also serve as electron acceptors for autotrophic bacteria such as *Crenothrix*, which aggravate staining and may reduce the capacity of pipelines. The precipitates formed range from reddish-brown to black, depending on the proportion of iron and manganese.

Sulfate, in association with magnesium or sodium, can have a pronounced laxative effect on people who are not accustomed to the water. Chloride in high concentration can contribute a salty taste to water and is sometimes an indication of sewage contamination, since the chloride concentration increases when water is used for domestic purposes. Neither is harmful in moderate concentration.

Bicarbonate and carbonate result from the dissolution of carbonate rocks and provide a very important buffer system in natural waters. *Alkalinity* consists of those chemical species in water which can neutralize acid. Its major constituents are hydroxyl (OH^-), carbonate (CO_3^{2-}), and bicarbonate (HCO_3^-) ions. The relative quantities of each are a function of pH. No significant concentration of hydroxyl ion exists below pH 10 and no significant carbonate concentration below pH 8.5. In most waters, alkalinity thus consists of the bicarbonate ion. The other species may be formed in water or wastewater treatment processes.

The quantity pH is a measure of the free hydrogen ion concentration in water. Water, and other chemicals in solution, will ionize to a greater or lesser degree. The ionization reaction of water may be written as

$$HOH \leftrightarrows H^+ + OH^- \tag{8-1}$$

This reaction has an equilibrium defined by

$$\frac{(H)(OH)}{(HOH)} = K_w \tag{8-2}$$

in which (H) is the chemical activity of the hydrogen ion, (OH) the chemical activity of the hydroxyl ion, and (HOH) is the chemical activity of the water. Since water is the solvent, its activity is defined as unity. In dilute solution, molar concentrations are often substituted for activities, yielding

$$[H][OH] = K_w = 10^{-14} \text{ (at 20°C)} \tag{8-3}$$

Taking logs of both sides gives

$$\log [H] + \log [OH] = -14 \tag{8-4}$$

Defining $-\log = \text{p}$ gives

$$\text{pH} + \text{pOH} = 14 \tag{8-5}$$

In neutral solutions [OH] = [H], hence pH = pOH = 7. Increasing acidity leads to higher values of [H], thus to lower values of pH. Low pH is associated with acidity, high pH with causticity. Its value is important in the operation of many water and wastewater treatment processes and in the control of corrosion.

Dissolved gases in water include all those to which it is exposed. Those which are commonly encountered are nitrogen, oxygen, carbon dioxide, hydrogen sulfide, and methane. *Hydrogen sulfide* is produced by reduction of sulfates, dissolution of pyrites, or anaerobic decomposition of organic matter. It has a disagreeable rotten-egg odor in low concentration, is poisonous in high concentration, and contributes to corrosion of metals and concrete. *Carbon dioxide* is dissolved from the atmosphere and produced by decomposition of organic matter and by subterranean geochemical activity. It is readily soluble in water, combining with the water to form carbonic acid. Carbonic acid exists in chemical equilibrium with bicarbonate alkalinity and can contribute to dissolution of subsurface minerals and corrosion of metals. *Oxygen* content is increased by transfer from the atmosphere and decreased by biological and chemical reactions. It is often absent in groundwater and badly contaminated surface waters and is generally added by water treatment processes either deliberately or inadvertently. It contributes to corrosion of metals under some conditions.

8-6 Water Chemistry

The various chemical reactions which occur in natural waters and in water and wastewater treatment can generally be considered to occur in dilute solution. This permits the use of simplified equilibrium equations in which molar concentrations are considered to be equal to chemical activities. The assumption of dilute conditions is not always justified, but the error introduced by the simplification is no greater than that which might be introduced by competing reactions with species which are not normally measured in water treatment. In this text, molar concentrations will be considered as equal to chemical activities.

Concentrations of different chemical species in water may be expressed in moles per liter, in equivalents per liter, or in mass per unit volume (typically milligrams per liter) in terms of the species itself or in terms of some other species. In the United States, units of pounds per million gallons (lb/Mg) are still used occasionally. The equivalent weight of a species is its molecular weight divided by the net valence or by the net change in valence in the case of oxidation-reduction reactions.

The number of equivalents per liter is the concentration divided by the equivalent weight. The number of equivalents per liter is called the normality. The number of moles per liter is called the molarity.

Example 8-1 500 mg of anhydrous $Ca(HCO_3)_2$ is dissolved in 750 mL of distilled water. What is the molar concentration, the normality, and the concentration in mg/L and lb/Mg expressed as $CaCO_3$?

Solution
The molecular weight of $Ca(HCO_3)_2$ is $40 + 2[1 + 12 + 3(16)] = 162$. The equivalent weight is the molecular weight divided by 2, or 81.

$$\text{Concentration} = \frac{500}{0.75} = 667 \text{ mg/L as } Ca(HCO_3)_2$$

$$= \frac{667 \text{ mg/L}}{162 \text{ g/mole}} = 4.12 \text{ mmoles/L}$$

$$= \frac{667 \text{ mg/L}}{81 \text{ g/equiv}} = 8.23 \text{ mequiv/L}$$

$$= 8.23 \text{ mequiv/L} \times 50 \text{ g } CaCO_3\text{/equiv}$$

$$= 412 \text{ mg/L as } CaCO_3$$

$$= 412 \text{ mg/L} \times 8.34 = 3432 \text{ lb/Mg}$$

The expression of concentrations of one species in terms of another (normally $CaCO_3$) is a convention in water and wastewater treatment. It involves multiplying the normality by the equivalent weight of the other species and is thus only a device for expressing normalities in terms of concentration of a common species.

An individual ion in aqueous solution may be involved in a large number of different chemical equilibria. Hydrogen ion, for example, is involved in the ionization of water, the bicarbonate-carbonate equilibria, and potentially, a great many other systems. There will be at any time, however, only one value of hydrogen ion concentration—which must simultaneously satisfy all the equilibria in which it is involved.

Example 8-2 A water has an alkalinity of 250 mg/L as $CaCO_3$ and a pH of 7.5. Determine the concentration of OH^-, CO_3^{2-}, HCO_3^- and CO_2.

Solution
The ionization constant for water is 1×10^{-14}. The two ionization constants for carbonic acid are

$$H_2CO_3 \leftrightarrows H^+ + HCO_3^-$$

$$K_1 = 4.3 \times 10^{-7}$$

$$HCO_3^- \leftrightarrows H^+ + CO_3^{2-}$$

$$K_2 = 4.7 \times 10^{-11}$$

The equilibrium equations corresponding to these reactions are

$$[H][OH] = 10^{-14}$$

$$\frac{[H][HCO_3]}{[H_2CO_3]} = 4.3 \times 10^{-7}$$

$$\frac{[H][CO_3]}{[HCO_3]} = 4.7 \times 10^{-11}$$

The terms in these equations, representing the molar concentrations, are algebraic quantities. Thus [H] equals $10^{-7.5}$, or 3.16×10^{-8}, wherever it appears. From Eq. (8-3),

$$[OH] = 3.16 \times 10^{-7}$$

The total alkalinity consists of OH^-, CO_3^{2-}, and HCO_3^- and is 250 mg/L ÷ 50 g/eq or 5 meq/L. Thus,

$$[OH] + 2[CO_3] + [HCO_3] = 5 \times 10^{-3}$$

The molar concentration of each species must be converted to equivalent concentration so that they can be added. Molarity and normality are equal for the monovalent species, but differ by a factor equal to the valence for multivalent species. We now have

$$2[CO_3] + [HCO_3] = 5 \times 10^{-3} - 3.16 \times 10^{-7} \approx 5 \times 10^{-3}$$

$$\frac{(3.16 \times 10^{-8})[CO_3]}{[HCO_3]} = 4.7 \times 10^{-11}$$

From which

$$[CO_3] = 0.007 \times 10^{-3}$$

$$[HCO_3] = 4.985 \times 10^{-3}$$

From this calculation we can note that the alkalinity is essentially all in the form of bicarbonate at this pH, as stated in Art. 8-5. The concentration of CO_2 is

$$[H_2CO_3] = \frac{(3.16 \times 10^{-8})(4.985 \times 10^{-3})}{4.3 \times 10^{-7}} = 0.366 \times 10^{-3}$$

Results expressed in mg/L as particular species are

$$H^+ = 3.16 \times 10^{-8} \text{ mg/L as H}$$

$$OH^- = 5.37 \times 10^{-6} \text{ mg/L as OH}$$

$$CO_3^- = 0.42 \text{ mg/L as } CO_3$$

$$HCO_3^- = 304.1 \text{ mg/L as } HCO_3$$

$$CO_2 = 16.1 \text{ mg/L as } CO_2$$

The dissolution of minerals and their precipitation are very important in water and wastewater treatment. Such reactions also have equilibria definable by equations similar to those used above. In such cases the equilibrium constants are called *solubility products*.

Example 8-3 Calcium is often removed from water by addition of lime. The lime reacts with bicarbonate alkalinity, converting it to carbonate alkalinity. The carbonate ion combines with calcium to form $CaCO_3$, which is quite insoluble. The solubility product of $CaCO_3$ is 5×10^{-9}. Determine the concentration of calcium when the CO_3 concentration is 10, 50, and 100 mg/L as $CaCO_3$.

Solution
The molar concentration of CO_3 in the three cases is 0.1×10^{-3}, 0.5×10^{-3}, and 1.0×10^{-3}. The solubility product equation is

$$[Ca][CO_3] = 5 \times 10^{-9}$$

since $[CaCO_3] = 1$. For the three alkalinity levels, $[Ca] = 5 \times 10^{-5}$, 1×10^{-5}, and 5×10^{-6} moles/L or 5, 1, and 0.5 mg/L as $CaCO_3$.

The reduced solubility of calcium in the presence of increased concentrations of carbonate is an example of what is called the *common ion effect*. The same phenomenon is important in removal of magnesium, iron, manganese, and phosphate.

Oxidation-reduction reactions are important in corrosion, disinfection, and iron and manganese removal. In such reactions there is a flow of electrons from one atom to another, resulting in a change of valence for both. The equivalent weight of chemical species involved in such reactions, as noted above, is the molecular weight divided by the net change of valence.

The application of chemical equilibrium theory to specific water treatment processes is treated in Chaps. 9 and 11. Further information concerning theoretical concepts may be found in References 10 and 11.

8-7 EPA Standards

Under the Safe Water Drinking Act of 1974 and subsequent legislation, the EPA has been charged with establishing drinking water standards for all "public water systems." A "public" system is one which serves more than 25 individuals or which has more than 15 connections; hence, nearly all water supplies other than private wells are included. The present standards shown in Table 8-1 include MCLs for those organic and inorganic chemicals known to have toxic or carcinogenic effects; for turbidity, for reasons discussed above; and for bacterial population. In addition, RCLs have been established for certain contaminants which are primarily of esthetic importance.

The water quality standards of the EPA have undergone continuous modification since they supplanted the earlier Public Health Service Standards, which applied only to waters used in interstate commerce. It is to be expected that, with new analytical techniques and more information concerning the health effects of water contaminants, the standards will continue to change. Water providers who are subject to the regulations of EPA must monitor the water in their system and notify the public if it fails to meet the MCLs of the regulations. States or individual communities are, of course, free to establish standards more restrictive than those of the EPA, and assurance of compliance is left to the states unless they fail to accept the responsibility.

8-8 Liability for Unsafe Water

The water sold to the public by cities, private companies, or individuals is not guaranteed to be pure or suitable for any purpose. Producers are required to test the water in their system and notify the public if those tests show it fails to meet current standards of EPA. Liability for health or other problems of the users normally hinges upon the demonstration of negligence on the part of the supplier. In the case of problems arising after notification to the public of failure to meet the standards, contributory negligence might be attributed to the user.

Negligence is a legal issue but could include such actions as failure to provide an adequate supply of required chemicals, failure to quickly correct known deficiencies in the treatment or distribution system, or failure to notify the public as required by law in the case of failure to meet EPA standards. In the past, when negligence has been demonstrated, water providers have been held liable for damages resulting from death or disease attributable to the water supply.

8-9 Characteristics of Waterborne Epidemics

Widespread outbreaks of disease are often attributed to water systems, although, as noted above, there are other media through which infectious diseases are spread. In order to assess the probability that the water system is at fault, it is desirable to consider the characteristics of a waterborne epidemic.

Such outbreaks are likely to be widely distributed. Exceptions to this general rule may occur if water from several sources is used in the system and only one source is contaminated or if the contamination results from a cross-connection. In these situations, the cases of disease may be grouped around the location of the cross-connection or the point where the contaminated source enters the system. All classes and ages of people are affected by waterborne epidemics.

Outbreaks of waterborne disease in the past have generally been associated with use of untreated water, deficiencies in treatment systems or deficiencies in the distribution system, particularly cross-connections. In the period from 1981 through 1985 there were 176 outbreaks of waterborne disease in the United States, which resulted in 32,807 cases of illness.[12]

8-10 Watershed and Reservoir Protection

Impounding reservoirs and their watershed area should be protected to the degree necessary to ensure that the water supply is not contaminated in a way that would render it unfit for use. EPA now requires that all public supplies drawn from surface sources be filtered unless very stringent criteria with regard to quality and protection are met.[13] If the reservoir is intended to provide water with no treatment other than disinfection, no recreational use should be permitted and use of the watershed for any purpose should be severely restricted. When a complete treatment plant is provided, there is no clear health reason for prohibiting limited recreational uses such as fishing, hunting, camping, and bathing.

In the absence of state regulations, the following rules might be applied:

1. Recreational use should be permitted only when there is a real need for such use and the need cannot be supplied by other bodies of water.
2. Use should be controlled by caretakers with police authority whose costs are paid by fees assessed against the recreational users of the lake.
3. Picnics and camping should be restricted to areas with garbage and toilet facilities.
4. Swimming and other water-contact sports should be restricted to areas at least 2 km (1.2 mi) distant from the intake.
5. Noncontact recreation such as fishing, boating, and hunting should be restricted to areas at least 200 m (600 ft) from the intake.

6. Any residential development within the drainage area should be provided with sewage treatment adequate to ensure protection of the resource.

The policy of the American Water Works Association[14] is that recreational use of reservoirs be prohibited for waters that have been treated and are ready for distribution or which are sufficiently high in quality to be distributed after disinfection alone. If the water requires treatment in addition to disinfection, recreational use may be permitted, subject to the control of the water utility which is responsible for the quality of the finished water.

8-11 Groundwater and Well Protection

Groundwater supplies and individual wells may be contaminated by surface water during floods and by percolation of waste material through the soil. Wells are protected against contamination from flooding by careful construction techniques which require that the casing be grouted down to the first impermeable stratum, that the casing extend above the surface of the ground, and that a concrete apron protect the area surrounding the casing. It is practically impossible to guard against contamination which enters the aquifer at points remote from the well. Leaking sewers, septic tanks, privies, and abandoned wells are all potential points of pollution, as are landfills and old dump sites. New wells should never be located in areas where such sources are known to exist.

8-12 Protection within the Treatment and Distribution Systems

Careful design of the treatment system will ensure that there are no actual or potential opportunities for mixing of potable and nonpotable water. The designer must avoid such errors as bypasses around individual treatment units which may also permit flow to bypass the plant and enter the distribution system, common wall construction between basins containing treated and untreated water which may permit contamination through small cracks, basins at low elevations which may be contaminated by flood waters, and certain filter designs which permit mixing of backwash and product water.

The designer must ensure that the plant has adequate chemical storage to provide for interruptions in supply resulting from weather, strikes, or natural disasters. An emergency power supply adequate to support the treatment facilities as well as the pumping plant should be provided. The operator, in turn, must ensure that ample supplies of required chemicals are maintained and that emergency systems are tested regularly.

All piping conveying potable water within the treatment plant should be of iron and be installed above ground. Treated water reservoirs must be covered and air inlets must be screened to exclude insects.

Protection within the distribution system should not depend on the maintenance of a residual disinfectant concentration in the water supply, although this is required as a final safety measure. The system itself is most likely to be contaminated by cross-connections and other plumbing defects in industrial, commercial, and residential facilities. Cross-connections to auxiliary supplies are surprisingly numerous in large cities and can be detected only by careful inspection of the premises of industrial and commercial users. Connections between such auxiliary supplies and public supplies are best protected by an air break. Air breaks must be carefully designed by competent engineers, since a vacuum in the supply line may draw water back, even across an air break if the gap is too small. As a rule of thumb, the gap should be at least equal to the diameter of the supply line.

A variety of older plumbing fixtures—toilets, glass washers, bedpan washers, and surgical instrument sterilizers—may have their inlets below the level of water in the fixtures and thus be particularly dangerous. Actual connections may exist between waste and water lines in the complicated plumbing systems of large buildings or waste lines may be above water tanks and have the opportunity to leak into them.

Within the distribution system itself, contamination may result from placing sewers and water lines in close proximity. Leaks in the distribution system can permit sewage to enter from the soil when the pressure is off or during periods of high demand when pressures are low. Whenever breaks occur in the distribution system, the mains affected should be thoroughly disinfected before they are put back in service. The procedures for disinfecting repaired sections of pipeline are discussed in Art. 7-10.

PROBLEMS

8-1. The solubility product of $Mg(OH)_2$ is 9×10^{-12}. Determine the solubility of Mg^{2+} at pH equal to 10, 11, and 12.

8-2. Phosphoric acid has a series of equilibria similar to those of carbonic acid. The three ionization reactions are

$$H_3PO_4 \rightarrow H^+ + H_2PO_4^-$$

$$H_2PO_4^- \rightarrow H^+ + HPO_4^{2-}$$

$$HPO_4^{2-} \rightarrow H^+ + PO_4^{3-}$$

with $K_1 = 7.5 \times 10^{-3}$, $K_2 = 6.2 \times 10^{-8}$, and $K_3 = 4.8 \times 10^{-13}$. If the total phosphate concentration is 20 mg/L as PO_4, determine the amounts of each species at pH values of 8, 10, and 12.

8-3. Phosphorus may be precipitated in the form of $Ca_3(PO_4)_2$ or $CaHPO_4$, which have solubility products of 1×10^{-27} and 3×10^{-7} respectively. Using the results of Prob. 8-2 and assuming the final calcium concentration is 50 mg/L as $CaCO_3$, determine the total phosphate content which will remain at pH 8, 10, and 12.

8-4. Iron may be precipitated as the hydroxide in either the ferrous or ferric state. The solubility products of the two species are 5×10^{-15} for $Fe(OH)_2$ and 6×10^{-38} for $Fe(OH)_3$. Determine the solubility of iron at pH 6, 8, and 10 for each species. What does this tell us about the desirability of oxidizing iron prior to its removal?

REFERENCES

1. Linden E. Witherall et al., "Investigation of *Legionella pneumophilia* in Drinking Water," *Journal of American Water Works Association*, **80**:2:87, 1988.
2. Fred P. Williams and Elmer W. Akin, "Waterborne Viral Gastroenteritis," *Journal of American Water Works Association*, **78**:1:34, 1986.
3. Joan B. Rose, "Occurrence and Significance of *Cryptosporidium* in Water," *Journal of American Water Works Association*, **80**:2:53, 1988.
4. William D. Rosenweig, Harvey Minnigh, and Wesley O. Pipes, "Fungi in Distribution Systems," *Journal of American Water Works Association,* **78**:1:53, 1986.
5. Raymond D. Letterman and Charles T. Driscoll, "Survey of Residual Aluminum in Finished Water," *Journal of American Water Works Association,* **80**:4:154, 1988.
6. Kevin L. Dixon and Ramon G. Lee, "Occurrence of Radon in Well Supplies," *Journal of American Water Works Association*, **80**:7:65, 1988.
7. Jerry D. Lowry and Sylvia B. Lowry, "Radionuclides in Drinking Water," *Journal of American Water Works Association*, **80**:7:50, 1988.
8. Robert M. Krill and William C. Sonzogni, "Chemical Monitoring of Wisconsin's Groundwater," *Journal of American Water Works Association*, **78**:9:70, 1986.
9. Gunther F. Craun and Leland J. McCabe, "Problems Associated with Metals in Drinking Water," *Journal of American Water Works Association*, **67**:593, 1975.
10. Clair N. Sawyer and Perry L. McCarty, *Chemistry for Environmental Engineers*, McGraw-Hill, New York, 1978.
11. Vernon L. Snoeyink and David Jenkins, *Water Chemistry*, John Wiley and Sons, New York, 1980.
12. Gunther F. Craun, "Surface Water Supplies and Health," *Journal of American Water Works Association*, **80**:2:40, 1988.
13. David E. Leland and Paul A. Berg, "Assessing Unfiltered Water Supplies," *Journal of American Water Works Association*, **80**:1:36, 1988.
14. "Recreational Use of Domestic Water Supply Reservoirs," *Journal of American Water Works Association*, **50**:5, 1958.

CHAPTER 9

CLARIFICATION OF WATER

9-1 Purpose of Clarification Processes

Water is treated for a variety of purposes, including removal of pathogenic microorganisms, tastes and odors, color and turbidity, dissolved minerals, and harmful organic materials. The product of water treatment may be suitable for general domestic purposes or may be produced to higher standards such as those required for high-pressure steam, manufacture of food or beverages, and other specialized industrial purposes.

The treatment processes include purely physical methods such as screening and simple sedimentation, purely chemical methods such as adsorption and ion exchange, and physicochemical techniques in which contaminants are altered chemically to enhance their removal by physical processes.

Clarification processes, in modern practice, are typically physicochemical techniques which are intended to remove microorganisms, turbidity, and color—including the humic materials associated with the formation of trihalomethanes. Depending on the concentration of contaminants to be removed, the process may include separate coagulation, sedimentation, and filtration stages or may incorporate all these in a single unit.

There are a number of older treatment plants in the United States which provide simple sedimentation processes intended solely for the removal of fine sand and silt. Such processes are not employed in modern plants, since most surface waters now contain far less sediment as a result of improved agricultural practices and the construction of upstream impoundments. Additionally, simple sedimentation is inefficient compared to coagulation-sedimentation and is inadequate by itself with regard to suspended solids removal.

Screens are used at surface water intakes to prevent the entrance of materials which might damage pumps or other mechanical equipment. The screens used at intakes have openings of 6 mm ($\frac{1}{4}$ in) or less and thus exclude leaves, twigs, and fish. A typical moving water screen is shown in Figure 9-1.

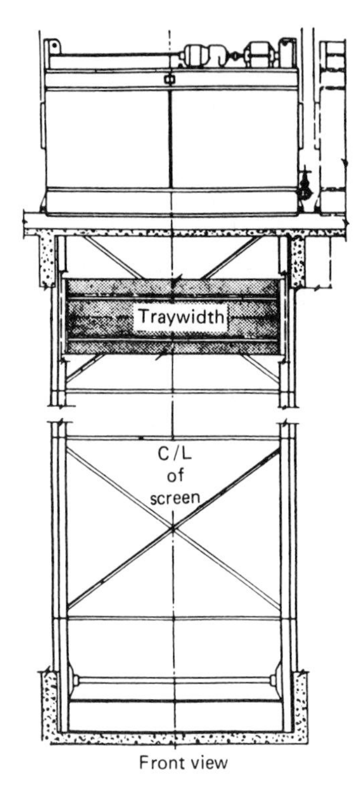

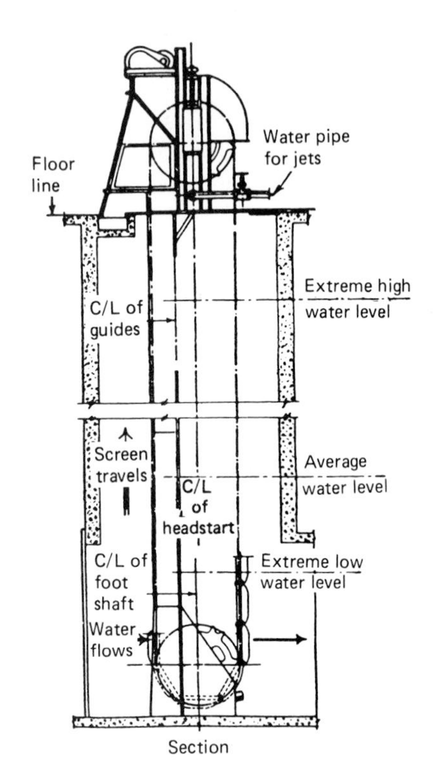

FIGURE 9-1
Traveling water screen. (*Courtesy FMC Corporation.*)

The screen may be advanced manually, on a timed basis, or automatically on development of a prescribed head loss. The accumulated debris is carried from the flow path and is dislodged by gravity, scraping, or water sprays. The head loss through screens depends on the details of their construction as well as the approach velocity, which is typically 0.3 to 0.6 m/s (1 to 2 ft/s). Equipment manufacturers may provide information concerning the losses through their products, but losses in clean screens are quite small and negligible in comparison to those which occur in operation when the screens are partially blocked.

9-2 Sedimentation of Discrete Particles

The sedimentation of discrete particles may be described by Newton's law, from which the terminal settling velocity of a spherical particle is found to be

$$v = \left[\frac{4g(\rho_s - \rho)d}{3C_D\rho}\right]^{1/2} \tag{9-1}$$

where v = terminal settling velocity
ρ_s = mass density of particle
ρ = mass density of fluid
g = gravitational constant
d = diameter of the particle

C_D is a dimensionless drag coefficient defined by

$$C_D = \frac{24}{N_R} + \frac{3}{N_R^{0.5}} + 0.34 \tag{9-2}$$

in which N_R is the Reynolds number, $vd\rho/\mu$, where μ is the absolute viscosity of the fluid and the other terms are as defined above. Equation (9-2) is applicable for Reynolds numbers up to 1000, which includes all situations of interest in water treatment. Where N_R is small (less than 0.5), the last terms of Eq. (9-2) may be neglected to yield

$$C_D = \frac{24}{N_R} = \frac{24\mu}{vd\rho} \tag{9-3}$$

which, when substituted in Eq. (9-1), gives

$$v = \frac{g}{18\mu}(\rho_s - \rho)d^2 \tag{9-4}$$

which is Stokes' law.

Particles in water are not spherical; however, the effect of irregular shape is not pronounced at low settling velocities. Most sedimentation processes are designed to remove small particles which settle slowly. Larger particles which settle at higher velocity will be removed in any event—whether or not they follow Stokes' or Newton's law.

The theoretical design of sedimentation processes is generally based on the concept of the ideal settling basin (Fig. 9-2). A particle entering the basin will have a horizontal velocity equal to the velocity of the fluid,

$$V = \frac{Q}{A} = \frac{Q}{w \times h} \tag{9-5}$$

and a vertical velocity equal to its terminal settling velocity defined by Stokes' or Newton's law. If a particle is to be removed, its settling velocity and horizontal velocity must be such that their resultant will carry it to the bottom of the basin before the outlet zone is reached. If a particle entering at the top of the basin is removed, all particles with the same settling velocity will be removed. Considering the slope of the velocity vector in Fig. 9-2, one may write

$$\frac{v_s}{V} = \frac{h}{L} \tag{9-6}$$

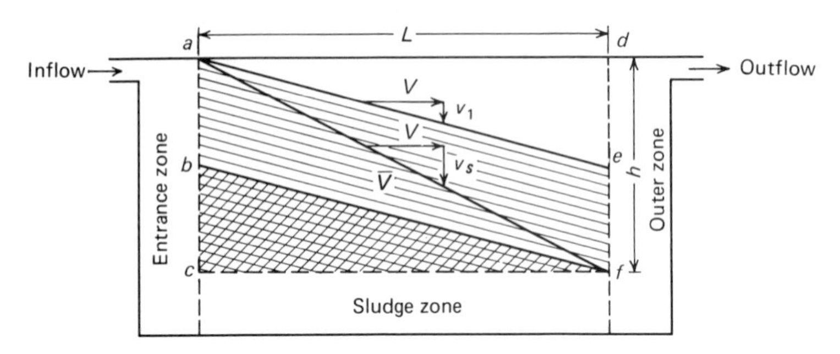

FIGURE 9-2
Ideal settling basin.

or

$$v_s = \frac{Vh}{L} = \frac{h}{L} \times \frac{Q}{wh} = \frac{Q}{wL} \tag{9-7}$$

Equation (9-7) defines the *surface overflow rate* (SOR), which is numerically equal to the flow divided by the plan area of the basin, but which physically represents the settling velocity of the slowest-settling particles that are 100 percent removed. Those particles which settle at velocities equal to or greater than the surface overflow rate will be entirely removed, while those which settle at lower velocities will be removed in direct proportion to the ratio of their settling velocity to v_s, assuming they are uniformly distributed on entering the basin.

In Fig. 9-2 one may note that a particle with settling velocity v_1 which enters at the top (point a) will settle only to point e and will thus enter the outlet zone in which higher velocities will carry it from the basin. An identical particle which enters at point b and settles at the same rate will be removed, as will all such particles which enter below point b. Considering the slope of the velocity vector and the geometry of the basin, one may equate the fraction X_r of particles with velocity v_1 that is removed to the vertical dimensions $(b - c)$ and $(a - c)$:

$$X_r = \frac{b - c}{a - c} = \frac{(v_1/V)L}{(v_s/V)L} = \frac{v_1}{v_s} \tag{9-8}$$

For actual suspensions of particles with a considerable variety of sizes and densities, prediction of the efficiency of a clarifier requires either a particle size distribution analysis or a settling column analysis. From either technique, a settling velocity cumulative frequency curve similar to Fig. 9-3 may be obtained. As noted above, all particles with settling velocity greater than v_s will be removed, together with a proportion of those with lesser settling velocity. The fraction of all particles removed is thus

$$F = (1 - X_s) + \int_0^{X_s} \frac{v}{v_s}\, dx \tag{9-9}$$

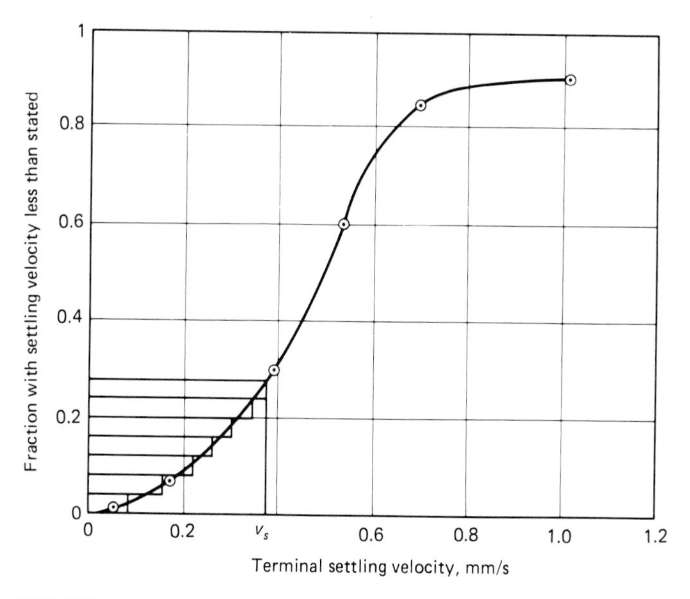

FIGURE 9-3
Cumulative distribution of particle settling velocity.

in which $(1 - X_s)$ is the fraction of particles with $v \geq v_s$ and the integral is the fraction of the particles with $v < v_s$, which is removed in the basin.

The actual calculation is performed by using a finite number of points on the distribution curve so that, in practice, Eq. (9-9) is approximated by

$$F = (1 - X_s) + \frac{1}{v_s}\Sigma v \Delta x \tag{9-10}$$

Example 9-1 A settling basin is designed to have a surface overflow rate of 32.6 m/day [800 gal/(ft² · d)]. Determine the overall removal obtained for a suspension with the size distribution given in the table below. The specific gravity of the particles is 1.2 and the water temperature is 20°C.

Particle size, mm	**0.10**	**0.08**	**0.07**	**0.06**	**0.04**	**0.02**	**0.01**
Weight fraction greater than size, %	10	15	40	70	93	99	100

Solution
From App. 3, at 20°C, $\mu = 1.0087$ and $\rho = 0.99823$. The settling velocity of the particles may be calculated from Stokes' law as follows:

$$v = \frac{g}{18\mu}(\rho_s - \rho)d^2 = \frac{9800}{18(1.0087)}(1.2 - 0.99823)d^2 = 108.91d^2$$

Representative values are tabulated below.

Weight fraction, %	**10.0**	**15.0**	**40.0**	**70.0**	**83.0**	**99.0**	**100**
v, mm/s	1.08	0.689	0.527	0.387	0.172	0.043	0.011
N_R	0.10	0.05	0.04	0.02	0.01	0.001	0.0001

Since the calculated Reynolds numbers are all less than 0.5, Stokes' law is applicable and the calculation of velocity is correct. From the calculated settling velocities, the cumulative distribution curve of Fig. 9-3 is drawn. All particles with settling velocities greater than the SOR will be removed. Thus, from the graph, $(1 - X_s) = 0.73$. The determination of $v\,\Delta x$ is tabulated below.

Δx	**0.04**	**0.04**	**0.04**	**0.04**	**0.04**	**0.04**	**0.027**
v	0.06	0.16	0.22	0.26	0.30	0.34	0.37
$v\,\Delta x$	0.0024	0.0064	0.0088	0.0104	0.0120	0.0136	0.0099

The overall removal is thus

$$F = (1 - X_s) + \frac{1}{v_s}\Sigma v\,\Delta x = 0.73 + \frac{1}{0.37}(0.0635) = 0.898 \qquad \text{or} \qquad 89.8\%$$

The preceding analysis shows that for the sedimentation of discrete particles, only the plan area of the basin is important. This theoretical conclusion is subject to certain practical modifications dictated by wind generated currents, density currents, and the need for uniform distribution of the flow. Actual sedimentation tanks have depths ranging from less than 3 to more than 6 m (10 to 20 ft). The depth is governed by the need to provide an undisturbed settling zone above sludge-removal equipment which is installed in the bottom and, in wide basins, by the need to minimize the effect of wind-generated mixing on the sludge accumulation zone. At the depths in common use, the *detention time* of standard clarifiers in water treatment plants ranges from 1 to 8 h. The detention time is defined as the tank volume divided by the flow.

The period individual particles of water remain in the basin depends on the details of its design—particularly the type of inlet and outlet structures which are used. Studies of clarifiers with salt solutions, dyes, and radioactive tracers indicate that the time elapsed to peak effluent concentration may be as little as 10 percent of the theoretical detention time, while the mean residence time may be only 50 to 60 percent of theoretical. The observed short-circuiting results from "dead" areas in the basins through which no flow occurs. This phenomenon increases the horizontal velocity, thus modifying the theoretical analysis above. The flow may be concentrated by locally high velocities at entrance and outlet and by differences in density between the liquid entering the basin and the basin contents. Differences in density can result from variations in either temperature or the concentration of suspended solids.

Better utilization of the basin volume might be obtained by subdividing it vertically by the addition of horizontal trays. If, for example, a tray were installed in the basin of Fig. 9-2 from *b* to *e*, the slower-settling particles with velocity v_1 would be removed entirely. Such a scheme, in effect, increases the surface area of the basin and reduces the surface overflow rate. Although basins with trays are seldom built because of the difficulty of removing accumulated solids, a number of proprietary devices such as tube settlers and plate clarifiers (Figs. 9-4 and 9-5) take advantage of the same principle. Such systems have the additional benefit of suppressing wind currents and of reducing the Reynolds number (and thus the turbulence) in the basin and permit more effective use of existing clarifiers as well as higher design loads on new systems.

The design of clarifiers is based primarily on the settling velocity of the particles which are to be removed and thus varies with the density and size of the suspension. The solids in a turbid river water have specific gravities ranging from 2.65 for sand to 1.03 for flocculated mud particles containing 95 percent water. Suspended water-logged vegetable matter has specific gravities ranging from 1.0 to 1.5. Chemical precipitates produced in coagulation have specific gravities ranging from 1.18 to 1.34, but the larger particles resulting from flocculation of the precipitates and other suspended matter contain a large volume of entrained water which may reduce their specific gravity to as little as 1.002.

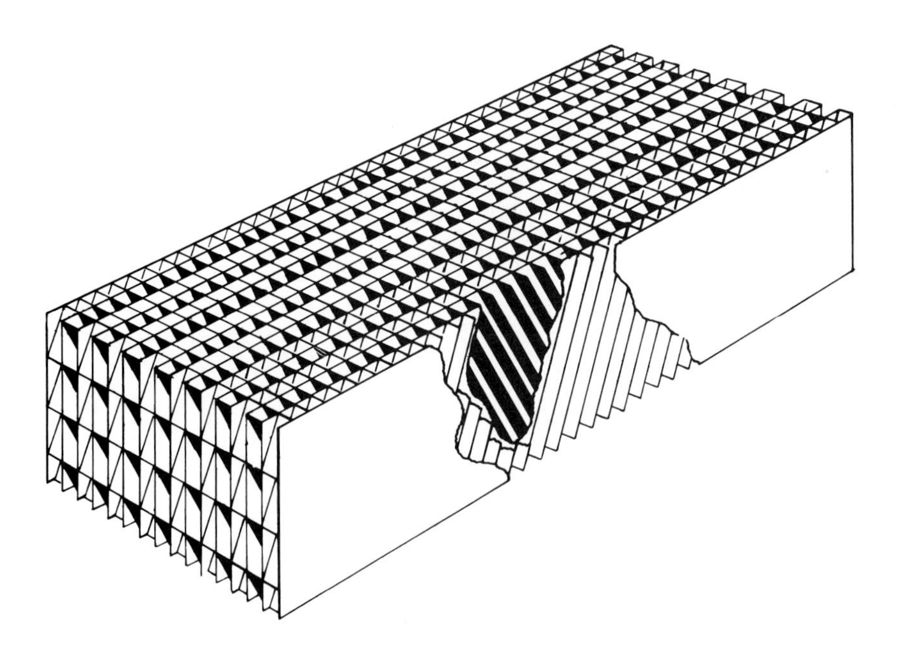

FIGURE 9-4
Tube settler. (*Courtesy Neptune Microfloc, Inc., a subsidiary of Neptune International.*)

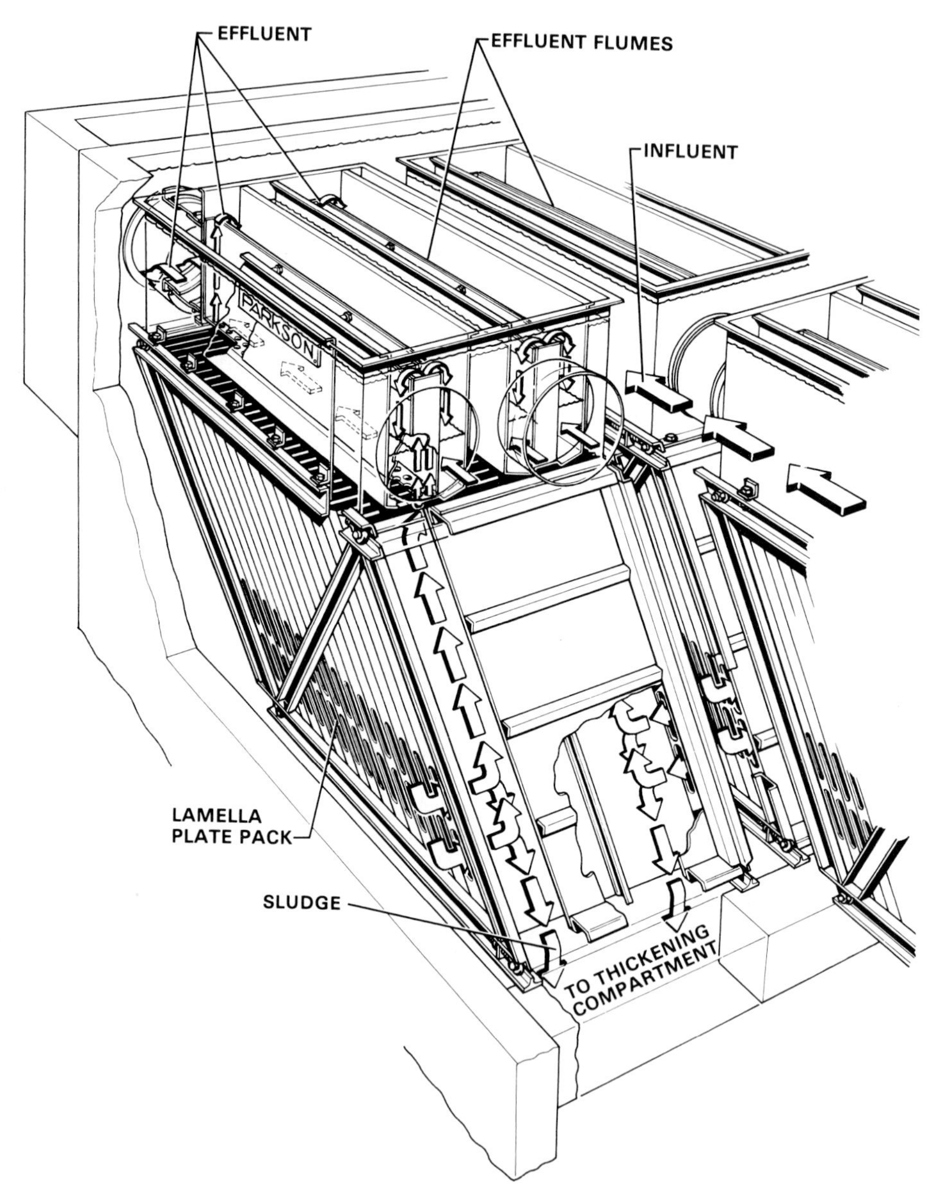

FIGURE 9-5
Lamella clarifier. (*Courtesy Parkson Corporation.*)

The precipitates formed in lime-soda softening processes consist principally of crystals of $CaCO_3$ with a size of 15 to 20 μm and a specific gravity of 2.7, which settle in clusters about 0.1 mm in size with an effective specific gravity of about 1.2. $Mg(OH)_2$, which may also be produced in softening, forms lighter suspensions which settle more slowly.

9-3 Sedimentation of Flocculant Suspensions

The chemical precipitates which are formed in coagulation and other destabilization processes tend to agglomerate while settling as a result of interparticle collisions. The density of the composite particles may decrease as a result of entrainment of water, but the overall result is generally an increase in settling velocity.

Settling analyses of such suspensions are performed in columns at least 300 mm in diameter with depth equal to that of the proposed clarifier. Samples are withdrawn at regular time intervals from multiple ports along the column and are analyzed to determine the reduction in suspended solids. Percent removals are plotted as numerical values versus depth and time as shown in Fig. 9-6. From this plot the removal obtained at various times may be predicted and a theoretical SOR can be established. The design SOR for the full-scale clarifier must be decreased by a factor of about 1.5 in order to obtain equivalent results.

Example 9-2 From the settling curves of Fig. 9-6, determine the theoretical efficiency of a sedimentation tank with a depth equal to the test cylinder and a detention time of 25 min. What surface overflow rate should be used in a full-sized clarifier in order to achieve equivalent results? The test cylinder has a depth of 3 m (9.8 ft).

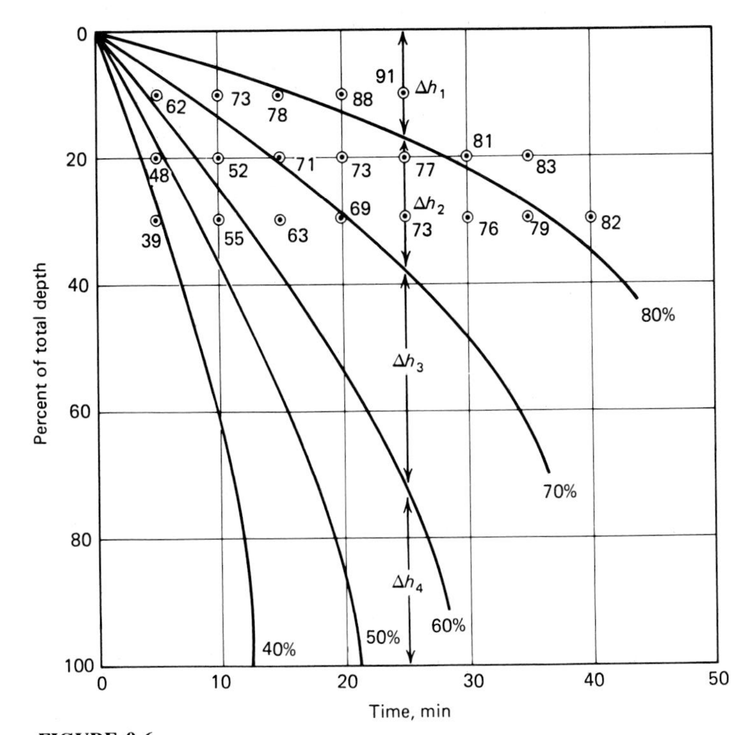

FIGURE 9-6
Settling of a flocculent suspension.

Solution
At t = 25 min, construct a vertical line as shown. In the volume of the basin corresponding to Δh_4, between 55 and 60 percent removal will occur. Similarly, in the volume corresponding to Δh_3, between 60 and 70 percent will be removed. In like fashion, the overall removal, assuming linear variation between contours, will be

$$F = \frac{55 + 60}{2} \times \frac{\Delta h_4}{h} + \frac{60 + 70}{2} \times \frac{\Delta h_3}{h} + \frac{70 + 80}{2} \times \frac{\Delta h_2}{h} + \frac{80 + 100}{2} \times \frac{\Delta h_1}{h}$$

Determining the Δh_i from Fig. 9-6, we get

$$F = 57.5(0.30) + 65(0.32) + 75(0.21) + 90(0.17) = 69\%$$

The clarifier has a detention time of 25 min and a depth of 3 m. $V/Q = 25$. $A(3)/Q = 25$.

$$\text{SOR} = \frac{Q}{A} = \frac{3}{25 \text{ m/min}} = 172.8 \text{ m/day}$$

Adjusting for full scale,

$$\text{SOR} = \frac{172.8}{1.5} = 115.2 \text{ m/day} \qquad (2826 \text{ gal/(ft}^2 \cdot \text{day)})$$

This rate is substantially higher than would be used in an actual clarifier, since the efficiency is much lower than is normally desired. Similar analyses at other detention times using the data of Fig. 9-6 would permit establishment of a relationship between efficiency and SOR which, in turn, could be used to determine the area required for any required efficiency.

9-4 Hindered Settling

Hindered settling occurs when high densities produce particle interactions and individual particles are so close to one another that the displacement of water produced by the settling of one affects the relative velocities of its neighbors. Such conditions may occur in sludge thickeners in water treatment plants. An estimate of the extent to which settling is hindered may be obtained from this equation:

$$\frac{v_h}{v} = (1 - C_v)^{4.65} \qquad (9\text{-}11)$$

where v_h is the hindered settling velocity, v is the free settling velocity, and C_v is the volume of particles divided by the total volume of the suspension. Equation (9-11) is valid provided the Reynolds number is less than 0.2, which is generally the case in hindered settling.

Example 9-3 A suspension of $CaCO_3$ with a mean particle size of 0.1 mm and a specific gravity of 1.1 is thickened by sedimentation to a concentration of 7 percent solids by mass. Find the final settling velocity and the required SOR for the thickener.

Solution
The original settling velocity, from Stokes' law, is

$$v = \frac{g}{18\mu}(\rho_s - \rho)d^2 = \frac{9800}{18(1.0087)}(1.1 - 0.99823)(0.1)^2$$

$$v = 0.55 \text{ mm/s}$$

At 7 percent solids, the solids concentration is 70,000 mg/L. At a specific gravity of 1.1, the volume occupied by the solids will be 70,000/1.1 = 63,636 mL.

$$C_v = \frac{63{,}636}{10^6} = 0.064$$

$$V_h = 0.55(1 - 0.064)^{4.65} = 0.40 \text{ mm/s}$$

The required SOR is thus 0.40 mm/s or 35 m/day [857 gal/(ft² · day)].

One may observe from the example above that, even at fairly high solids concentrations, the reduction in settling velocity is not particularly great. Hindered settling in clarifiers handling biological solids is analyzed by a different technique which may be found in Chap. 22.

9-5 Scour

The horizontal velocity in sedimentation basins must be limited to a value less than that which will carry the particles in traction along the bottom. The horizontal velocity just sufficient to cause scour has been defined as

$$V = \left[\frac{8\beta(s - 1)gd}{f}\right]^{1/2} \tag{9-12}$$

where V = horizontal velocity
s = specific gravity of particle
β = dimensionless constant ranging from 0.04 to 0.06
f = Darcy-Weisbach friction factor (usually 0.02 to 0.03)

The other terms are as defined earlier.

In most sedimentation tanks, the horizontal velocity is well below that required to cause scour. In some grit chambers (Chap. 20), scour is an important design parameter.

Example 9-4 A rectangular clarifier is designed to remove particles with a diameter of 0.2 mm and a specific gravity of 1.005. Determine the SOR and the depth required to prevent scour.

Solution
From Stokes' law, the SOR is

$$v = \frac{9800}{18(1.0087)}(1.005 - 0.99823)(0.2)^2 = 0.146 \text{ mm/s}$$

$$= 12.6 \text{ m/day } [310 \text{ gal/(ft}^2 \cdot \text{day)}]$$

The scour velocity, at worst, is

$$V = \left[\frac{8(0.04)(1.005 - 1)(9800)(0.2)}{0.03}\right]^{1/2} = 10.2 \text{ mm/s} = 883 \text{ m/day}$$

Since $v = 12.6 = Q/wL$ and $V = 883 = Q/wh$,

$$\frac{h}{L} = \frac{12.6}{883} = 0.0143$$

A clarifier with the minimum normal depth of 3 m could thus be 210 m long without producing scour. Since clarifiers with such lengths are not practical, it is clear that scour is not important with this type of suspension.

9-6 Coagulation Processes

A large portion of the suspended particles in water are sufficiently small that their removal in a sedimentation tank is impossible at reasonable surface overflow rates. Additionally, humic materials, which are the most important precursors of the trihalomethanes formed by disinfection processes, are not removed at all by simple sedimentation.

Colloidal particles, as a result of their small size, have a very large ratio of surface area to volume. For example, 1 cm^3 of material, if divided into cubes 0.1 mm on a side (the size of fine sand) would have a surface area of 0.06 m^2, while if divided into cubes 10^{-5} mm on a side (the midpoint of the colloidal range) would have a surface area of 600 m^2. As a result of this immense area, surface chemical phenomena are very important. Preferential adsorption of ions from solution onto the colloidal surface and ionization of chemical groups on the surface produce net charges on the particles. A schematic representation of the resulting colloidal state is presented in Fig. 9-7. Most colloidal particles in water are negatively charged as shown. The stationary charged layer on the surface is surrounded by a bound layer of water in which ions of opposite charge drawn from the bulk solution produce a rapid drop in potential. This drop within the bound-water layer is called the *Stern potential*. A more gradual drop, called the *zeta potential*, occurs between the shear surface of the bound-water layer and the point of electroneutrality in the solution.

The surface charge on colloidal particles is the major contributor to their long-term stability. Particles which might otherwise settle or coalesce are mutually repelled by their like charge. *Coagulation* is a chemical technique directed toward destabilization of colloidal particles. *Flocculation*, in engineering usage, is a slow mixing technique which promotes the agglomeration of destabilized particles.

Although other techniques are possible, the coagulation of water generally involves the addition of chemicals—either hydrolyzing electrolytes or organic polymers. The action of metallic coagulants is complex, involving the dissolution of the salt (which may reduce the zeta potential by altering the ionic concentration in the bound layer), the formation of complex hydroxyoxides of

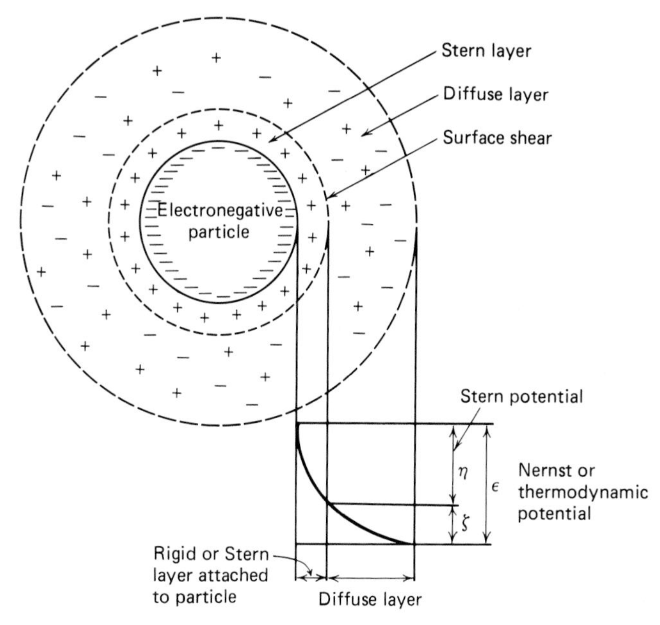

FIGURE 9-7
Guoy-Stern colloidal model.

the metal which may be highly charged, and the entrapment of individual particles in the chemical precipitate formed. The processes are very complex[1,2] but may be described as follows:
Dissolution:

$$Al_2(SO_4)_3 \rightarrow 2Al(H_2O)_6^{+3} + 3SO_4^{-2} \tag{9-13}$$

Hydrolysis:

$$\begin{aligned} Al(H_2O)_6^{+3} + H_2O &\rightarrow Al(H_2O)_5\,OH^{+2} + H^+ \\ Al(H_2O)_5OH^{+2} + H_2O &\rightarrow Al(H_2O)_4(OH)_2^{+1} + H^+ \\ Al(H_2O)_4(OH)_2^{+1} + H_2O &\rightarrow Al(H_2O)_3(OH)_3 + H^+ \\ Al(H_2O)_3(OH)_3 + H_2O &\rightarrow Al(H_2O)_2(OH)_4^{-} + H^+ \end{aligned} \tag{9-14}$$

Polymerization:
The products of the hydrolysis combine to form a variety of molecules including:

$Al_6(OH)_{15}^{+3}$
$Al_7(OH)_{17}^{+4}$
$Al_8(OH)_{20}^{+4}$
$Al_{13}(OH)_{34}^{+5}$

with a molecular structure of the form

```
                  H              H              H
                  O     H2O      O     H2O      O     H2O
           H      |  H   |   H   |  H   |   H   |  H   |   H
           O      |  O   |   O   |  O   |   O   |  O   |   O
(H2O)4Al       Al       Al      Al      Al      Al      Al       Al(H2O)4
           O      |  O   |   O   |  O   |   O   |  O   |   O
           H      |  H   |   H   |  H   |   H   |  H   |   H
                 H2O     O      H2O     O      H2O     O
                         H              H              H
```

Hydrolysis of iron salts is somewhat different than that of aluminum, but results in the formation of similar polymeric species. The net effect of addition of a metallic coagulant is seen to be the formation of large, insoluble, positively charged particles and production of free hydrogen ion from the water involved in the hydrolysis. This complex process is frequently represented by the simplified equation

$$Al_2(SO_4)_3 + 6H_2O \rightarrow 2Al(OH)_3 + 3H_2SO_4 \qquad (9\text{-}15)$$

The polymeric species formed and the effectiveness of coagulation depend on both pH and the concentration of coagulant applied. For any water, there is an optimum pH range and optimum coagulant concentration. Curves such as that presented in Fig. 9-8 can be obtained for particular waters and coagulants by means of laboratory jar tests. The practical control of coagulation is dependent on such analyses. The optimum chemical dosage produces a maximum mean particle size, a minimum count of small particles, and minimum turbidity prior to settling.[4] Dosages which are substantially too high or too low will be ineffective and may, in fact, produce colloidal suspensions of the coagulant itself.

The chemicals commonly used in coagulation include alum (aluminum sulfate), ferric chloride, ferric sulfate, sodium aluminate, polyaluminum chloride, ferrous sulfate and lime, and chlorinated copperas. The choice is dictated by relative cost and effectiveness in particular waters. Alum is by far the most commonly used, although it is not always the most effective on a molar basis[5] and it is not desirable to increase the concentration of aluminum in treated water.[6]

Removal of specific contaminants in coagulation processes may be affected by such factors as temperature, pH, alkalinity, and the choice of coagulant. Reduction of suspended solids and turbidity is adversely affected by low temperature[7]; however, removal of total organic carbon (TOC), which includes some of the precursors of trihalomethanes, is not.[8] Trihalomethane (THM) formation itself is less at low temperatures, but the reduction in formation is not a result of improved removal of the precursors in coagulation but rather of a reduction in the rate of THM formation in disinfection processes.

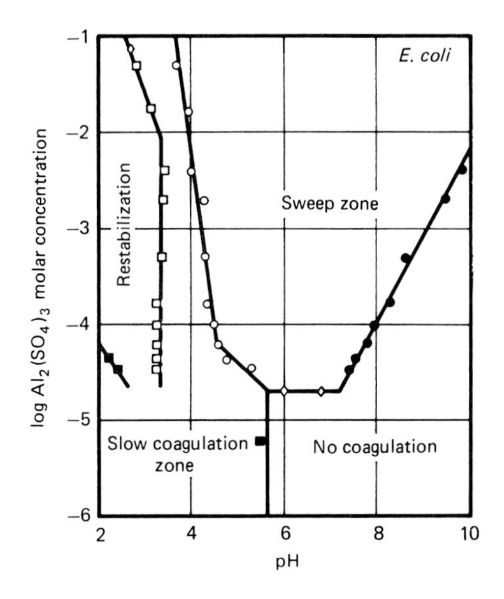

FIGURE 9-8
Aluminum sulfate domain of stability.[3] (*Reprinted from Journal of American Water Works Association, **62**, by permission of the Association. Copyright 1976 by the American Water Works Association, Inc., 6666 W. Quincy Avenue, Denver, CO 80235.*)

Removal of *Giardia* in coagulation is closely associated with removal of turbidity. Reported removal in coagulation and sedimentation ranges from 65 to over 90 percent and is approximately equal to the reduction in turbidity.[9] Virus removal, on the other hand, is not clearly associated with turbidity. Removals of hepatitis A virus and rotavirus in coagulation and sedimentation are typically in excess of 90 percent.[10]

Most of the metallic coagulants react with water to produce free hydrogen ions as shown in Eqs. (9-14) and (9-15). Since nearly all natural waters contain alkalinity, the hydrogen ions released will react with this, reducing the variation in pH. If a water contains insufficient alkalinity, the addition of a metallic coagulant may depress the pH below the range in which the particular salt is effective. In such circumstances, an alkaline salt must be added to increase the buffer capacity of the solution. The adequacy of the alkalinity can be estimated from the following simplified equations, which represent the approximate overall reactions.

Aluminum sulfate (alum):

$$AL_2(SO_4)_3 \cdot 18H_2O + 3Ca(HCO_3)_2 \rightarrow 2Al(OH)_3 + 3CaSO_4 + 18H_2O + 6CO_2$$

Ferric chloride:

$$2FeCl_3 + 3Ca(HCO_3)_2 \rightarrow 2Fe(OH)_3 + 3CaCl_2 + 6CO_2$$

Ferric sulfate:

$$Fe_2(SO_4)_3 + 3Ca(HCO_3)_2 \rightarrow 2Fe(OH)_3 + 3CaSO_4 + 6CO_2$$

Ferrous sulfate and lime:

$$FeSO_4 \cdot 7H_2O + Ca(OH)_2 \rightarrow Fe(OH)_2 + CaSO_4 + 7H_2O$$

followed by, in the presence of oxygen,

$$4Fe(OH)_2 + O_2 + 2H_2O \rightarrow 4Fe(OH)_3$$

Chlorinated copperas:

$$3FeSO_4 \cdot 7H_2O + 1.5Cl_2 \rightarrow Fe_2(SO_4)_3 + FeCl_3 + 21H_2O$$

followed by

$$Fe_2(SO_4)_3 + 3Ca(HCO_3)_2 \rightarrow 2Fe(OH)_3 + 3CaSO_4 + 6CO_2$$

and

$$2FeCl_3 + 3Ca(HCO_3)_2 \rightarrow 2Fe(OH)_3 + 3CaCl_2 + 6CO_2$$

The optimum pH range for each of the metallic coagulants is tabulated below.

Coagulant	pH
Alum	4.0 to 7.0
Ferrous sulfate	8.5 and above
Chlorinated copperas	3.5 to 6.5 and above 8.5
Ferric chloride	3.5 to 6.5 and above 8.5
Ferric sulfate	3.5 to 7.0 and above 9.0

Polymeric coagulants or *polyelectrolytes* are long-chain, high-molecular-weight molecules which bear a large number of charged groups. The net charge on the molecule may be positive, negative, or neutral. Representative molecular structures are

Anionic:

$$\begin{array}{cccccccccccc} -CH_2- & CH & -CH_2- & CH & -CH_2- & CH- \\ & | & & | & & | \\ & COO^- & & COO^- & & COO^- \end{array}$$

Cationic:

$$\begin{array}{cccccc} -CH_2- & CH & -CH_2- & CH & -CH_2- & CH- \\ & | & & | & & | \\ & C_6H_4 & & C_6H_4 & & C_6H_4 \\ & | & & | & & | \\ & N^+ & & N^+ & & N^+ \\ & R & & R & & R \end{array}$$

Ampholytic:

$$\begin{array}{ccccccccc} -NH- & CH & -CO-NH- & CH & -CO-NH- & CH & -CO- \\ & | & & | & & | & \\ & (CH_2)_4 & & (CH_2)_2 & & (CH_2)^4 & \\ & | & & | & & | & \\ & NH_3^+ & & COO^- & & NH_3^+ & \end{array}$$

Although it might appear that cationic polymers would be most effective in coagulation of the negatively charged colloids found in water, this is not always the case.[2] The chemical groups on the polymer are thought to combine with active sites on the colloid. Such interaction of a single molecule with a large number of particles produces a bridging effect, combining them into a larger particle which may settle under the action of gravity. Both molecular weight of the polymer and the charge density influence the effectiveness of polyelectrolytes; however, recent research indicates that charge density is the most important single factor.[11,12]

Polyelectrolytes are excellent coagulants which may be used alone or in conjunction with metallic coagulants. A large number of such products have been approved for use in treating public water supplies. Permissible dosages range from 1 to 150 mg/L, most being less than 10 mg/L. The selection of an appropriate coagulant requires determination of the necessary dosage through jar tests and comparison of all relevant costs—including subsequent management of the sludge produced.

Coagulant aids, properly speaking, do not aid in coagulation, but rather in the subsequent flocculation of the destabilized particles. Agents include oxidizers such as chlorine and weighting agents such as clay and activated silica.

Oxidizing agents are thought to improve the coagulation-flocculation process by the destruction or alteration of organic compounds which might otherwise interfere. When chlorine has been used, the dosage has been that required to reach the breakpoint (see Art. 11-1). Such addition of chlorine prior to coagulation may result in increased production of trihalomethanes; hence, other oxidizers such as ozone may be preferable.

Weighting agents are sometimes used in the coagulation of waters of low initial turbidity. It is a curious phenomenon in water treatment that highly turbid waters are more easily clarified than those that are relatively limpid, although removal of disease-causing microorganisms does not seem to be related to initial turbidity. The addition of materials such as bentonite clay increases the particle density and the average weight of the suspension and provides a considerable surface for the adsorption of organic compounds. Dosages of clay typically range from 10 to 50 mg/L. Other weighting agents/adsorbents include activated carbon, powdered silica, and limestone. These have potential effects in addition to those associated with clay.

Activated silica consists of a preparation of colloidal sodium silicate which can act as a coagulant itself, as a coagulant aid in association with alum, or as a flocculating agent. The preparation of activated silica must be continuous, since it ages rapidly and may gel within a matter of hours. Its preparation involves the neutralization of approximately 80 percent of the alkalinity of a 1.5 percent solution of sodium silicate with any available acid. The solution is then diluted after being aged for about 10 min and is fed to the water being treated. Further details of the process may be found in Ref. 13.

Under optimum conditions, activated silica will increase the rate of coagulation and flocculation, reduce the coagulant dosage, broaden the pH range of effective coagulation, produce larger and tougher floc particles, and increase the removal of both color and colloidal material. Dosages are on the order of 10 percent of the alum dose, with the optimum being determined by jar tests. Polyelectrolytes have generally replaced activated silica in modern practice, since they can produce the same effects and are somewhat easier to handle.

9-7 Flocculation Processes

As noted earlier, the destabilized particles and chemical precipitates resulting from coagulation may still settle very slowly. *Flocculation* is a slow mixing process in which these particles are brought into contact in order to promote their agglomeration. The rate of agglomeration or flocculation depends on the the number of particles present, the relative volume which they occupy, and the velocity gradient G in the basin. The mean velocity gradient is a measure of power input in mixing processes and is equal to

$$G = \left(\frac{P}{\mu V}\right)^{1/2} \tag{9-16}$$

where P = power dissipated
μ = absolute viscosity
V = volume to which the power is applied

The theory of flocculation is quite complex and may be found in Refs. 14 and 15. Practical design factors include the use of detention times of 20 to 30 min and values of G between 25 and 65 s^{-1}. Flocculation is generally effected by slowly rotating large-diameter mixers. Older treatment plants may contain separate flocculation basins such as those in Fig. 9-9, in which large paddle

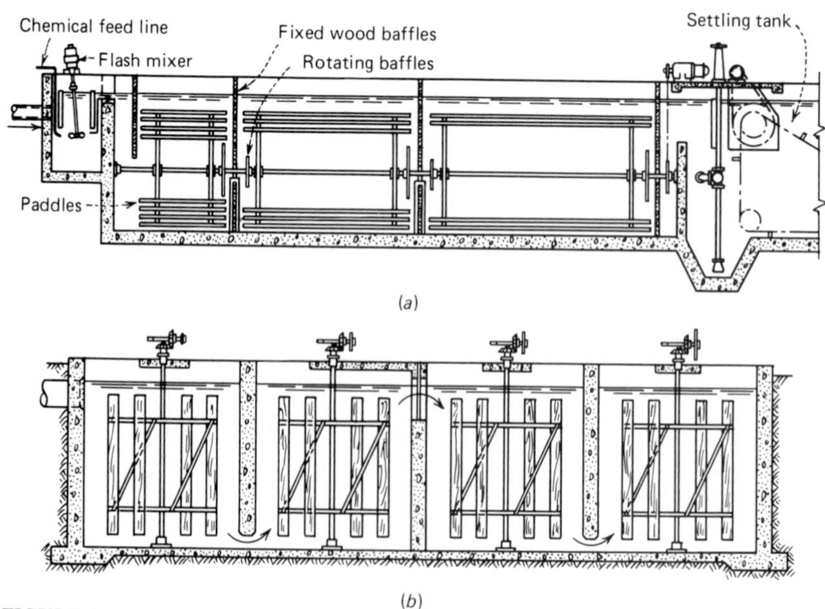

FIGURE 9-9
Flocculation basins. (*a*) Horizontal paddles; (*b*) vertical paddles.

wheel mixers are mounted either vertically or horizontally. Newer plants are more likely to incorporate dispersion of the coagulant (flash mixing), flocculation, and sedimentation in a single unit called a *contact clarifier*.

Contact clarifiers may be rectangular or circular in plan. Typical systems are illustrated in Figs. 9-10 and 9-11. The water enters in the center, where the chemical addition and rapid mixing occur, flows downward through the central area under the skirt, where flocculation occurs, and then flows upward through the bulk of the basin which serves as a sedimentation tank. As the water moves upward, its velocity will decrease since the flow area increases. Particles borne by the water will be carried upward until they reach the point at which their settling velocity is equal to the upward velocity of the fluid. As the number of particles so suspended increases, a sludge blanket will be formed which will act in a sense as a filter—straining out or providing additional opportunities for flocculation of particles which might otherwise be carried out by their upward velocity. As the particle density in the sludge blanket increases, the water velocity through the blanket will increase, resulting in upward movement of the suspended mass to a new equilibrium point. Large particles which are formed either in flocculation or in the sludge blanket may fall to the bottom of the basin. The sludge blanket itself will continue to rise and, in time, would be carried over the effluent weirs if a portion of the accumulated mass were not

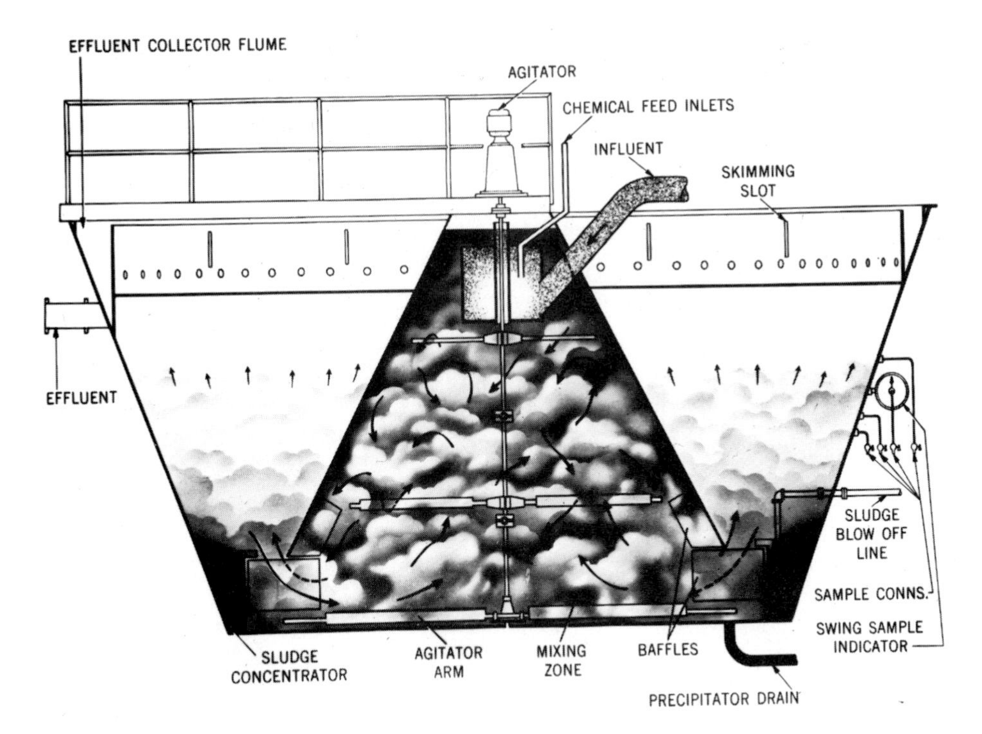

FIGURE 9-10
Suspended solids contact clarifier. (*Courtesy Permutit Co., Inc.*)

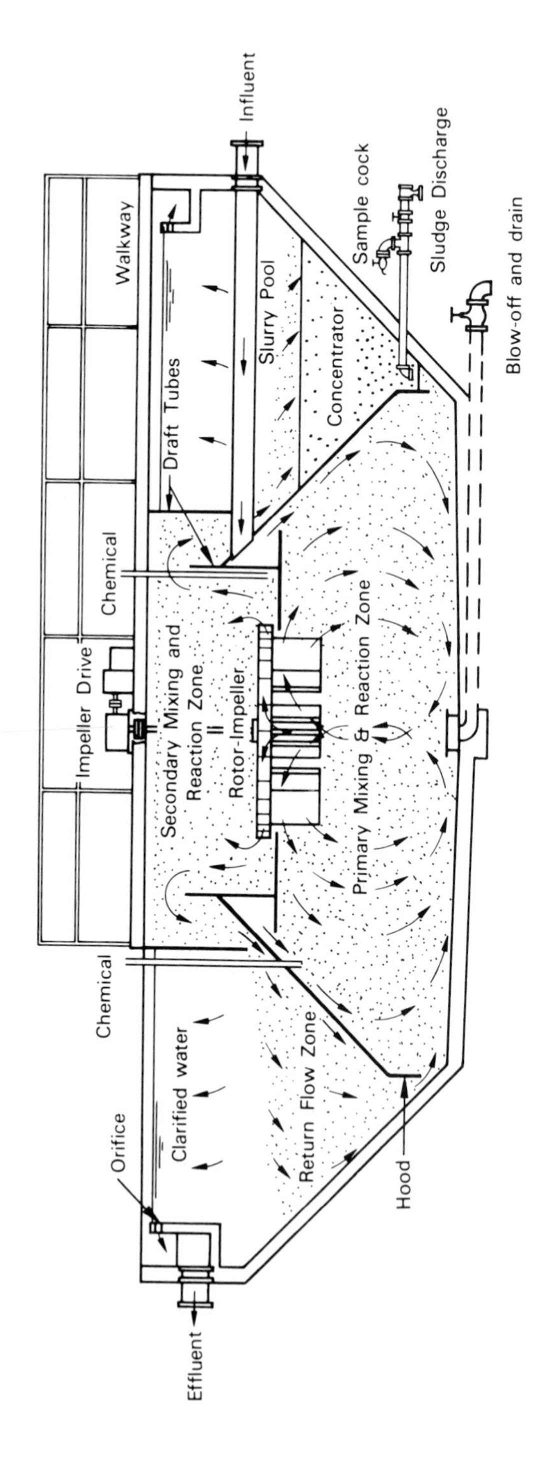

FIGURE 9-11
Suspended solids contact clarifier. (*Courtesy Infilco-Degremont, Inc.*)

wasted periodically to maintain a reasonable elevation. Sludge wasting from the blanket may be done automatically on a timed basis or, more commonly, by observation of the elevation of its upper surface.

9-8 Chemical Feeding Methods

Chemical feeding includes handling, storage, measurement, and proportioning and conveying of the proper amounts to the proper point of application. The methods and equipment used depend to some extent on the size of the plant. Metallic salts which are fed in relatively large amounts (such as alum, lime, and some of the iron salts) are most economically purchased as dry chemicals. These are generally stored in elevated towers, much like silos, from which they can be withdrawn by gravity. The bulk chemicals are delivered to the plant in trucks or railcars and are moved to storage by bucket elevators, screw conveyers, or high-velocity-air conveyers. The conveying equipment must include dust suppression and air pollution control features.

Alum may be purchased in a variety of commercial grades with densities ranging from 600 to 1100 kg/m^3 (40 to 70 lb/ft^3). The angle of repose is about 40° for the "rice" and powdered grades which are most commonly used. Hopper sections at the bottom of the storage containers should have a minimum slope of about 60° to prevent arching. Alum dust is irritating to the eyes and mucous membranes and operators should wear respirators, eye protection, and gloves when handling this material. Alum can be stored in mild steel or concrete containers, but should be protected against excessive humidity since it is mildly hygroscopic, forms acid solutions, and has a slight tendency to lump when moist.

Sodium aluminate has a bulk density ranging from 640 to 800 kg/m^3 (40 to 50 lb/ft^3). It can be stored in mild steel containers but deteriorates on exposure to the atmosphere, hence it is shipped in bags rather than in bulk. The powder forms alkaline solutions and the skin, eyes, and respiratory systems of workers must be protected.

Ferric sulfate has a bulk density of 1000 to 1550 kg/m^3 (63 to 100 lb/ft^3). It may be stored in steel or concrete but must be protected from moisture, since it is very hygroscopic and may lump or cake. In solution, it is acidic and aggressive. Workers require protection as noted for aluminum salts.

Ferrous sulfate has a bulk density approximately equal to that of water. Handling requirements are similar to those for ferric sulfate. In moist warm air it will cake, oxidize to the ferric state, and hydrate further. It is acidic in solution and workers must be protected from dust.

Lime is available as *quicklime* (CaO) and *slaked lime* [$Ca(OH)_2$]. Quicklime has a bulk density ranging from 880 to 1120 kg/m^3 (55 to 70 lb/ft^3). It is stored in airtight steel or concrete bins with 60° slope on the hopper bottom. Workers must wear skin and eye protection since, in solution, quicklime is very caustic. Slaked or hydrated lime has a bulk density about one-half that of water, tends to flow readily, and may arch in storage bins. Machinery should be provided in or on the bins to break arches which may form. Slaked lime is an irritant to mucous membranes and tends to dry the skin.

Soda ash has a bulk density ranging from 640 to 1000 kg/m^3 (40 to 63 lb/ft^3). It may be stored in steel bins and does not arch or flow like slaked lime. The solution is alkaline. It and the dust are irritating to skin and mucous membranes.

Dry metallic salts are proportioned and fed by a variety of commercially available gravimetric and volumetric feeders. The feeders may employ screw conveyers, rotating disks, oscillating pans, vibrating troughs, or weighing belts. The dry chemical is fed to a solution tank (a slaking tank in the case of quicklime), from which it flows or is pumped to the point of application.

Alum may also be purchased as a liquid and stored in stainless steel, fiberglass, PVC, or rubber-lined containers. The solution will crystallize at temperatures slightly below freezing (depending on the concentration). Sodium aluminate in liquid form may be stored in mild steel but must also be protected against low temperatures.

Ferric chloride and *ferrous chloride* are available in bulk only in liquid form. The solutions are corrosive and have densities of about 1.4 and 1.2 kg/L (90 and 75 lb/ft^3) respectively. Workers should wear rubber gloves and aprons and eye protection. The liquids can be stored in fiber glass or rubber or plastic-lined steel. Depending on the strength of the solution, ferric chloride may crystallize at temperatures above freezing. The minimum crystallization temperature occurs at about 33 percent $FeCl_3$. As normally shipped, crystals of $FeCl_3 \cdot 6H_2O$ may begin to form at temperatures of about 7°C (45°F).

Materials which are suitable for the formation and conveyance of solutions of metallic salts are as follows:

Alum. Type 316 stainless steel, fiber glass, plastics, and rubber.

Sodium aluminate. Mild steel, type 304 stainless steel, iron, concrete, or plastics.

Ferric/ferrous chloride. Rubber or Saran-lined steel, hard rubber, fiber glass, or plastic. Valves and other appurtenances must be of plastic or be lined with Saran or Teflon.

Ferric/ferrous sulfate. Rubber, fiber glass, type 316 stainless steel, or plastics.

Lime. Rubber, iron, steel, concrete, or plastics

Soda ash. Rubber, iron, steel, or plastics

Lime is not fed as a solution since its solubility is very low. On mixing with water, quicklime will be slaked with the release of considerable heat. The water-to-lime ratio may range from as little as 2:1 to as much as 6:1 by weight. The slaking time varies from 5 to 10 min. The lime, once slaked, is fed as a suspension called *milk of lime*, which has a tendency to deposit solids at changes in direction and, to a degree, throughout the conveyance system. The system should thus consist, as much as possible, of open channels. Closed pipe sections should be designed so that they can be readily disassembled for cleaning.

Polyelectrolytes are available as both dry chemicals and liquids, although not all types are obtainable in both forms. Specific information concerning the characteristics of individual polymers should be obtained from the manufacturer.

Dry polymers must be wetted and dispersed in water to form a solution which is often aged briefly before being further diluted and fed to the treatment process. Wetting and dispersion often requires specialized equipment and a considerable energy input to prevent lumping of the polymer and formation of "fish-eyes," which are impossible to disperse or dissolve.

Liquid polymer is readily dispersed in water and solutions are then fed in the same way as dry polymer. Polymer solutions are commonly made in batches sufficient to meet the demand for a day or two. Polymer solutions may be very viscous, hence the designer must consider this factor carefully in selecting pipes and pumps for conveying this type of material (see Chap. 3).

The coagulants must be thoroughly dispersed in the water to be treated. This *rapid mixing* step is a critical part of all coagulation and flocculation processes[11] and requires a high degree of turbulence and power dissipation. Rapid mixing may be provided by static mixers (baffled piping, baffled channel, or hydraulic jump) or mechanical mixers (paddle, turbine, or propeller) which may be installed either in-line or in a separate small basin. Detention times in flash mixers of 10 to 20 s are usually adequate, but some regulatory agencies may employ other standards. Power required is 2 to 5 $kW/(m^3 \cdot min)$ [0.1 to 0.2 $hp/(ft^3 \cdot min)$]. The mean velocity gradient G in rapid mixers is on the order of 500 s^{-1} and the optimum product of gradient and detention time, GT, is reportedly on the order of 2×10^5.[11]

9-9 Sedimentation Basin Design

Sedimentation basins, as noted, will always have a detention time somewhat less than the nominal value and a surface overflow rate somewhat higher than nominal as a result of nonideality of the flow pattern. Design of sedimentation basins is directed toward reducing the degree of nonideality.

Sedimentation tanks may be rectangular or square. In *rectangular basins*, the flow is directed along the long axis. This flow pattern minimizes the effect of inlet and outlet disturbances. Sludge removal equipment in such basins consists of horizontal scrapers which drag the solids to a hopper at one end, from which they are removed intermittently or continuously by gravity or augers. Typical designs are shown in Figs. 9-12 and 9-13. Vacuum or siphon devices may be used to remove sludge from clarifiers, but such devices are best suited to very light flocculant sludges such as those encountered in biological waste treatment processes. Rectangular basins offer certain economies in construction if common-wall design is used. Square basins are occasionally used

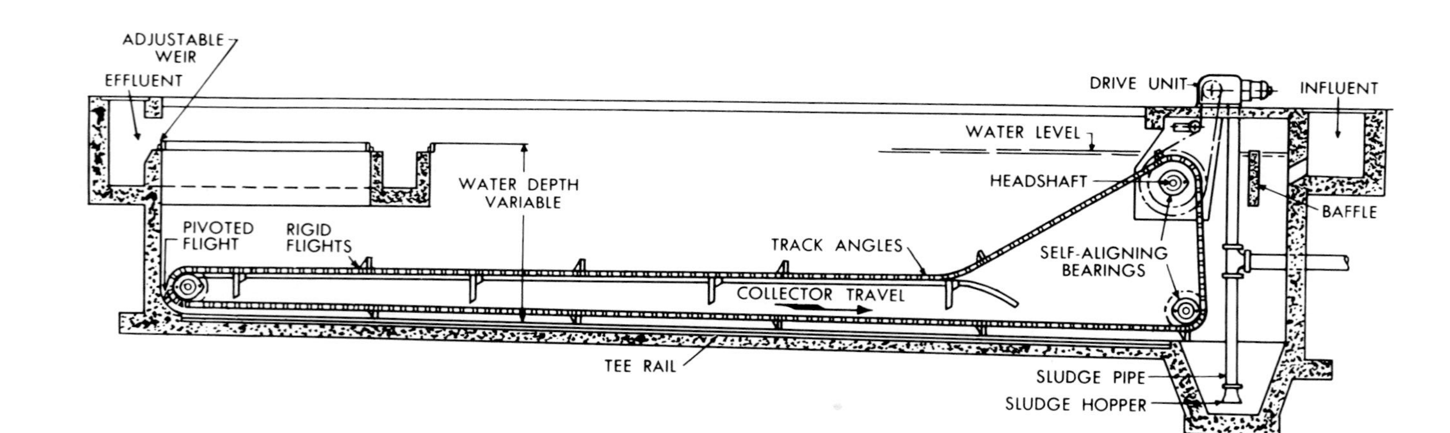

FIGURE 9-12
Rectangular clarifier. (*Courtesy Envirex, a Rexnord Company.*)

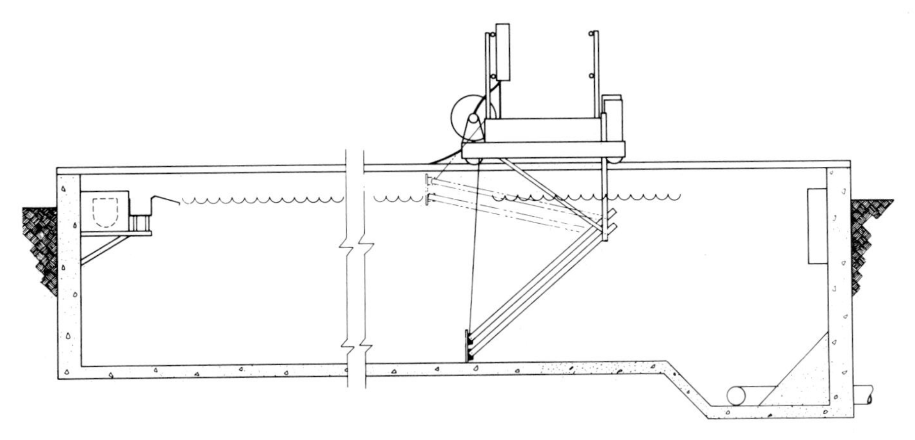

FIGURE 9-13
Rectangular clarifier. (*Courtesy Aqua-Aerobic Systems, Inc.*)

for clarifiers. Their flow pattern is not as desirable as that in rectangular designs, and the sludge removal equipment is more complicated. Square basins generally employ rotating scrapers similar to those in circular clarifiers with an additional corner sweep mechanism similar to that shown in Fig. 9-14.

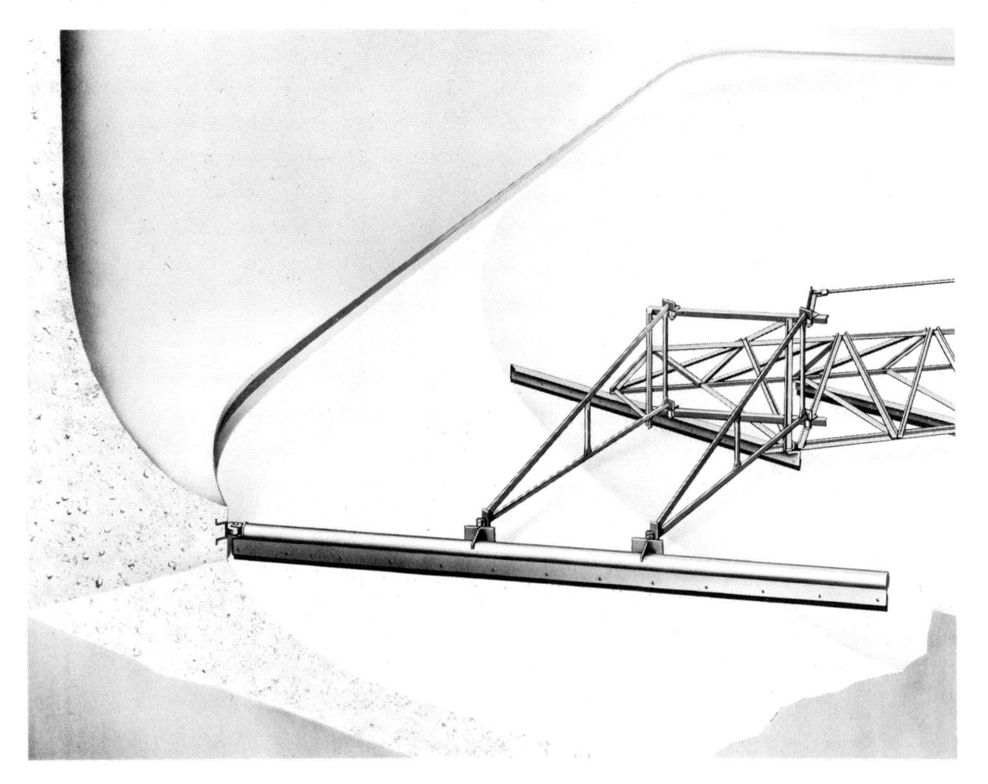

FIGURE 9-14
Corner sweep mechanism. (*Courtesy Envirex, a Rexnord Company.*)

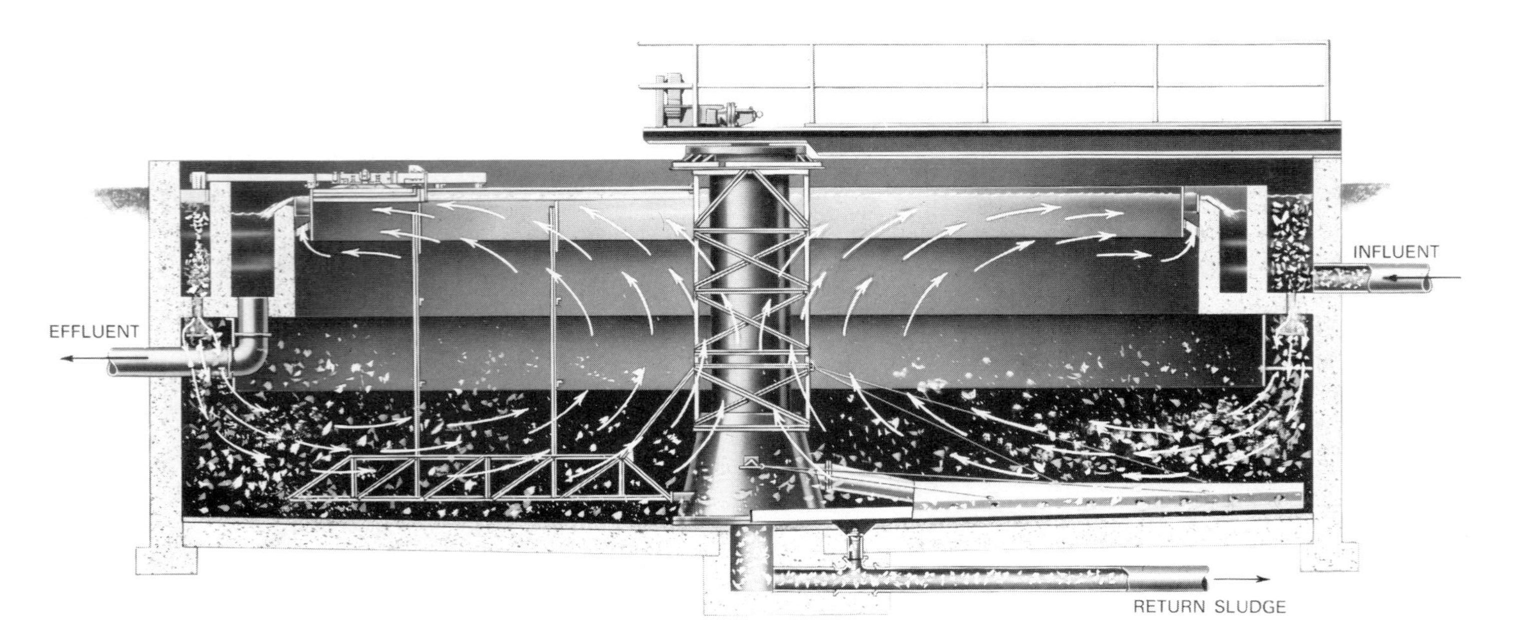

FIGURE 9-15
Peripheral-flow circular clarifier. (*Courtesy Envirex, a Rexnord Company.*)

In *circular basins*, the flow may enter around the perimeter, as in Fig. 9-15, or at the center, as in Fig. 9-16. The flow pattern is more complicated than in rectangular basins, and there is more opportunity for short-circuiting. Studies of the flow pattern in circular clarifiers have indicated that the average detention time is greater in peripherally fed basins than in those in which the flow enters in the center. Cleaning equipment in circular basins usually consists of scraper blades mounted on radial arms. The bottom of the basin is sloped toward the central hopper, and the rotating blades push the sludge into a series of windrows which are gradually worked to the center. Circular basins have smaller wall area for a given plan area but do not permit common-wall construction.

Careful design of inlets and outlets is very important to the proper operation of clarifiers. The *ideal inlet* reduces the entrance velocity to prevent development of currents toward the outlet, distributes the water as uniformly as possible across the basin, and mixes it with water already in the tank to prevent density currents. A near-perfect inlet consists of a very large number of very small openings distributed across the width and depth of the basin in which the head loss through the openings is large with respect to the variation in head between different openings. Some typical designs which offer a compromise between simplicity and function are illustrated in Fig. 9-17. Poorly designed inlets are the most common cause of poor clarifier performance.

Outlets of clarifiers usually consist of weirs which skim the clarified water from the surface and are sufficiently long to reduce the local velocity in their vicinity to levels which will not resuspend solids. The design of weirs is based on a weir loading or weir overflow rate expressed in flow per unit length. Effluent weirs are placed as far from the inlet as possible—at the opposite end of rectangular basins, around the perimeter of center-fed circular tanks, and toward the center and along the radii of peripherally fed basins. The weirs with their associated effluent channels may cover a substantial portion of the area of

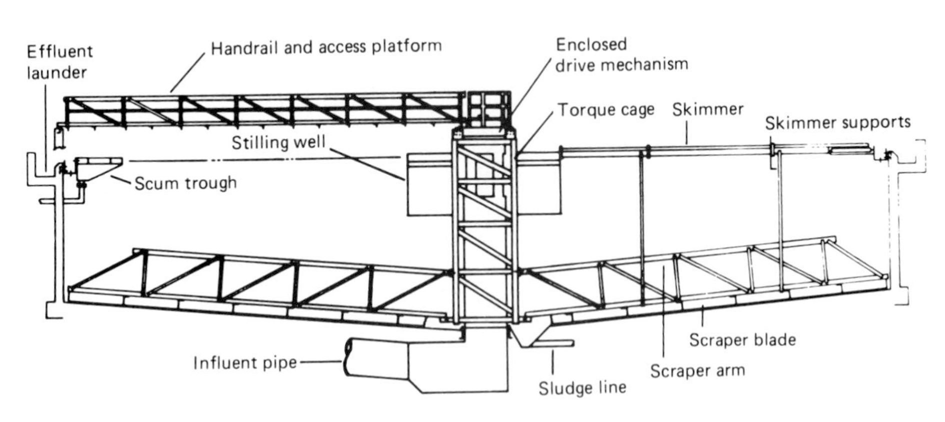

FIGURE 9-16
Center-flow circular clarifier. (*Courtesy Environmental Elements Corporation.*)

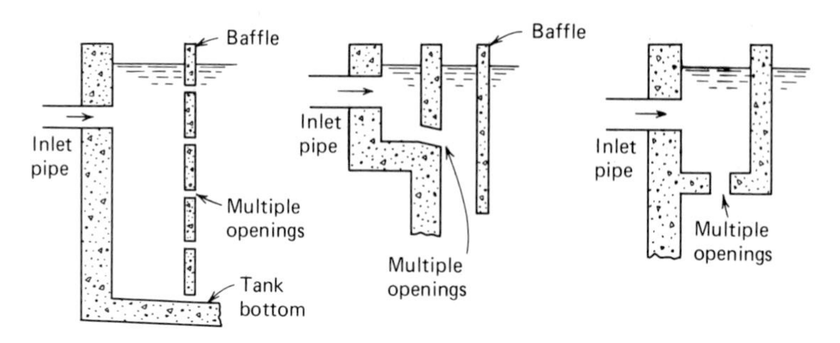

FIGURE 9-17
Typical sedimentation tank inlets.

the basin. The area so covered is still an effective part of the clarifier and is not subtracted in determining the SOR.

The weirs must be level and have a free discharge if the flow is to be uniform over their length. Typical weirs consist of 90° V notches approximately 50 mm (2 in) deep placed from 100 to 300 mm (4 to 12 in) on centers. The length calculated from the weir overflow rate is the total length, not the length over which flow occurs. The channel to which the weirs discharge must be designed so that the water surface therein does not restrict free flow. The hydraulics of this situation is described in Art. 3-22.

A compilation of typical surface overflow rates, weir overflow rates, and detention times which have been used in water treatment are presented in Table 9-1. These values are provided for purposes of comparison, not as recommended design standards. Individual states normally establish recom-

TABLE 9-1
Typical water treatment clarifier design details

Type of basin	Detention time, h	Weir overflow rate		Surface overflow rate	
		$m^3/(m \cdot day)$	$gal/(ft \cdot day)$	m/day	$gal/(ft^2 \cdot day)$
Presedimentation	3–8				
Standard basin following:					
Coagulation and flocculation	2–8	250	20,000	20–33	500–800
Softening	4–8	250	20,000	20–40	500–1000
Upflow clarifier following:					
Coagulation and flocculation	2	175	14,000	55	1400
Softening	1	350	28,000	100	2500
Tube settler following:					
Coagulation and flocculation	0.2				
Softening	0.2				

mended design criteria which the engineer may alter by demonstrating that they are not applicable to the particular water which is being treated or the process being used. Design of water treatment systems should be based on laboratory evaluation of the systems which are proposed.

9-10 Mechanical Equipment Selection

There are a large number of manufacturers of equipment for water treatment, most of whom can provide a variety of designs for basins of different types and shapes. Specifications ideally should address performance rather than details of the design, but it is often impossible to predict performance from manufacturers' literature. Existing installations of equipment under consideration offer the best opportunity to evaluate its performance and durability. Factors which must be considered are availability of replacement parts, accessibility of running gear for maintenance and repair, and resistance to corrosion.

In drafting specifications the engineer should be careful not to include proprietary equipment such as gears with unusual pitches or patented elements of a larger assembly unless it is intended to exclude other manufacturers.

PROBLEMS

9-1. A water treatment plant designed for a flow of 12,000 m^3/day is expected to use alum at a rate of 20 mg/L. Determine the storage volume required to provide a minimum of 1 month's supply if deliveries can be expected on a biweekly basis.

9-2. A flow of 20,000 m^3/day which contains 100 mg/L of suspended solids is coagulated with 50 mg/L of alum. If 90 percent of the solids and chemical precipitates is removed in sedimentation estimate:

(*a*) Total dry mass of sludge produced per day.

(*b*) Total wet mass produced per day, assuming the sludge is 5 percent solids by weight.

(*c*) Total volume produced per day, assuming the specific gravity of the wet sludge is 1.05.

9-3. A sedimentation basin is to be designed so that it will remove 100 percent of all particles which have a settling velocity of 0.3 mm/s.

(*a*) For a flow of 10 m^3/min, determine appropriate dimensions for a rectangular basin in which the length is 4 times the width. The detention time is 2 h.

(*b*) Determine the total weir length required if the weir overflow rate is 250 $m^3/(m \cdot day)$. Show how you would locate this weir on a sketch of the basin.

9-4. Describe the effect you would expect with regard to solids removal if the basin of Prob. 9-3 were kept at a detention time of 2 h but made only half as deep. Why would this effect occur? What other factors might influence your choice of basin depth.

9-5. A clarifier with an area of 150 m^2 treats a flow of 3000 m^3/day. The water entering the clarifier contains a substantial number of clay agglomerates with a specific gravity of 1.05 and a diameter of 0.05 mm. What percentage of these particles will be removed in the clarifier?

9-6. A water treatment plant with a flow of 10,000 m^3 has a normal concentration of suspended solids of 80 mg/L which can be reduced to nearly zero by treatment with either 25 mg/L alum $[Al_2(SO_4)_3 \cdot 18H_2O]$ or 3 mg/L of a polyelectrolyte. Determine the total mass of sludge produced per day by treatment with each coagulant. If the bulk density of the sludge is 1.02 for the alum and 1.025 for the polymer, determine the volume produced per day by each coagulant.

9-7. A clarifier designed for a flow of 6000 m^3/day has an area of 300 m^2. What is the settling velocity of the particles this basin is intended to remove? What diameter sand particles (ρ = 2650 kg/m^3) will be entirely removed? What diameter flocculated mud particles (ρ = 1030 kg/m^3) will be removed?

9-8. A circular clarifier has a surface overflow rate of 35 m/day and a weir overflow rate of 250 m^2/day. What is the maximum diameter which will meet these standards if only a single peripheral weir is used?

9-9. A rapid mixer disperses chemicals in a flow of 20,000 m^3/day. The detention time is 30 s and G is 700 s^{-1}. What is the power required for this mixer?

REFERENCES

1. Stephen K. Dentel and James M. Gossett, "Mechanisms of Coagulation with Aluminum Salts," *Journal of American Water Works Association*, **80**:4:187, 1988.
2. Walter J. Weber, *Physicochemical Processes for Water Quality Control*, Wiley-Interscience, New York, 1972.
3. George P. Hanna and A. J. Rubin, "Effect of Sulfate and Other Ions in Coagulation with Aluminum (III)," *Journal of American Water Works Association*, **62**:315, 1970.
4. Gregory D. Reed and Patricia C. Mery, "Influence of Floc Size Distribution on Clarification," *Journal of American Water Works Association*, **78**:8:75, 1986.
5. Johannes Haarhoff and John L. Cleasby, "Comparing Aluminum and Iron Coagulants for In-line Filtration of Cold Water," *Journal of American Water Works Association*, **80**:4:187, 1988.
6. Raymond D. Letterman and Charles T. Driscoll, "Survey of Residual Aluminum in Filtered Water," *Journal of American Water Works Association*, **80**:4:154, 1988.
7. Juli K. Morris and William R. Knocke, "Temperature Effects on the Use of Metal-Ion Coagulants for Water Treatment," *Journal of American Water Works Association*, **76**:3:74, 1984.
8. William R. Knocke, Sara West, and Robert C. Hoehn, "Effects of Low Temperature on the Removal of Trihalomethane Precursors by Coagulation," *Journal of American Water Works Association*, **78**:4:189, 1986.
9. Gary S. Logsdon et al., "Evaluating Sedimentation and Various Filter Media for Removal of *Giardia* Cysts," *Journal of American Water Works Association*, **77**:2:61, 1985.
10. V. Chalapati Rao et al., "Removal of Hepatitis A Virus and Rotavirus by Drinking Water Treatment," *Journal of American Water Works Association*, **80**:2:59, 1988.
11. Rong-Jin Leu and Mriganka Ghosh, "Polyelectrolyte Characteristics and Flocculation," *Journal of American Water Works Association*, **80**:4:159, 1988.
12. Richard E. Hubel and James K. Edzwald, "Removing Trihalomethane Precursors by Coagulation," *Journal of American Water Works Association*, **79**:7:98, 1987.
13. J. M. Cohen and S. A. Hannah, "Coagulation and Flocculation," in *Water Quality and Treatment*, 3d ed., McGraw-Hill, New York, 1971.
14. Gordon M. Fair, John C. Geyer, and Daniel A. Okun, *Water and Wastewater Engineering*, vol. 2, John Wiley and Sons, New York, 1968.
15. R. S. Gemmel, "Mixing and Sedimentation," in *Water Quality and Treatment*, 3d ed., McGraw-Hill, New York, 1971.

CHAPTER 10

FILTRATION OF WATER

10-1 Slow Sand Filters

The slow sand filter was developed in Great Britain in the early nineteenth century and has been widely used throughout the world. In the United States, such filters were found to be unsuitable for the turbid clay-bearing waters of the major rivers, and the rapid sand filter, or American filter, was developed in response to this special need. Slow sand filter systems have, nevertheless, been used in the United States and are enjoying a new popularity in certain applications both here and abroad.[1,2]

Slow sand filters are typically applied in the United States to waters which are very low in turbidity and thus require no pretreatment. In other countries it is not uncommon, however, to pretreat the influent to slow filters by a variety of techniques including upflow and downflow filtration through very coarse media,[2] rapid sand filtration, sedimentation alone, or coagulation and sedimentation processes.

The filtration rate on slow filters is normally less than 0.4 m/h [0.16 gal/(min · ft^2)], although rates up to 1.5 m/h [0.60 gal/(min · ft^2)] have been used recently on some experimental systems.[3,4] "Ripening," or the development of a gelatinous surface coat of biological growth, is important to the proper operation of the filter. Optimum operation is not obtained for a period ranging from 6 h to as much as 30 days,[5,6] during which this surface coat, called *schmutzdecke*, is accumulated. The coat is reportedly most effective when it consists primarily of filamentous algae,[3] while it offers most resistance to flow and thus requires most frequent cleaning when unicellular algae predominate.[7] The filters are operated with the product water being wasted at the beginning of the run until the requisite quality is achieved.

The filter medium is normally somewhat finer and less carefully graded than that used in rapid filters (Art. 10-4). Effective sizes of 0.1 to 0.3 mm and uniformity coefficients of 2 to 3 are commonly employed. Although fine sand is considered preferable, coarser materials, which may be locally available, have also proven satisfactory.[4,8] The thickness of the medium may initially be as much as 1.4 m (4.5 ft), although 1 m (3.3 ft) is more common. The depth is gradually reduced during use, since the bed is cleaned by removal of successive layers. The minimum satisfactory depth is reported to be about 0.5 m (1.5 ft).

Slow sand filters may be of any shape, although rectangular designs are most common (Fig. 10-1). Individual units may be as large as 6000 m^2 (1.5 acre). The size of individual units is governed by the need to provide the required flow while at least one unit is out of service for cleaning or other maintenance. Sidewalls may be vertical or sloped and are constructed of concrete or masonry. The sand is supported by a layer of gravel about 0.3 m (1 ft) thick which is graded from an effective size of about 5 mm ($\frac{3}{16}$ in) at the top to 50 mm (2 in) at the bottom. Underdrains, normally constructed of perforated pipe, are placed within the lower portion of the support gravel to collect the filtrate. The lateral pipes are spaced about 2.5 to 3.5 m on centers (8 to 12 ft). Frictional losses in the underdrains should be kept low in order that the flow distribution on the bed be uniform. In cold climates, slow sand filters may be covered to minimize the chance of freezing.

Control methods for slow filters can be quite simple. The influent flow can be measured by simple V-notch weirs and regulated with gate valves or shear gates. Minimum depth on the filter can be maintained by an effluent weir set slightly above the maximum level of the sand— a device which also prevents development of negative head within the bed. It is desirable to provide some means of draining the water from above the filter bed when cleaning is required. This can be done easily with an adjustable overflow which can be lowered to the minimum level of the sand surface. Provision should be made to fill the filter bed from below with clean water when it is first put in service and after it has been refilled with sand. Sophisticated controls are not necessary and are undesirable in the situations in which these filters are most likely to be used in modern practice.

During operation, the level of the water above the filter surface will gradually increase as the upper layers of the sand become plugged. When the depth has increased to 1.25 to 2 m (4 to 7 ft), the filters are cleaned. The adjustable overflow mentioned above is very desirable to speed the dewatering

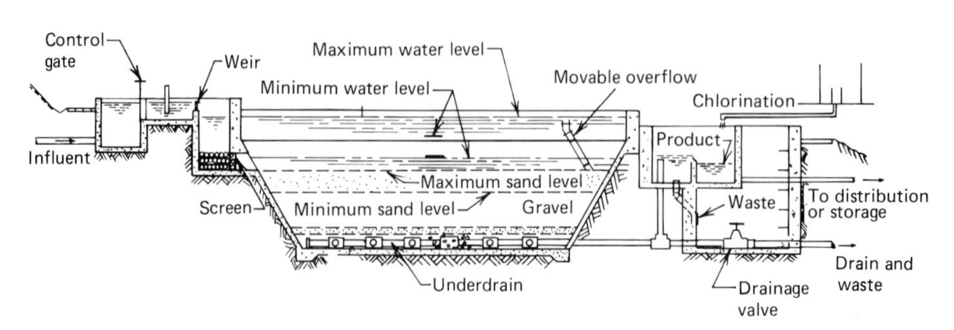

FIGURE 10-1
Cross section of a slow sand filter. (*After Ref. 2*).

of the filter prior to cleaning. It may take more than a day for a 2-m depth to drain through a dirty filter. The duration of filter runs depends on the character of the raw water and may range from as little as 1 week to as long as 6 months. Typical filter runs in the United States are 40 to 60 days in length.[1] Cleaning is effected by removal of the top layer of sand, the amount removed depending upon the depth of penetration of the *schmutzdecke*. Typical thicknesses removed range from 12 to 50 mm (0.5 to 2 in). The cleaning operation is most often done manually, although some operators have adapted light machinery to this task. The work requires little skill and can be done by casual labor recruited for short periods or by municipal employees normally occupied with other tasks.

The sand can be cleaned of the accumulated material by agitation in a continuously flowing stream of water. The cleaned sand is then stored until the thickness of the bed has been reduced to about 0.5 m (1.5 ft), when the bed is refilled, either by hand or hydraulically.

Modern applications of slow sand filtration to water treatment in the United States are limited to surface waters of very low turbidity, typically upland waters drawn from undeveloped areas which, until recently, were considered suitable for use without treatment or with no treatment other than disinfection. Recent outbreaks of waterborne giardiasis (see Chap. 8) among the users of such supplies have demonstrated the need for some technique for removal of *Giardia* cysts, which may persist in low-temperature waters of low turbidity for many months and which are very resistant to the action of disinfectants. Slow sand filtration has been clearly demonstrated to be capable of complete removal of *Giardia* after ripening,[8,9] and has the advantages of requiring little labor, little skill in operation, no chemical addition, and relatively low capital outlay. Although other techniques, discussed below, can also remove *Giardia*, for waters with extremely low turbidity slow sand filtration is more dependable.[7] In addition to the certain removal of *Giardia*, slow sand filters provide excellent removal of coliforms, other bacteria, and viruses.

In other countries, the advantages of slow sand filtration cited above have led to its application to waters which are quite high in turbidity. In these cases, pretreatment is required in order to provide reasonably long filter runs. The pretreatment may be as simple as passage through a series of coarse filters containing gravel ranging from 6 to 20 mm ($\frac{1}{4}$ to $\frac{3}{4}$ in). Recent studies in Colombia[2] have demonstrated that such systems can reduce the turbidity from over 100 to less than 10 ntu (nephelometric turbidity units), a level which permits reasonably long runs on slow sand filters.

10-2 Rapid Filtration

Rapid filtration generally implies a process which includes coagulation, flocculation, clarification, and disinfection. In treating low-turbidity waters and industrial process waters, the clarification step may be omitted and the coagulation and flocculation may occur within the conduit supplying flow to the filter.[10,11]

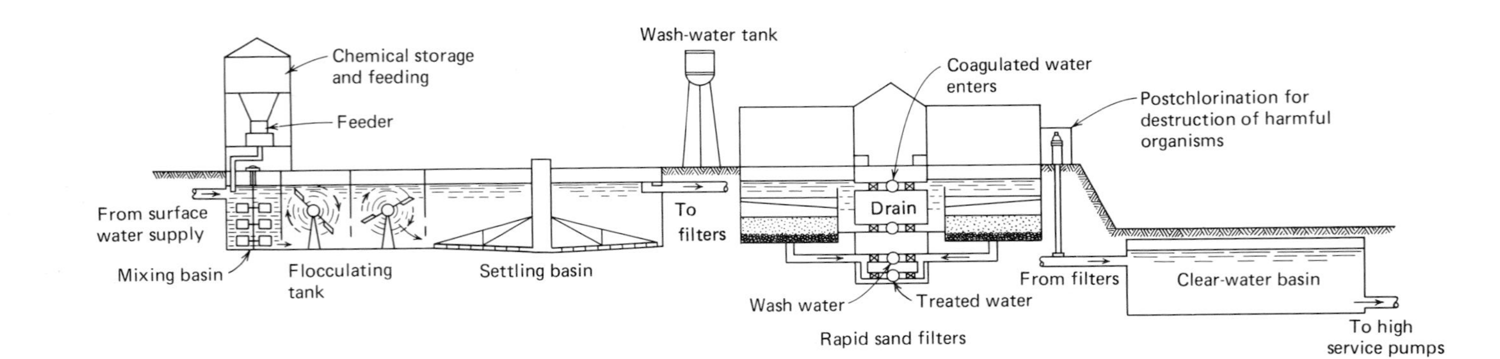

FIGURE 10-2
Cross section of a complete rapid filtration plant.

Figure 10-2 shows a cross section through a standard rapid filter plant containing separate coagulation, flocculation, and sedimentation units in advance of the rapid filters. As has been noted, many modern plants combine these processes in a single unit or, in the case of waters of low turbidity, eliminate them entirely. Figure 10-3 shows a cross section of a single rapid filter employing a rate-of-flow controller. As discussed below, rapid filters may employ a variety of flow-regulation techniques, media other than sand, and many different underdrain and backwash systems. The essential characteristics of a rapid filter system include careful pretreatment of the water before filtration, a high filtration rate—typically 5 to 10 m/h [2 to 4 gal/(min · ft^2)] or more, and cleaning of the bed by reverse flow of previously filtered water.

In order to assure satisfactory performance of rapid filters, the proper application of coagulants to destabilize colloidal suspensions is critical.[12,13] Additionally, as with slow filters, the initial period of filtration following cleaning normally produces water of inferior quality which may have to be wasted. The duration of the ripening period in rapid filters with properly pretreated

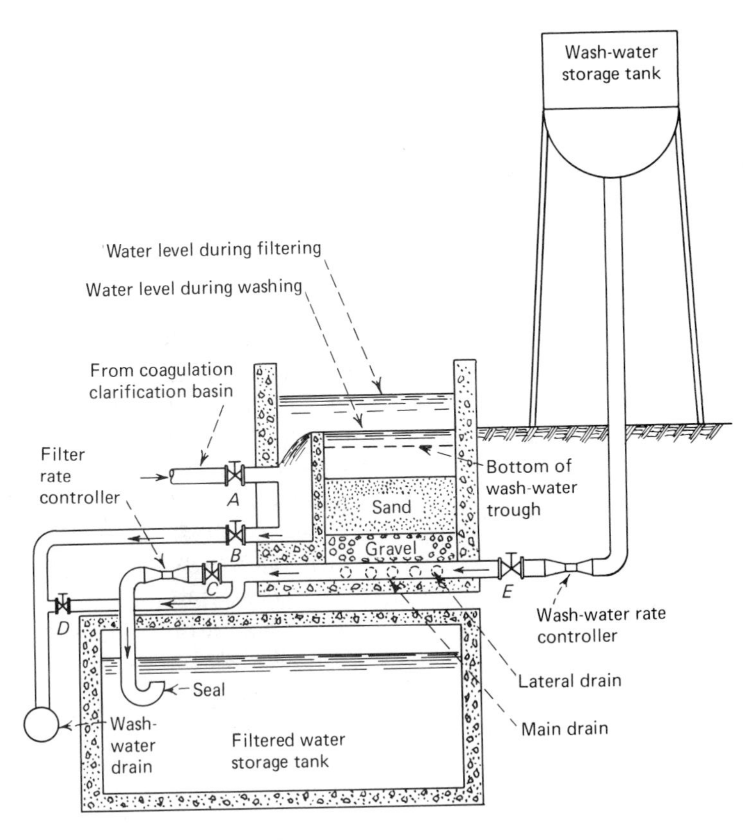

FIGURE 10-3
Schematic section of a rapid filter.

water is much shorter than in slow filters, ranging from 5 to 10 min or 1 to 2 empty bed detention times (EBDT).[14,15]

10-3 Theory of Filtration in Rapid Filters

A rapid filter contains a bed of a coarse medium, such as sand, ranging in depth from 300 mm (1 ft) to several meters. The kinetics of removal of particles smaller in size than the pore openings of the bed is extremely complex and very dependent on proper chemical conditioning. The removal itself has been described as consisting of a transport and an attachment process.[16,17] The *transport* to the surface of the filter medium may be produced by diffusion, interception, settling, impingement, or hydrodynamic carriage. The transport process is aided by flocculation in the interstices of the filter and by the short travel required for removal by sedimentation.

Attachment of the particles after their contact with the medium is chemical in nature and is influenced by pH, ionic composition of the water, age of the floc, nature and dosage of the coagulant, and the composition and surface condition of the medium.[17,18,19] Both the filter medium and the suspended particles in the influent may bear surface charges which can either aid or inhibit attachment. These surface charges are altered by changes in pH and by addition of coagulants.

Removal of particulate matter in a granular filter results from the transport and attachment processes described above. Removal tends to occur initially in the upper strata of the bed with particles penetrating deeper as the run progresses. As the interstices of the bed are filled, the superficial velocity of the water increases, resulting in resuspension of some particles, increased pressure loss in the bed, and creation of relatively large flow channels in the upper surface. If the head loss in the bed becomes too large, a partial vacuum may be created, resulting in the formation of bubbles from gases drawn from solution. This phenomenon, called *air binding*, further restricts the area of flow, increases the velocity and head loss, and may cause particles to be carried through the filter.

Large, strong floc particles will tend to be removed in the upper layers of the filter, producing high head loss with little penetration of the bed, particularly if the medium is fine. Smaller floc particles, particularly on coarse media, will penetrate farther into the bed, distributing the reduction in flow capacity and producing lower head loss for equivalent removal over equal time periods. Coarse-medium filter beds are normally deeper to provide for the greater penetration of the floc which is expected.

In order to achieve deep penetration without carrying solids through the filter, the floc must be small and relatively tough. This can be provided by using short flocculation periods and relatively high velocity gradients. The importance of proper pretreatment is difficult to overstate. The length of the filter run, the head loss developed, and even the question of whether the process will work as intended all depend on chemical conditioning.

10-4 Filter Media

The ideal filter medium should be of such a size that it will provide a satisfactory effluent, retain a maximum quantity of solids with minimum head loss, and be readily cleaned with a minimum quantity of water.[18]

The size and uniformity of filter media are specified by the *effective size* and the *uniformity coefficient*. The effective size is the sieve size in millimeters which permits 10 percent of the medium by weight to pass. The uniformity coefficient is the ratio between the sieve size which permits 60 percent by weight to pass and the effective size. Fine materials produce higher head loss but provide better protection against passage of small particles. Uniform materials permit deeper penetration of floc and better utilization of the storage capacity of the bed. Coarse materials require higher backwash velocities for fluidization (if this is produced by the water flow) but are less likely to form large agglomerates called *mudballs* during backwash.

Filter beds may be constructed of a single medium, two media of different characteristics, or a mixture of more than two media. These are called *monomedium*, *dual-media*, and *multimedia*, respectively.[20]

Sand is normally the cheapest filter medium. The sand used in rapid filters should be free from dirt, be hard and resistant to abrasion, and preferably be quartz or quartzite. It should not lose more than 5 percent by weight after 24 h immersion in 40 percent hydrochloric acid. In present practice, sand with an effective size of 0.45 to 0.55 mm and a uniformity coefficient from 1.2 to 1.7 is used in rapid filters.

Anthracite is a hard coal which has been used as a substitute for sand in monomedium filters and is often used as a component of dual-media and multimedia designs. Crushed anthracite used in filters has an effective size of 0.7 mm or more and a uniformity coefficient of less than 1.75.

Garnet sand and *ilmenite* are particularly dense materials [specific gravity (s.g.) $\approx$ 4.2] which may be used as components of multimedia filters. The relatively high cost and limited availability of these materials make their use as a monomedium impractical.

Other materials which may be locally available, such as crushed glass, slag, metallic ores, and even shredded coconut husks and burned rice husks,[21] have been used as filter media, as have activated carbon and pelletized metals.[22] The latter media have effects on water quality other than those associated with simple filtration processes. The materials of major engineering interest are limited to sand and anthracite.

Ordinary granular filters, upon being backwashed (Art. 10-6), will settle with the finest particles on top and the coarsest on the bottom. This gradation is unfavorable in that suspended particles will tend to be removed in the upper, finer strata, producing higher head loss than if they were more uniformly distributed through the bed. The gradation can be reversed, to a degree, by employing two or more media of different densities, so selected that the coarser particles will settle more slowly than the finer and will thus be on top after the fluidized bed reconsolidates following backwash. Mixed-media filters usually employ anthracite (s.g. ≈ 1.5), silica sand (s.g. ≈ 2.6), and garnet or ilmenite (s.g. ≈ 4.2). It can be shown[18] that, for equal settling velocities, the required particle sizes for media of different density can be calculated from

$$\frac{d_1}{d_2} = \left(\frac{\rho_2 - \rho_w}{\rho_1 - \rho_w}\right)^{2/3} \tag{10-1}$$

Example 10-1 Determine the particle sizes of anthracite and ilmenite which have settling velocities equal to that of sand with a diameter of 0.5 mm.

Solution
For the anthracite,

$$d_1 = (0.50)\left(\frac{2.6 - 1}{1.5 - 1}\right)^{2/3} = 1.1 \text{ mm}$$

For the ilmenite,

$$d_1 = (0.50)\left(\frac{2.6 - 1}{4.2 - 1}\right)^{2/3} = 0.3 \text{ mm}$$

Thus anthracite smaller than 1.1 mm would remain above the sand and grains of ilmenite larger than 0.3 mm would remain below it.

Since the individual media are unlikely to be completely uniform, there will be some degree of intermixing, with the coarser coal penetrating the finer sand and the coarser sand mixing with the finer ilmenite. Nevertheless, mixed-media filters are not true depth filters, but rather provide two or three filter surfaces with progressively smaller openings. This permits effective use of a larger portion of the volume of the filter, since coarse particles can be removed at the upper surface while the finer penetrate deeper into the bed. Filter runs are proportionately longer and head losses lower than those in monomedium filters. Since solids can be expected to accumulate on planes within the filter as well as on the upper surface, backwash systems should be selected to provide for removal of these accumulations. Scour systems which have proven satisfactory for multimedia filters are discussed in Art. 10-6.

10-5 The Underdrain System

The filter medium in rapid filters is underlain by a system which serves as a support, as a collector of filtered water, and as a distributor of backwash flow. This underdrain, depending on its design, may employ pipes and gravel or any of a large variety of proprietary systems.

Gravel, when used as a part of the collection and backwash distribution system, is normally placed in five or six layers, with the finest material on top. The upper layer is of such a size that its openings are smaller than the diameter of the filter medium. Succeeding layers are graded to support the layer immediately above. The finest material is typically 2.5 to 5 mm and the coarsest 40 to

60 mm. Individual layers may be 60 to 200 mm thick and the total depth of gravel, 400 to 600 mm.

In order to assure that the filtered water passes through the entire area of the bed, that backwash flows are evenly distributed, and that the filter bed is not disturbed by locally high velocities, the underdrain system must be designed to provide uniform pressure over the entire bottom of the filter and to prevent jetting of backwash flow. While it is not possible that the pressure be truly identical at all points, uniform collection and distribution can be reasonably assured by providing a system that has head losses through the orifices which are large relative to those in the main distribution and carriage system. Systems which provide such conditions include pipe laterals with nozzles or orifices, concrete teepee laterals, ceramic tile block, plastic block, and concrete plenums with nozzles, porcelain spheres, or porous plates. These systems may also employ a gravel layer to minimize jetting if the backwash enters the filter with a significant vertical velocity component.

An assortment of such systems is illustrated in Fig. 10-4.

10-6 The Backwash Process

Washing consists of fluidizing the filter media either with water, air, or a combination of the two so that the individual grains will be suspended, be subjected to abrasion by their contact with each other, and thus be cleansed of the material which has been accumulated during the filter run. Washing may be begun at a predetermined head loss through the filter, on a timed basis, or on evidence of increasing turbidity of the product water.

The process may consist of (1) simple fluidization with or without auxiliary scour, (2) surface wash, air scour, and partial fluidization, or (3) combinations of the two. The backwash velocity must be sufficiently great to carry out the suspended matter removed by the filter, yet not so great as to wash out the filter medium. As a practical matter, this means, for most waters, that the backwash rate must exceed 0.3 m/min [7 gal/(min · ft^2)] but be less than $V_t = 10D_{60}$ m/min for sand and less than $V_t = 4.7D_{60}$ m/min for anthracite, where D_{60} is the 60 percent size in millimeters.[23] The velocities given above are approximations for 20°C and specific gravities of 2.65 and 1.55, respectively, for sand and coal. The actual hydraulics of fluidization of granular filters is quite complicated and depends upon the densities of medium and fluid, viscosity of the fluid, the void ratio of the medium, and the sphericity of the particles.[24] Since both the density and viscosity of the water will vary with temperature, it is clear that some provision should be made for variations in expansion with season. The velocity which just begins to fluidize the bed may be calculated from[23]

$$V_b = V_t \times f^{4.5} \tag{10-2}$$

where V_b = minimum fluidization velocity
V_t = terminal velocity to wash medium from bed, as above
f = porosity of medium

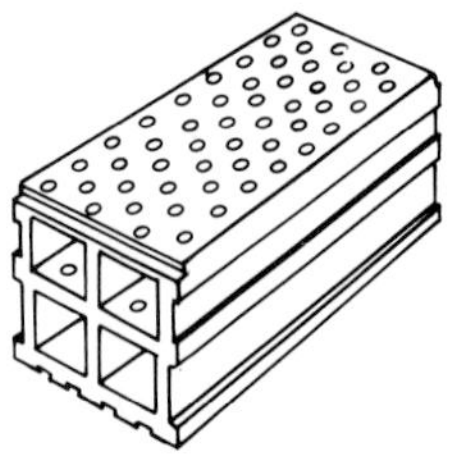

Tile underdrain

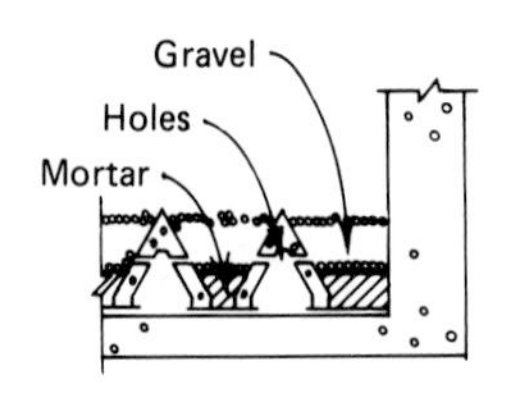

Teepee underdrain

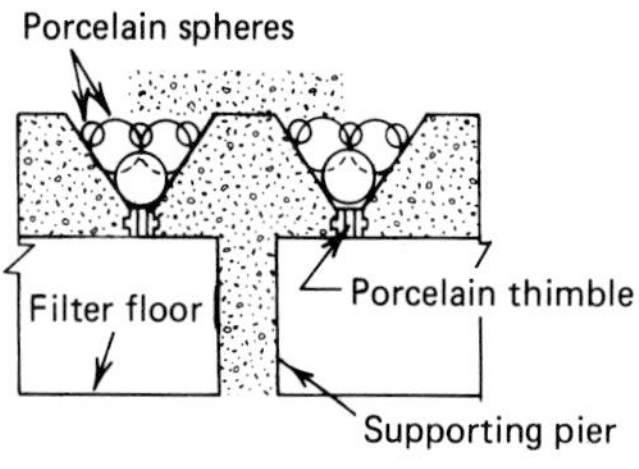

Wheeler underdrain

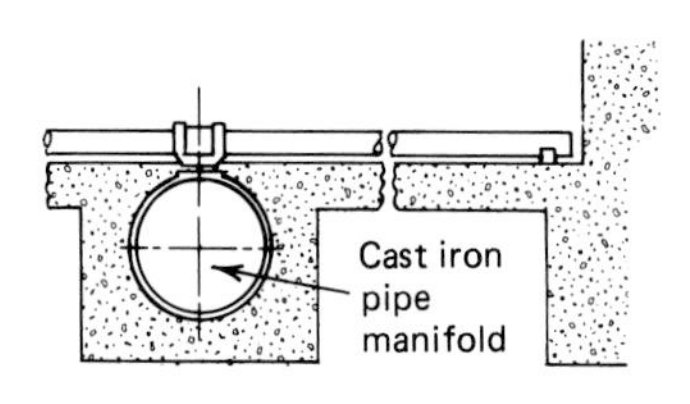

Pipe underdrain

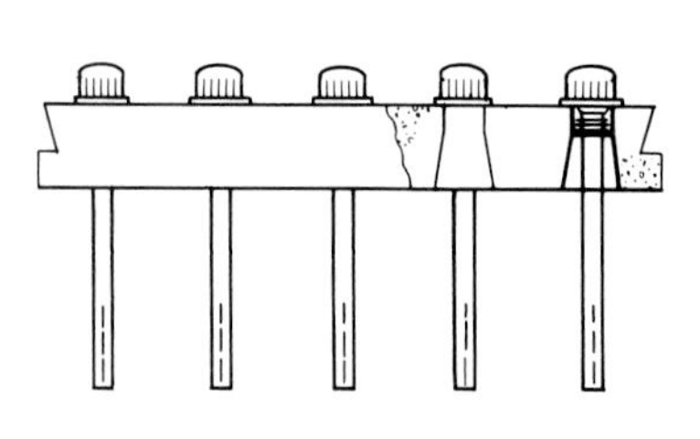

Infilco air/water nozzle

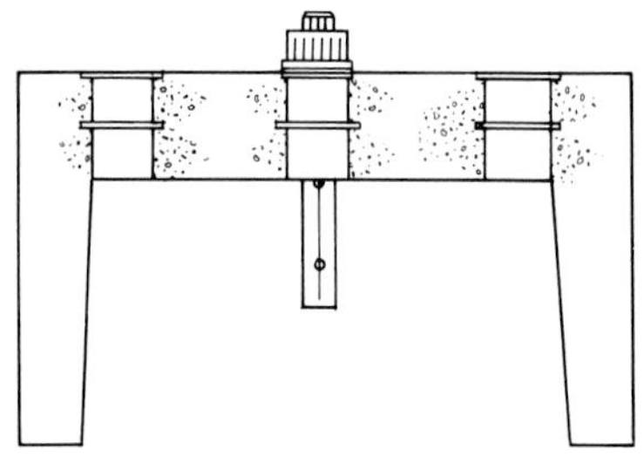

Camp air/water nozzle

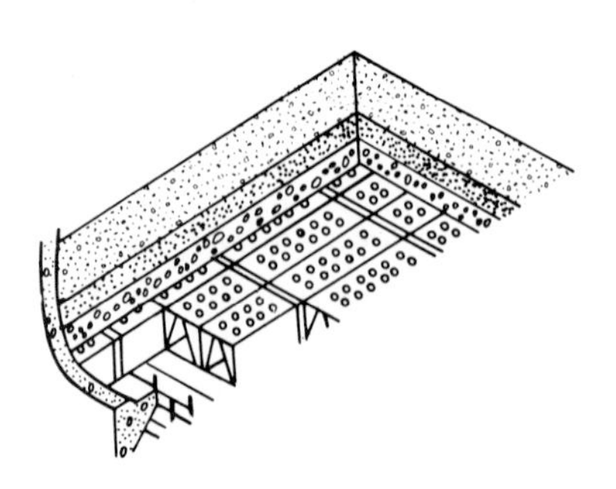

Fiber-glass underdrain

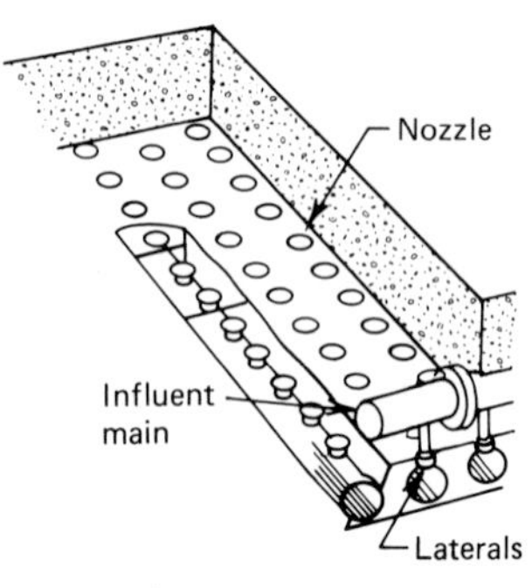

Air/water pipe underdrain

FIGURE 10-4
Typical filter underdrain designs.

Example 10-2 Calculate the terminal velocity and the minimum fluidization velocity of filter sand with an effective size of 0.55 mm, a uniformity coefficient of 1.5, a specific gravity of 2.65, and a porosity of 0.45.

Solution

$$V_t = 10(1.5 \times 0.55) = 8.25 \text{ m/min}$$

$$V_b = 8.25 \times 0.45^{4.5} = 0.23 \text{ m/min}$$

Thus a backwash velocity of 0.23 m/min [5.6 gal/(min · ft^2)] would fluidize but not expand the bed.

The cleansing of the medium is a result of shear produced by the water passing the suspended grains and of abrasion resulting from interparticle contacts. The maximum abrasion occurs when[23]

$$V_b = 0.1V_t \tag{10-3}$$

Thus, for sand

$$V_b = D_{60} \tag{10-4}$$

and for anthracite

$$V_b = 0.47D_{60} \tag{10-5}$$

in which V_b is in meters per minute and D_{60} is in millimeters. These rates are for a temperature of 20°C but can be corrected for other temperatures by

$$V_{b(T)} = V_{b(20)} \times \mu_T^{-1/3} \tag{10-6}$$

in which μ_T is the viscosity in centipoise at the temperature in question.

Example 10-3 Determine the appropriate backwash rate for a sand medium with an effective size of 0.5 mm and a uniformity coefficient of 1.5 at temperatures of 5 and 35°C.

Solution

$$D_{60} = 0.5 \times 1.5 = 0.75 \text{ mm}$$

$$V_b(20) = D_{60} = 0.75 \text{ m/min}$$

$$V_{b(5)} = 0.75 \times 1.52^{-1/3} = 0.65 \text{ m/min}$$

$$V_{b(35)} = 0.75 \times 0.71^{-1/3} = 0.84 \text{ m/min}$$

The design of a backwash system requires consideration not only of the characteristics of the medium but also the type and arrangement of the underdrains; the number, size, and location of the wash troughs above the filter; the head loss through the various parts of the system; the type of control system; and the type and capacity of the auxiliary scour or surface wash.

The *head loss* during backwash is the sum of the head lost in the expanded bed, the gravel (if present), the underdrain, and the pipe conveying the wash water to the filter. This may be written as

$$H = h_f + h_g + h_u + h_p \qquad (10\text{-}7)$$

The term h_f is the head lost in the expanded bed in meters of water and is equal to the weight of the medium in water:

$$h_f = L(1 - f)(\rho_s - \rho) \qquad (10\text{-}8)$$

where L = depth of unexpanded bed
f = porosity
ρ = density of water
ρ_s = density of medium

The term h_g is the head lost in the gravel in meters of water and may be approximated by

$$h_g = 0.03 L_g V_b \qquad (10\text{-}9)$$

where L_g is depth of gravel layer, m, and V_b is backwash velocity, m/min. The actual loss in the gravel is also influenced by additional factors which do not appear in Eq. (10-9), but the additional loss is relatively small and is very difficult if not impossible to calculate exactly.

The term h_u is the head lost in the underdrains and depends on the specific design. In general, the loss in the underdrain may be calculated from

$$h_u = \frac{1}{2g}\left(\frac{V_b}{\alpha\beta}\right)^2 \qquad (10\text{-}10)$$

in which α is an orifice coefficient and β is the ratio of the orifice area to the filter bed area (normally 0.2 to 0.7 percent).[23]

The head loss in the pipe supplying the backwash flow can be calculated by the techniques presented in Chap. 3.

Example 10-4 Determine the head loss during backwash for a filter bed consisting of 0.6 m of sand with $f = 0.45$, 0.4 m of gravel, and a pipe lateral underdrain with circular orifices having $\alpha = 0.8$ and $\beta = 0.005$. $V_b = 0.8$ m/min.

Solution

$$h_f = 0.6(1 - 0.45)(2.65 - 1) = 0.54 \text{ m}$$

$$h_g = 0.03(0.4)(0.8) = 0.01 \text{ m}$$

$$h_u = \frac{1}{2 \times 9.8}\left(\frac{0.8/60}{0.8(0.005)}\right)^2 = 0.57 \text{ m}$$

$$H = 0.54 + 0.01 + 0.57 = 1.12 \text{ m (3.7 ft)}$$

When backwash is provided from an elevated tank, the system should be designed so that the head loss is large compared to the change in head in the tank, otherwise an adjustable-rate-of-flow controller must be provided to ensure that the backwash rate will not change significantly. Backwash may also be provided by pumping from storage or directly from the distribution system. In the latter case, a pressure-reducing valve is required, since the pressure in the distribution system is normally far higher than that required for backwash.

The amount of water required for backwash depends on the design of the filter and the quality of the water being filtered. Filter runs on rapid filters may range from a few hours to several days. The washing cycle requires from 15 to 30 min to lower the water level in the bed, begin surface wash, expand and wash the medium until it is clean, allow it to reconsolidate, and return the filter to service. The actual washing lasts 5 to 10 min. For typical systems, backwash use amounts to 1 to 5 percent of the flow produced.

Surface wash or *air wash* should be incorporated in all rapid-filter designs to ensure thorough cleaning of the media and prevent the formation of mudballs (Art. 10-9). Surface wash is provided by directing jets of water downward against the surface of the filter before and during the first few moments of backwash. The jets may be mounted on a fixed network of pipe above the bed or on rotating arms which sweep the entire filter (Fig. 10-5). The fixed nozzles apply water at a rate of 0.2 m/min [5 gal/(min · ft^2)] to the entire bed and the surface wash is normally continued while the bed is expanded. The rotating design of Fig. 10-5 uses water at the rate of 0.02 to 0.04 m/min [0.5 to 1 gal/(min · ft^2)], based on the entire bed area. The surface wash begins 1 min before the backwash and continues during the expansion of the bed. The high-pressure jets in this system coupled with the stirring provided by its rotary motion provide good scour of surface deposits which might otherwise contribute to the formation of mudballs. Multimedia filters which provide more than one filtering surface must include surface wash for the internal surfaces as well. The devices cannot rotate until the bed has been fluidized, hence backwash of multimedia filters with surface wash requires fluidization without expansion, application of surface wash, and subsequent increase of backwash rate to provide expansion

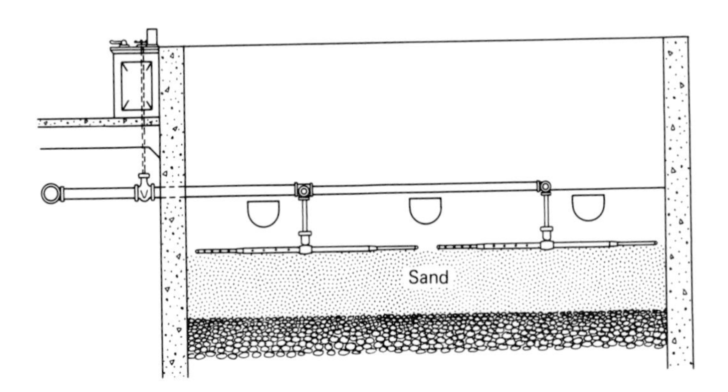

FIGURE 10-5
Rotating-arm surface wash system.

and sweep out the accumulated material which has been scoured from the media.

Air wash systems can be employed with plenum or pipe-type underdrains using nozzles. Air is introduced into the underdrain at rates ranging from 0.3 to 1.5 m/min (commonly 0.9 to 1.2 m/min), based on the filter plan area. After a few minutes of air agitation, water is added to the underdrain at rates ranging from 0.3 to 0.5 m/min [7 to 12 gal/(min · ft^2)]. The air is left on in some designs until the rising flow reaches the effluent structure or troughs and, in others, throughout the washing process. The air wash fluidizes the bed without expanding it, and the violent rolling action provided by the air scours deposits both on the upper surface and internal surfaces such as those in multimedia filters. Air wash is widely used in Europe and has become increasingly popular in the United States as engineers have become familiar with its benefits.[20] Air velocities in the underdrains are generally less than 50 m/s (150 ft/s), while water velocity is in the range of 0.3 to 0.6 m/s (2 to 3 ft/s).

During backwash, the rising water and the accumulated contaminants scoured from the filter must be removed from the top of the bed. In Europe, the common practice is to direct the backwash flow to a gullet at one side of the filter box. The elevation of the overflow structure is set so that it is above the highest level of the expanded bed. In the United States, it has been more common to use troughs placed above the bed, usually at a spacing less than 2 m (7 ft). The height of the lip of the trough above the fluidized bed should be approximately one-third of the center-to-center spacing in order to prevent washing of the filter medium from the bed.[25] The side weirs used in Europe appear to provide better retention of the filter medium but limit the width of the filter box. Designs with widths up to 4.5 m (15 ft) are reported to be satisfactory.[20] The hydraulic design of gullets and effluent troughs is treated in Chap. 3.

10-7 Filter Controls

The original rapid-filter designs incorporated rate-of-flow controllers to maintain constant filtration rates despite variations in head loss within the filter during a run. Such controllers, while still available, are not the most desirable method of regulating filter flow rates. It is generally accepted[23,26] that declining-rate filters produce better quality and larger volumes of water between backwash cycles, although some investigators have not found these differences to occur[27] and not all state regulatory agencies have accepted the advantages of such designs.[28]

The designer must recognize that filter systems are not likely to be operated at a constant rate from the beginning to the end of a filter run. Municipal water consumption rates vary widely from the average, and, while storage at the filter plant and in the system helps to equalize production rates, the available storage would have to be very large to equalize demand over a filter run which might last for 2 days.[29] It is very undesirable to subject filters to sudden changes in flow rate, since this may dislodge material and carry it through the bed,[12] hence the control system should be designed to prevent such variations even when changes in production rate occur.

A variety of classifications of filter control systems have been offered. In this text, controls are divided into constant or variable head and constant or variable head loss systems.

Constant head loss, constant head systems employ weirs at the filter entrances which provide equal flow to each unit. A level sensor in each filter sends a signal to a modulating valve which opens or closes to maintain a constant head. As the head loss in the filter medium increases, the valve opens, maintaining a constant net head loss (medium plus valve) across the system. When one filter in such an array is taken out of service for backwashing, the flow will automatically redistribute to the other units and the control valve will open to maintain a constant head at the increased flow. The same response—a rapid increase in flow rate through the filter—occurs when the operator alters the plant flow rate.

Variable head loss, variable head—uniform flow systems split the incoming flow uniformly among the filters in operation in the manner described above. The water level at the beginning of the filter run is just above the top of the medium—held at that level either by an effluent weir or a restriction in the effluent piping. As the filter clogs, the level of water in the individual filter boxes will increase to provide the additional head required. Changes in flow upon individual units produced by backwashing or variations in production will cause changes in depth accompanied by gradual changes in velocity through the filter.

In an alternative variable head loss, variable head—uniform flow design, the elevation in all the filters of a group and in their common influent structure is at the same level and changes as the filters become clogged. The flow through the individual units is held constant by a rate-of-flow controller which measures the effluent flow, compares it to a preset value, and adjusts a modulating valve to maintain a constant rate. Changing the rate on the filters during backwash or to accommodate changes in production requires resetting the control valves. Changing the valve setting produces an immediate increase in flow rate, and control systems of this type sometimes "hunt," which leads to surges of flow.

Variable head loss, variable head—declining flow systems provide a common head on all filters and in their common supply structure. The minimum head is just above the filter medium, normally held at that level by an effluent weir structure. The flow through the individual units is not regulated. Since all operate at the same total head loss, the cleanest filter will carry the highest flow. As the filters become plugged, the level in the system will increase to maintain a constant flow. When the flow is altered to accommodate backwash or variation in production, the level will gradually change, producing a gradual change in flow.

An alternative method of operating declining-rate filters involves reducing the water level in a common effluent structure as the filters become plugged. In such designs, the influent level is constant. As the filters plug, the flow will tend to decrease. If the pumping rate from the effluent structure is constant, its level will decrease, increasing the net head on the filters and gradually increasing the flow. The same phenomenon will occur during changes in production rate.

A common modification in the declining-flow design incorporates an orifice in either the influent or effluent line of each unit to limit the flow rate immediately after backwash.

The various filter control methods have as their common intent the minimization of surges and high filtration velocities which might carry contaminants through the bed. Designs which require a change in water elevation in order to produce a change in flow are best suited to this common purpose, since changing the depth requires measurable time and changes in velocity must thus occur slowly. Declining-rate filters appear to be more logical in that the highest rates are applied to clean filters, making breakthrough less likely.

A modification of the declining-rate filter which employs a common underdrain plenum and a common outlet structure is illustrated in Fig. 10-6. In this system, the individual units are so proportioned that each unit may be backwashed by the flow through the others. During backwash the flow into the system is maintained, but the common outlet is closed. At the same time, the inlet to the unit to be cleaned is closed and its drain is opened. The incoming water will enter those filters still in service (gradually increasing the depth), will pass through them to the underdrain, and will then flow upward through the filter being cleaned. The backwash flow increases slowly, thus preventing possible upsets of the bed. Such designs require minimum mechanization and control but have certain drawbacks as well, in that the common outlet prevents location of problems in a single unit and the entire battery of filters must be operated in order to provide adequate wash water.

Filter performance may be monitored by a variety of techniques, not all of which are required in all designs. *Head loss* through the media can be measured by simple pressure gauges and will give the operator a clear indication of the need for backwash. Head loss is commonly limited to prevent the production of negative head and air binding (Art. 10-9). In some designs, of course, the head loss is indicated by the level of the water above the medium. Additional pressure gauges within the filter will permit the operator to determine *where* the head loss is occurring and thus optimize the pretreatment to distribute the head loss through the bed and achieve longer runs.

Turbidity sensors should be included in all filter designs and are commonly connected to an alarm system which is triggered by increases beyond 1 ntu. Turbidity is a very important indicator of the biological quality of a water and has been shown to be closely correlated with the presence of both bacteria and *Giardia*.[30] Ideally, each unit should have its own turbidity sensor, and in some designs backwash may be initiated automatically at a preset effluent turbidity.

Flow meters are not required on individual filters but should be provided for the total influent and effluent flow. Measurement of backwash flow is

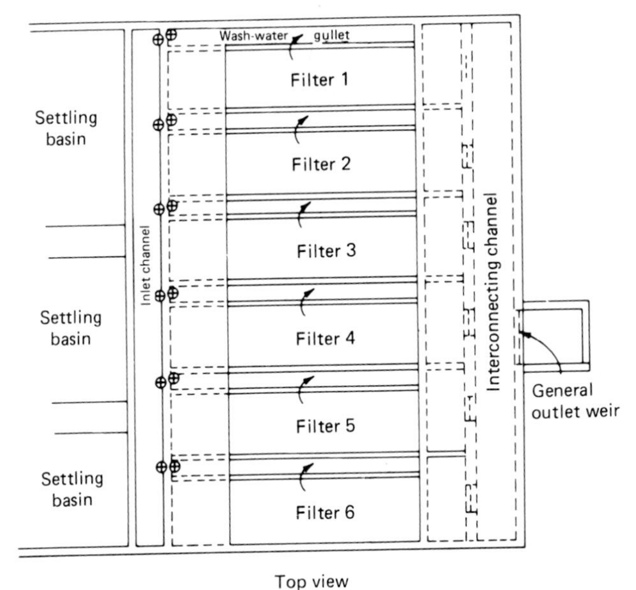

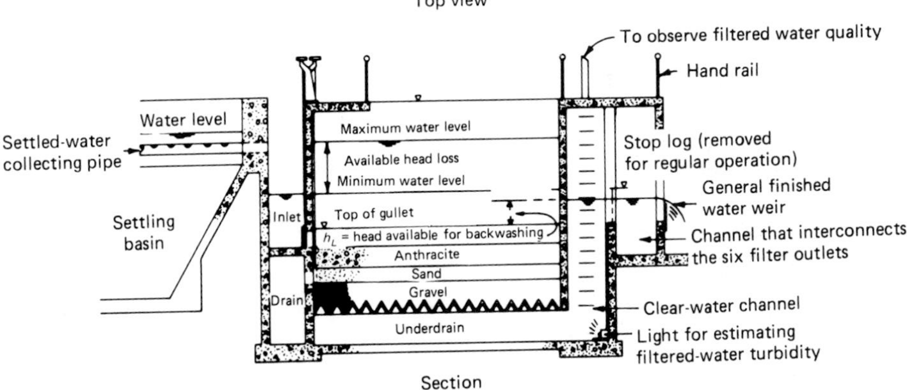

FIGURE 10-6
Declining-rate filter with common underdrain. (*Reprinted from Journal of American Water Works Association, **66**, by permission of the Association. Copyright 1974 by the American Water Works Association, Inc., 6666 W. Quincy Avenue, Denver, CO 80235.*)

desirable unless the system is so designed that the flow is constant. When air scour systems are used, the air flow should be measured unless it is provided by a positive displacement compressor which is preset at the desired rate.

10-8 Piping in Filtration Plants

Pipe galleries are unnecessary in some declining-rate and variable-head filter designs. In those cases where individual filter controls and pumped or elevated backwash are required, a pipe gallery provides a common location for pipes, controls, valves, and other fittings. Galleries should be uncrowded and be provided with good lighting, ventilation, and drainage. Some leakage and condensation will occur even in the best of circumstances, and poorly designed pipe galleries can be most unpleasant places to work.

Air-release valves should be provided at high points in the piping, which is usually iron with either flanged or victaulic joints. Drain lines carrying the effluent backwash water may be of concrete or clay and are usually encased in concrete. Influent lines to the filters are sometimes open channels which provide a uniform head to influent weirs or to the filters themselves.

Velocities in influent lines should be low, 0.3 to 0.6 m/s (2 to 3 ft/s), to prevent breakup of floc and to minimize head differences between individual filters. Velocity in the backwash influent line is usually about 3 m/s (10 ft/s), since the flow is great and the line would otherwise be quite large. The backwash waste line, for similar reasons, is designed to have a velocity of up to 3.5 m/s (12 ft/s). The treated effluent line to the clear well is designed for velocities in the range of 1.2 to 1.5 m/s (4 to 5 ft/s). Water which is filtered to waste at the beginning of a filter run (Art. 10-9) is conveyed at velocities similar to those in the backwash waste line. In some cases, low available head may require lower velocities and larger pipe sizes.

Valves in filter-pipe galleries may be operated pneumatically, electrically, hydraulically, or manually. Manually operated valves are usually provided with long stems which pass to stands on the operating floor. Filter plants use both gate and butterfly valves. The latter are more commonly used in new construction, particularly in larger sizes.

10-9 Operating Difficulties

The turbidity of the effluent from rapid filters at the beginning of a run, like that from slow sand filters, is usually higher than that obtained after the filter has been in service for some time. This initially poor quality is attributed to the dilution of the incoming coagulated and flocculated water by the backwash water left in the filter, which contains no coagulant. Such dilution can cause restabilization of the colloidal material and permit its passage through the filter.

The most common method of handling the initial poor quality is to filter to waste, that is, discharge the filtrate to a sewer for the first period of operation. The duration of such wasting is typically 5 to 10 min.[14,15]

As an alternative to wasting the initial flow, one may operate the filter at a lower rate. This procedure, called *slow-start*, was developed in Great Britain. The time required before full flow is achieved depends upon the characteristics of the conditioned water. As an approximation, one should expect that 2 to 3 empty bed volumes will have to pass through the filter before the full flow rate is permissible.

A third option, somewhat more complicated, involves the addition of polymer or other coagulant to the backwash flow during the last few moments of the process.[31] The volume of water treated should be at least equal to that left in the bed, or about 1 filter volume. With proper application of this technique, the initial quality of the filtrate can be greatly improved and no wasting of flow or delay in reaching full rate is required.

As has been noted in the discussion above, there will be a gradual increase in head loss through the filter medium as it becomes plugged. If the loss in the upper layers exceeds the head above the surface, the medium below can act as a draft tube and produce a partial vacuum in the bed. Whether this is physically possible depends upon the details of the design; filters with effluent weirs set above the upper level of the sand, for example, cannot develop *negative head*. If negative head is produced, gases can be drawn from solution and form bubbles within the bed. In addition to interfering with the flow during filtration, such accumulations of air may be released violently when backwash is initiated, causing locally high water velocities and possible displacement of the gravel underlying the media. *Air binding* can best be controlled by adjustment of pretreatment to permit deeper penetration of floc into the filter. This will distribute the head loss throughout the depth and provide higher net pressure within the bed.

Mud and the products of coagulation may accumulate on the filter surface, forming a dense mat. During filtration and as washing begins, there is lateral pressure within the bed which tends to compact these deposits and, as the filter begins to expand, forms them into lumps or balls. Depending on the density of the mudballs, they may either sink within the bed or remain at or near the surface. As their size increases in subsequent filter cycles, the balls will tend to settle to the gravel-sand interface where they interfere with the backwash flow. The diverted backwash flow can displace the gravel, permitting sand to enter the underdrain. Additionally, the sand above the mudball accumulations is not properly cleaned and the problem is thus aggravated. Mudball formation can be controlled most easily by provision of proper surface wash or air scour equipment. Multimedia filters can form mudballs on interior planes, hence such filters must include cleaning equipment for these surfaces as well.

Sand incrustation may result in water softening plants (Chap. 11) in which inadequate stabilization of the treated water is provided. If water which is supersaturated with calcium carbonate enters the filters, the mineral may come out of solution, coating individual grains and cementing them together. Prevention involves pH adjustment with acid or addition of polyphosphates to complex with the calcium and keep it in solution.

Filters following iron and manganese oxidation-precipitation processes (Chap. 11) may serve as a niche for the growth of autotrophic bacteria which obtain their energy by reducing the metals in the influent. The reduction of the metals results in their redissolution and a subsequent deterioration in water quality. The problem can generally be controlled by more frequent backwashing and maintenance of a chlorine residual in the filter.

10-10 The Clear Well and Plant Capacity

The maximum capacity of a filter plant depends on the water consumption rate and the storage available for treated water. The greater the storage, the closer the capacity may be to the annual average consumption. The striking of the optimum balance between storage and treatment capacity is a problem in engineering economy, but, in any event, some storage will be required at the plant to permit operation without frequent changes in rate. Storage at most plants is in the range of one-quarter to one-third of the daily capacity. The clear well may be below the filters or may be a separate structure. The stored water must be covered and protected against the possibility of chance or deliberate contamination.

Selecting the number and the size of individual filter units also involves balancing cost and ease of operation. A minimum of three filters is required to assure continued production while one unit is out of operation for repair and a second is being backwashed. A large number of filters provides more flexibility to the operator, but requires greater expense for controls and other appurtenances. Usually all filters will be of the same size and will be constructed in two parallel rows on either side of a common pipe gallery. The depth of the filter box depends on the media and the control system and is typically 3 m (10 ft) or more. Interior filter walls are usually roughened to prevent streaming of water between the walls and the sand.

It is generally not desirable to build much excess filter capacity, since expansion is comparatively easy. At some plants, extra filter boxes may be constructed without installing underdrains, piping, media, and other appurtenances until the extra capacity is required.

10-11 Other Filtration Processes

A number of media which provide effects on water quality not offered by sand and anthracite have been employed in filtration processes.

Granular *activated carbon* has been shown to be an effective filtration medium insofar as removal of turbidity is concerned.[32,33] Whether the combination of filtration and removal of organic contaminants in a single process is desirable is, of course, a different question. Carbon systems and their operational problems are treated in Chap. 11.

With respect to the use of activated carbon in filters, it is necessary to note that the carbon will remove chlorine from the influent and this, coupled with its very large surface area and the availability of adsorbed organic material, is likely to enhance the growth of bacteria and other microorganisms within the bed. Activated carbon usually has a larger uniformity coefficient than standard filter materials and is less dense, but its density increases with adsorption of organics. Most carbons are more subject to particle size reduction by abrasion than either anthracite or sand. These factors somewhat complicate the use of carbon as a filter medium and require that adjustments be made in anticipated filter run duration and backwash rate.

A mixture of *coal and metallic aluminum* has been used as a filter medium[22] without a previous coagulation step or the addition of any coagulant aids. The system is most applicable to waters of low alkalinity, since adjustment of the pH to levels less than 6 is required. The intimate mixture of dissimilar materials creates a galvanic cell and galvanic corrosion. The galvanic corrosion produces electrical fields within the bed which destabilize colloidal particles. Aluminum is dissolved, but does not appear to be precipitated at the low pH and low alkalinity levels involved. Filter run lengths are comparable to those with other media (up to 20 h), as are filtration rates (up to 20 m/h).

In-line filtration processes which employ either single- or two-stage filtration without prior treatment other than addition of a coagulant in the influent line have been used successfully on waters of low turbidity.[10,15,34,35] These processes offer an alternative to slow sand filtration for waters of low turbidity which are contaminated by *Giardia*. The location of chemical addition is critical to the proper operation of these systems, and coagulant aids such as clay may be required. The choice of an appropriate treatment process for such waters appears to depend on the size of the community, since small communities may not have employees capable of operating sophisticated systems requiring careful control and large communities may not have sufficient land for slow sand filters.

Pressure filters are rapid filters contained in a pressure vessel as shown in Fig. 10-7. The filter cross section and medium are similar to those in standard rapid filters. Chemical dosages are generally applied in-line as shown in the figure. The application of pressure filters is primarily in removal of precipitated iron and manganese, in treating industrial process waters, and in filtration of recirculated swimming-pool water. Backwash is commonly effected by directing the influent to the bottom of the filter and wasting the flow from the top of the bed. When the backwash is done with untreated water, such units are not used for production of potable water, although with proper provision for backwash with treated water, there is no reason why safe water could not be provided. The advantage of pressure filters over gravity units is that more pressure is available to drive the water through the filter and runs are proportionately longer.

Diatomaceous earth filters were developed by the army for water treatment under combat conditions. Since the units are small, they have also been applied to swimming-pool use and have been shown to be effective in removing *Giardia*.[36]

The filter consists of either rigid elements which are covered by closely wound wire or fabric or cylinders of fused alumina or ceramic. A precoat of diatomaceous earth amounting to 0.3 kg/m^2 of filter area is applied by placing it in the first water applied after backwash. Water is wasted until the filter element is coated. The diatomaceous earth is added continuously in a slurry to the raw water to form what is called a body coat and which gradually increases the thickness of the layer on the filter. The body coat is added at a rate of 2 to 3 mg/L per turbidity unit in the raw water.

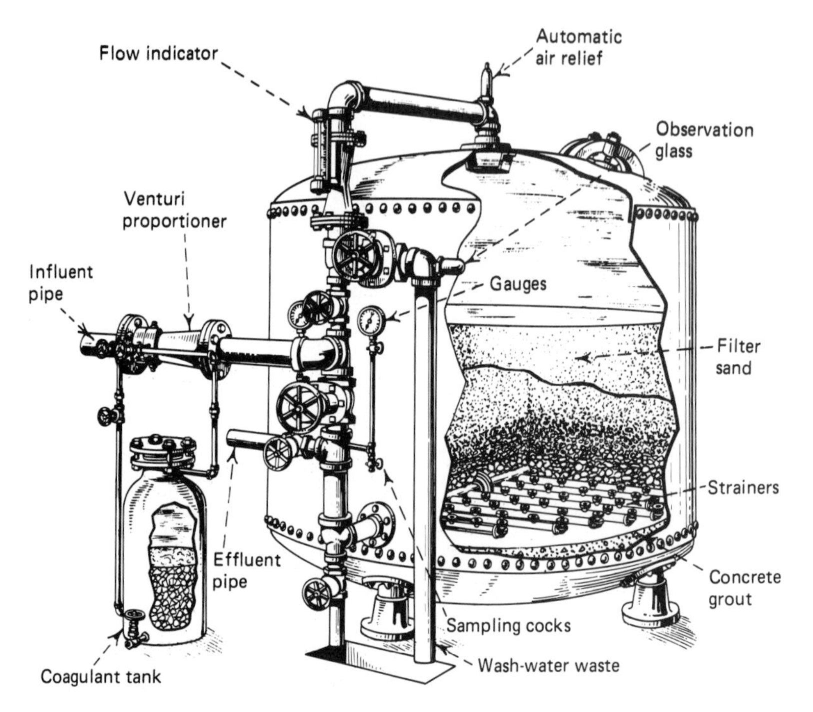

FIGURE 10-7
Pressure filter. (*Courtesy Infilco-Degremont, Inc.*)

The filters are backwashed when the head loss is about 70 kPa (10 lb/in^2). Backwash is effected by reversing the flow and passing treated water through the screen. Some manufacturers provide an "air bump" to help dislodge the accumulated material. Runs may extend for as much as 2 days, depending on the water quality and the operator's skill. Filtration rates are typically 2.5 m/h [1 gal/(min · ft^2)].

Diatomite filters are not suitable for waters of high turbidity and are not permitted at all for potable water production in some states. Where their use is permitted, they are normally limited to waters with turbidities less than 30 ntu. The reluctance of state officials to approve these systems results from the lack of protection against breaks in the thin filter layer and the potential for mixing contaminated and product water during and after backwash.

Some industrial diatomite filters utilize vertical plate structures with water applied from either side. As the filter plugs, the flow is reversed, washing the filter and then filtering in the direction of the wash until the head loss increases again. The backwash water may be collected and the diatomite recovered in designs of this type.

In *upflow filters*, the direction of flow is the same as that of the backwash, hence the water is filtered by progressively finer layers as it passes through the bed. Such a design permits maximum utilization of the storage capacity of the bed and thus longer filter runs. In order to prevent undesired expansion of the bed at filtration rates greater than the minimum fluidization velocity of the medium, the bed is restrained by a grid which produces arching in the fine grains. Expansion is provided during backwash by breaking the arches with air. Once the arches are broken, the bed will expand as though the grid were absent.

Upflow filters have been loaded at rates equaling or exceeding the maxima used on standard filters and have the potential of producing potable waters, although their use in the United States is limited to industrial water treatment. They are generally used with in-line coagulation. Flocculation, sedimentation, and filtration all occur within the filter medium.

A *biflow filter* developed in the Soviet Union employs a modification of the upflow principle in which the expansion of the bed is restrained by directing a portion of the flow downward. Effluent is collected through a screened pipe a few centimeters below the upper surface of the sand. Approximately 80 percent of the flow is directed upward and 20 percent downward in this system.

PROBLEMS

10-1. Design a slow sand filter system for a community which has a projected population of 3000 and an average per capita usage of 300 L/day. Use the Goodrich formula of Chap. 2 to estimate the peak flow rates, and size the system so it can produce water at the maximum weekly rate with one unit out of service. The design filtration rate is 6 m/day.

10-2. A rapid filter plant is to produce a maximum flow of 23,000 m^3/day. The nominal filtration rate is 120 m/day and is not to exceed 180 m/day with one filter being backwashed nor 240 m/day with one filter being backwashed and one filter out of service. Determine the number of units required and the individual filter area. How much water is required to wash one filter if the backwash rate is 1 m/min and the duration is 10 min?

10-3. Calculate the backwash velocity necessary to expand a sand filter with an effective size of 0.6 mm and a uniformity coefficient of 1.3. This medium (or any other, for that matter) can be cleaned at a backwash velocity of 0.3 m/min if fluidization is provided by air scour. Calculate the volume of water saved during a 10-min backwash of a 2 by 4 m filter if air scour is used. If the cost of water production is \$0.25/m^3, how much money would be saved per year by air fluidization if the filter runs average 2 days in length?

10-4. A filter system similar to that of Fig. 10-6 contains four filter units operated at a nominal rate of 180 m/day. During backwash the total flow is unchanged, but the rate on the units in service increases with all the flow being used to backwash the fourth. Assuming a temperature of 20°C and a uniformity coefficient of 1.7, determine the maximum effective size which can be used if the medium is (*a*) sand and (*b*) anthracite. Which medium would you select? Why?

10-5. If the filter system illustrated in Fig 10-6 contains sand with an effective size of 0.6 mm and a uniformity coefficient of 1.2, what is the minimum nominal filtration rate with all units in operation if one cell is to be backwashed at a time with no reduction in influent flow rate and all flow used for washing? Is this filtration rate within the normal range?

10-6. Rework Prob. 10-5 assuming the medium is anthracite with an effective size of 0.95 mm and a uniformity coefficient of 1.4.

10-7. Determine the variation in backwash velocity required to expand an anthracite medium with an effective size of 1 mm and a uniformity coefficient of 1.3 if the water temperature varies from 3 to 30°C.

10-8. An anthracite medium with the size given in Prob. 10-7 and a depth of 1 m is supported on a proprietary nozzle underdrain. The nozzles are installed at a density of 11 per square meter (about 1 per square foot). Each nozzle is reported to have a head loss of 0.3 m of water at a flow rate of 30 L/min, and the head loss varies as the square of the flow. Determine the head loss through the nozzles and the medium when the medium is expanded during backwash. $f = 0.4$.

10-9. A dual-media filter contains anthracite with an effective size of 1.2 mm and a uniformity coefficient of 1.5. What must be the size of the underlying sand layer if the anthracite is to remain on top?

10-10. In Prob. 10-9, what backwash velocity is necessary to expand the anthracite? Is this also suitable for expansion of the sand?

10-11. A rapid filter plant contains 10 units, each 5 m long and 2.5 m wide. The filter medium is sand with an effective size of 0.55 mm, a uniformity coefficient of 1.4, and a depth of 1 m. The filters are each washed once per 24-h period. The washing process involves 10 min of fluidization and requires that the filter be out of service 30 min. The treated flow is wasted for the first 30 min of each run. Determine the volume of treated water produced per day at a nominal filtration rate of 180 m/day and the percentage of the product water required for backwash.

REFERENCES

1. Lloyd A. Slezak and Ronald C. Sims, "The Application and Effectiveness of Slow Sand Filtration in the United States," *Journal of American Water Works Association*, **76**:12:38, 1984.
2. Gerardo Galvis and Jan Teun Visscher, "Filtración Lenta en Arena y Pretratamiento," *Seminario Internacional sobre Tecnología Simplificada para Potabilización de Agua*, Cali, Colombia, 1987.
3. Daniel R. McNair et al., "Schmutzdecke Characterization of Clinoptilite-Amended Slow Sand Filtration" *Journal of American Water Works Association*, **79**:12:74, 1987.
4. Timothy J. Seelaus, David W. Hendricks, and Brian A. Janones, "Design and Operation of a Slow Sand Filter," *Journal of American Water Works Association*, **78**:12:35, 1986.
5. Thomas R. Cullen and Raymond D. Letterman, "The Effect of Slow Sand Filter Maintenance on Water Quality," *Journal of American Water Works Association*, **77**:12:48, 1985.
6. Kim R. Fox et al., "Pilot-Plant Studies of Slow-Rate Filtration," *Journal of American Water Works Association*, **76**:12:62, 1984.
7. John L. Cleasby, David J. Hilmoe, and Constantine J. Dimitracopoulos, "Slow Sand and Direct In-Line Filtration of a Surface Water," *Journal of American Water Works Association*, **76**:12:44, 1984.
8. William D. Bellamy, David W. Hendricks, and Gary S. Logsdon, "Slow Sand Filtration: Influences of Selected Process Variables," *Journal of American Water Works Association*, **77**:12:62, 1985.

9. William D. Bellamy et al., "Removing *Giardia* Cysts With Slow Sand Filtration," *Journal of American Water Works Association*, **77**:2:52, 1985.
10. Keith Craig, "Direct Filtration: An Australian Study," *Journal of American Water Works Association*, **77**:12:56, 1985.
11. John R. Bratby, "Optimizing Direct Filtration in Brasília," *Journal of American Water Works Association*, **78**:7:106, 1986.
12. Appiah Amirtharajah, "Some Theoretical and Conceptual Views of Filtration," *Journal of American Water Works Association*, **80**:12:36, 1988.
13. John E. Tobiason and Charles R. O'Melia, "Physicochemical Aspects of Particle Removal in Depth Filtration," *Journal of American Water Works Association*, **80**:12:54, 1988.
14. Ron R. Mosher and David W. Hendricks, "Rapid Rate Filtration of Low Turbidity Water Using Field-Scale Pilot Filters," *Journal of American Water Works Association*, **78**:12:42, 1986.
15. James B. Horn et al., "Removing *Giardia* Cysts and Other Particles from Low Turbidity Waters Using Dual-Stage Filtration," *Journal of American Water Works Association*, **80**:2:68, 1988.
16. Charles R. O'Melia and W. Stumm, "Theory of Water Filtration," *Journal of American Water Works Association*, **59**:1393, 1967.
17. P. Logonathan and W. J. Maier, "Some Surface Chemical Aspects in Turbidity Removal by Sand Filtration," *Journal of American Water Works Association*, **67**:336, 1975.
18. Susuma Kawamura, "Design and Operation of High Rate Filters—Part 1," *Journal of American Water Works Association*, **67**:535, 1975.
19. Russell H. Boyd and Mriganka M. Ghosh, "An Investigation of the Influences of Some Physicochemical Variables on Porous Media Filtration," *Journal of American Water Works Association*, **66**:94, 1974.
20. Robert D. G. Monk, "Design Options for Water Filtration," *Journal of American Water Works Association*, **79**:9:93, 1987.
21. Richard J. Frankel, "Series Filtration Using Local Media," *Journal of American Water Works Association*, **66**:124, 1974.
22. Anthony G. Collins and Robert L. Johnson, "Reduction of Turbidity by a Coal-Aluminum Filter," *Journal of American Water Works Association*, **77**:6:88, 1985.
23. Susuma Kawamura, "Design and Operation of High Rate Filters—Part 2," *Journal of American Water Works Association*, **67**:653, 1975.
24. A. H. Dharmarajah and John L. Cleasby, "Predicting the Expansion Behavior of Filter Media," *Journal of American Water Works Association*, **78**:2:66, 1986.
25. Jonathan A. French, "Flow Approaching Filter Washwater Troughs," *Journal of Environmental Engineering Division, American Society of Civil Engineers*, **107**:EE2:359, 1981.
26. Jorge Arboleda, "Hydraulic Control Systems of Constant and Declining Flow Rate in Filtration," *Journal of American Water Works Association*, **66**:87, 1974.
27. David J. Hilmoe and John L. Cleasby, "Comparing Constant-Rate and Declining-Rate Filtration of a Surface Water," *Journal of American Water Works Association*, **78**:12:26, 1986.
28. David A. Cornwell, Mark M. Bishop, and Howard J. Dunn, "Declining-Rate Filters: Regulatory Aspects and Operating Results," *Journal of American Water Works Association*, **76**:12:55, 1984.
29. Committee Report, "Comparison of Alternative Systems for Controlling Flow through Filters," *Journal of American Water Works Association*, **76**:1:91, 1984.
30. Mohammed Y. Al-Ani et al., "Removing *Giardia* Cysts from Low Turbidity Waters by Rapid Rate Filtration," *Journal of American Water Works Association*, **78**:5:66, 1986.
31. Kelly O. Cranston and Appiah Amirtharajah, "Improving the Initial Quality of a Dual-Media Filter by Coagulants in Backwash," *Journal of American Water Works Association*, **79**:12:50, 1987.

32. Sandra L. Graese, Vernon L. Snoeyink, and Ramon G. Lee, "Granular Activated Carbon Filter-Adsorber Systems," *Journal of American Water Works Association*, **79**:12:64, 1987.
33. Robert G. Hyde et al., "Replacing Sand With GAC in Rapid Gravity Filters," *Journal of American Water Works Association*, **79**:12:33, 1987.
34. Keith E. Carns and Judith Dickson Parker, "Using Polymers with Direct Filtration," *Journal of American Water Works Association*, **77**:3:44, 1985.
35. Susuma Kawamura, "Two-Stage Filtration," *Journal of American Water Works Association*, **77**:12:42, 1985.
36. Kelly P. Langé et al., "Diatomaceous Earth Filtration of *Giardia* Cysts and Other Substances," *Journal of American Water Works Association*, **78**:1:76, 1986.

CHAPTER 11

MISCELLANEOUS WATER TREATMENT TECHNIQUES

11-1 Disinfection

Water contains many microorganisms, some of which cause disease. In the water treatment processes already discussed—coagulation, flocculation, sedimentation, and filtration—most, at times all, of these are removed. Typical bacterial reductions in coagulation-flocculation-sedimentation are 60 to 70 percent and addition of a filtration process increases the overall removal to close to 99 percent. Disease-causing microorganisms are normally removed more effectively than bacteria in general, hence reductions of close to 100 percent can normally be expected for these. The possibility remains, however, that some harmful organisms might pass through the treatment plant or might enter the water after treatment. For this reason, public water supplies in the United States and most of the world are treated with disinfectants to provide additional protection against transmission of disease.

Disinfection is the killing of disease-causing microorganisms. In the process, coliform bacteria and other indicator species will be killed as well and the total bacterial count will be substantially reduced. Complete sterilization of water is not ordinarily sought nor achieved in disinfection processes.

Chlorine, in a variety of chemical forms, has been the disinfectant most commonly used in the United States. A number of equilibria affect the form and the effectiveness of chlorine in water. Chlorine itself will combine with water to form hypochlorous and hydrochloric acids:

$$Cl_2 + H_2O \rightleftarrows HOCl + HCl \tag{11-1}$$

in a reaction which is nearly complete to the right. Molecular chlorine is found in measurable concentration only at pH levels less than 3.0. The hypochlorous acid, HOCl, ionizes to form hypochlorite in a reversible reaction, the equilibrium of which is governed by pH:

$$HOCl \rightleftarrows H^+ + OCl^- \tag{11-2}$$

where the equilibrium is defined by

$$\frac{[H][OCl]}{[HOCl]} = K \tag{11-3}$$

Chlorine may also be added in the form of salts such as sodium or calcium hypochlorite which will ionize in water:

$$Ca(OCl)_2 \rightarrow Ca^{2+} + 2OCl^- \tag{11-4}$$

releasing hypochlorite which will then enter into the equilibrium described by Eqs. (11-2) and (11-3). Both hypochlorite ion and hypochlorous acid are disinfectants, but hypochlorous acid is the more effective. Since the balance between the two forms is a function of hydrogen ion concentration, pH has considerable influence on the rate and effectiveness of disinfection with chlorine. The total of the molecular chlorine, hypochlorous acid, and hypochlorite ion in water is called *free available chlorine*.

Chlorine is a very strong oxidizing agent and, when added to water in the molecular form, will react with both organic and inorganic materials which are present. On reaction with ammonia or amines, it will form *chloramines* which, while still providing reasonable disinfecting action, are less likely to react with other materials that may be present. Chlorine in the form of chloramines is called *combined available chlorine*. The reaction of chlorine with ammonia occurs as follows:

$$NH_3 + HOCl \rightarrow H_2O + NH_2Cl \text{ (monochloramine)} \tag{11-5}$$

$$NH_2Cl + HOCl \rightarrow H_2O + NHCl_2 \text{ (dichloramine)} \tag{11-6}$$

$$NHCl_2 + HOCl \rightarrow H_2O + NCl_3 \text{ (trichloramine or nitrogen trichloride)} \tag{11-7}$$

The proportion of monochloramine, dichloramine and trichloramine formed depends on the molar ratio of chlorine to ammonia and the pH of the water. Typical proportions are 1 part ammonia to 4 parts chlorine, but the dosage must be adjusted to suit the particular water. The di- and trichloramines produce unpleasant taste and odor in water at low concentration, hence their formation is undesirable. Maintaining the pH above 7.5 will inhibit the production of dichloramine, and maintaining it above 5 will minimize the amount of trichloramine.

The use of combined chlorine as a disinfectant has been encouraged by the evidence that free chlorine contributes to the production of THMs (Chap. 8) and that chloramines, being less reactive, are less likely to create these compounds. Chloramines have also been used for many years as a means of maintaining a residual chlorine concentration within distribution systems,[1] where their reduced reactivity contributes to their persistence. Not all states permit the use of chloramines as primary disinfectants, and some states do not permit their use at all.[2] As with any chemical in water, there are potential undesirable effects. Chloramines may induce hemolytic anemia in dialysis patients, are mutagenic to bacteria, and are toxic to fish.[3] The health risk of chlorine and chlorinated by-products in water is not clear, and the potential health risks of the alternative disinfectants discussed below have not been studied thoroughly. Conservative practice lies in reducing the amount of chlorine used as much as possible while still providing protection, and in being prepared to change technique if clear evidence becomes available that another procedure is safer.[4]

The mechanism by which chlorine kills disease-causing organisms is uncertain. It is likely that the chlorine destroys the extracellular enzymes of the bacteria and possible that it actually passes through the cell wall to attack intercellular systems. The concentrations used are not sufficiently high to destroy the cell by oxidation. The efficiency of chlorine is adversely affected by high pH [Eq. (11-3)] and low temperature.

Chlorine may be added in a variety of ways and at different points within the system. Waters which are of excellent quality may employ chlorine as the sole treatment process. This is only practicable when the water is very low in turbidity and contains no *Giardia* cysts. Dosages in *plain chlorination* may be as much as 0.5 mg/L in order to maintain residual protection within the distribution system.

Prechlorination is the addition of chlorine prior to any other treatment. Although it improves coagulation and may reduce tastes and odors, it also enhances the production of THMs and other halogenated organics. The potential adverse health impact of these materials makes prechlorination undesirable, and this method is no longer in common use. Chlorine is still sometimes added before filtration, but not before coagulation and sedimentation, which provide a substantial reduction in the humic materials which are the precursors of THMs. The dosage is proportioned to maintain a small residual (0.1 mg/L) in the filter effluent and is helpful in reducing bacterial growth in the filter as well as in disinfection.

Postchlorination is the addition of chlorine subsequent to filtration. The point of addition is usually in the filter effluent pipe so that the water will be exposed to the chlorine for some time in the clear well before it is pumped into the distribution system. Dosages of 0.25 to 0.5 mg/L are usually adequate to maintain a residual in the clear well. Higher dosages or subsequent addition within the distribution system may be required to maintain residual protection until the water reaches the user. Residual chlorine concentration in distribution systems is reduced by reaction with organic or inorganic matter in the water or with the pipes themselves.

Breakpoint chlorination involves addition of chlorine in an amount sufficient to react with any ammonia and readily oxidizable organics which are present. During addition of chlorine to waters containing such oxidizable materials, the residual chlorine concentration will increase at a rate less than the rate of addition. At some point a further increase in dosage will produce a decrease in residual concentration—perhaps to zero. Further increases in dosage will eventually cause the residual to increase again. The point at which the concentration begins to increase again is called the breakpoint, and the dosage required to reach that point is called the breakpoint dosage. Beyond the breakpoint dosage, the residual concentration normally increases at the same rate as the dosage. Below the breakpoint dosage, the available chlorine is primarily combined chlorine, while above the breakpoint, it is primarily free chlorine.

Dechlorination can be effected with a variety of reducing agents (sulfur dioxide, bisulfites, sulfites, or thiosulfates), by aeration, and by activated carbon. In present practice, dechlorination is not likely to be required, since excessive chlorine dosages should not be used. The removal of chlorine is an undesirable side effect of the use of activated carbon for other purposes.

Chlorine gas is the most commonly used form of chlorine. It may be purchased in pressurized cylinders ranging in capacity from 45 to 1000 kg (100 to 2200 lb) and in railroad tank cars with capacities from 15,000 to 50,000 kg (16 to 55 tons). The size used depends on the consumption rate at the plant as well as the availability of rail service. There is a maximum rate at which chlorine can be withdrawn from a pressurized container without causing freezing of the container and/or the connecting piping. When necessary, small cylinders can be connected in parallel to reduce the rate of withdrawal per container. The maximum rate of removal for a 1000-kg cylinder is about 10 kg/h at 4°C (500 lb/day at 40°F) but drops to 1 kg/h at −27°C (50 lb/day at −17°F). For a 68-kg cylinder, the corresponding rates are 2 kg/h and 0.1 kg/h at 4 and −27°C, respectively.

Gas chlorinators must be able to feed at adjustable rates, but be capable of maintaining a constant rate despite pressure changes in the cylinder resulting from temperature variations. Most modern chlorination systems dissolve the gas in water in close proximity to the cylinder and convey the solution to the point of application. This technique reduces the possibility of leaks of chlorine gas, since the dissolution and conveyance system never contains gas under pressure. Figure 11-1 shows a typical gas chlorination system. A water aspirator creates a vacuum which draws chlorine from the cylinder and through a control valve and rate indicator. The rate of withdrawal can be set by adjusting the control valve. When the gas reaches the aspirator, it is dissolved in the water to form a solution of hypochlorous acid [Eq. (11-1)]. If the vacuum applied by the aspirator should fail, the inlet safety valve closes automatically. If the safety valve should fail to close, the chlorine pressure forces the seal of the vent and the gas is discharged outside the building. In large installations the rate controller and ejector may be enclosed in a cabinet with electrical displays and, perhaps, remote control from a central operating room.

Chlorine use, and the amount remaining in the containers, is monitored by weighing the cylinders. It is not possible to judge the amount remaining from the pressure, since this is a function only of temperature. Duplicate systems are

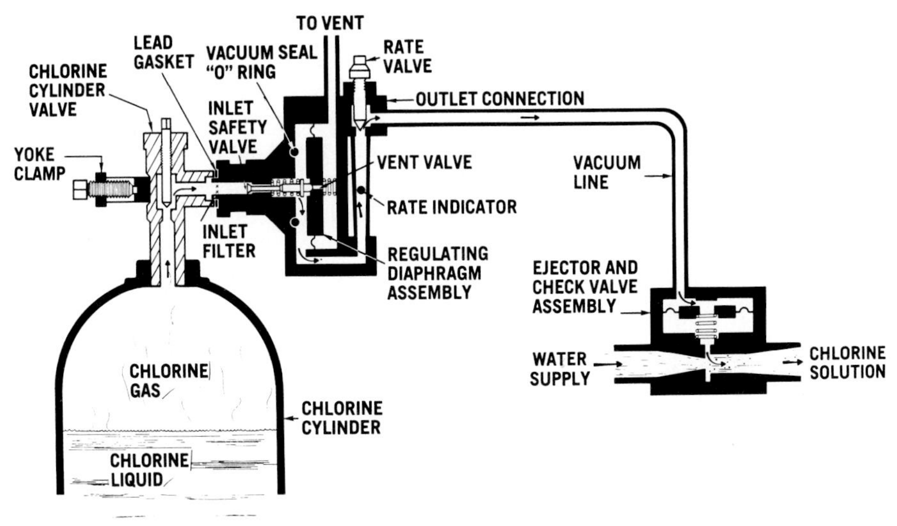

FIGURE 11-1
Gas chlorinator. (*Courtesy Capital Controls Company.*)

provided to eliminate the possibility of interruption. A minimum supply adequate for 2 weeks should always be on hand. Chlorine cylinders and the connecting piping should be kept in heated rooms in all but the warmest climates. Even in the southernmost United States, it is necessary to heat the connecting piping if the cylinders are stored outdoors. Chlorine gas may reliquefy after leaving the cylinder if it comes in contact with cold surfaces and, at temperatures below 10°C, will combine with moisture in the lines to form a crystalline chlorine hydrate called *chlorine ice*. Either the liquefied chlorine or the crystalline solid may interfere with flow of the gas by plugging the control valve or the orifice.

Chlorine, chlorinators, and piping carrying gaseous chlorine under pressure should always be located in a separate room. Positive ventilation to the outside should be provided at a rate equal to 1 room volume per minute. The air outlet should be near the floor, since chlorine is heavier than air. An automatic chlorine detector connected to an alarm is required, and switches for the exhaust fan and lights should be located outside the room. In addition, a gas mask must be stored outside the room in case of major leaks. Minor leaks can be detected by moving an open bottle of ammonia along the piping; white fumes will be created by the reaction between chlorine and ammonia. The room should be heated to maintain a temperature of about 15°C. Substantial heat may be required simply to evaporate the chlorine in the cylinders in addition to that required to raise the ambient temperature.

When gaseous ammonia is used for forming chloramines, the equipment is very similar to that used for chlorine. The ammonia should be stored in a room separate from the chlorine and should be added to the water before the chlorine. Protection devices are similar to those for chlorine, with the exception that the ventilation outlet should be near the ceiling.

Hypochlorination can be effected by addition of sodium hypochlorite or calcium hypochlorite. Sodium hypochlorite is the active ingredient in ordinary household bleach and can be obtained in carboys and large containers at a strength of 12 to 15 percent available chlorine. Calcium hypochlorite is a solid sold in cake or powder form under a variety of trade names and primarily used for swimming pool treatment. It has a chlorine content of 70 percent and may be fed as a dry chemical or dissolved in water and fed as a solution. Solution feeders for hypochlorite commonly employ small positive-displacement diaphragm pumps which have adjustable speed and stroke. The pumps may be driven by a turbine-type water meter and thus provide automatic flow proportioning.

Hypochlorination is primarily employed under emergency conditions and in very small plants. It is much less dangerous than chlorine gas, but the solution will attack the eyes and skin, so a certain degree of care is still required. Chloramination with hypochlorites is usually provided by adding ammonia as a water solution (ammonium hydroxide) or as a dry chemical (ammonium chloride or ammonium sulfate). These require less care than the gas, but ammonium solutions are caustic and skin and eye protection is necessary. Chlorine and ammonia solutions are usually conveyed in plastic pipe.

Additional oxidizing disinfectants include other halogens, chlorine dioxide, and ozone. Disinfection can also be provided by ultraviolet irradiation, extreme values of pH, heat, ultrasonic waves, and certain metals.

Two other *halogens*, bromine and iodine, are also effective germicidal agents with chemical equilibria similar to those of chlorine. *Iodine*, unlike bromine and chlorine, does not react with ammonia or organic nitrogen compounds to form amines and thus persists as hypoiodous acid and molecular iodine. It is a good disinfectant which has been used in swimming pools, but is unlikely to be widely used in water treatment because of its physiological effects on thyroid activity and its relatively high cost.[5] A mixture of halogens, monochloramine, and iodide has been found to provide more rapid disinfection than chloramine alone, presumably as a result of the formation of hypoiodous acid. The iodine-iodate system is a better disinfectant at neutral pH than chloramines, but the combination proved to be more effective than either chloramine or iodine alone.[6] *Bromine*, while an effective disinfectant, is more expensive than chlorine; it also forms THMs, and there is little experience with its use. For these reasons it is unlikely to be commonly used.

Chlorine dioxide is especially useful since it aids in destruction of taste- and odor-causing compounds, promotes improved coagulation, particularly in colored waters, and reduces the concentrations of the precursors of THMs and other organic halides.[7,8] It is produced by the chlorination of sodium chlorite in a ratio of 1 mole chlorine to 2 moles chlorite. The resulting compound, ClO_2, does not react with ammonia, is unaffected by pH within the normal range encountered in water treatment, and is a powerful oxidizing agent and germicide. Chlorine dioxide and its degradation products chlorite (ClO_2^-) and chlorate (ClO_3^-) are potentially harmful in that they are known to cause hemolytic anemia in animals. Chlorine dioxide is sometimes used prior to any other treatment, with residual protection in the distribution system being provided by postchlorination. The total dosage of chlorine dioxide should be limited to about 1.2 mg/L since the degradation products are stable, amount to about 70 percent of the dosage, and should be kept below 1.0 mg/L to avoid health impacts on consumers.[8]

Ozone (O_3), like chlorine dioxide, has effects other than disinfection which make it particularly attractive in water treatment. Since it is a strong oxidizing agent, it may be applied in any circumstance in which chlorine is effective. It improves the effectiveness of subsequent coagulation,[9] apparently by polymerization of metastable organics; aids in removal of musty, earthy, fishy, and muddy tastes and odors;[10,11] reduces subsequent chlorine demand; reduces the concentration of the precursors of THMs; and produces fewer halogenated organics than other common oxidizing disinfectants.[12] Ozone, unlike chlorine, requires very little contact time for effective disinfection. Contact times even for virus inactivation are reportedly as little as 2 min. The action of ozone is not particularly sensitive to pH in the range of pH 5 to 8, but is quite sensitive to temperature. As temperature increases, both the rate and degree of completion of oxidation reactions are likely to increase, hence ozone demand may increase with temperature. The spontaneous breakdown of ozone to molecular oxygen also increases with temperature, while its solubility decreases. The combination of these factors may require large increases in dosage at high temperatures. Ozone has been widely used in Europe for many years and its use in the United States is increasing.

While ozone is preferable to chlorine or chlorine dioxide in many respects, it is not a perfect solution to the problems associated with organics in water. Ozone does produce some halogenated organics and under some circumstances has been reported to actually increase THM formation.[11] The removal of taste and odor is associated with chemical changes into substances which have odors which are not objectionable—often reported as being fruity.[10] These odors are associated with aliphatic and aromatic aldehydes and ketones. Ozone has been observed to interfere with the ability of activated carbon to remove volatile organic compounds, evidently by producing other low-molecular-weight organics which compete with the VOCs for the adsorption sites on the carbon.[13]

Ozone is manufactured by electrical discharge into cooled, dried air, high-purity oxygen, or oxygen-enriched air. In the case of air, approximately 1 percent of the oxygen content is converted to O_3 at an energy consumption of 0.025 kWh per gram O_3. Increased yield is provided by oxygen enrichment, but the economy of the process depends upon the cost of producing the oxygen. The mixture of gases is transferred to water either by bubbling it through the bulk solution or by permitting droplets of water to fall through a rising column of gas. Ozone which is not transferred should not be discharged to the atmosphere, since it is an air pollutant. The off-gas from the head space of the transfer reactor can be recycled to the ozone generator if high-purity oxygen is used. Otherwise the excess ozone must be removed by a chemical reduction process before the air is discharged.

Ozone decays spontaneously to oxygen, hence its production at locations remote from the point of use is impractical. This decay phenomenon also prevents maintenance of a residual ozone concentration which might guard against subsequent contamination of the distribution system. For this reason, ozonation is normally followed by chlorination in modest dosages, either as free or combined chlorine. The ozonation step provides disinfection and reduction in the concentration of THM precursors as well as improved coagulation. Subsequent chlorination can then provide residual protection with minimum formation of halogenated organics.

Ultraviolet irradiation is effective in killing all types of bacteria and viruses through the probable mechanism of destruction of nucleic acids. The ultraviolet (uv) rays are generated by mercury-vapor quartz lamps which have an efficiency of about 30 percent, a wavelength of 253.7 nm, and an intensity of 50 uv W/m^2 at a distance of 50 mm.[14] Minimum retention times are on the order of 15 s with water films less than 120 mm thick. The advantages of ultraviolet disinfection include ease of automation; no chemical handling; short retention time; no effect on chemical characteristics, taste, or odor of the water; little maintenance; and no ill effect from overdoses. The disadvantages lie primarily in the lack of residual protection, relatively high cost, and ineffectiveness on turbid waters in which the rays cannot penetrate. The process is used for the most part in industrial applications and small private water systems.

Extreme values of pH, either high or low, can provide good bacterial kills. Precipitation of magnesium (associated with a pH of 11) can give coliform reductions of more than 99.9 percent. Mineral acids, if they are present, can produce similar reductions. Neither caustics nor acids are added to water for the purpose of disinfection. Reduction of bacterial numbers by pH variation is incidental to other processes and offers no residual protection if the pH is later adjusted to near neutral values.

Heat can be used to disinfect water, but the method is impractical on a large scale. Continuous-flow pasteurization has been used successfully on small-scale systems but is high in cost and provides no residual protection.

Ultrasonic waves at frequencies of 20 to 400 kHz have been demonstrated to provide complete sterilization of water at retention times of 60 min and very large reduction in bacterial numbers at retention times as low as 2 s. Costs have been excessive, but combinations of short-term sonation and ultraviolet light might be economically attractive.[15] No residual protection is provided against subsequent contamination.

Metallic ions such as silver, copper, and mercury exhibit disinfecting action. Silver is effective in concentrations less than those which are harmful to human health (0.05 mg/L). The advantages of silver include its low dosage, ease of application, and residual protection. The disadvantages include the possibility of sorption of silver on colloidal materials, the possibility of inhibition or precipitation of silver by other chemicals, reduced efficiency at reduced temperature, and relatively high cost.

Potassium permanganate, which is sometimes used in iron and manganese removal or for taste and odor control, is a strong oxidizing agent and exhibits germicidal properties. Its disinfectant action is incidental to its use for other purposes.

11-2 Organic Contaminants

As has been discussed above, water contains a variety of naturally occurring and artificial organic materials. These contribute to taste and odor problems (which may be aggravated by some disinfection techniques), may be precursors of the THMs which are formed by some disinfection techniques, or may be toxic in themselves. Some of these chemicals may be favorably modified or be removed by water treatment processes such as aeration, ozonation, application of chlorine dioxide, coagulation, flocculation, sedimentation, and filtration. Others may be unaffected by these techniques and require additional treatment if their concentration would otherwise exceed the limits set by the EPA or the states.

Tastes and odors in water may be produced by either organic or inorganic materials. The perception of taste and odor is very subjective, and measurement techniques are thus imprecise. Threshold odor number (TON) has been widely used to express the concentration of odor-causing materials but may be replaced by instrumental methods or a relatively new procedure called *flavor profile analysis* (FPA) which identifies the character of the taste or odor as well as its intensity.[16] TON is the dilution factor required to produce a solution in which the odor is just perceptible. It is customarily assessed by a group of trained analysts in order to minimize the variability inherent in a subjective determination.

Many of the organic materials associated with earthy or musty taste or odor in water are nonvolatile metabolic products of actinomycetes and blue-green algae. As noted earlier, these materials may be modified by oxidizing agents such as chlorine, chlorine dioxide, ozone, and potassium permanganate. Since their volatility is low, they are not effectively removed by aeration.[17] Oxidative techniques are unlikely to destroy these materials, but rather modify them so that their odors are no longer perceived as unpleasant. *Geosmin* (trans-1,10-dimethyl-trans-9-decalol) and certain methoxy pyrazines produced by algae may be converted to aliphatic and aromatic aldehydes and ketones with sweet, fruity odors.

Tastes and odors associated with dissolved gases and some volatile organic chemicals may be removed by aeration processes (Art. 11-6). Materials which are not favorably altered or removed by aeration or oxidative techniques can generally be adsorbed on activated carbon (Art. 11-4).

Humic and fulvic acids encompass a group of acidic randomly polymerized macromolecules which constitute the major organic constituent of natural waters. These materials are the chief precursors of THMs, which are produced by their partial oxidation by chlorine and some other oxidizing disinfectants. As has been noted earlier, these precursors can be removed by coagulation techniques and be altered by ozonation or chlorine dioxide so that THMs are not formed by subsequent chlorination in modest amounts. Although the optimum technique for removal of THM precursors appears to be coagulation either alone or following treatment with ozone or chlorine dioxide, these materials can also be removed by adsorption (Art. 11-4). In fact, they may compete successfully for adsorption sites with other materials which are inherently more undesirable.

Volatile organic compounds include a number of THMs as well as materials primarily of industrial origin such as benzene and vinyl chloride. MCL levels ranging from 0.02 to 2.0 μg/L have been established for eight such chemicals, and it is likely that others may be added to the list in the future. These chemicals are not altered or removed by coagulation, sedimentation, or filtration but can be removed by aeration (Art. 11-6) or adsorption processes (Art. 11-4). If aeration is used, the exhaust gases must be treated to prevent simply transferring the problem from the water to the air.

Pesticides which enter water supplies from either agricultural or domestic use are, in general, quite resistant to removal by ordinary water treatment techniques.[18] Exceptions to this general rule include carbofuran, which will undergo hydrolysis at pH levels encountered in softening, and metribuzin, which is oxidizable by chlorine and chlorine dioxide. Activated carbon (Art. 11-4) in either powdered or granular form has been used successfully in reducing the concentration of these materials.

11-3 Algae Control

Algae are microscopic photosynthetic plants which, under certain circumstances, may produce very heavy growths called "blooms" in lakes and reservoirs used for water supply. The algal cells themselves can be removed, albeit with some difficulty, by ordinary water treatment processes; however, certain products of the algal metabolism create problems which are more difficult to resolve. As has been noted above, certain taste- and odor-causing materials are associated with algal blooms. In addition, some cyanobacteria can produce neurotoxins or hepatotoxins which are not effectively removed by ordinary water treatment. These toxins have been associated with livestock deaths and, more recently, human illness. While both the taste- and odor-causing materials and the toxins can be removed by activated carbon,[19] it is possible under some circumstances to prevent the algal blooms from occurring in the first place.

Both copper sulfate and chlorine have been used to control algae. Dosages which have been found adequate to kill various species of algae are shown in Table 11-1. Smaller concentrations are sometimes effective in preventing growth from beginning, which is desirable since killing large growths of algae can precipitate the very problems (taste and odor, toxicity) which it is desired

TABLE 11-1
Dosages of copper sulfate and chlorine for various organisms

Organisms	Odor	Copper sulfate, mg/L	Chlorine, mg/L
Chrysophyta:			
Asterionella	Aromatic, fishy, geranium	0.10	0.5–1.0
Melosira		0.30	2.0
Synedra	Earthy	1.00	1.0
Navicula		0.07	
Chlorophyta:			
Conferva		1.00	
Scenedemus		0.30	
Spirogyra		0.20	0.7–1.5
Ulothrix		0.20	
Volvox	Fishy	0.25	0.3–1.0
Xygnema		0.70	
Coelastrum		0.30	
Cyanophyta:			
Anabaena	Moldy, grassy, vile	0.10	0.5–1.0
Clathrocystis	Grassy, vile	0.10	0.5–1.0
Oscillaría		0.20	1.1
Aphanizomenon	Moldy, grassy, vile	0.15	0.5–1.0
Protozoa:			
Euglena		0.50	
Uroglena	Fishy, oily	0.05	0.3–1.0
Peridinium	Fishy	2.00	
Chlamydomonas		0.50	
Dinobryon	Aromatic, violet, fishy	0.30	0.3–1.0
Synura	Cucumber, fishy, bitter	0.10	0.3–1.0
Schizomycetes:			
Beggiatoa	Putrefactive	5.00	
Crenothrix	Putrefactive	0.30	0.5

to prevent. Excessive dosages of copper sulfate can be toxic to certain fish as shown in Table 11-2, hence reservoirs which support populations of fish must be dosed with considerable care. The use of chlorine to prevent algal growth is probably not desirable, since the waters in reservoirs and lakes can be expected to contain the precursors of the THMs.

Copper sulfate may be applied by dragging sacks containing the chemical through the water behind boats which follow parallel paths about 5 m (15 ft) apart. The amount of chemical added is calculated on the basis of treating the top 2 m (6 ft) of the basin. An alternative method of application consists of blowing the crystals onto the surface from a boat. The larger particles (about 1 mm) will penetrate as much as 5 m (16 ft) before dissolving completely. In very alkaline waters, copper may be precipitated as the carbonate. This can be prevented by addition of sodium citrate, which will complex the copper in a soluble form.[20]

TABLE 11-2
Tolerance of various fish for copper sulfate

Fish	mg/L
Trout	0.14
Carp	0.30
Suckers	0.30
Catfish	0.40
Pickerel	0.40
Goldfish	0.50
Perch	0.75
Sunfish	1.20
Black bass	2.10

11-4 Chemistry of Activated Carbon

Activated carbon can be prepared from virtually any organic solid (coal, lignite, wood, etc.) by a destructive distillation process which drives off the volatile components of the material, leaving behind a porous carbon skeleton which has a very large surface area per unit volume. This skeleton is then "activated" by steam in an oxygen-depleted atmosphere at temperatures ranging from 750 to 950°C, often with the addition of dehydrating agents such as zinc chloride or phosphoric acid.

Molecules in solution are attracted to, and may be held by, a surface in contact with the solution. The forces holding adsorbed molecules to the surface may result from either chemical bonding or van der Waals attraction. Adsorption will remove gases, liquids, and solids from solution. The rate and degree of completion of the reaction depend on pH, temperature, initial concentration, and molecular size, weight, and structure.

Adsorption is greatest at low pH because then activated carbon is positively charged by adsorption of hydrogen ions and most of the colloidal material and all ionized polar groups on organic molecules are negatively charged. Although the adsorption reaction is exothermic, it proceeds more rapidly at high temperature as a result of increased diffusion of molecules into the fine pores of the carbon. The amount removed at equilibrium, if equilibrium is attained, will, however, be lower at high temperature.

The rate of removal increases with increasing concentration but decreases with increasing molecular size, molecular weight, and complexity of molecular structure. In mixtures of pollutants, a degree of preferential adsorption may occur, with smaller, lighter molecules being more completely removed. Formation of short-chain molecules by some oxidation processes may thus result in decreased removal of other species.

The relative capacity of different carbons is best assessed by trial with the water to be treated. For screening purposes, many manufacturers test their products against standard solutions to yield iodine, molasses, or phenol numbers. These values indicate, respectively, the ability of a particular carbon to adsorb small, large, and complex molecules. The number represents the number of milligrams of a particular material adsorbed per gram of carbon at a specified equilibrium concentration. Values of these indices are typically greater than 1000.

Activated carbon is used in water treatment in the form of both powder and granules. Granular activated carbon (GAC) has an effective size of the same order of magnitude as filter media, while powder activated carbon (PAC) is generally less than 0.075 mm in size. PAC, being finely divided, has an extremely high ratio of surface area to volume. Since adsorption is a surface phenomenon, this increases its effectiveness, but also makes it slow to settle and difficult to remove once added. The carbon may be added before coagulation, just before filtration, or in both locations. Dosage prior to the filters may be applied at an increased rate after backwashing until a layer of carbon has been deposited atop the bed. Dosages are determined by testing the water and typically range from 0.25 to 8 mg/L.

PAC is fed by the same general type of dry feed machines used for metallic coagulants. The feed machine must include some device to prevent *flooding*, which is the uncontrolled running of powder through the mechanism. Finely divided powders exhibit some of the characteristics of fluids and positive control must be provided. Carbon is an abrasive material and in aqueous solution is acidic. Conveyance systems must be made of materials which are resistant both to abrasion and low pH and must be designed to prevent deposition of carbon in the conduits and to permit cleaning if deposits do occur. Steel pipe is normally satisfactory, provided the use is not continuous. Linings of rubber or plastic and stainless steel pipe may be justified if use is continuous or replacement would be difficult. Valves should be ball, cone, or plug types, since these are less likely than others to plug or wear in a manner which would interfere with their function. Dust-suppression devices are necessary to prevent air pollution when PAC is handled.

PAC is used primarily as a short-term treatment to correct seasonal taste and odor or pesticide problems. When the need for carbon treatment is intermittent, GAC systems are seldom justifiable. Although PAC can be recovered and regenerated, this is not common. If the carbon is to be recovered from the clarifiers or filter backwash, polyelectrolytes must be used as the sole coagulant to minimize the accumulation of inerts. Wet oxidation systems (Chap. 23) have been adapted to the regeneration of powdered carbon slurries.

GAC systems ordinarily employ either closed containers like pressure filters or open boxes similar to rapid filters containing a bed of granular carbon. Systems of this kind are used both in wastewater treatment and in water treatment when adsorption is necessary on a continuous or near-continuous basis. The choice between GAC and PAC is an economic decision involving assessment of the relative costs for capital and operation. GAC systems can be less expensive overall if carbon use is substantial, since granular carbon can be regenerated fairly easily. The capital cost of the fixed beds and the carbon handling and regeneration system can be justified only if use is at least near-continuous and carbon consumption is fairly high.

Although activated carbon has been used as a replacement for other media in rapid filters,[21,22] the apparent economy of combining two processes in a single unit (filtration and adsorption) is illusory. Activated carbon changes in effective density as adsorption proceeds, hence spent carbon is denser than clean carbon. Since filter beds, whatever the medium, must be fluidized for removal of accumulated suspended solids, a filter containing GAC must be fluidized as well. When fluidization occurs, the spent carbon will tend to move toward the bottom of the bed, where it will no longer be in equilibrium with the water which surrounds it and the adsorbed contaminants may desorb and reenter the water. It appears to be preferable to separate the filtration and adsorption processes, removing suspended solids first so that backwashing of the adsorber will be minimized.

The beds are operated in both upflow and downflow modes, with operation continuing until the effluent quality begins to deteriorate. At this point the capacity of the bed must be restored by replacing part or all of the carbon. Downflow units must be taken out of service for emptying and refilling while upflow units can be operated on a more or less continuous basis. Upflow rates on the order of 0.2 to 0.4 m/min [5 to 10 gal/(min · ft^2)], depending on the carbon size, will expand the bed by about 10 percent. In this condition the spent carbon will migrate to the bottom, maintaining a countercurrent operation in which the adsorptive capacity of the carbon is most fully utilized and in which the spent carbon can be readily removed from the process for regeneration.

Spent granular carbon is normally regenerated thermally in a three-step process consisting of drying at 100°C for about 15 min, pyrolysis of adsorbates in which the temperature increases from 100 to 800°C in a 5-min period, and reactivation with steam in a reduced atmosphere at temperatures above 800°C for about 10 min. The regenerated carbon is then quenched in water and moved to storage or replaced in the adsorbers. A certain amount of carbon is lost in each cycle as a result of irreversible adsorption, mechanical attrition, combustion, and loss of fines in the carriage water. The net loss is typically 5 to 10 percent per cycle from all causes. Contactors are constructed of concrete or of steel lined with rubber or plastic.

Conveyance systems for GAC are similar to those discussed above for PAC. The cost of GAC treatment of potable water ranged from \$0.025 to \$0.25 per cubic meter (\$0.10 to \$1.00 per 1000 gal) in 1988,[23] depending on flow rate, materials of construction, mode of operation, carbon life, and other factors. This is a very significant addition to the other costs of water treatment.

11-5 Iron and Manganese Removal

Although iron and manganese are most commonly found in groundwaters, surface waters may also contain significant amounts at times. Concentrations in excess of 0.3 mg/L may produce detectable taste and odor; red-colored water which may stain clothes, cooking utensils, and plumbing fixtures; accumulations of precipitated iron in the distribution system; and growth of *Crenothrix* (an iron bacterium) in the pipes. The bacterial growth can produce additional taste and odor problems.

Iron and manganese contribute to hardness and are removed by water softening (Art. 11-7); however, it is not always desirable to soften water solely for the purpose of removing relatively small quantities of these metals. Their removal can be enhanced by oxidizing them to a higher valance state in which their solubility is reduced. The oxidation reactions of interest are

$$Fe^{2+} \rightarrow Fe^{3+} + e^- \tag{11-8}$$

and

$$Mn^{2+} \rightarrow Mn^{4+} + 2e^- \tag{11-9}$$

The oxidizing agent can be atmospheric oxygen, chlorine, chlorine dioxide, ozone, permanganate, or any other oxidant which will not leave an unwanted residue.

Iron alone in groundwaters which contain little or no organic material can be removed by simple aeration followed by detention/sedimentation and filtration. The aeration can be effected by any of the techniques of Art. 11-6, but is enhanced in towers containing slats or trays which are either seeded with catalytic material such as pyrolusite (MnO_2) or which are permitted to accumulate deposits of Fe_2O_3. The latter deposits also serve to catalyze the oxidation of iron. The detention/sedimentation basin common to such systems is unlikely to accumulate much in the way of solids. Its purpose is primarily to provide sufficient time for the oxidation reaction to reach completion and for the iron to come out of solution. The precipitated iron is removed by the filter. A number of manufacturers provide packaged iron removal systems which incorporate an aeration tower, a small detention tank, and a filter in a single unit. This type of system can be completely satisfactory for small flows with the characteristics described above.

If *both iron and manganese* are present or if the water contains organic material such as humic or fulvic acid, aeration is sufficiently rapid only if it is catalyzed by pyrolusite or by accumulation of oxidation products (Fe_2O_3 and MnO_2) on a porous bed such as coke. Simple aeration will not provide oxidation and precipitation within a reasonable time, although elevation of the pH will increase the rate substantially. *Manganese* is much more slowly oxidized than iron—in fact, the rate is negligible at pH levels below 9. *Organic material* interferes with removal of iron and manganese by peptization, i.e., by complexing with the metal without complete neutralization of molecular charge, by forming soluble complexes with both the reduced and oxidized metals, by reduction of the oxidized metals, or by a combination of these.[24]

The application of strong oxidizing agents such as chlorine, ozone, chlorine dioxide, or potassium permanganate can serve to oxidize iron and manganese more rapidly and can also modify or destroy the organic materials present so they do not interfere with the reaction. Such oxidizing agents, of course, have the potential of forming THMs from humic and fulvic acid and thus must be applied with care. Permanganate in this application will function adequately at neutral pH. The other oxidants require pH levels above 8.5 for satisfactory manganese removal. Iron is oxidized satisfactorily at neutral pH, but its removal is enhanced under more alkaline conditions. Removal of the oxidized metals is effected by rapid filtration.

Bacterial growth in filters used to remove iron and manganese can lead to reduction of the oxidized metals and their return to solution. Autotrophic bacteria may find such filters to be an excellent environment and may quickly remove all dissolved oxygen entering the bed and then turn to the oxidized metals as a substitute hydrogen acceptor. The problem can be reduced by frequent backwash (head loss development is typically quite slow, since little material is removed) and by maintaining a chlorine residual through the filter. If a heavy biological growth has developed on the media, it may be removed by concentrated dosages of chlorine or permanganate.[25]

Manganese and iron can be removed from solution by adsorption on a packed bed of granular pyrolusite. This system is called the *manganese zeolite* process, although, unlike other so-called zeolite processes, it does not involve an ion-exchange reaction (Art. 11-9). The adsorbed iron and manganese is periodically oxidized by dosing the bed with permanganate. Systems employing this process are widely used in individual home treatment units.

The problems associated with iron and manganese in water are primarily esthetic, and in some circumstances may be best handled by sequestering the metals with polyphosphates or silicates. Polyphosphates in dosages of about 5 parts polyphosphate per part metal will form stable complexes with iron and manganese in concentrations less than about 1 mg/L. The near simultaneous addition of sodium silicate and chlorine provides a similar effect at similar dosages. The silicate evidently forms a colloidal sol with the oxidized metal and has been successfully applied at total metal concentrations in excess of 1 mg/L.[26] High concentrations of other multivalent ions (such as calcium) reduce the time of stability of the sol and require that the dosage be increased.[27]

There is some evidence that groundwater containing iron and manganese may be treated in situ by pumping oxygenated water into a metal-rich aquifer.[28] The iron and manganese are evidently oxidized and precipitated within the aquifer and are deposited upon the sand in a manner which does not cause clogging. The injection wells must be carefully located to ensure that only the required volume is treated. Whether such a technique is less expensive than the more usual methods depends on the particular circumstances.

11-6 Aeration

Aeration is used in water treatment to alter the concentration of dissolved gases, to strip volatile organics, and to reduce tastes and odors. The last generally involves removal of dissolved gases (such as hydrogen sulfide or chlorine) or volatile organic materials. The tastes and odors associated with algal growth are not appreciably reduced by aeration.

Aeration techniques include spray, cascade, diffused-air, multiple-tray, and packed-column systems. The object of all designs is to maximize the area of contact between the water and the air and to produce motion of one fluid relative to the other so that exchange can be enhanced by maximizing the concentration gradient.

Spray nozzles provide a large air-water surface area but exposure time is short, 2 s or less. Nozzles require considerable head and so much land area that housing is difficult and cold weather operation may be impossible. Operating pressures are typically 70 to 140 kPa (10 to 20 lb/in^2) and the height of the spray is about 2 m (6 ft).

A *cascade* consists of a stairlike assembly over which the water flows in a thin film, falling from one level to the next. The head required depends on the height of the structure, which is typically 1 to 3 m (3 to 10 ft). Cascades are less effective than other aeration devices but are also less likely to freeze in cold weather.

Diffused-air aerators consist of concrete tanks with depth ranging from 3 to 5 m (10 to 18 ft), width ranging from 3 to 10 m (10 to 33 ft), and length adequate to provide a detention time of 5 to 30 min. Air is applied along one side of the basin through the same sort of diffusers used in sewage treatment (Chap. 22). The air produces a spiral flow of the water within the tank which increases the contact time between the bubbles and the liquid. Air flow ranges from 0.04 to 1.5 m^3 per cubic meter of water, depending on the application. Diffused aeration systems also provide some mixing and flocculation and do not offer an opportunity for freezing. They are reasonably effective in stripping VOCs.

Multiple-tray aerators consist of a series of trays formed of wooden slats, perforated plates, or screens which are spaced about 500 mm (20 in) apart. The individual trays sometimes contain a 200- to 300-mm- (8- to 12-in-) deep layer of coke, slag, stone, or ceramic balls 50 to 150 mm (2 to 6 in) in size. Tray aerators with thick layers of such media approximate packed-bed aerators. The water is applied at the top of the structure through either sprays or a perforated pan. Application rates are 40 to 200 L/min per square meter of total tray area [1 to 5 gal/(min · ft^2)], depending on the application. Air flow through the trays is provided by open louvers on the sides, normally with natural draft. In other cases, forced-draft systems are used. These have the advantage of permitting a design with balanced gas-liquid flow rates but have higher first cost and operating cost. The air flow should be countercurrent to the water for optimum efficiency. Tray aerators are subject to freezing and require housing in some parts of the United States.

Packed-bed aerators can be used for simple aeration, but are more likely to be applied in stripping operations, where their greater efficiency is particularly important. Tower aeration, in fact, has been shown to be by far the most economical technique for removal of VOCs.[29]

The exchange of volatile compounds between a gas phase and a liquid phase is described by Henry's law, whether the material being exchanged is a gas or an organic liquid with high vapor pressure. The rate of exchange is a function of the difference between the partial pressure of the substance in the gas and liquid phases. For particular materials it is possible to calculate an optimum gas-liquid ratio based on Henry's constant and the assumption that the concentration in the influent air stream and the effluent liquid stream are both equal to zero.[30] This theoretical ratio must be modified to account for variations in temperature, the presence of mixtures of volatile materials, and the finite length of a real aeration tower.

Temperature has a pronounced effect on the vapor pressure of volatile materials in both liquid and gas phases. Vapor pressures of both phases increase with increasing temperature, but not necessarily in equal amounts. If a mixture of contaminants is present, which is the normal case, each will affect the partial pressures of the others and thus their rate of removal. The assumption of zero concentration in the influent gas stream and the effluent liquid requires an infinite contact period and thus a tower of infinite height. Since real towers may be 5 to 10 m (15 to 33 ft) in height, the air flow must be increased to obtain equivalent performance. Theoretical air-water ratios for typical VOCs range from 1:1 to 20:1 while average operating ratios may be from 6:1 to nearly 200:1.[30]

The towers are packed with a material which will force the water to move in a thin film over a large surface area and to follow a tortuous path which will provide a long contact time. The air, similarly, must move through the media in the opposite direction. Media include randomly placed plastic shapes, ceramic cylinders called Raschig rings, and natural materials such as rock. Forced ventilation is required to provide the necessary air-water ratio. This ratio is a function not only of the variables cited earlier, but also of the effectiveness of the tower design.

Packed-bed towers may freeze in cold climates if the stripping air is not heated. The air leaving the tower will contain the VOCs removed from the water, and these materials are air pollutants as well as water pollutants. Depending on the quantity of VOCs being removed from the water and the local air-quality standards, treatment of the head gases may be required before their discharge.

11-7 Water Softening

Hardness consists principally of calcium and magnesium salts but includes all divalent cations. These, while not undesirable from a health standpoint, may make the water less suitable for some nonpotable uses. Benefits of softening to domestic users include reduction in soap use, longer life for water heaters, and less incrustation of pipes. If the water is softened by addition of lime, additional benefits include removal of iron and manganese, coprecipitation of humic and fulvic acid, and reduction in suspended solids—including bacteria and viruses.

Industrial users of municipal water benefit through the lower cost of producing process and boiler waters from softened water.

There is no evidence that hardness is harmful to health and some evidence that consumption of very soft waters may contribute to heart disease and accumulation of certain heavy metals in the liver.[31] Ordinary lime-soda softening will not remove all of the hardness and is usually operated to produce a residual hardness of about 100 mg/L as $CaCO_3$. Greater reductions are uneconomic and may have adverse health consequences as well.

The chemical dosages required to precipitate calcium and magnesium are most easily calculated by presenting the ionic balance of the water on a bar diagram like that of Fig. 11-2 . The concentrations of the various ions present are expressed in meq/L and drawn to scale as shown. The species which are important are Ca^{2+}, Mg^{2+}, HCO_3^-, CO_3^{2-} and CO_2. The remaining ions may be shown as "other species."

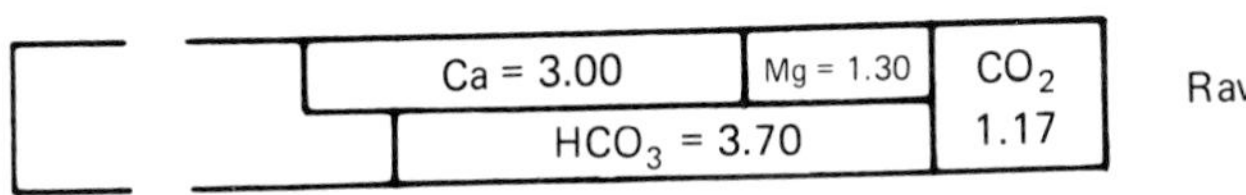

Lime dosage = 1.17 + 3.00 = 4.17 meq/L:

Ca = 7.17
Mg = 1.30
CO2 1.17
HCO3 = 3.70
OH = 4.17

Following reaction of OH⁻ and CO2:

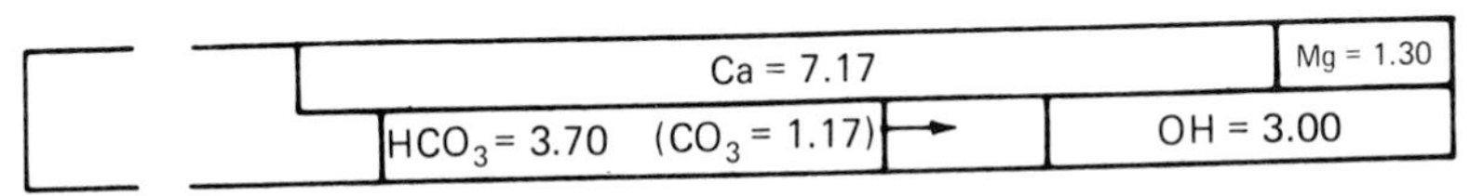

Following reaction of OH⁻ and HCO3⁻:

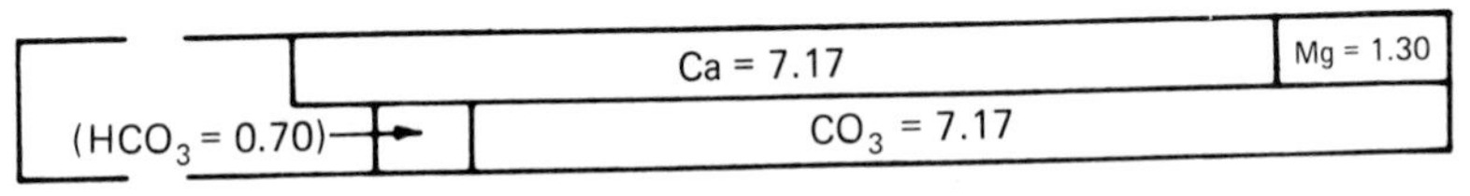

Following precipitation of $CaCO_3$:

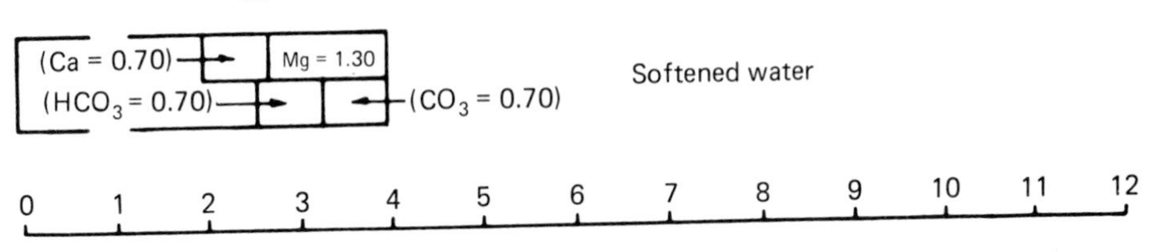

FIGURE 11-2
Bar diagrams for softening analysis. *(Reprinted from Journal of American Water Works Association, 68, by permission of the Association. Copyright 1976 by the American Water Works Association, Inc., 6666 W. Quincy Avenue, Denver, CO. 80235.)*

Calcium is precipitated as the carbonate in accord with the equation

$$Ca^{2+} + CO_3^{-} \rightarrow CaCO_3 \tag{11-10}$$

Provision of a carbonate ion concentration equal to that of the calcium will permit nearly complete combination. The theoretical solubility of $CaCO_3$ at equimolar concentrations of calcium and carbonate is approximately 0.14 meq/L; however, the practical solubility at reasonable detention time is 0.7 meq/L or more.

The carbonate ion can be added directly, but it is generally more economical to produce it by reaction of lime with the bicarbonate alkalinity. If the quantity available from this reaction is inadequate, soda ash (Na_2CO_3) is added to supply the deficit. If a final calcium hardness greater than 0.7 meq/L is desired, this can be achieved by providing a carbonate ion concentration 0.7 meq/L greater than the required removal of calcium. This is not precisely correct either theoretically or practically but will give a good approximation of the required dosage—which will require adjustment in any event.

The chemical equilibria of the carbonate-bicarbonate system are complex. The pH of maximum carbonate ion concentration is a function of the total alkalinity and temperature, as shown in Fig. 11-3, and of the total dissolved solids concentration.[32] The total alkalinity and dissolved solids, it should be noted, are increased in proportion to the quantity of chemical added and change during the softening reactions. A reasonable approximation of the required dosages can be calculated from the bar diagram, with the dosage then being adjusted to yield optimum results.

Example 11-1 Calculate the necessary amounts of lime and soda ash to soften a water which contains a total hardness of 215 mg/L as $CaCO_3$, alkalinity of 185 mg/L as $CaCO_3$, Mg^{2+} 15.8 mg/L as Mg^{2+}, Na^+ 8 mg/L as Na^+, SO_4^{2-} 28.6 mg/L as SO_4^{2-}, Cl^- 10.0 mg/L as Cl^-, NO_3^- 1.0 mg/L as N, CO_2 25.8 mg/L as CO_2, and has a pH of 7.07.

Solution

A portion of the data given is not required. The pertinent values are

Total hardness = 215 mg/L as $CaCO_3$ = 4.30 meq/L
Magnesium = 15.8 mg/L as Mg^{2+} = 1.30 meq/L
Calcium = total hardness − magnesium = 3.00 meq/L
CO_2 = 25.8 mg/L as CO_2 = 1.17 meq/L
Alkalinity = 185 mg/L as $CaCO_3$ = 3.70 meq/L

Since the pH is close to 7, it may be concluded that the alkalinity is entirely bicarbonate (at pH 7, $CO_3 \approx 0.001\ HCO_3$). The bar diagram can then be constructed as in Fig. 11-2.

If the final hardness desired is 100 mg/L as $CaCO_3$ (2 meq/L), it is evident that magnesium removal will not be required. This may be concluded because the calcium concentration can be reduced to 0.7 meq/L and this plus the magnesium

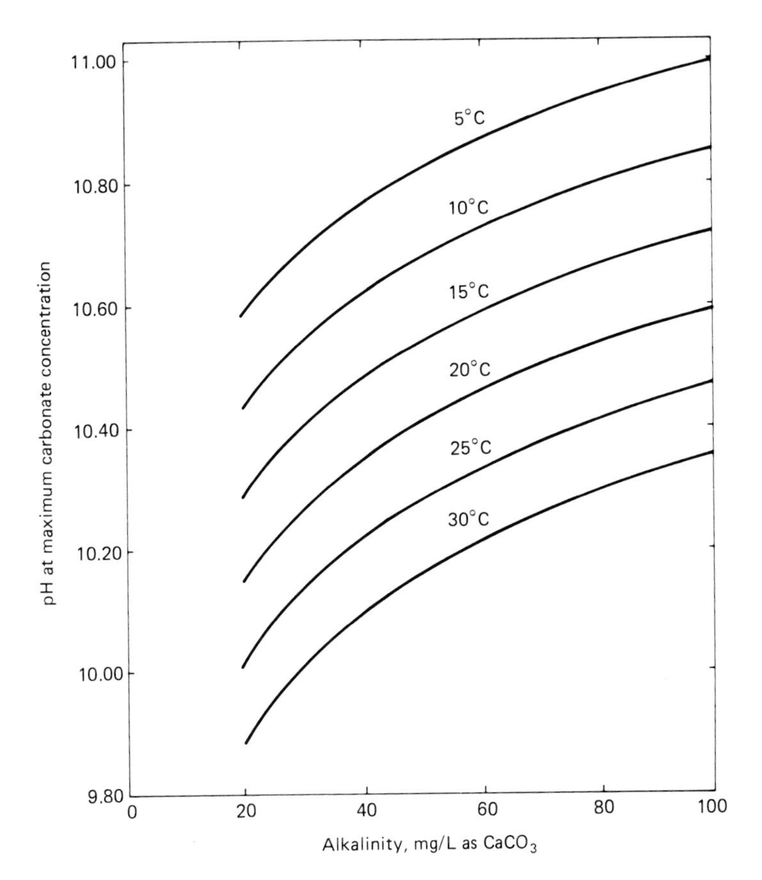

FIGURE 11-3
Alkalinity versus pH at maximum carbonate ion concentration at various temperatures for 200 mg/L total dissolved solids. (*Reprinted from Journal of American Water Works Association, **67**, by permission of the Association. Copyright 1975 by the American Water Works Association, Inc., 6666 W. Quincy Avenue, Denver, CO, 80235.*)

will meet the required standard. Lime will be added to convert a portion of the HCO_3 to CO_3 and to neutralize the CO_2, which in water forms carbonic acid. The carbonate ion concentration required to remove 2.3 meq/L of calcium is 3.0 meq/L (the required removal plus 0.7 meq/L). This can be provided entirely from the HCO_3, since the latter exceeds 3.0 meq/L. If it did not, the additional CO_3 would be provided by adding soda ash.

The required lime dosage is thus equal to the CO_2 concentration plus the necessary CO_3 production, or $1.17 + 3.00 = 4.17$ meq/L. The effect of this addition is shown in the additional bar diagrams of Fig. 11-2. The first reactions occur rapidly while the precipitation is slow. The concentrations shown on the last diagram are those which would be obtained after mixing and sedimentation. The final pH, based on the CO_3 and HCO_3 concentrations, may be calculated to be 10.3.

Magnesium is precipitated as the hydroxide in accord with

$$Mg^{2+} + 2OH^- \rightarrow Mg(OH)_2 \tag{11-11}$$

In order to obtain a satisfactory rate of reaction, excess hydroxyl ion must be added, usually about 1 meq/L. As with calcium, the calculated dosages must be adjusted somewhat. With an excess hydroxyl ion concentration of 1.0 meq/L, the practical solubility of $Mg(OH)_2$ is 0.2 meq/L. If final magnesium concentrations higher than 10 mg/L as $CaCO_3$ are desired, they are obtained by treating only a portion of the flow and bypassing the remainder. The portion to be treated depends on the influent and effluent concentrations and may be calculated by a simple mass balance.

Example 11-2 Assume that the water of the example above is to be softened to a final hardness of 2 meq/L with the additional restriction that the final magnesium concentration is not to exceed 0.6 meq/L.

Solution

The proportion of flow to be treated for magnesium removal may be calculated from the knowledge that the effluent concentration is to be 0.6 meq/L, the influent concentration is 1.30 meq/L, and the fraction treated, X, will contain 0.2 meq/L. A mass balance for this condition yields

$$(1 - X)(1.30) + X(0.2) = 0.6$$

Solving for X gives $X = 0.636$.

The lime dosage is that required to react with the CO_2, that required to convert the HCO_3 to CO_3, that required to combine with the magnesium, plus an excess. No free hydroxyl ion will be present until both the CO_2 and HCO_3 have been converted to CO_3.

$$\text{Lime dosage} = 1.17 + 3.70 + 1.30 + 1.00 = 7.17 \text{ meq/L}$$

This dosage is added only to the 63.6 percent of the flow being treated for magnesium removal. The ionic distribution in the water after the addition and the subsequent reactions are shown in Fig. 11-4. The initial chemical reactions are rapid while the precipitation is slow. The effluent from the magnesium softening process will have the constituents shown in the fifth diagram of Fig. 11-4. This flow is then mixed with the bypassed flow, which still has the makeup shown in the first diagram. The constituents of the mixed flow are calculated as follows:

	Concentration in		
Ion	**Bypass flow**	**Mg flow**	**Net concentration in mixed flow**
Mg^{2+}	1.30	0.20	1.30(0.364) + 0.20(0.636)
Ca^{2+}	3.00	10.17	3.00(0.364) + 10.17(0.636)
CO_2	1.17	0	1.17(0.364) + 0
HCO_3^-	3.70	0	3.70(0.364) + 0
CO_3^{2-}	0	8.57	0 + 8.57(0.636)
OH^-	0	1.20	0 + 1.20(0.636)

Ca = 3.00 | Mg = 1.30 | CO_2
HCO_3 = 3.70 | 1.17

Raw water

Lime dosage = 1.17 + 3.70 + 1.30 + 1.00 = 7.17 meq/L:

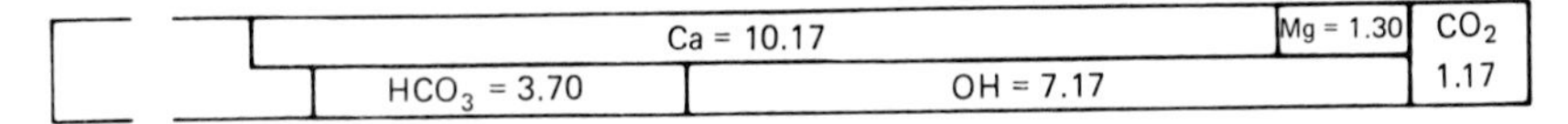

Following reaction of OH^- and CO_2:

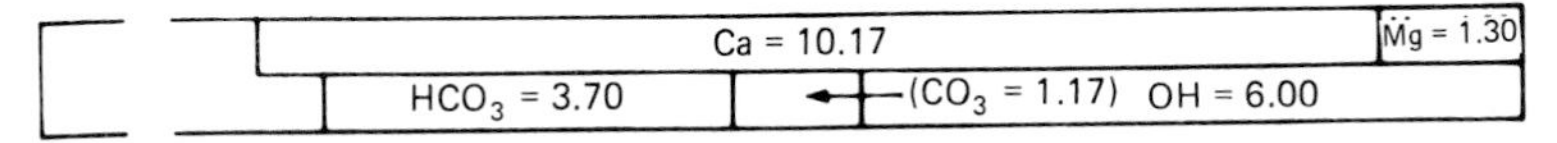

Following reaction of OH^- and HCO_3^-:

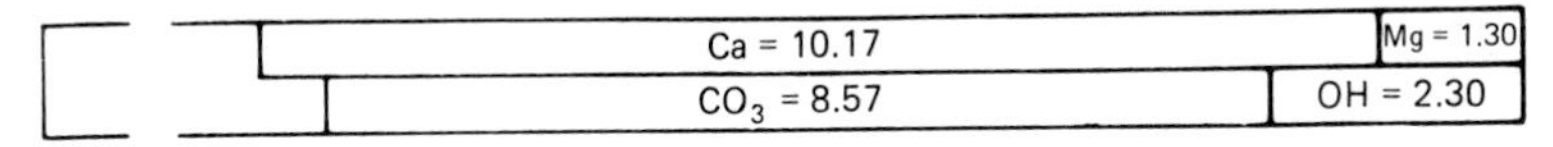

Following precipitation of $Mg(OH)_2$:

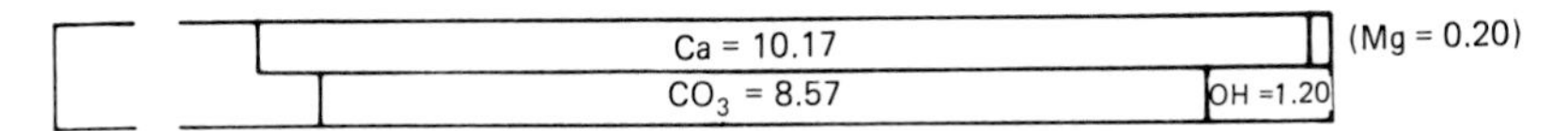

Following mixing of bypassed flow:

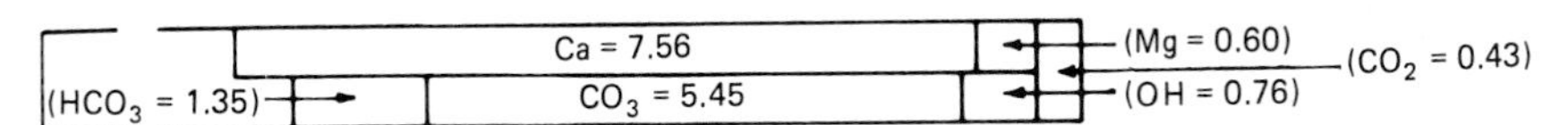

Following reaction of OH^- and CO_2:

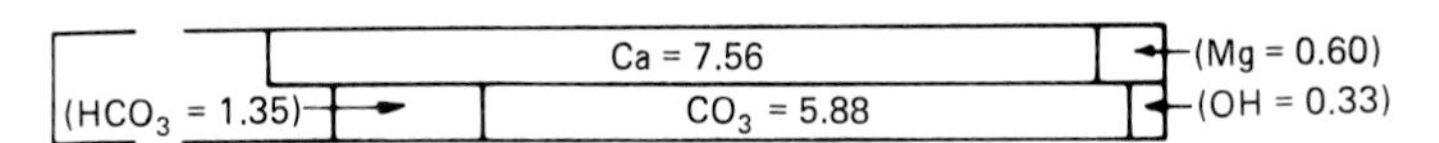

Following reaction of OH^- and HCO_3^-:

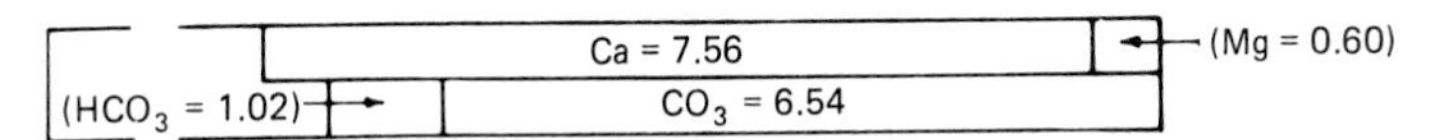

Lime dosage = 0.32 meq/L for reaction with HCO_3^-:

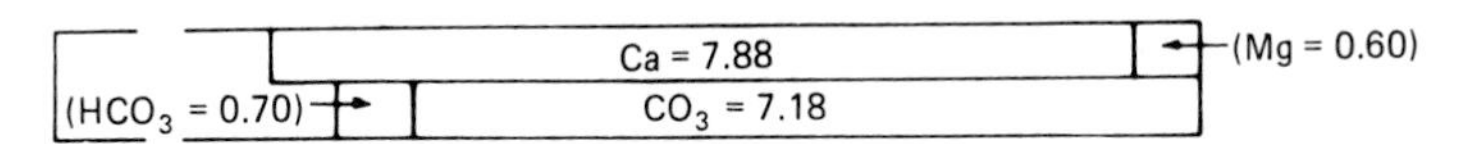

Following precipitation of $CaCO_3$:

FIGURE 11-4
Split-treatment softening analysis. (*Reprinted from Journal of American Water Works Association, 68, by permission of the Association. Copyright 1976 by the American Water Works Association, Inc., 6666 W. Quincy Avenue, Denver, CO. 80235.*)

The concentration in the mixed flow is shown in the sixth diagram, with subsequent reactions below, as the hydroxyl ion from the treated flow reacts with the bypassed flow. The resulting water in the eighth diagram must then be considered as an additional problem in calcium removal. In order to reduce the calcium concentration to 1.4 meq/L (for a final total hardness of 2.0 meq/L), the CO_3 concentration must be made 0.7 meq/L greater than the required removal.

The removal required is 7.56 − 1.4 = 6.16, thus the CO_3 concentration must be 6.16 + 0.7 = 6.86 meq/L. Since 6.54 meq/L is already available, 0.32 meq/L is required, and this can be produced from the HCO_3 by adding 0.32 meq/L of lime.

If no HCO_3 remained, the CO_3 would be provided by adding soda ash. The effect of the lime addition and the subsequent reactions are shown. The resulting water has a final hardness of 2.0 meq/L as in Example 11-1, but the proportions of calcium and magnesium are changed. The total lime dosage is 7.17 meq/L added to 63.6 percent of the flow plus 0.32 meq/L to the total flow, or 7.17(.636) + 0.32 = 4.87 meq/L.

This may be compared to the 4.17 meq/L required to obtain the same final hardness by calcium removal alone, and illustrates the economy possible by not removing magnesium.

Facilities required for storage and handling of lime and soda ash are described in Art. 9-8. The softening system includes the chemical proportioning and feeding equipment, a rapid mixing process in which the chemicals are dispersed in the flow, a flocculation stage in which the precipitates come out of solution and particle growth is facilitated, a sedimentation step in which the precipitates are removed, and a recarbonation step in which the pH of the effluent is adjusted to provide a stable water which is neither corrosive nor depositing (Art. 11-8). Upflow clarifiers or suspended solids contact units (Art. 9-7) are often used in softening plants. The dispersion, flocculation, and sedimentation steps are thus contained within a single unit. A softening plant will also include filters, and a certain amount of precipitated calcium and magnesium can be expected to reach the filter medium, where it may create incrustation if the water has not been stabilized.

The advantages of lime-soda softening include a reduction in the total mineral content of the water, removal of suspended solids and precursors of THMs, removal of iron and manganese, and a reduction in color and bacterial numbers. The process, however, produces large quantities of sludge which must be handled further, requires careful operation, and, if the pH is not properly adjusted, may create problems with filter media and within the distribution system.

An alternative softening process utilizes caustic soda (NaOH) to react with the alkalinity and produce carbonate ion for reaction with calcium.[33] The calcium carbonate is precipitated in a fluidized bed upon seed grains of sand, marble chips, steel grit, or similar dense material. As particles grow in size by deposition of $CaCO_3$, they migrate toward the bottom of the fluidized bed from which they are periodically wasted. The detention time in this process is very short (5 to 10 s) and no sludge, in the usual sense, is produced. The waste material is in the form of pellets of calcite which can be moved by solids-handling equipment. The process is reportedly effective on all waters which do not contain orthophosphate concentrations greater than 0.5 mg/L. If the alkalinity is low, a similar reaction can be produced by direct addition of soda ash. The effluent pH is lower than in other softening processes (typically less than 9.0), so subsequent stabilization may not be required. The crystallization process can also be achieved with lime as the source of alkali, but in this case is quite sensitive to low temperature—a factor which limits its applicability. When caustic soda is used, even near freezing, temperature effects are quite small.

11-8 Stabilization

Stabilization is the adjustment of the ionic condition of a water so that it will neither corrode the pipes through which it passes nor deposit encrusting films (generally of calcium carbonate). Some waters are naturally corrosive, while others become corrosive through pH reduction occasioned by addition of metallic coagulants or chlorine. Other waters which have been softened by the lime-soda process may be alkaline and supersaturated with calcium carbonate. Considering the example problems of Art. 11-7, one may note that the effluent in either case will have a pH of about 10.3 and the flow will contain 0.7 to 1.4 meq/L of calcium and 0.7 meq/L of carbonate ion. Since the true solubility of $CaCO_3$ is much less than 0.7 meq/L, the excess will gradually come out of solution and will tend to coat whatever surfaces it contacts. This can cement the individual grains of a rapid filter together or deposit a gradually increasing layer of material inside the pipe system, gradually decreasing the diameter of the lines. These phenomena can be prevented by adjustment of the pH through addition of carbon dioxide or a mineral acid, in the case of softened water, or addition of an alkali (lime or soda ash), in the case of acid water.

The pH of stability for a water may be approximated from Langlier's index:

$$SI = pH - (pK_2' - pK_S' + pCa + pAlk) \tag{11-12}$$

where the symbol p indicates the negative log. K_2' and K_S' are empirical constants dependent on temperature and ionic strength, Ca and Alk denote the molar concentration of calcium and alkalinity, and SI is Langlier's index. If the index is greater than zero the water will be depositing; if less than zero, corrosive; if equal to zero, stable. The magnitude of SI does not indicate the rate of corrosion or deposition nor is the prediction always accurate in this abbreviated form.

The *marble test* for measuring stability is a practical method which clearly indicates the condition of the water. A water of known alkalinity is placed in contact with powdered calcium carbonate for 24 h and the alkalinity is re-measured. If the alkalinity has decreased, the water is depositing and may cause incrustation of filters or pipes. If the alkalinity has increased, the water is corrosive and may damage the distribution system. If the alkalinity is unchanged, the water is stable. The water may then be adjusted to achieve a stable pH. The pH of stability will not change rapidly during routine operation of the plant.

Stabilization can also be achieved by addition of sodium hexametaphosphate. The phosphate will complex with calcium to form stable suspensions which will not cause incrustation. Phospate will also aid in removing deposits which have accumulated in the system. Dosages following softening typically range from 0.5 to 5.0 mg/L.

11-9 Ion Exchange

Ion exchange is a chemical phenomenon in which materials in solution are removed by interchange with other ions immobilized within a solid matrix through which the flow is passed. Ion-exchange processes look much like filters in that they contain a bed of granular material, an influent structure which distributes the flow, and an effluent structure which collects the product. The exchange media include natural inorganic aluminum silicates (sometimes called *zeolites* or *greensands*), bentonite clay, synthetic siliceous gels, sulfonated coal, and synthetic organic resins and are generally prepared in the form of particles approximately 0.5 mm in diameter. Most processes currently in use employ artificial organic resins.

Cation exchange is used primarily in softening and employs sodium-based resins in which the divalent ions are replaced by sodium:

$$\underset{\text{Hard water}}{\begin{matrix}Ca\\Mg\end{matrix}\left\{\begin{matrix}(HCO_3)_2\\SO_4\\Cl_2\end{matrix}\right.} + \underset{\text{Sodium exchange bed}}{Na_2X} = \underset{\text{Soft water}}{Na_2\left\{\begin{matrix}(HCO_3)_2\\SO_4\\Cl_2\end{matrix}\right.} + \underset{\text{Exhausted exchange bed}}{\left.\begin{matrix}Ca\\Mg\end{matrix}\right\}X}$$

Regeneration:

$$\underset{\text{Exhausted exchange bed}}{\left.\begin{matrix}Ca\\Mg\end{matrix}\right\}X} + \underset{\text{Salt solution}}{2NaCl} = \underset{\text{Sodium exchange bed}}{Na_2X} + \underset{\text{Waste water}}{\left.\begin{matrix}Ca\\Mg\end{matrix}\right\}Cl_2}$$

The rate of application depends on the depth of the bed, since the contact time is the important design factor. The empty bed detention time is typically about 10 min. When the exchange capacity of the resin has been exhausted, it may be regenerated by a concentrated solution of sodium chloride.

The exchange capacity of modern synthetic resins is as high as 1600 equivalents per cubic meter. The resin should not be exhausted completely before it is regenerated, since this increases the amount of regenerant required. In large-scale plants, the hardness of the product water can be monitored; when it increases, the regeneration cycle is begun. One may also calculate the volume which can be treated between regenerations and regenerate either on a time or flow basis.

Example 11-3 Calculate the flow which can be treated between regeneration cycles and the time between regenerations for a flow of 5000 m³/day with a total hardness of 215 mg/L as $CaCO_3$, which is softened by a resin with a total capacity of 1200 eq/m³. Not more than 75 percent of the capacity is to be used before regeneration. The empty bed detention time is 10 min at average flow.

Solution
The volume of resin is 5000 × 10/1440 = 34.7 m³. This would be divided among several beds. The total usable capacity is 34.7 × 1200 × 0.75 = 31,200 eq.

The total hardness is 4.3 meq/L or 4.3 eq/m³. The volume which can be treated is thus

$$V = \frac{31{,}200}{4.3} = 7260 \text{ m}^3$$

The time between regenerations at a flow of 5000 m³/day would be about 35 h.

If the water were being treated for municipal purposes, all the flow would not be passed through the ion-exchange process, since water of zero hardness is not desirable for domestic purposes. The regeneration would thus be at greater intervals. The regeneration is accomplished by filling the bed with a saturated or nearly saturated salt solution and allowing it to remain for about 15 min. The salt solution is then drained and product water is wasted until the hardness falls to about 20 mg/L.

Example 11-4 Calculate the quantity of salt required to regenerate the bed of Example 11-3 if a 100 percent excess is required. Calculate the concentration of sodium chloride if the entire salt dosage is to be applied in one filling of the bed and the void ratio is 0.45.

Solution
The bed, at regeneration, has exchanged 31,200 eq. The salt dosage required is thus 2 × 31,200 = 62,400 eq or 3650 kg of NaCl. The void volume of the bed is 34.7 × 0.45 = 15.6 m³. The salt concentration is 3650/15.6 = 234 kg/m³. The solubility of NaCl ranges from 225 to 243 kg/m³ over the temperature range 0 to 100°C, hence this is at or near the solubility limit of the salt.

Cation-exchange resins will remove iron and manganese, but care must be taken to ensure that these remain in the reduced state. If they should be oxidized either deliberately or inadvertently, they will precipitate upon the resin, coating the grains and reducing the exchange capacity. It is preferable to remove these metals prior to ion exchange.

Ion-exchange softening processes are compact and produce no sludge, but do produce concentrated brines which may be equally difficult to manage. Home water softeners which operate on this principle often discharge the brine to a sewer, but this option is normally not available to industrial or municipal treatment plants.

Hydrogen-cycle resins exchange hydrogen ion for *all* the cations present in the water. Typical reactions in this process are

$$\begin{matrix}\text{Raw water} & & \text{Hydrogen exchange bed} & & \text{Treated water} & & \text{Exhausted exchange bed}\\ \left.\begin{matrix}Ca\\Mg\\Na_2\\Fe\end{matrix}\right\}\left\{\begin{matrix}(HCO_3)_2\\Cl_2\\SO_4\end{matrix}\right. & + & H_2X & = & H_2\left\{\begin{matrix}CO_3\\SO_4\\Cl_2\end{matrix}\right. & + & \left.\begin{matrix}Ca\\Mg\\Na_2\\Fe\end{matrix}\right\}X\end{matrix}$$

Regeneration:

$$\begin{matrix}\text{Exhausted exchange bed} & & \text{Dilute acid} & & \text{Hydrogen exchange bed} & & \text{Waste water}\\ \left.\begin{matrix}Ca\\Mg\\Na_2\\Fe\end{matrix}\right\}X & + & H_2SO_4 & = & H_2X & + & \left.\begin{matrix}Ca\\Mg\\Na_2\\Fe\end{matrix}\right\}SO_4\end{matrix}$$

The product water contains both carbonic acid and a variety of strong acids, depending on the particular anions in the raw water. The carbonic acid can be removed by air stripping, and the mineral acids can be neutralized, to a degree, by mixing with untreated water. It is also possible to pass the acidic effluent through *anionic-exchange* resins which will exchange the sulfate, chloride, nitrate, etc. for hydroxyl ion:

$$\begin{matrix}\text{Cation exchange effluent} & & \text{Anion exchange bed} & & \text{Treated water} & & \text{Exhausted exchange bed}\\ H_2\left\{\begin{matrix}SO_4\\Cl_2\\(NO_3)_2\\SiO_3\\CO_3\end{matrix}\right. & + & A(OH)_2 & = & H_2O & + & A\left\{\begin{matrix}SO_4\\Cl_2\\(NO_3)_2\\SiO_3\\CO_3\end{matrix}\right.\end{matrix}$$

Regeneration:

$$\begin{matrix}\text{Exhausted exchange bed} & & \text{Caustic soda} & & \text{Regenerated exchange bed} & & \text{Waste water}\\ A\left\{\begin{matrix}So_4\\Cl_2\\(NO_3)_2\\SiO_3\\CO_3\end{matrix}\right. & + & 2NaOH & = & A(OH)_2 & + & Na_2\left\{\begin{matrix}SO_4\\Cl_2\\(NO_3)_2\\SiO_3\\CO_3\end{matrix}\right.\end{matrix}$$

The effluent from the two-stage process is comparable to distilled water in quality and costs less to produce. While demineralized water is not required or desired for municipal use, it is useful in food manufacture, high-pressure boilers, and other industrial processes.

The hydrogen- and hydroxyl-cycle resins are regenerated with strong acids and bases. Sulfuric acid, which is shown as the regenerant for cation exchange, is not always suitable, since the solubility of $CaSO_4$ is only about 1250 mg/L and thus $CaSO_4$ may form insoluble precipitates on the resin. Hydrochloric acid is preferable from this standpoint, but is more expensive. There is no similar problem with the anionic bed, since the only cation present in the influent is hydrogen.

The waste regenerants from these beds are either acid or caustic and are highly mineralized. If the two streams are carefully mixed, a more or less neutral brine is obtained. The strongly basic exchange medium will remove carbonic and silicic acids as well as the strong mineral acids. This process is thus particularly desirable for high-pressure boiler waters where silica can create operational problems.

11-10 Reverse Osmosis

The principle of *osmosis* can be readily understood by considering a semipermeable membrane separating two bodies of water which contain differing salt concentrations. A semipermeable membrane has pore openings which are larger than the molecular size of water but smaller than those of the salts. In the situation described, the water will flow through the membrane from the lower to the higher concentration, attempting to equalize the salt concentrations. If a pressure is applied to the concentrated solution, the flow can be halted. The pressure which is just adequate to stop the flow is called the *osmotic pressure*. If a greater pressure is applied, the natural flow direction will be reversed and water will flow from the higher to the lower concentration. The pressure applied in practical systems ranges from 5 to 50 times the osmotic pressure.

Reverse osmosis systems consist of the membrane, a support structure, a pressure vessel, and a pump. The optimum membrane configuration is a hollow fiber which has an area-to-volume ratio of up to 30,000 (compared to 300 to 3000 for other designs), requires no support structure, and has reasonable water flux rates. Modern membranes are manufactured of aromatic polyamides which have more desirable mechanical and chemical properties than earlier cellulose acetate designs.

The influent to reverse osmosis systems often requires pretreatment to prevent fouling of the membrane by bacteria or other suspended solids. Since the membranes are not completely efficient in excluding molecules other than water, the product water is not completely demineralized. Dissolved gases such as carbon dioxide pass through the membrane, producing increased pH on the reject side and somewhat acidic conditions on the product side. The alkaline conditions on the reject side may cause precipitation of salts, as may the high concentration of materials of low solubility such as $CaSO_4$. The reject water, like the brines from ion exchange, may require special handling.

11-11 Treatment of Brackish and Saline Waters

Increasing water consumption and depletion of existing water resources has led to considerable interest in conversion of saline and brackish waters. Although the cost of potable water production from brackish or saline water is substantially greater than that of treating fresh water, where adequate fresh water is not available it may be more economical to treat locally available saline waters than to import fresh water. It has been predicted that continued development of desalination techniques may reduce the cost to levels comparable to freshwater treatment,[34] although increasing energy costs make the achievement of this goal less likely. Desalination systems can be separated into those which employ a phase change, like distillation or freezing, and those which separate water and dissolved minerals within the aqueous phase, like ion exchange, electrodialysis, and reverse osmosis.

Evaporators, generally in the form of multiple-effect systems, can be used to distill fresh water from salty or brackish water. Problems associated with evaporators include accelerated corrosion and scaling due to the high temperatures involved. First costs and operating costs are high. Estimated costs of large-scale multistage flash distillation of seawater exceeded $1.40/m^3 ($5.40/1000 gal) in 1983.[34]

Solar stills employ shallow basins covered with glass or plastic upon which the water vaporized by the sun can condense. The capital costs are comparable to multiple-effect evaporators, but energy costs are substantially less. The process is only applicable in areas with a high percentage of sunny days and varies in capacity with season and weather. Multiple-effect evaporators have also been operated with geothermal power as an energy source.[35]

Freezing for desalination is effected by application of vacuum processes in which evaporation of a portion of the water or of a miscible secondary refrigerant freezes the flow. The influent to the process is cooled in a heat exchanger by product and waste streams until it is close to 0°C. The evaporator then withdraws the refrigerant, dropping the temperature and producing ice crystals which are removed and rinsed with a portion of the product water. The ice crystals are used to condense the vapor stream from the evaporator, or to cool it after its compression, and this, in turn, melts the ice. A portion of the product is used to rinse the ice crystals and the rinse water, product, and brine are used to chill the incoming flow. Freezing, because of the low temperature and absence of direct surface heat exchange, is generally less subject to maintenance problems than distillation. It is also more economical overall, since it requires substantially less energy transfer to freeze than to vaporize water.

Ion exchange has been applied to desalting by using the hydrogen- and hydroxyl-base resins discussed in Art. 11-9. Ion-exchange systems are simple to operate and have moderate capital costs and few operating problems, but they require costly regenerants and produce troublesome waste streams. The net production ranges from 50 to 80 percent of the influent, depending on the total dissolved solids concentration. A number of novel industrial systems which reduce regenerant requirements have been developed. The Sul-biSul process, for example, employs an anionic-exchange resin in which sulfates are converted to bisulfates. The bed can then be regenerated by using the raw water. Ion exchange is more applicable to brackish (total dissolved solids less than 5000 mg/L) than to saltwater. Recent costs have been approximately one-fourth to one-fifth those of distillation processes.

Electrodialysis employs electrical energy to drive dissolved ionized solids across semipermeable membranes. The system consists of cathodic and anionic semipermeable membranes and two electrodes. Cations and anions are driven across their respective membranes (which reject oppositely charged or neutral molecules), leaving demineralized water behind. Manufacturers have claimed yields of as much as 85 to 90 percent with costs equivalent to ion-exchange processes. Experience in Florida[36] has shown that electrodialysis is less than satisfactory and is not consistently capable of producing water which meets the EPA's secondary standards.

Reverse osmosis is the best-demonstrated technology for saline water conversion. The principles of the process are described in Art. 11-10. There were 85 municipal reverse-osmosis systems either in use or under construction in Florida alone in 1983[36] with capacities as high as 19,000 m^3/day (5 mgd). This process is particularly useful in coastal communities. Estimated costs of potable water production by this technique were in the range of \$0.50/m^3 (\$1.80/1000 gal) in 1983.

11-12 Fluoridation and Defluoridation

It has been established that fluoride is helpful in reducing the incidence of dental caries and that modest amounts of this ion in drinking water will provide the degree of protection which is desirable. The therapeutic dosage is thought to be about 1 mg/L as F, while slightly higher concentrations have been associated with mottling and loss of teeth. EPA specifies variable upper limits for fluoride in drinking water (0.6 to 2.4 mg/L), with the variation being based on air temperature.

Fluoride is added in the form of sodium fluoride, sodium silicofluoride, or hydrofluosilicic acid. Sodium fluoride, which is most commonly used, is commercially available in purities of 90 to 98 percent NaF. It is soluble in concentrations of up to 4 percent. Some states require that the chemical, which is white, be tinted blue to distinguish it from other water treatment chemicals. Saturated solutions of NaF can be easily fed in the proportions necessary to obtain the concentration desired in the water. The water used to make the solution may need to be softened, since CaF_2 and MgF_2 are practically insoluble.

Sodium silicofluoride (Na_2SiF_6) is the least expensive of the chemicals used in fluoridation but is somewhat difficult to dissolve. Slurries should ordinarily not be used, since deposition and subsequent dissolution may create variations in concentration in the treated water.

Hydrofluosilicic acid (H_2SiF_6) is available commercially in concentrations of 22 to 30 percent. It is most easily fed as delivered without prior dilution. Hydrofluosilicic acid is the most corrosive of the commonly used fluoride compounds, but all require the use of fiber-glass, plastic, or stainless steel tanks and piping.

Waters which naturally contain fluoride in excess of the recommended concentration may be treated for its removal. Three methods of removal have been demonstrated—two involving ion exchange and one precipitation. Bone char and an artificial bone char consisting of calcium hydroxyphosphate (apatite) have been used as specific ion-exchange media for fluoride. When the exchange capacity has been exhausted, the medium is regenerated with a 1 percent solution of caustic soda. Activated alumina will also remove fluoride by an exchange reaction and may be regenerated with dilute caustic.

Fluoride is removed by adsorption on or coprecipitation with magnesium hydroxide. Its concentration is thus reduced by magnesium softening, and a possible fluoride removal process is therefore the addition of a soluble magnesium salt (such as dolomitic lime, magnesia, or magnesium sulfate) followed by its precipitation as the hydroxide. Approximately 50 mg of magnesium must be precipitated per milligram of F^- removed.

11-13 Water Treatment Wastes

The production of potable water from surface or ground supplies usually results in a variety of waste streams which may not be suitable for discharge to the environment. These include brines from ion exchange, reverse osmosis, and other desalination techniques; filter backwash water; and coagulation, softening, and other chemical sludges.

The wastes from water treatment include the impurities which have been removed from the product water as well as any contaminants which may have been present in the chemicals used or which may have been produced by the treatment itself. Spent activated carbon, for example, may carry pesticides or other materials which are considered hazardous or toxic. Coagulation and softening sludges contain humic materials from the raw water and heavy metals present in the treatment chemicals. Filter backwash, similarly, contains all those materials retained by or produced within the bed. Brines produced in desalting processes, in addition to specific contaminants, are very high in dissolved solids content. Some of these materials would be characterized as hazardous or toxic under the current regulations of the EPA and the states if they were produced in manufacturing processes. As the list of regulated chemicals is expanded and new test protocols for evaluating toxicity are established, it is likely that some water treatment plant wastes will become subject to the regulations of the Resource Conservation and Recovery Act of 1976 (RCRA).

Brines have been managed by discharge to deep wells or saline surface waters or by lagooning to obtain evaporation. Discharge to sewers is not a satisfactory solution, since this simply moves the problem to the sewage treatment plant which has no facilities for treating this type of contaminant and which may have its treatment processes adversely affected by high dissolved solids content. Most desalination plants are located in coastal areas or arid climates in which the other options are feasible.

Deep wells which penetrate brackish or saline aquifers have been used for disposal of hazardous materials as well as brines. There is some danger of contamination of fresh water through discontinuities in aquicludes or through imperfections in the well casing. The aquifer may plug because of precipitation in the vicinity of the well, increasing the injection pressure required and increasing the likelihood that overlying strata may be fractured. Careful monitoring of pressure and pretreatment of the waste stream to ensure that it is entirely soluble are important. A sudden drop in pressure normally indicates failure of the system, although it makes disposal of the waste easier.

Lagooning permits the evaporation of brines in arid climates. The evaporation leaves behind a salt residue which must be managed as a solid waste and which may be hazardous. Lagoons used for such a purpose must be watertight or the brine will contaminate underlying groundwater. It is extremely difficult to construct a truly watertight impoundment of any size. If these wastes are regulated under RCRA, shallow monitoring wells would be required to monitor the amount and character of any leakage. If the contents were classified as hazardous, double liners and a leachate collection and treatment system would be required.

Discharge to saline surface waters in coastal environments appears to have the least potential for adverse environmental effects. Nevertheless, the actual contents of the waste stream must be monitored carefully, since its character may be substantially different from that of the receiving waters. Effects on wildlife, chemical reactions between waste and receiving water, and introduction of contaminants from regeneration chemicals must all be considered.

The waste brine from some cation-exchange processes may be treated by addition of soda ash. If the water being treated contains principally calcium, the brine will be mostly $CaCl_2$. Addition of Na_2CO_3 will precipitate calcium carbonate, leaving a sodium chloride brine which can be reused as a regenerant. Other ions such as magnesium will build up in the brine, making it necessary to waste a portion each time it is treated. The calcium carbonate sludge requires further handling similar to that described below for softening sludges.

Filter backwash water is the largest single waste flow in water treatment plants. It is typically managed by directing it to a surge tank and then returning it at a low constant rate to the head of the plant for recycle. Polymers may be added in the surge tank to aid in removal of fines in subsequent processes. Iron and manganese removed in the filters are likely to be reduced and returned to solution in the surge tank or in subsequent clarifiers. There may thus be a buildup in the concentration of these materials throughout the plant and an increase in treatment costs.

Coagulation sludges contain microbial, organic, and inorganic contaminants derived from the water, the metallic or polymeric coagulants, and any contaminants which the latter may contain. The solids content of these sludges averages 8 to 10 percent, but may be substantially higher or lower in specific cases. Alum sludges have been concentrated by lagooning, drying on sand beds, gravity thickening, vacuum or pressure filtration, centrifugation, solvent extraction, and freezing. Once the sludge has been reduced to a more manageable volume it *may* be disposable in sanitary landfills or on agricultural or forest land.[37] Alum sludge may tend to complex phosphate, making it less available for plant growth. Additionally, the sludge itself may contain extractable hazardous or toxic materials which may leach into the soil.

Alum sludge concentration in centrifuges or pressure filters may produce solids contents as high as 40 percent. Permitting or causing the sludge to freeze will separate the water. On thawing, the resulting slurry will settle to about 20 percent solids within 5 h and may be further dewatered on sand beds or by mechanical processes. *Aliphatic amine solvents* are miscible with water at low temperature (18°C), but separate when warmed (55°C). The solids are readily separable from the solvent-water mixture, and the solvent can then be extracted from the water by heating the mixture.

Alum may be recovered from coagulation sludges by addition of sulfuric acid. The resulting liquid alum solution can be reused as a coagulant in either water or wastewater treatment plants. Drawbacks to this process include the gradual accumulation of acid-soluble contaminants in the treatment chemicals. The concentrations of iron, manganese, heavy metals, and some organic contaminants are likely to be higher in water treated with recovered alum.[38]

Lime softening sludges can be dewatered by centrifuges to 50 or 60 percent solids. Centrifugation is particularly useful for lime sludges, since it tends to classify the waste to a degree, leaving the bulk of the Mg, Si, Fe, and Al in the centrate.[39] Calcium carbonate sludge can be recovered by recalcining, normally following removal of magnesium. Magnesium can be extracted from lime softening sludge without dissolving calcium by addition of carbon dioxide, provided the pH is kept above 7.5. The $CaCO_3$ sludge is then heated to 900 to 1200°C, in which temperature range CO_2 is driven off, leaving quicklime. The process is economical in some cases and has been applied at some large treatment plants. The cost of lime produced by this method is about the same as if it were purchased, but the reduction in solid waste may make the method attractive. One should note that this technique does not eliminate the waste sludge, but only reduces the amount. Two moles of calcium carbonate are produced per mole of lime added, hence even with reuse, there is a net production of sludge.

Accumulation of undesirable contaminants is possible in reused lime as well as in reused alum. The softening sludge itself will contain humic and other organic materials, heavy metals, and similar contaminants which may fall under the regulations of RCRA. Recovered lime and alum have been used in wastewater treatment, both to enhance suspended solids removal and for precipitation of phosphorus. Reuse in this manner may be less likely to create problems since the sludges produced in wastewater treatment require careful management in any event.

PROBLEMS

11-1. The ionization constant for hypochlorous acid is $K = 3 \times 10^{-8}$. Determine the percentage of hypochlorous acid present at pH values of 3, 5, 7, 9 and 11 [see Eq. (11-2)].

11-2. The following data were obtained in breakpoint chlorination of a water. Plot the data and determine the breakpoint dosage. The breakpoint is best determined by projecting the decreasing and increasing slopes on either side to their point of intersection.

Dosage, mg/L	1.00	2.00	3.00	4.00	5.00	6.00	7.00
Residual, mg/L	0.80	1.55	1.95	1.25	0.50	0.85	1.95

11-3. A water treatment plant is designed to treat a flow of 20,000 m^3/day. The chlorine dosage is 1 mg/L. What size containers should be used in this plant? What is the minimum weight of chlorine which should be kept on hand?

11-4. If chlorine for the plant in Prob. 11-3 can be provided only in 68-kg (150-lb) cylinders and the minimum temperature is 4°C, how many cylinders must be connected to provide the required supply?

11-5. Estimate the power required to provide 2 mg/L of ozone from dry air for the plant of Prob. 11-3. If the power cost is $0.08/kWh, what is the annual power cost for ozone generation?

11-6. Estimate the energy cost to disinfect the flow of Prob. 11-3 using ultraviolet radiation. The water film is 100 mm thick.

11-7. A water contains 200 mg/L Ca, 75 mg/L Mg, 180 mg/L HCO_3, and 150 mg/L CO_2, all expressed as $CaCO_3$. Determine the chemical dosages required to soften this water as much as possible without removing magnesium. What will the final hardness be?

11-8. If the water of Prob. 11-7 is to be softened to a final hardness of 85 mg/L as $CaCO_3$ and is not to contain more than 25 mg/L Mg as $CaCO_3$, find the required chemical dosages.

11-9. Design a cation-exchange process which will produce a water meeting the criteria of Prob. 11-8. Assuming the salt must be provided in an excess of 100 percent, how much salt would be used per year in a plant with a capacity of 5000 m^3/day?

11-10. An ion-exchange resin has a capacity of 1600 eq/m^3. The manufacturer recommends that the resin be regenerated when 75 percent of its capacity has been utilized. The regeneration cycle takes 1 h for draining, regeneration, and flushing the bed. The contact time in the bed(s) is to be 10 min when one bed is out of service and one bed is being regenerated. The plant is to reduce the hardness of a flow of 7000 m^3/day from 250 to 100 mg/L as $CaCO_3$. Determine the total volume of resin required and the number of beds to be used. What is the frequency of regeneration which would be required?

11-11. Sodium fluoride has a solubility of 4 percent over the temperature range commonly encountered in water treatment. A water containing 0.3 mg/L F is to have its concentration increased to 1.2 mg/L F. For a flow of 10,000 m^3/day, determine the weight of sodium fluoride used per day. If the chemical is fed as a 4 percent solution, what is the appropriate flow rate?

11-12. A water plant with a flow of 20,000 m^3/day uses 25 mg/L alum as $Al_2SO_4 \cdot 18H_2O$. The sludge produced consists of the alum precipitates plus 20 mg/L of other material removed by the process. Determine the dry mass of sludge produced per day.

11-13. The sludge of Prob. 11-12 is 10 percent solids and has an effective specific gravity of 1.02. Determine the wet mass and volume of sludge produced per day.

REFERENCES

1. Jack C. Dice, "Denver's Seven Decades of Experience with Chloramination," *Journal of American Water Works Association*, **77**:1:34, 1985.
2. David J. Hack, "State Regulation of Chloramination," *Journal of American Water Works Association*, **77**:1:46, 1985.
3. Peter Kreft et al., "Converting from Chlorine to Chloramines: A Case Study," *Journal of American Water Works Association*, **77**:1:38, 1985.
4. Gunther F. Craun, "Surface Water Supplies and Health," *Journal of American Water Works Association*, **80**:2:40, 1988.
5. Morris J. Carrell, "Chlorination and Disinfection: State of the Art," *Journal of American Water Works Association*, **63**:769, 1971.
6. Riley N. Kinman and Ronald F. Layton, "New Method for Water Disinfection," *Journal of American Water Works Association*, **68**:298, 1976.
7. W. J. Masschelein, "Experience with Chlorine Dioxide in Brussels, Part 3: Operational Case Studies," *Journal of American Water Works Association*, **77**:1:73, 1985.
8. Karen S. Werdehoff and Phillip C. Singer, "Chlorine Dioxide Effects on THMFP, TOXFP and the Formation of Inorganic By-Products," *Journal of American Water Works Association*, **79**:9:107, 1987.
9. Domenic Grasso and Walter J. Weber, Jr., "Ozone-Induced Particle Destabilization," *Journal of American Water Works Association*, **80**:8:73, 1988.
10. Cristophe Anselme, I. H. Suffet, and Joel Mallevialle, "Effects of Ozonation on Tastes and Odors," *Journal of American Water Works Association*, **80**:10:45, 1988.
11. Wilfred L. LePage, "A Treatment Plant Operator Assesses Ozonation," *Journal of American Water Works Association*, **77**:8:44, 1985.
12. Benjamin W. Lykins, Jr., Wayne E. Koffskey, and Robert G. Miller, "Chemical Products and Toxicological Effects of Disinfection," *Journal of American Water Works Association*, **78**:11:66, 1986.
13. Stephen W. Maloney et al., "Ozone-GAC Following Conventional US Drinking Water Treatment," *Journal of American Water Works Association*, **77**:8:66, 1985.
14. Robert C. Hoehn, "Comparative Disinfection Methods," *Journal of American Water Works Association*, **68**:302, 1976.
15. Edmund J. Laubusch, "Chlorination and Other Disinfection Processes," *Water Quality and Treatment*, 3d ed., McGraw-Hill, New York, 1971.
16. Jeroen H. M. Bartels, Gary A. Burlingame, and I. H. Suffet, "Flavor Profile Analysis: Taste and Odor Control of the Future," *Journal of American Water Works Association*, **78**:3:50, 1986.
17. Shahla Lalezary et al., "Air Stripping of Taste and Odor Compounds from Water," *Journal of American Water Works Association*, **76**:3:83, 1984.
18. Richard J. Miltner et al., "Treatment of Seasonal Pesticides in Surface Waters," *Journal of American Water Works Association*, **81**:1:43, 1989.
19. Ian R. Falconer et al., "Using Activated Carbon to Remove Toxicity from Drinking Water during Cyanobacterial Blooms," *Journal of American Water Works Association*, **81**:2:102, 1989.
20. Marcia Headstream, Dan M. Wells, and Robert M. Sweazy, "The Canyon Lakes Project," *Journal of American Water Works Association*, **67**:125, 1975.

21. Sandra L. Graese, Vernon L. Snoeyink, and Ramon G. Lee, "Granular Activated Carbon Filter-Adsorber Systems," *Journal of American Water Works Association*, **79**:12:64, 1987.
22. Robert G. Hyde et al., "Replacing Sand with GAC in Rapid Gravity Filters," *Journal of American Water Works Association*, **79**:12:33, 1987.
23. Jeffrey Q. Adams and Robert M. Clark, "Cost Estimates for GAC Treatment Systems," *Journal of American Water Works Association*, **81**:1:35, 1989.
24. Timothy J. Cromley and John T. O'Connor, "Effect of Ozonation on the Removal of Iron from Groundwater," *Journal of American Water Works Association*, **68**:315, 1976.
25. John T. O'Connor, "Iron and Manganese," *Water Quality and Treatment*, 3d ed., McGraw-Hill, New York, 1971.
26. F. James Dart and Paul D. Foley, "Silicate as Fe, Mn Deposition Preventative in Distribution Systems," *Journal of American Water Works Association*, **64**:244, 1972.
27. R. Bruce Robinson, Roger A. Minear, and Joseph M. Holden, "Effects of Several Ions on Iron Treatment by Sodium Silicate and Hypochlorite," *Journal of American Water Works Association*, **79**:7:116, 1987.
28. Committee Report, "Research Needs for the Treatment of Iron and Manganese," *Journal of American Water Works Association*, **79**:9:119, 1987.
29. Robert M. Clark, Richard G. Eilers, and James A. Goodrich, "VOCs in Drinking Water: Cost of Removal," *Journal of Environmental Engineering Division, American Society of Civil Engineers*, **110**:EE6:1146, 1984.
30. O. Thomas Love, Jr., and Richard G. Eilers, "Treatment of Drinking Water Containing Trichlorethylene and Related Industrial Solvents," *Journal of American Water Works Association*, **74**:413, 1982.
31. Robert S. Ingols and T. F. Craft, "Analytical Notes—Hard vs. Soft Water Effects on the Transfer of Metallic Ion from Intestine," *Journal of American Water Works Association*, **68**:209, 1976.
32. Paul M. Schierholz, John D. Stevens, and John L. Cleasby, "Optimum Calcium Removal in Lime Softening," *Journal of American Water Works Association*, **68**:112, 1976.
33. Cornelis van der Veen and Anthonie Graveland, "Central Softening by Crystallization in a Fluidized-Bed Process," *Journal of American Water Works Association*, **80**:6:51, 1988.
34. Millard P. Robinson, Jr., Garret P. Westerhoff, and Thomas M. Leahy III, "Desalting—A Water Supply Alternative for Virginia Beach," *Journal of American Water Works Association*, **75**:109, 1983.
35. James J. O'Brien, "Geothermal Resources as a Source of Water Supply," *Journal of American Water Works Association*, **64**:649, 1972.
36. Glenn M. Dykes, "Desalting Water in Florida," *Journal of American Water Works Association*, **75**:104, 1983.
37. Robert J. Grabarek and Edward C. Krug, "Silvicultural Application of Alum Sludge," *Journal of American Water Works Association*, **79**:6:84, 1987.
38. Mark M. Bishop et al., "Testing of Alum Recovery for Solids Reduction and Reuse," *Journal of American Water Works Association*, **79**:6:76, 1987.
39. Michael A. Burris, Kenneth W. Cosens, and David M. Mair, "Softening and Coagulation Sludge—Disposal Studies for a Surface Water Supply," *Journal of American Water Works Association*, **68**:247, 1976.

CHAPTER 12

SEWERAGE—GENERAL CONSIDERATIONS

12-1 Definitions

Sewerage refers to the collection, treatment and disposal of liquid waste. *Sewerage works* or *sewage works* include all the physical structures required for that collection, treatment, and disposal.

Sewage is the liquid waste conveyed by a sewer and may include domestic and industrial discharges as well as storm sewage, infiltration, and inflow. *Domestic* or *sanitary sewage* is that which originates in the sanitary conveniences of dwellings, commercial or industrial facilities, and institutions. *Industrial waste* includes the liquid discharges from industrial processes such as manufacturing and food processing. *Storm sewage* is flow derived from rainfall events and deliberately introduced into sewers intended for its conveyance. *Infiltration* is water which enters the sewers from the ground through leaks. *Inflow* is water which enters the sewers from the surface, during rainfall events, through flaws in the system, or through connections to roof or basement drains.

A *sewer* is a pipe or conduit, generally closed, but normally not flowing full, which carries sewage. A *common sewer* serves all abutting properties. A *sanitary sewer* carries sanitary sewage and is designed to exclude storm sewage, infiltration, and inflow. Industrial waste may be carried in sanitary sewers, depending upon its characteristics. A *storm sewer* carries storm sewage and any other wastes which may be discharged into the streets or onto the surface of the ground. A *combined sewer* carries both domestic and storm sewage. A system composed of combined sewers is called a *combined system*, while one which segregates the storm water is called a *separate system*.

A *house sewer* is a pipe conveying wastewater from an individual structure to a common sewer or other point of disposal. A *lateral sewer* is a common sewer with no tributary flow except from house sewers. A *submain sewer* collects flow from one or more laterals as well as house sewers. A *main* or *trunk sewer* collects flow from several submains as well as laterals and house sewers. *Force mains* are pressurized sewer lines which convey sewage from a pumping station to another main or to a point of treatment or disposal. An *intercepting sewer* intersects other sewers to separate the dry weather flow from storm-water flow which they may carry. A *relief sewer* is a sewer which has been built to carry a portion of the flow in a system with otherwise inadequate capacity. An *outfall sewer* is a sewer which carries the collected waste to a point of treatment or disposal.

Sewage treatment includes any process which may be used to favorably modify the characteristics of the wastewater. *Sewage disposal* refers to the discharge of liquid wastes to the environment. Normally, but not always, disposal implies some degree of treatment prior to discharge.

12-2 General Considerations

Provision of sewerage for an urban area requires careful design. The sewers must be adequate in size and slope so that they will contain the maximum flow without being surcharged and maintain velocities which will prevent deposition of solids. Before design can begin, the flow and the variations in flow must be estimated. In addition, any subterranean features, including other utilities, which could interfere with construction must be located.

The sewage, once it enters the sewers, becomes the responsibility of the community. The potential danger to health and to the environment posed by sewage creates both legal and financial liabilities for the community and makes the construction of a sewerage system something of a mixed blessing in many cases. Small communities which have managed in the past with on-site disposal systems (Chap. 25) may be well-advised to avoid modernization.

Sewage is normally treated in some manner before being discharged to the environment. In the United States, the EPA requires, as a minimum, that the waste receive the equivalent of *secondary treatment*. This, in most cases, involves a reduction in waste strength of about 85 percent. The actual degree of treatment provided in the United States depends on the water quality standards of the receiving stream and the flow and characteristics of both the stream and waste. In some cases no discharge of pollutants may be permitted to particular bodies of water.

The required degree of treatment may normally be achieved in a variety of ways by combining different processes. Selection of the optimum combination of processes for a particular situation requires preliminary design and evaluation of costs of those alternatives which can produce the necessary effluent quality. The following chapters cover design of collection, treatment, and disposal systems for liquid wastes. Estimation of sanitary sewage quantity is discussed in Chap. 2.

12-3 Combined and Separate Sewers

Modern sewer systems are typically separate systems. Exceptions to this general rule are found in some larger, older cities where combined sewers were constructed in the past and where new additions follow the existing practice. In most instances these communities were densely populated and had constructed storm sewers before the need for sanitary sewers was commonly accepted. In the early 1800s, the discharge of household wastes into the sewerage system was actually prohibited in some cities, but household sewers were later connected to the storm sewers, converting them to combined sewers. Subsequent additions were then designed as combined sewers, often including provision for separation of dry weather flows from the storm water flow.

As urban areas develop, the first need is for sanitary sewers. Storm water can be handled by street and gutter flow to natural watercourses until development is fairly extensive. Since storm sewers are larger and substantially more expensive than sanitary sewers, their construction is frequently deferred. New sanitary sewers are designed specifically to exclude storm water, with strict specifications regarding allowable infiltration at joints and elimination of potential sources of inflow.

Although storm water may be heavily polluted, particularly at the beginning of a runoff event, the flow is usually so great that it cannot be economically contained and/or treated. The receiving streams are also likely to be otherwise contaminated at the same time by flows from nonurban sources. Hence, requiring treatment of urban storm water appears unreasonable.

This text treats sanitary and storm sewers as separate systems, corresponding to present practice in the United States. Combined sewers, if they should be required, can be provided by constructing a storm sewer system and connecting household drains to these lines. The sanitary sewage flow is negligible with respect to storm water flow.

12-4 Liability for Damages Caused by Sewage

Decisions in state courts concerning liability for damages associated with sewerage works are far from uniform; however, certain principles appear to be commonly accepted.

A city cannot be required to furnish sewerage, nor can it be held liabile if it is not furnished. Once such service is provided, however, the community and its officials assume certain responsibilities for damages to health, property, or the environment which may result from unsatisfactory design, construction, or operation of the system.

Deficiencies in design or construction may permit the city to involve the engineer or contractor in any legal difficulties which result. The city, however, as owner of the facilities, may bear the ultimate responsibility if the other parties are unable to pay for the damages.

Storm drains normally concentrate flows at fewer points and increase both the total discharge and the peak rate of flow. Such changes in normal drainage patterns may cause downstream flooding or erosion for which the city can be held liable. Many political subdivisions in the United States now require that storm drainage systems be designed so as to prevent the flow after development from exceeding that which existed before.

Damages resulting from poor operation or maintenance are always the responsibility of the city. Failure to respond quickly to known deficiencies is normally sufficient to establish legal liability. State and federal agencies such as the EPA may assess fines against communities which fail to meet relevant discharge standards. In addition, private citizens or other legal persons whose use of the contaminated water is adversely affected may be able to sue at common law.

CHAPTER 13

STORM-WATER FLOW

13-1 Urban Hydrology

In order to design a storm sewer system, it is necessary to determine the flow which each segment must carry. This is not always a straightforward matter. As noted in Chap. 4, rainfall events are predictable only in a statistical sense, and, whatever return period may be selected as a basis for design, some storms will occur which will overtax the system. The actual flow which exists in the sewers will be a function not only of the statistical frequency of the design storm, but also the distribution of the storm in time (the hyetograph), the antecedent conditions, the season of the year, and the physical design of the collection system. The last factor is very important. As will become clear, provision of storm drainage in developing urban areas, while reducing flooding in the area served, increases both the total amount of runoff and the magnitude of the peak flow. This can result in flooding of downstream areas. Many communities now require that storm sewers be designed so as to prevent the discharge from the area served from exceeding that which existed prior to development.

The relationship among rainfall intensity, duration, and frequency, as noted in Chap. 4, is obtainable from compilations of data of the National Weather Service.[1,2] The data may be used to produce both synthetic hyetographs and intensity-duration curves for different return periods. From the isohyetal maps of Ref. 1, for example, one may obtain for the area of New Orleans the following data for a 10-year return period:

Duration, h	0.5	1	2	3	6	12	24
Rainfall, in	2.6	3.3	4.4	5.0	6.2	7.5	9.0

Similarly, from Ref. 2, using both isohyetal maps and equations based on Gumbel's fitting of the Fisher-Tippett Type I frequency distribution, one may obtain for New Orleans and a 10-year return period:

Duration, min	5	10	15	30	60
Rainfall, in	0.68	1.19	1.54	2.43	3.35

The values from Refs. 1 and 2 coincide, more or less, for the 60-min rainfall, but differ for the 30-min duration. As noted in Chap. 4, the values of Ref. 2 should be used for durations of less than 1 h. The combined data set is presented in Table 13-1.

The data of Table 13-1 may then be plotted as shown in Fig. 13-1, and similar curves can be obtained for other return periods. It is important to remember that curves of the sort shown in Fig. 13-1 are not hyetographs and do not represent an actual or hypothetical storm pattern. Rather, the curve tells us what the maximum average intensity will be during a rainfall of specified duration with a return period equal to that stipulated.

Figure 13-1 or Table 13-1 tell us, among other things, that the average intensity during the maximum 5 min of a 10-year storm in New Orleans is 207 mm/h. We may also calculate, since the maximum 10 min has an average intensity of 181 mm/h, that the second highest 5-min intensity is 155 mm/h. This procedure can be continued for each period of a storm. What is unknown (and unknowable) is precisely when these intensities will occur during a storm of, for example, 12 h duration. One might resolve this question by mimicking actual storm hyetographs or by selecting a pattern designed to maximize the peak runoff. The latter procedure is often used in design of urban drainage systems.

In considering what occurs during a rainfall-runoff event, certain factors are evident in light of the discussion in Chap. 4. The initial rainfall will produce

TABLE 13-1
Rainfall and rainfall intensity for a 10-year return period in New Orleans

		Average intensity	
Duration	**Net rainfall, in**	**in/h**	**mm/h**
5 min	0.68	8.16	207
10 min	1.19	7.14	181
15 min	1.54	6.16	156
30 min	2.43	4.86	123
60 min	3.35	3.35	85
2 h	4.40	2.20	56
3 h	5.00	1.67	42
6 h	6.20	1.03	26
12 h	7.50	0.63	16
24 h	9.00	0.38	10

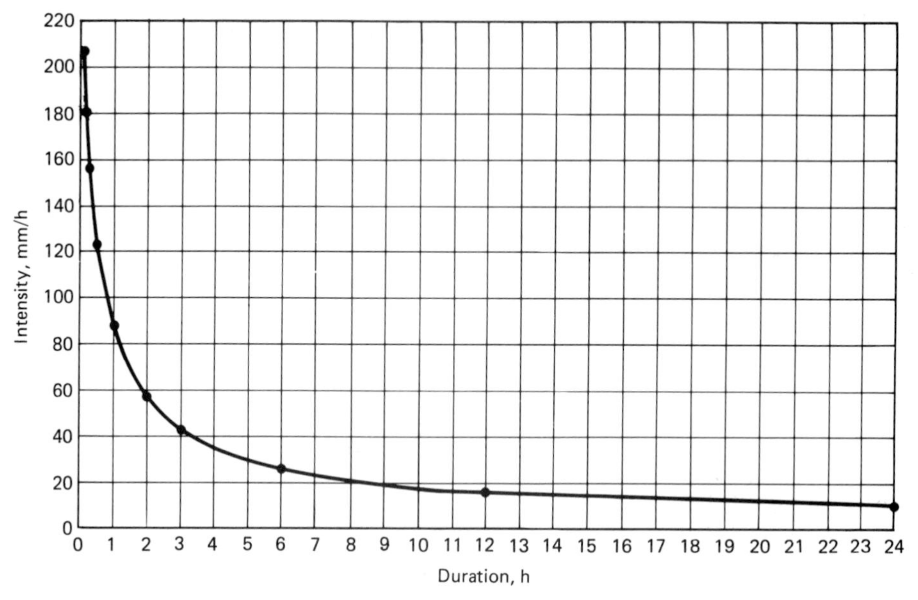

FIGURE 13-1
Intensity-duration curve for 10-year return period in New Orleans.

little or no runoff, since it will be lost to evaporation, depression storage, infiltration, or other abstractive processes. This will be true whether the intensity is high or low. Once the abstractions have been largely satisfied, it is reasonable to assume that the runoff rate will be more or less related to the rainfall intensity. A storm of given return period will thus produce maximum runoff if the rainfall is distributed so that maximum intensities occur relatively late in the event. A pattern which is often assumed is that the maximum quarter will be the third, the second largest the second quarter, the third largest the fourth quarter, and the smallest the first quarter. Within each quarter the intensities are distributed to yield a gradual increase to a peak at the beginning of the third quarter followed by a gradual decrease. The technique of constructing such a hyetograph may be illustrated by example.

Example 13-1 Using the data of Table 13-1 and Fig. 13-1, construct a synthetic hyetograph for a 12-h storm with a 10-year return period in New Orleans. Use a 15-min interval.

Solution
From the data, the storm will have an average intensity of 16 mm/h, which means the total rainfall will be $16 \times 12 = 192$ mm. The maximum 15-min period will have an average intensity of 156 mm/h, the maximum 30 min an average intensity of 123 mm/h, the maximum 45 min an average intensity of 100 mm/h, the maximum hour an average intensity of 85 mm/h, and so on. Tabulating these values, one may calculate the intensity i during each 15-min period of the storm:

Duration	Average intensity, mm/h	*i*, mm/h
0.25	156	156
0.50	123	90
0.75	100	54
1.00	85	40
1.25	75	36
.	.	.
.	.	.
.	.	.
11.25	16.9	.
11.50	16.6	3.1
11.75	16.3	2.5
12.00	16.0	1.9

The plotting positions are selected to ensure that the peak 3-h period (with an average intensity of 42 mm/h) occurs in the third quarter, that the second highest 3-h period (with an average intensity of 10 mm/h) occurs in the second quarter, and so on. The resulting hyetograph is shown in Fig. 13-2.

It should be noted that the peak intensity is not a function of the storm duration, but only of the return period. Shortening the duration of the design storm will have little effect on the peak flow, but will reduce the total discharge.

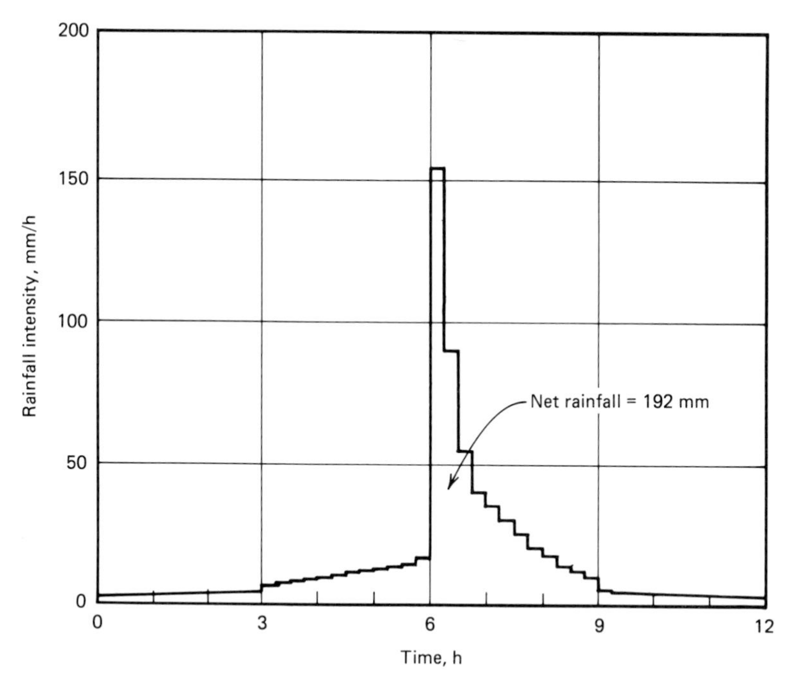

FIGURE 13-2
Synthetic hyetograph for 12-hour duration 10-year storm return in New Orleans.

13-2 The Rational Method

The rational method is the simplest of the methods used for storm sewer design. Although it was widely used in the past, this method should not be applied to areas larger than 3 km^2. The procedure calculates the flow as the product of rainfall intensity, drainage area, and a coefficient which reflects the combined effects of surface storage, infiltration, and evaporation. The total volume of water which falls upon an area A per unit time under a rainfall of intensity i is

$$Q = iA \tag{13-1}$$

Of this total, a portion will be lost as a result of the abstractions discussed in Chap. 4. This portion is not constant and tends to decrease during a rainfall event. For given conditions of antecedent moisture, soil character, ground slope, and level of development, the fraction of incident precipitation which appears as runoff is defined as C. If C is known, then the discharge will be

$$Q = CiA \tag{13-2}$$

C for an area is not invariant, but tends to increase as the rainfall continues. Research conducted during the early 1900s led to the development of the following equations. For impervious surfaces:

$$C = 0.175t^{1/3} \tag{13-3}$$

or

$$C = \frac{t}{8 + t} \tag{13-4}$$

For improved pervious surfaces:

$$C = \frac{0.3t}{20 + t} \tag{13-5}$$

in which t is the duration of the storm in minutes. Average values of C which have been commonly used for various surfaces are presented in Table 13-2.

An effective runoff coefficient for a composite drainage area can be obtained by estimating the percentage of the total which is covered by roofs, paving, lawns, etc., multiplying each fraction by the appropriate coefficient and then summing the products.

Example 13-2 Determine the runoff coefficient for an area of 0.2 km^2, of which 3000 m^2 is covered by buildings, 5000 m^2 by paved driveways and walks, and 2000 m^2 by portland cement streets. The remaining area is flat, heavy soil covered by grass.

Solution

From Table 13-2, one may obtain values of C for each area:

Roofs 0.70 to 0.95

Driveways and walks 0.75 to 0.85

Street 0.80 to 0.95

Lawn 0.13 to 0.17

The fraction of the area with each surface is

Roofs 3000/200,000 = 0.015

Driveways and walks 5000/200,000 = 0.025

Street 2000/200,000 = 0.010

Lawn 190,000/200,000 = 0.95

The average value of C, depending on the specific values chosen for the individual areas, will thus lie between 0.16 and 0.21.

Conservative practice, in the absence of local knowledge, would be to use the higher values in the table, resulting in a runoff coefficient of 0.21. Some engineers in the past have used the values presented in Table 13-3 for composite areas of specific character.

The rational method may still be useful for designing drainage improvements for restricted areas such as small subdivisions. In no case should it be used for areas larger than 3 km^2 (1 mi^2), and for areas larger than 0.5 km^2 (0.2 mi^2) the other techniques discussed below are more suitable.

Both simple and more sophisticated methods may utilize the concept of *time of concentration* or *time to equilibrium flow*. This is the time required to permit flow from the entire tributary area to reach the point under consideration. Izzard[4] developed an equation for time to equilibrium flow:

TABLE 13-2
Runoff coefficients for various surfaces

Type of surface	C
Watertight roofs	0.70–0.95
Asphaltic cement streets	0.85–0.90
Portland cement streets	0.80–0.95
Paved driveways and walks	0.75–0.85
Gravel driveways and walks	0.15–0.30
Lawns, sandy soil	
2% slope	0.05–0.10
2–7% slope	0.10–0.15
> 7% slope	0.15–0.20
Lawns, heavy soil	
2% slope	0.13–0.17
2–7% slope	0.18–0.22
> 7% slope	0.25–0.35

TABLE 13-3
Runoff coefficients for different areas[3]

Description of area	C
Business	
Downtown area	0.70–0.95
Neighborhood area	0.50–0.70
Residential (urban)	
Single-family area	0.30–0.50
Multiunits, detached	0.40–0.60
Multiunits, attached	0.60–0.75
Residential (surburban)	0.25–0.40
Apartment areas	0.50–0.70
Industrial	
Light	0.50–0.80
Heavy	0.60–0.90
Parks, cemeteries	0.10–0.25
Playgrounds	0.20–0.35
Railroad yards	0.20–0.40
Unimproved areas	0.10–0.30

$$t_e = 526.76kL^{1/3}i_e^{-2/3} \tag{13-6}$$

in which k is given by

$$k = \frac{2.76 \times 10^{-5}i_e + c}{s^{1/3}} \tag{13-7}$$

where L = distance of flow, m
s = slope
c = retardance coefficient (Table 13-4)
i_e = excess rainfall, mm/h

The experimental verification of this formula is limited to $i_e \times L < 3800$.

TABLE 13-4
Izzard's retardance coefficient

Surface	c
Very smooth asphalt pavement	0.0070
Tar and sand pavement	0.0075
Concrete pavement	0.012
Tar and gravel pavement	0.017
Closely clipped sod	0.046
Dense bluegrass sod	0.060

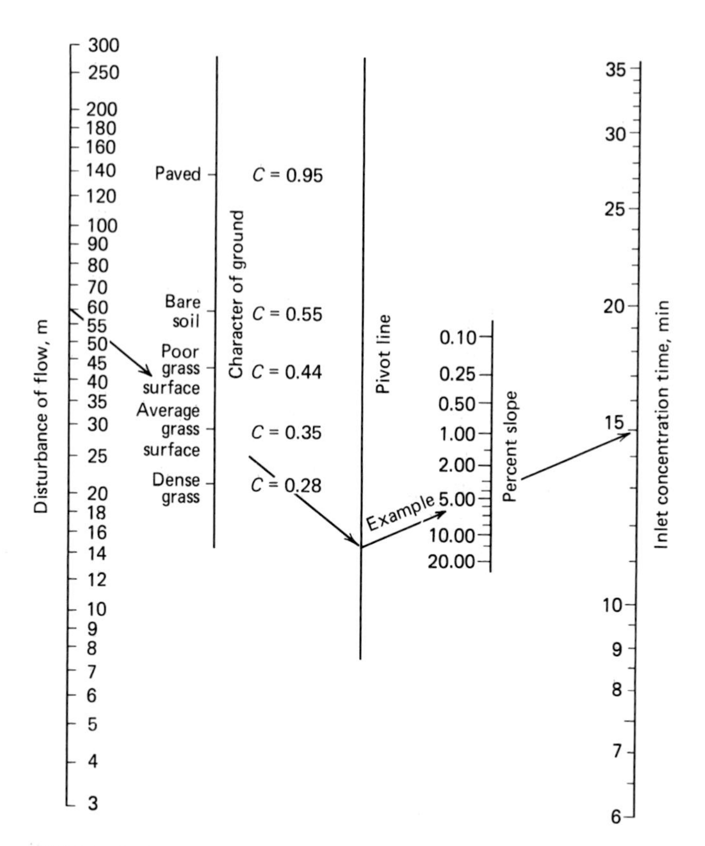

FIGURE 13-3
Overland flow time. (*Modified from a figure in Data Book for Civil Engineering, Vol. I, Design, 3d ed., by Elwyn E. Seelye. Copyright 1960. With permission of John Wiley and Sons, Inc.*)

Simpler methods of estimating the time of concentration include assuming a standard value, perhaps 5 to 10 min, or using a nomograph such as that of Fig. 13-3. In the example shown on the figure, a flow distance of 60 m over an ordinary grass surface with a slope of 4 percent yields a time of concentration of 15 min. This procedure neglects the effect of rainfall intensity but is adequate for the very small areas to which the rational method may be appropriately applied.

The maximum rate of runoff from a given rainfall intensity will occur when the rainfall has continued for a period sufficient to permit flow to reach the outlet from the most remote point in the drainage area. In Fig. 13-4, if it is assumed that it takes 5 min for water to flow from the boundary of one zone to the next, it is clear that at the end of 5 min only zone *A* will contribute flow at point *I*. Similarly, after 10 min only zones *A* and *B* will contribute, and 15 min will be required before the entire area is contributing flow. If the rainfall

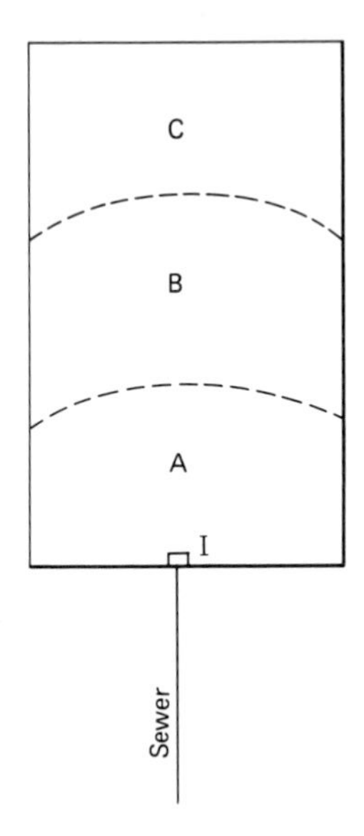

FIGURE 13-4
Diagram illustrating time of concentration.

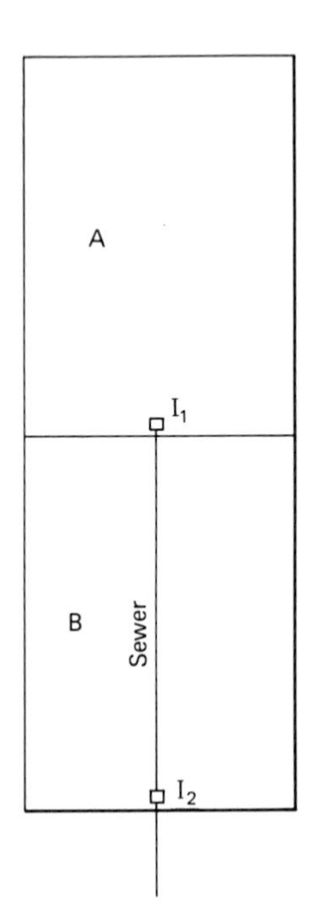

FIGURE 13-5
Diagram illustrating time of concentration.

continues for only 10 min, the water arriving from zone C during the period from 10 to 15 min would be offset, in part, by declining flow from area A. The same logic may be applied to the drainage area of Fig. 13-5. The water from A enters the sewer at I_1 and that from B at I_2. The time of concentration at I_2 is either the time of concentration for area B or the time of concentration for area A plus the time of flow in the sewer from I_1 to I_2, whichever is greater.

The time of concentration for each sewer is determined in similar fashion, by comparing the time of concentration for the area immediately tributary to the sewer inlet and the time of concentration plus time of flow for upstream tributary areas. When there is more than one upstream area, the time of concentration is the longest of those possible.

In determining rainfall intensity to be used in the rational method, it must be recognized that the shorter the duration of a rainfall event, the greater the expected average intensity will be. The critical duration of rainfall will be that which produces maximum runoff, and this will be that which is sufficient to produce flow from the entire drainage area. Shorter periods will provide lower flows, since the entire area is not involved, and longer periods will produce lower average intensities. The time of concentration is used in conjunction with curves like that of Fig. 13-1 to determine the appropriate intensity. This intensity is then substituted into Eq. (13-2) to determine the design flow for the sewer.

Example 13-3 A sewer drains a single-family residential area with $C = 0.35$. The distance of flow from the most remote point is 60 m over ordinary grass with a slope of 4 percent. The area drained is 10,000 m^2 and the data of Fig. 13-1 are used to determine intensity. Determine the design flow.

Solution
From Fig. 13-3, the time of concentration is found to be 15 min. From Fig. 13-1, i = 156 mm/h. Substituting values in Eq. (13-2) gives

$$Q = 0.35 \times 0.156 \text{ m/h} \times 10{,}000 \text{ m}^2$$

$$= 546 \text{ m}^3\text{/h} = 0.15 \text{ m}^3/s$$

13-3 The SCS Technique

The SCS technique was originally developed by the Soil Conservation Service (SCS) of the U.S. Department of Agriculture for use in rural areas. The procedure has been modified to permit its application to urban areas[5] and has been further adapted to a computerized simulation technique[6] which permits actual generation and routing of hydrographs.

The SCS technique hinges on determination of a *curve number*, CN, which depends primarily upon soil type, but which may be modified to account for the degree of development and antecedent moisture conditions. The curve number is a runoff coefficient of sorts which includes the effects of infiltration and detention storage. It is not only that, however, since its value is also influenced by the duration of excess rainfall. The peak runoff from an area with CN = 70 will not, in general, be seven-eighths that from an equal area with CN = 80, although it may approach that value in very large storms.

Curve numbers for areas of different descriptions are presented in Table 13-5. The hydrologic soil groups are identified in Ref. 5 for all soil series in the United States. Group A consists chiefly of deep, well- to excessively drained sands or gravels; group B consists chiefly of moderately deep to deep, moderately well to well-drained soils with moderately fine to moderately coarse texture; group C consists chiefly of soils with a layer which impedes downward motion of water or soils with moderately fine to fine textures; and group D consists chiefly of clay soils with a high swelling potential, soils with a permanent high groundwater table, soils with a claypan or clay layer at or near the surface, and shallow soils over nearly impervious materials.

For given rainfall depths and curve numbers, one may then determine the total depth of runoff from Table 13-6. The values of CN in Table 13-5 are based on the conditions specified in the table. If the area being studied does not meet the criteria in Table 13-4, a composite curve number can be developed in a manner analogous to that used for composite runoff coefficients in the rational method.

TABLE 13-5
Runoff curve numbers for selected agricultural, suburban, and urban land use[5]

Land use description	Hydrologic soil group			
	A	B	C	D
Cultivated land*:				
Without conservation treatment	72	81	88	91
With conservation treatment	62	71	78	81
Pasture or range land:				
Poor condition	68	79	86	89
Good condition	39	61	74	80
Meadow: good condition	30	58	71	78
Wood or forest land:				
Thin stand, poor cover, no mulch	45	66	77	83
Good cover†	25	55	70	77
Open spaces, lawns, parks, golf courses, cemeteries, etc.:				
Good condition: grass cover on 75% or more of the area	39	61	74	80
Fair condition: grass cover on 50–75% of the area	49	69	79	84
Commercial and business areas (85% impervious)	89	92	94	95
Industrial districts (72% impervious)	81	82	91	93
Residential‡ average lot size; average % impervious§				
500 m^2 or less; 65%	77	85	90	92
1000 m^2; 38%	61	75	83	87
1350 m^2; 30%	57	72	81	86
2000 m^2; 25%	54	70	80	85
4000 m^2; 20%	51	68	79	84
Paved parking lots, roofs, driveways, etc.¶	98	98	98	98
Streets and roads:				
Paved with curbs and storm sewers¶	98	98	98	98
Gravel	76	85	89	91
Dirt	72	82	87	89

* For a more detailed description of agricultural land use curve numbers refer to *National Engineering Handbook*, Sec. 4, "Hydrology," Chap. 9, August 1972.

† Good cover is protected from grazing. Litter and brush cover soil.

‡ Curve numbers are computed assuming the runoff from the house and driveway is directed toward the street with a minimum of roof water directed to lawns where additional infiltration could occur.

§ The remaining pervious areas (lawn) are considered to be in good pasture condition for these curve numbers.

¶ In some warmer climates of the United States, a curve number of 95 may be used.

Example 13-4 Determine the runoff from 125 mm of rain for a 4×10^6 m^2 watershed which is to be developed on soils of group C. The proposed land use is 50 percent detached houses on 1000-m^2 lots; 10 percent townhouses with 500-m^2 lots; 25 percent streets with curbs and gutters, schools, parking lots, and plazas; and 15 percent open space, parks, schoolyards, etc. with good grass cover.

Solution

Find the composite CN:

TABLE 13-6
Runoff depth in millimeters as a function of CN and rainfall (after Ref. 5)

	Curve number (CN)*								
Rainfall, mm	**60**	**65**	**70**	**75**	**80**	**85**	**90**	**95**	**98**
25	0	0	0	0.7	2.0	4.8	8.0	14.0	19.8
30	0	0	0.8	1.8	3.8	7.0	11.5	18.5	24.8
35	0	0.5	1.5	3.3	6.0	9.8	15.3	23.0	29.5
40	0.3	1.3	2.8	5.0	8.5	13.0	19.0	27.8	34.5
45	0.8	2.3	4.3	7.3	11.0	16.3	23.3	32.3	39.5
50	1.5	3.5	6.0	9.5	14.0	20.0	27.3	37.0	44.3
65	4.3	7.5	11.5	16.3	22.3	29.5	38.3	49.0	56.8
75	8.3	12.8	18.0	24.0	31.3	39.8	49.5	61.3	69.5
100	19.0	25.8	33.3	41.8	51.0	61.5	73.0	85.8	94.3
125	32.5	41.3	51.0	61.3	72.3	84.3	97.0	110.5	119.0
150	48.0	58.8	70.0	82.0	94.5	107.8	121.3	135.3	144.0
175	65.0	77.5	90.5	103.8	117.3	131.5	145.3	160.3	169.0
200	83.3	97.5	111.8	126.0	140.5	155.3	170.3	185.0	194.0
225	102.5	118.0	133.5	148.8	164.3	197.8	194.8	210.0	219.0
250	122.5	139.3	155.8	172.0	188.0	204.0	219.5	235.0	244.0
275	143.0	161.0	178.3	195.5	212.0	228.5	244.3	259.8	269.0
300	164.0	183.0	201.3	219.0	236.3	253.0	269.0	284.8	294.0

* For CN and rainfalls not in table use arithmetic interpolation.

Land use	Percent	CN	Product
Houses	50	83	4150
Townhouses	10	90	900
Streets, etc.	25	98	2450
Open space, etc.	15	74	1110

The sum of the products is 8610 and the composite curve number is 8610/100 = 86. From Table 13-6, interpolation for CN equal to 86 yields a runoff of 86.8 mm. The total runoff is thus $0.0868 \times 4 \times 10^6 = 347{,}200\ m^3$.

Example 13-5 Determine the CN for a development on soils consisting of 50 percent group B and 50 percent group C. The proposed land use is

40 percent residential which is 30 percent impervious
12 percent residential which is 65 percent impervious
18 percent paved roads with curbs and storm sewers
16 percent open land with 50 percent fair cover and 50 percent good cover
14 percent parking lots, plazas, schools, etc.

Solution
Tabulate the calculations as follows:

	Soil group					
	B			C		
Land use	**%**	**CN**	**Product**	**%**	**CN**	**Product**
Residential	20	72	1440	20	81	1620
Residential	6	85	510	6	90	540
Roads	9	98	882	9	98	882
Open land:						
Fair cover	4	69	276	4	79	316
Good cover	4	61	244	4	74	296
Parking lots, etc.	7	98	686	7	98	686
	50		4038	50		4340

$$\text{CN} = \frac{1}{2}\left(\frac{4038}{50} + \frac{4340}{50}\right) \approx 84$$

Although the SCS technique can be used to obtain synthetic hydrographs, there are better techniques available if such information is required. The SCS technique is very useful, however, for determining the total discharge (as above) and the peak flow.

The peak flow (except for parts of the Pacific coast) may be determined for agricultural areas from Figs. 13-6 through 13-8. The peak values obtained from these figures may then be adjusted for various degrees of urbanization as detailed below.

Figures 13-6 through 13-8 refer to flat, moderate, and steep slopes. These mean, specifically, 1 percent, 4 percent, and 16 percent slopes. For slopes other than these, the values obtained from the graphs must be multiplied by the factors in Table 13-7. In addition, the values from the peak discharge curves must be adjusted to reflect the percent of impervious area in the developed watershed and the percentage of the hydraulic length which has been modified. This latter factor refers to modification in the natural condition of the main channel. No distinction is made with respect to the nature of the modification. Figures 13-9 and 13-10 may be used to determine appropriate adjustment factors.

Example 13-6 A 1,200,000 m^2 watershed is to be developed. The CN for the proposed development is 80, and 60 percent of the hydraulic length will be modified by gutters and storm drains; 30 percent of the area will be impervious. The average slope is 5 percent. Compute the present and future peak runoff from a 125-mm rainfall. The present CN is 75.

Solution
From Table 13-6, the total runoff for the present and future conditions will be 61.3 and 72.3 mm, respectively. From Fig. 13-7 (moderate slope), the present and

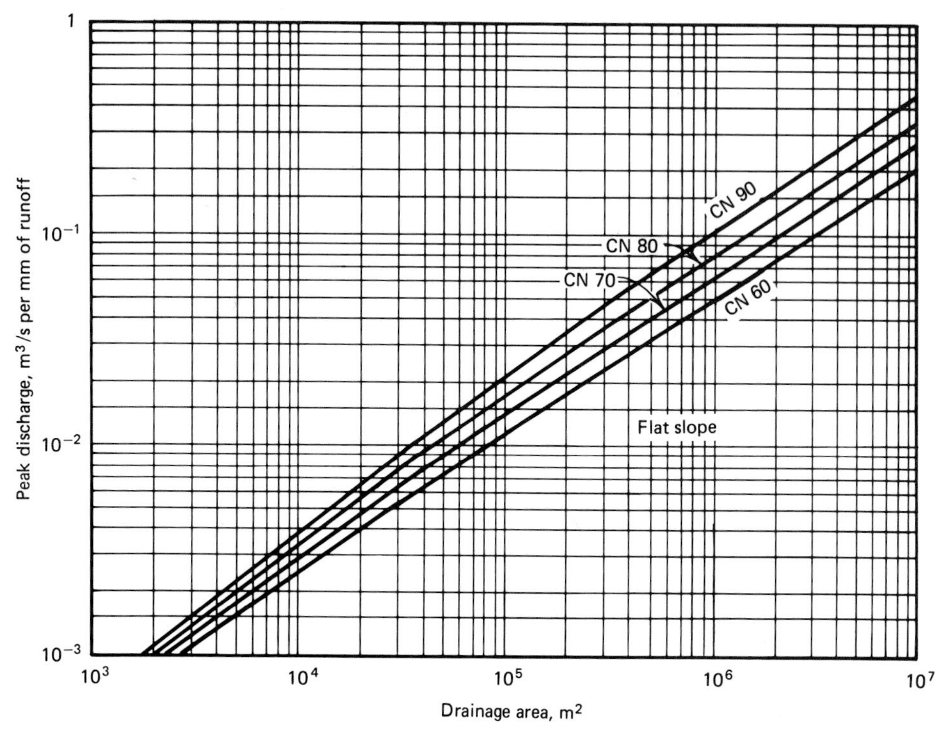

FIGURE 13-6
Peak rate of discharge for small watersheds—flat slope.

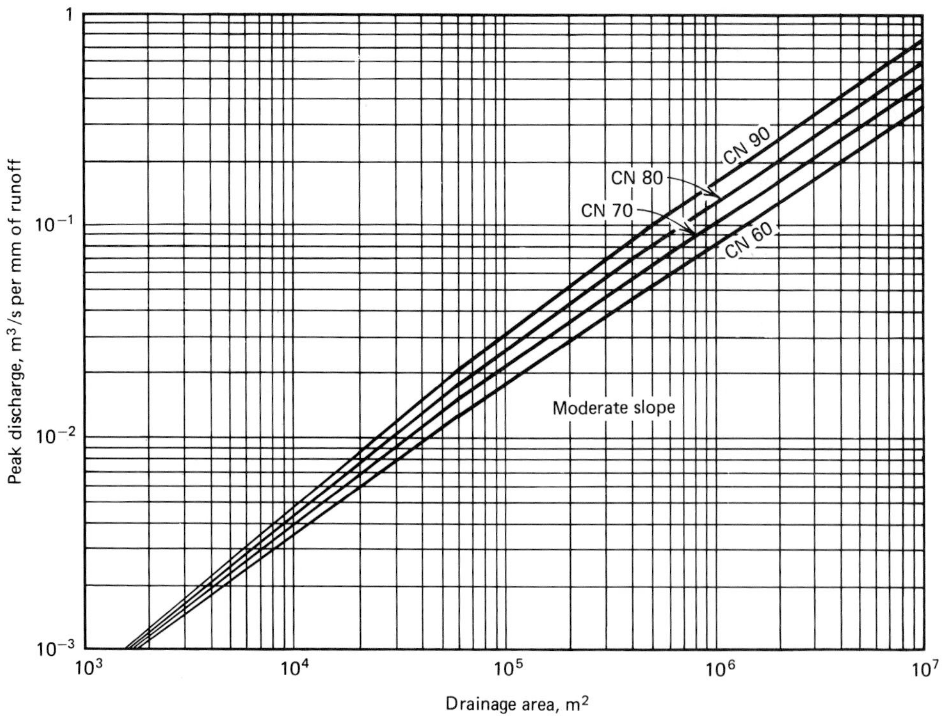

FIGURE 13-7
Peak rate of discharge for small watersheds—moderate slope.

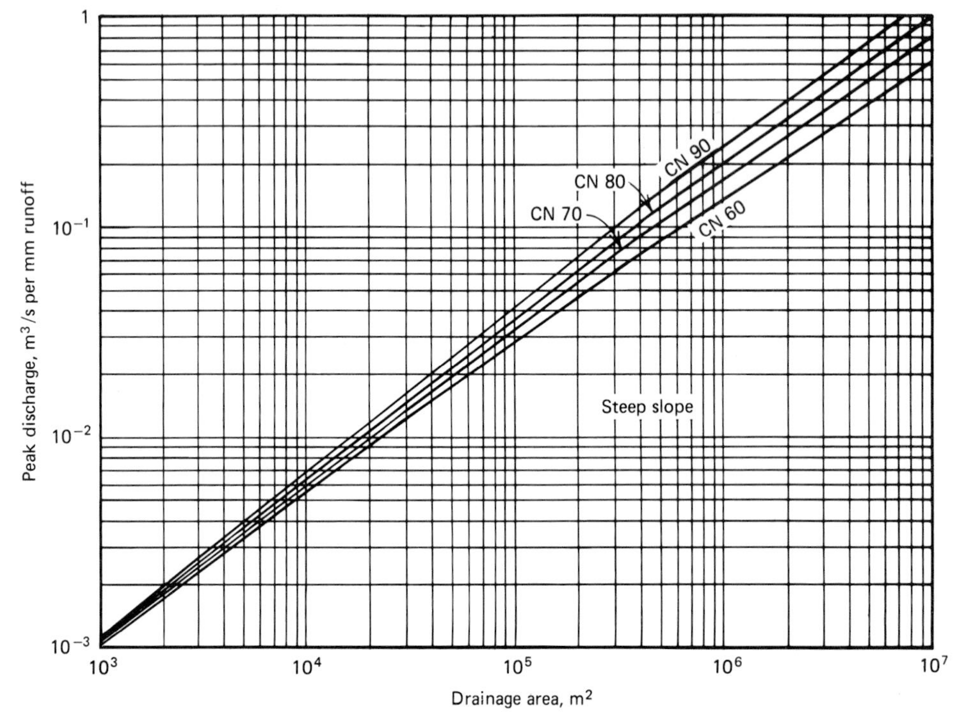

FIGURE 13-8
Peak rate of discharge for small watersheds—steep slope.

future peaks for agricultural land are 1.3×10^{-1} and 1.5×10^{-1} m³/s per millimeter. For 5 percent slope, these values must be multiplied by 1.08 (Table 13-7).

The present peak discharge from the design storm is thus

$$Q = 61.3 \times 1.3 \times 10^{-1} \times 1.08 = 8.6 \text{ m}^3/\text{s}$$

The future development's flow will be increased by factors of 1.16 (Fig. 13-9) and 1.40 (Fig. 13-10) as a result of increased imperviousness and improved hydraulic capacity.

The future peak is thus

$$Q = 72.3 \times 1.5 \times 10^{-1} \times 1.08 \times 1.16 \times 1.40 = 19.0 \text{ m}^3/\text{s}$$

The proposed development will increase the peak discharge by a factor of about 2.2.

The values obtained from Figures 13-6 through 13-8 are premised on approximately uniform surface flow. Where significant ponding occurs, the peaks will be reduced. Table 13-8 presents factors which may be used to adjust peak discharges for storms of different frequency. The table is for ponding areas which are centrally or uniformly distributed throughout the watershed. Peaks will be somewhat lower where the ponding areas are concentrated

TABLE 13-7
Slope adjustment factors for peak discharge (after Ref. 5)

	Area, 10^3 m							
Slope, %	**40**	**80**	**200**	**400**	**800**	**2000**	**4000**	**8000**
				Flat slopes				
0.1	0.49	0.47	0.44	0.43	0.42	0.41	0.41	0.40
0.2	0.61	0.59	0.56	0.55	0.54	0.53	0.53	0.52
0.3	0.69	0.67	0.65	0.64	0.63	0.62	0.62	0.61
0.4	0.76	0.74	0.72	0.71	0.70	0.69	0.69	0.69
0.5	0.82	0.80	0.78	0.77	0.77	0.76	0.76	0.76
0.7	0.90	0.89	0.88	0.87	0.87	0.87	0.87	0.87
1.0	1.00	1.00	1.00	1.00	1.00	1.00	1.00	1.00
1.5	1.13	1.14	1.14	1.15	1.16	1.17	1.17	1.17
2.0	1.21	1.24	1.26	1.28	1.29	1.30	1.31	1.31
				Moderate slopes				
3	0.93	0.92	0.91	0.90	0.90	0.90	0.89	0.89
4	1.00	1.00	1.00	1.00	1.00	1.00	1.00	1.00
5	1.04	1.05	1.07	1.08	1.08	1.08	1.09	1.09
6	1.07	1.10	1.12	1.14	1.15	1.16	1.17	1.17
7	1.09	1.13	1.18	1.21	1.22	1.23	1.23	1.24
				Steep slopes				
8	0.92	0.88	0.84	0.81	0.80	0.78	0.78	0.77
9	0.94	0.90	0.86	0.84	0.83	0.82	0.81	0.81
10	0.96	0.92	0.88	0.87	0.86	0.85	0.84	0.84
11	0.96	0.94	0.91	0.90	0.89	0.88	0.87	0.87
12	0.97	0.95	0.93	0.92	0.91	0.90	0.90	0.90
13	0.97	0.97	0.95	0.94	0.94	0.93	0.93	0.92
14	0.98	0.98	0.97	0.96	0.96	0.96	0.95	0.95
15	0.99	0.99	0.99	0.98	0.98	0.98	0.98	0.98
16	1.00	1.00	1.00	1.00	1.00	1.00	1.00	1.00
20	1.03	1.04	1.05	1.06	1.07	1.08	1.09	1.10
25	1.06	1.08	1.12	1.14	1.15	1.16	1.17	1.19
30	1.09	1.11	1.14	1.17	1.20	1.22	1.23	1.24
40	1.12	1.16	1.20	1.24	1.29	1.31	1.33	1.35
50	1.17	1.21	1.25	1.29	1.34	1.37	1.40	1.43

downstream and substantially higher where the ponding is concentrated upstream. Reference 5 should be consulted for further information on these conditions.

13-4 Hydrograph Techniques

The *unit hydrograph* has been defined as the hydrograph of surface runoff resulting from an effective rain (that is, one producing runoff) falling for a unit of time. The unit of time may be any value less than the time of concentration.

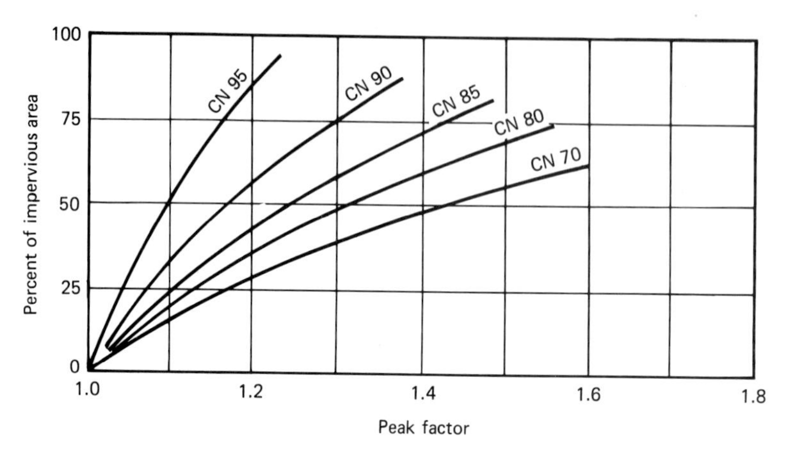

FIGURE 13-9
Factors for adjusting peak discharge for percentage imperviousness (value of CN for future conditions).

The usefulness of the unit hydrograph is based on the observation that all single storms on a watershed which have equal duration will produce runoff during equal lengths of time. For example, all storms on a watershed with a duration of 12 h may result in runoff extending over 5 days. Furthermore, the ordinates of the runoff hydrograph will be proportional to the rainfall excess (net rainfall minus abstractions).

A unit hydrograph may be constructed from existing records of rainfall and stream flow. Figure 13-11*a* from *A* to *C* represents the hydrograph of a storm of unit duration, assumed as 12 h in this case. The time intervals in the figure are also 12 h. The runoff curve before point *A* results from antecedent precipitation, and the recession from point *C* is a continuation of the base flow.

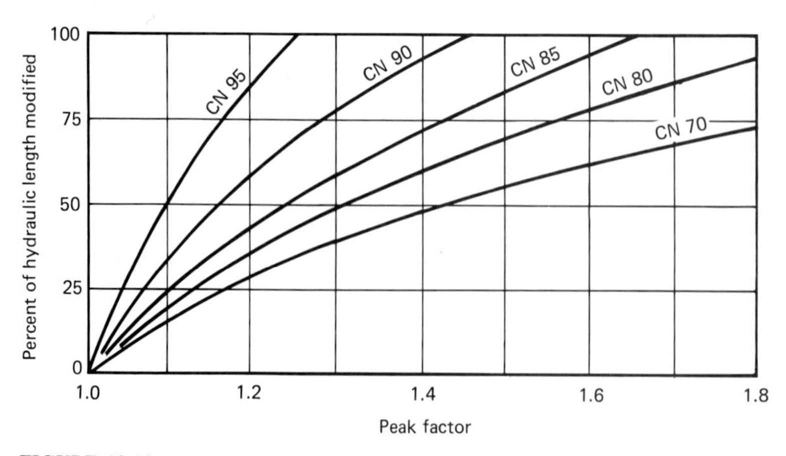

FIGURE 13-10
Factors for adjusting peak discharge for percentage of hydraulic length modified (value of CN for future conditions).

TABLE 13-8
Adjustment factors for central ponding[5]

Percentage of ponding and swampy area	Storm frequency, years					
	2	5	10	25	50	100
0.2	0.94	0.95	0.96	0.97	0.98	0.99
0.5	0.88	0.89	0.90	0.91	0.92	0.94
1.0	0.83	0.84	0.86	0.87	0.88	0.90
2.0	0.78	0.79	0.81	0.83	0.85	0.87
2.5	0.73	0.74	0.76	0.78	0.81	0.84
3.3	0.69	0.70	0.71	0.74	0.77	0.81
5.0	0.65	0.66	0.68	0.72	0.75	0.78
6.7	0.62	0.63	0.65	0.69	0.72	0.75
10.0	0.58	0.59	0.61	0.65	0.68	0.71
20.0	0.53	0.54	0.56	0.60	0.63	0.68
25.0	0.50	0.51	0.53	0.57	0.61	0.66

Figure 13-11*b* is obtained by subtracting the effects of antecedent rainfall and the base flow and dividing the ordinates by the total runoff. It represents a 12-h unit hydrograph for the drainage basin.

The concept of the unit hydrograph has been applied to small urban areas (0.04 to 38 km^2).[7] In this technique, parametric equations have been developed

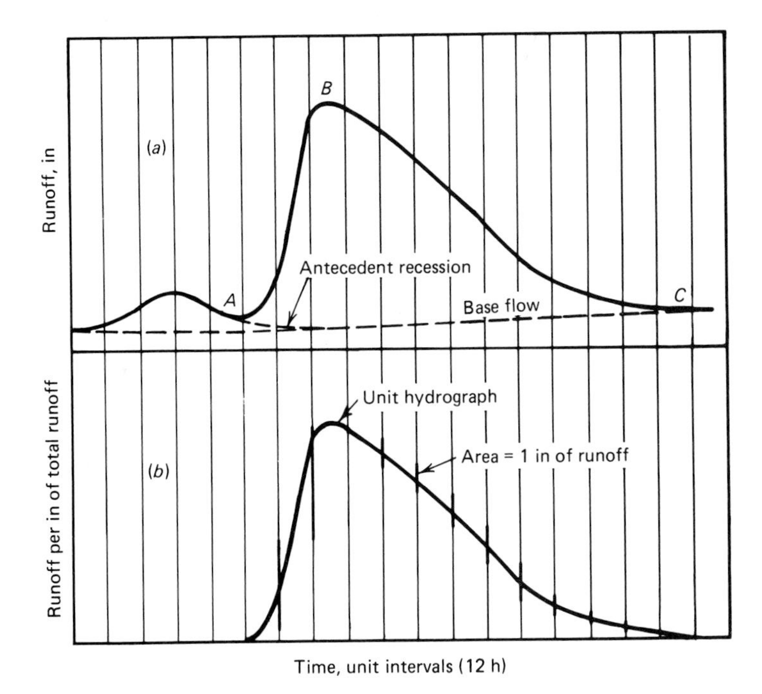

FIGURE 13-11
Construction of unit hydrograph.

which permit definition of the hydrograph shape based upon physical characteristics of the catchment. The equations are

$$T_R = 4.1L^{0.23}S^{-0.25}I^{-0.18}\Phi^{1.57} \tag{13-8}$$

$$Q = 13.27A^{0.96}T_R^{-1.07} \tag{13-9}$$

$$T_B = 71.21AQ^{-0.95} \tag{13-10}$$

$$W_{50} = 12.08A^{0.93}Q^{-0.92} \tag{13-11}$$

$$W_{75} = 7.21A^{0.79}Q^{-0.78} \tag{13-12}$$

where T_R = rise time of hydrograph, min
Q = peak discharge, m^3/s per mm net rain
T_B = time base of hydrograph, min
W_{50} = width of hydrograph, min, at 50% Q
W_{75} = width of hydrograph, min, at 75% Q
L = total distance along main channel, m
S = main channel slope (generally neglecting flatter 20% of upstream length)
I = imperviousness, %
Φ = dimensionless conveyance factor (0.6–1.3)
A = watershed area, km^2

These equations are based upon a 10-min unit time. The technique whereby a hydrograph is generated from a particular hyetograph is illustrated below.

Example 13-7 A 60-min storm has a net rainfall of 33 mm with 2.5 mm falling in the first 10 min, 6.5 in the second 10 min, 10 in the third, 9 in the fourth, 4 in the fifth and 1 in the sixth. Using the 10-min hydrograph equations, find the hydrograph resulting from the storm on a drainage area for which L = 4575 m, I = 30 percent, A = 5.5 km^2, Φ = 1.0, and S = 1 percent.

Solution
The unit hydrograph is defined by Eqs. (13-8) through (13-12). Substituting values gives

$$T_R = 4.1(4575)^{0.23}(0.01)^{-0.25}(30)^{-0.18}(1)^{1.57} = 49$$

$$Q = 13.27(5.5)^{0.96}(48.8)^{-1.07} = 1.06 \text{ m}^3\text{/s per mm}$$

$$T_B = 71.21(5.5)(1.06)^{-0.95} = 370 \text{ min}$$

$$W_{50} = 12.08(5.5)^{0.93}(1.06)^{-0.92} = 56 \text{ min}$$

$$W_{75} = 7.21(5.5)^{0.79}(1.06)^{-0.78} = 26 \text{ min}$$

The resulting unit hydrograph is shown in Fig. 13-12. Each 10-min rainfall will produce such a hydrograph with ordinate equal to the unit hydrograph value times the net rain. The six hydrographs for the storm can then be added (taking into

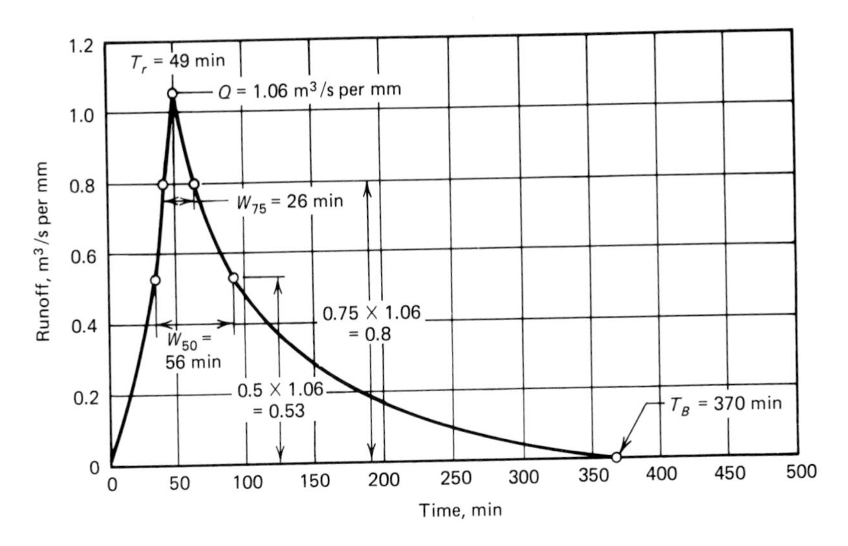

FIGURE 13-12
10-min unit hydrograph for an urban area.

account their various times of beginning) to produce the storm hydrograph of Fig. 13-13. The peak discharge is 27.2 m³/s and the total flow may be obtained from the area under the upper curve (approximately 180,000 m³).

13-5 Computer Simulation Techniques

Since the late 1940s, considerable attention has been directed to the development of computer-based models of urban drainage processes. These models require more or less complete definition of the hydraulic and hydrologic factors which affect the discharge and are capable of producing (when properly calibrated) a great deal of information concerning the response of a drainage system to any selected rainfall pattern. The various models vary widely in the level of detail required and produced and in the sophistication with which the hydraulic and hydrologic factors are modeled.

The most useful readily available technique for accurate routing in urban drainage systems involves solution of the one-dimensional equations of fluid motion called the Saint Venant equations. These consist of the *continuity equation*:

$$A\frac{\partial V}{\partial x} + V\frac{\partial A}{\partial x} + B\frac{\partial y}{\partial t} = 0 \tag{13-13}$$

and the *momentum equation*:

$$S_f = S_0 - \frac{\partial y}{\partial x} - \frac{V}{g}\frac{\partial V}{\partial x} - \frac{1}{g}\frac{\partial V}{\partial t} = \frac{V^2}{C^2 R} \tag{13-14}$$

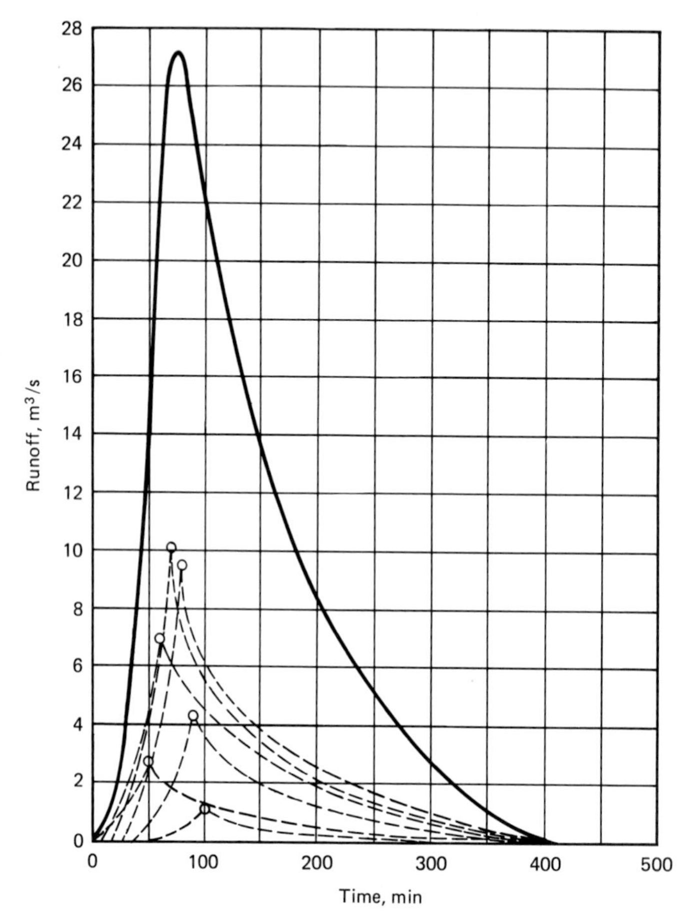

FIGURE 13-13
Unit hydrographs and storm hydrograph for an urban area.

where A = cross-sectional area of flow
V = velocity
B = width of water surface
S_f = friction slope from Manning's or Chézy's equation
S_0 = channel slope
y = water depth
x = distance along channel

These equations may be solved by a variety of finite-difference techniques, both explicit and implicit, provided the boundary conditions are properly described. The time step used in explicit solutions must generally be rather short, which leads to substantial computer use. These techniques do, however, yield the best solutions.

A number of approximate routing techniques are used which minimize computer time, at some cost in accuracy. The *kinematic wave* equation drops all the slope terms in the momentum equation save S_0. This neglects the dynamic terms, and flow variations will therefore propagate only in a downstream direction. As a result, downstream controls cannot be directly accounted for, although some models incorporate iterative procedures which attempt to adjust for this weakness.

Storage routing techniques based upon the continuity equation and the assumption that flow occurs at normal depth are used in many urban-drainage models. These procedures neglect all kinematic effects, including both the variable slope which occurs during the passage of a flood wave and backwater effects of downstream controls.

One of the first techniques to utilize the computer in analysis of urban-drainage problems was the *Chicago hydrograph method*.[8] The procedure is particularly tailored to the Chicago area and may not be useful in other communities without substantial modification. A design hyetograph, similar in shape to that of Fig. 13-12, with a rise time of 67.5 min and a base of 180 min, was selected on the basis of local records. Horton's equation (Chap. 4) was used to model infiltration, and, since a standard rainfall pattern was used, the equation could be directly integrated. Depression storage was considered to vary according to the ogee function of Fig. 4-6 and to total 6 mm on pervious areas and 1.5 mm on pavement. Overland flow (rainfall less infiltration and depression storage) was routed according to a modification of Izzard's overland flow equation (Eq. 13-6) and a storage routing procedure. Gutter routing was also done by using storage routing, but lateral sewers were modeled simply by lagging the hydrographs. For Chicago, this was found to give results equivalent to storage routing procedures.

ILLUDAS[9] is the U.S. version of the RRL method developed in Great Britain. The model requires that the drainage basin be broken down into subareas tributary to collection system inlets. Each subarea is subdivided into directly connected paved areas, contributing grassed areas, and supplemental grassed areas.

The program calculates inlet hydrographs from a hyetograph and descriptive data for the basin. Losses are based on an initial abstraction plus Horton's equation. Overland flow times are calculated from Manning's equation for paved areas and Izzard's overland flow hydrograph for grassed areas. Required data include length and slope of the subbasin, coefficients for Horton's equation, and percentages of paved, grassed, and supplemental paved area. In general, the required data may be obtained from aerial photographs and a physical inspection of the area.

Routing within the collection system may be effected by three different procedures: lagging without storage, storage routing with explicit solution of the continuity equation based on the assumption that the depth is uniform throughout each segment, and storage routing with implicit solution of the continuity equation in a manner analogous to the modified Puls method.

The hydraulic routing procedure is the weakest feature of ILLUDAS. The program has a design capability which will size required conduits. In the analysis mode, however, there is no carryover for water which is unable to enter the system. Flooding is thus not definable. The program will indicate that a conduit is inadequate but will not give any measure of the extent of flooding. Water which is "lost" at the inlets is lost from the simulation, and thus nonconservative errors can be introduced in subsequent computations.

HEC-1 is one of a family of hydrologic/hydraulic computer models developed by the U.S. Army Corps of Engineeers. In its most recent version,[10] it has the ability to model the components of urban drainage systems.

HEC-1 offers a number of internal options for generating hydrographs and for routing the flow. The data required include a hyetograph, definition of an infiltration/detention function, and sufficient hydraulic detail to permit routing by the selected technique. The program includes a number of storage routing techniques and can model pumping from one basin to another.

The hyetograph may be a historical record or may be synthesized from a duration-intensity curve (Art. 13-1). Interception-infiltration-detention can be modeled by any of four techniques: initial abstraction plus uniform loss rate, exponential loss rate, the SCS curve number, or the Holtan loss rate. Only the Holtan option permits consideration of multiple storms. Surface flow may be determined by any of three unit hydrograph techniques or a kinematic wave approximation.

The three unit hydrograph techniques are the Clark unit hydrograph, requiring three input parameters; a modification of the Snyder unit hydrograph, requiring two parameters; and the SCS dimensionless unit hydrograph, requiring a single parameter. All the hydrograph techniques require historical data for calibration of the model.

The kinematic wave approximation requires definition of the hydraulic characteristics of all system elements: overland flow, collector channels, and main channel. Values for roughness, hydraulic radius, area, and slope must be introduced into the program. The calculation procedure involves a solution of the finite-difference approximation to the kinematic wave equation. The solution is not necessarily correct, since the kinematic wave approach does not permit modeling of backwater effects. The model may be considerably simplified by neglecting minor collectors.

STORM is a Corps of Engineers model which was developed originally to assist in estimating pollutant loadings resulting from urban runoff.[11] STORM is a continuous simulation model which permits analysis of multiple storm events. The hydraulic routing procedure is far simpler than that in HEC-1 and therefore requires less data and yields less definitive results.

Runoff may be computed in STORM by three methods: a coefficient method similar to the rational method with an adjustment for depression storage; the SCS method, including adjustment for antecedent precipitation; and a combination of the two using the coefficient method for impervious areas and the SCS method for pervious areas.

Flows are routed through each subbasin by the SCS triangular unit hydrograph.[5] In order to use this procedure, the time of concentration and the ratio of recession time to peak time must be entered in the program.

STORM does not permit detailed modeling of hydraulic features of the drainage system. Its major intended use is in planning-level analysis of storm water pollution. The 1977 version does not permit channel routing or combination of subbasins.

SWMM[12] has been developed over an extended period of time under the auspices of the EPA and, like some of the other models discussed here, has substantial capabilities beyond simple storm-water modeling.

Rainfall is introduced in the form of hyetographs, which can be varied for different subareas. Abstractions are based on detention storage and infiltration. When rainfall exceeds losses, overland flow is calculated from Manning's equation. The calculation requires specification of a *characteristic width* for each subcatchment as well as roughness, slope, and area. Overland flow is routed using the continuity equation in storage form, which implies that the depth changes uniformly over the subcatchment. SWMM is the first of the methods considered here that considers storage in routing overland flow.

The overland flow may be routed to gutters, or these may be neglected, with flow being routed directly to the sewers. The gutter routing, if used, employs a quasi-steady-state method based on Manning's equation and the continuity equation.

The output from the RUNOFF module of the program is then normally routed through major collectors using a computational block called TRANSPORT, which employs a nonlinear kinematic wave procedure using a four-point implicit difference scheme and a dynamic wave approximation. The hydraulic calculations are superior to those used in HEC-1, but overall backwater effects are not modeled. An optional routing routine known as EXTRAN is available which will solve the dynamic equations for gradually varied flow. This technique is very versatile, but requires substantial computer time. EXTRAN, when used, does not permit simulation of storm-water quality, which is otherwise possible.

The program incorporates storage and pumping simulation capabilities like those of HEC-1 and can operate in either design or evaluation mode. A significant feature of SWMM is its ability to keep track of flows in excess of conduit capacity. If TRANSPORT is used, excess flows are "stored" at the inlet until the system can accommodate them. The program will print out messages at each time step for each location at which surcharging occurs, reporting the volume of water stored. EXTRAN, on the other hand, assumes that flows in excess of conduit capacity are discharged from the system at the location where the flow is too great. A summary table presents total discharge from each node. Discharges from interior nodes indicate inadequacies in the system.

SWMM has the capability of modeling quality as well as quantity of flow in storm sewers and combined sewerage systems. Subroutines which introduce sanitary sewage and industrial waste flows and predict storm-water quality based upon land use and antecedent conditions may be used, as may treatment routines. The latter are rather rudimentary, but do include subroutines for estimating treatment costs. The various blocks of SWMM can be, and often are, used alone. There is no need to simulate quality if one is interested only in the hydraulics of the collection system.

The *Illinois urban storm runoff method* (IUSR)[13] is one of the most sophisticated models available. Like SWMM and HEC-1, the model has capabilities beyond simple hydraulic analysis of storm sewers.

Hyetographs in the model supply water both to subcatchments consisting of linear strips and to gutters. All surface flows are modeled by the continuity equation and the kinematic wave approximation of the momentum equation. The difference equations are solved numerically in a manner similar to that used for major conduits in SWMM and HEC-1. The technique, while sophisticated, is not necessarily superior to use of the Manning or Izzard equations. No technique used for surface flow in these models can simulate backwater effects.

IUSR models inlets explicitly as weirs, orifices or a combination thereof. Other models do not include inlet effects. Flow in excess of inlet or conduit capacity is either stored at the inlet or bypassed to the next inlet downstream.

The model requires substantial data, including geometric characteristics of the subcatchments, gutters, and inlets. Like SWMM, it permits the use of different hyetographs for different subareas and may be operated in either design or analysis mode. In either case, the hydraulic calculations are conducted in a fashion which employs the Saint Venant equations and a method of successive approximations. This procedure permits modeling of backwater effects to a degree, but does not yield an exact solution.

IUSR requires substantially more data than SWMM, HEC-1 or ILLUDAS. The data required may not be readily available, and assuming values for the necessary parameters will tend to offset the potentially greater precision of the calculations. There are other models available which are not discussed here. Some are proprietary, while others were developed for particular communities and are not readily applicable elsewhere. The techniques discussed above, while not all-inclusive, define the range of systems available. Modern storm sewer design, save perhaps for small subdivisions, requires the use of procedures more sophisticated than the rational method. ILLUDAS and SWMM are particularly useful, and their minimum data requirements are not much more extensive than those of the rational method.

PROBLEMS

13-1. A residential urban area has the following proportions of different land use: roofs, 25 percent; asphalt pavement, 14 percent; concrete sidewalk, 5 percent; gravel driveways, 7 percent; grassy lawns with average soil and little slope, 49 percent. Compute an average runoff coefficient using the values in Table 13-2.

13-2. An urban area of 100,000 m^2 has a runoff coefficient of 0.45. Using a time of concentration of 25 min and the data of Fig. 13-1, compute the peak discharge resulting from a 10-year storm.

13-3. Using Fig. 13-5, find the time of concentration for an area in which the flow must travel 50 m over average grass on a slope of 5 percent.

13-4. A 700,000 m^2 watershed is 30 percent agricultural and 70 percent urban. The agricultural land is 40 percent cultivated with conservation treatment, 35 percent meadow in good condition, and 25 percent forest land with good cover. The urban area is residential: 60 percent on 1350 m^2 lots, 25 percent on 1000 m^2 lots, and 15 percent streets and roads with curbs and storm sewers. The soil group is B. Determine a composite CN and the total discharge from a 120-mm rainfall.

13-5. A watershed of 4,000,000 m^2 has a present CN of 70 and an average slope of 3 percent. Development will modify 70 percent of the hydraulic length, increase the impervious area to 40 percent, and increase CN to 80. Compute the present and future peak discharge from a 100-mm 24-h storm.

13-6. A developer proposes to construct a new housing project. The relevant data are

Factor	Present value	Future value
Area, m^2	3.88×10^6	3.88×10^6
CN	80	85
% impervious	—	50
% modified	—	65

Determine the present and future peak discharge and total runoff from a 150-mm 24-h storm. If all flow in excess of the present flow must be stored, how much reservoir capacity is required? The average slope is 1 percent.

13-7. Plot the hydrograph which would result from a 1-h storm event with 10-min net rainfalls of 10, 15, 20, 25, 15, and 5 mm. The drainage area has a hydraulic length of 10,000 m, an area of 15 km^2, I = 25 percent, Φ = 0.80, and s = 2.5 percent. What is the peak discharge? When does it occur?

REFERENCES

1. "Rainfall Frequency Atlas of the United States," Technical Paper No. 40, U.S. Department of Commerce, 1961.
2. "Five to 60-Minute Precipitation Frequency for the Eastern and Central United States," NWS Hydro-35, National Oceanic and Atmospheric Administration, Washington, D.C., 1977.
3. "Design and Construction of Sanitary and Storm Sewers," Manual of Practice 37, American Society of Civil Engineers, New York, 1969.
4. C. F. Izzard, "Hydraulics of Runoff from Developed Surfaces," *Proceedings of Highway Research Board*, **26**:129, 1943.
5. "Urban Hydrology for Small Watersheds," Technical Release No. 55, U.S. Department of Agriculture, 1975.
6. "Computer Program for Project Formulation—Hydrology (1982 Version)," Technical Release No. 20, 2d ed., U.S. Department of Agriculture, 1983.
7. William H. Espey Jr., Duke G. Altman, and Charles B. Graves, "Nomographs for Ten-Minute Unit Hydrographs for Small Urban Watersheds," *Urban Runoff Control Planning*, American Society of Civil Engineers, New York, 1977.
8. A. L. Tholin and C. J. Kiefer, "Hydrology of Urban Runoff," *Journal Sanitary Engineering Division, American Society of Civil Engineers*, **85**:SA2, 1959.

9. M. L. Terstriep and J. B. Stall, "The Illinois Urban Drainage Simulator—ILLUDAS," Bulletin 58, Illinois State Water Survey, Urbana, 1974.
10. "HEC-1 Flood Hydrograph Package Users Manual," 723-X6-L2010, U.S. Army Corps of Engineers, 1981.
11. "Storage, Treatment, Overflow, Runoff Model Users Manual," 723-S8-L7520, U.S. Army Corps of Engineers, 1977.
12. "Stormwater Management Model Users Manual Version III," U.S. Environmental Protection Agency, Cincinnati, 1982.
13. "Urban Stormwater Runoff: Determination of Volumes and Flowrates," U.S. Environmental Protection Agency, Cincinnati, 1976.

CHAPTER 14

SEWER MATERIALS

14-1 Precast Sewers

The pipe materials which are used to transport water (Chap. 6) may also be used to collect wastewater. It is more usual, however, to employ less expensive materials since sewers rarely are required to withstand any internal pressure. Iron and steel pipe are used to convey sewage only under unusual loading conditions or for force mains in which the sewage flow is pressurized.

The most commonly used sewer material is *clay pipe*, which is made of clay or shale that has been ground, wetted, molded, dried, and burned in a kiln. The burning produces a fusion or vitrification of the clay, making it very hard and dense and resistant to biological and chemical attack. Clay pipe was formerly glazed, producing a glasslike surface, but the glazing process is no longer provided since it contributed to air pollution.

Clay pipe is manufactured with integral bell-and-spigot ends and in a plain end configuration. The bell and spigot pipe is fitted with polymeric rings of various designs within the bell and a polymeric sleeve on the spigot.[1] The two ends are joined in a tight press fit with the aid of a lubricant. Plain-end pipe has a polymeric sleeve cast on either end and is joined by a plastic corrugated ring into which the lubricated pipe ends are pressed.[2] The polymeric ends provide a much tighter and more flexible joint than was available with older designs.

Dimensions and strengths of standard and extra-strength clay pipe are presented in Tables 14-1 and 14-2. Fittings are available in the forms illustrated in Fig. 14-1, and other shapes may be made on special order. Wyes and tees should be used for joining house sewers to common sewers and should be installed in the sewer when it is constructed—even if the adjoining property is not yet developed. The open ends of unused connections can be closed with stoppers or mortar until they are needed.

Failure to provide wyes or tees in common sewers invites builders to break the pipe to make new connections. Such breaks are seldom properly

TABLE 14-1
Minimum crushing strength of clay pipe (*reprinted by permission of the American Society for Testing and Materials, copyright 1977*)

Nominal size, in (mm)	Extra strength clay pipe		Standard strength clay pipe	
	lb/linear ft	kg/linear m	lb/linear ft	kg/linear m
4 (100)	2,000	2,980	1,200	1,790
6 (150)	2,000	2,980	1,200	1,790
8 (200)	2,200	3,270	1,400	2.080
10 (250)	2,400	3,570	1,600	2,380
12 (305)	2,600	3,870	1,800	2,680
15 (380)	2,900	4,320	2,000	2,980
18 (460)	3,300	4,910	2,200	3,270
21 (530)	3,850	5,730	2,400	3,570
24 (610)	4,400	6,550	2,600	3,870
27 (690)	4,700	6,990	2,800	4,170
30 (760)	5,000	7,440	3,300	4,910
33 (840)	5,500	8,190	3,600	5,360
36 (915)	6,000	8,930	4,000	5,950
39 (990)	6,600	9,820		
42 (1,070)	7,000	10,410		

closed and can be a major source of infiltration. City inspectors should check all new connections before the contractor is allowed to close the trench. When unplanned connections are necessary, wye or tee saddles may be installed in a carefully cut opening and be sealed with mortar.

Concrete pipe is sometimes used for sanitary sewers in locations where grades, temperatures, and sewage characteristics prevent corrosion (Art. 14-6). It is, of course, also used for storm drains—an application in which it is generally preferable to clay pipe.

Precast concrete pipe is manufactured with the dimensions and strengths shown in Table 14-3. Joints may be made with compression rings and gaskets or mortar or mastic packing, as shown in Fig. 14-2. For sanitary sewer applications, the compression rings or gaskets should be used. The dimensional tolerances are closer for pipe manufactured for use with such joints, and the joints themselves are less prone to leakage. Fittings such as tees and wyes may be readily obtained for concrete pipe, but are usually manufactured to order rather than kept in stock. Connections can also be made by carefully cutting the pipe, inserting the mating piece, and mortaring the joint.

All concrete pipe made in sizes larger than 610 mm (24 in) is *reinforced*, and reinforced pipe can also be obtained in sizes as small as 310 mm (12 in), as shown in Table 14-4. Joints are either bell-and-spigot or tongue-and-groove in

TABLE 14-2

Dimensions of clay pipe (*reprinted by permission of the American Society for Testing and Materials, copyright 1977*)

Nominal size, in (mm)	Laying length: Min., ft (m)	Laying length: Limit of minus variation, in/ft (mm/m)	Difference in length of two opposite sides, max., in (mm)	Outside diameter of barrel, in (mm): Minimum	Outside diameter of barrel, in (mm): Maximum	Inside diameter of socket at $\frac{1}{2}$ in (13 mm) above base, min, in (mm)*†
4 (100)	2 (0.61)	$\frac{1}{4}$ (20)	$\frac{5}{16}$ (8)	$4\frac{7}{8}$ (124)	$5\frac{1}{8}$ (130)	$5\frac{3}{4}$ (146)
6 (150)	2 (0.61)	$\frac{1}{4}$ (20)	$\frac{3}{8}$ (9)	$7\frac{1}{16}$ (179)	$7\frac{7}{16}$ (189)	$8\frac{3}{16}$ (208)
8 (200)	2 (0.61)	$\frac{1}{4}$ (20)	$\frac{7}{16}$ (11)	$9\frac{1}{4}$ (235)	$9\frac{3}{4}$ (248)	$10\frac{1}{2}$ (267)
10 (250)	2 (0.61)	$\frac{1}{4}$ (20)	$\frac{7}{16}$ (11)	$11\frac{1}{2}$ (292)	12 (305)	$12\frac{3}{4}$ (324)
12 (305)	2 (0.61)	$\frac{1}{4}$ (20)	$\frac{7}{16}$ (11)	$13\frac{3}{4}$ (349)	$14\frac{5}{16}$ (364)	$15\frac{1}{8}$ (384)
15 (380)	3 (0.91)	$\frac{1}{4}$ (20)	$\frac{1}{2}$ (13)	$17\frac{3}{16}$ (437)	$17\frac{13}{16}$ (452)	$18\frac{5}{8}$ (473)
18 (460)	3 (0.91)	$\frac{1}{4}$ (20)	$\frac{1}{2}$ (13)	$20\frac{5}{8}$ (524)	$21\frac{7}{16}$ (545)	$22\frac{1}{4}$ (565)
21 (530)	3 (0.91)	$\frac{1}{4}$ (20)	$\frac{9}{16}$ (14)	$24\frac{1}{8}$ (613)	25 (635)	$25\frac{7}{8}$ (657)
24 (610)	3 (0.91)	$\frac{3}{8}$ (30)	$\frac{9}{16}$ (14)	$27\frac{1}{2}$ (699)	$28\frac{1}{2}$ (724)	$29\frac{3}{8}$ (746)
27 (690)	3 (0.91)	$\frac{3}{8}$ (30)	$\frac{5}{8}$ (16)	31 (787)	$32\frac{1}{8}$ (816)	33 (838)
30 (760)	3 (0.91)	$\frac{3}{8}$ (30)	$\frac{5}{8}$ (16)	$34\frac{3}{8}$ (873)	$35\frac{5}{8}$ (905)	$36\frac{1}{2}$ (927)
33 (840)	3 (0.91)	$\frac{3}{8}$ (30)	$\frac{5}{8}$ (16)	$37\frac{5}{8}$ (956)	$38\frac{15}{16}$ (989)	$39\frac{7}{8}$ (1013)
36 (915)	3 (0.91)	$\frac{3}{8}$ (30)	$\frac{11}{26}$ (17)	$40\frac{3}{4}$ (1035)	$42\frac{1}{4}$ (1073)	$43\frac{1}{4}$ (1099)
39 (990)	5 (1.52)	$\frac{1}{4}$ (20)	$\frac{3}{4}$ (19)	$45\frac{3}{8}$ (1152)	$47\frac{1}{4}$ (1200)	$48\frac{1}{2}$ (1232)
42 (1070)	5 (1.52)	$\frac{3}{8}$ (30)	$\frac{7}{8}$ (23)	$48\frac{1}{2}$ (1232)	51 (1295)	$52\frac{1}{2}$ (1333)

Nominal size, in (mm)	Depth of socket*†: Nominal, in (mm)	Depth of socket*†: Min., in (mm)	Thickness of barrel, Extra strength: Nominal, in (mm)	Thickness of barrel, Extra strength: Min., in (mm)	Thickness of barrel, Standard strength: Nominal, in (mm)	Thickness of barrel, Standard strength: Min., in (mm)	Thickness of socket at $\frac{1}{2}$ in (13 mm) from outer end†: Nominal, in (mm)	Thickness of socket at $\frac{1}{2}$ in (13 mm) from outer end†: Min., in (mm)
4 (100)	$1\frac{3}{4}$ (44)	$1\frac{1}{2}$ (38)	$\frac{5}{8}$ (16)	$\frac{9}{16}$ (14)	$\frac{1}{2}$ (13)	$\frac{7}{16}$ (11)	$\frac{7}{16}$ (11)	$\frac{3}{8}$ (9)
6 (150)	$2\frac{1}{4}$ (57)	2 (51)	$\frac{11}{16}$ (17)	$\frac{9}{16}$ (14)	$\frac{5}{8}$ (16)	$\frac{9}{16}$ (14)	$\frac{1}{2}$ (13)	$\frac{7}{16}$ (11)
8 (200)	$2\frac{1}{2}$ (64)	$2\frac{1}{4}$ (57)	$\frac{7}{8}$ (22)	$\frac{3}{4}$ (19)	$\frac{3}{4}$ (19)	$\frac{11}{16}$ (17)	$\frac{9}{16}$ (14)	$\frac{1}{2}$ (13)
10 (250)	$2\frac{5}{8}$ (67)	$2\frac{3}{8}$ (60)	1 (25)	$\frac{7}{8}$ (22)	$\frac{7}{8}$ (22)	$\frac{13}{16}$ (21)	$\frac{5}{8}$ (16)	$\frac{9}{16}$ (14)
12 (305)	$2\frac{3}{4}$ (70)	$2\frac{1}{2}$ (64)	$1\frac{3}{16}$ (30)	$1\frac{3}{16}$ (27)	1 (25)	$\frac{15}{16}$ (24)	$\frac{3}{4}$ (19)	$\frac{11}{16}$ (17)
15 (380)	$2\frac{7}{8}$ (73)	$2\frac{5}{8}$ (67)	$1\frac{1}{2}$ (38)	$1\frac{3}{8}$ (35)	$1\frac{1}{4}$ (31)	$1\frac{1}{8}$ (29)	$\frac{15}{16}$ (24)	$\frac{7}{8}$ (22)
18 (460)	3 (76)	$2\frac{3}{4}$ (70)	$1\frac{7}{8}$ (48)	$1\frac{3}{4}$ (44)	$1\frac{1}{2}$ (38)	$1\frac{3}{8}$ (35)	$1\frac{1}{8}$ (29)	$1\frac{1}{16}$ (27)
21 (530)	$3\frac{1}{4}$ (83)	3 (76)	$2\frac{1}{4}$ (57)	2 (51)	$1\frac{3}{4}$ (44)	$1\frac{5}{8}$ (41)	$1\frac{5}{16}$ (33)	$1\frac{3}{16}$ (30)
24 (610)	$3\frac{3}{8}$ (86)	$3\frac{1}{8}$ (79)	$2\frac{1}{2}$ (64)	$2\frac{1}{4}$ (57)	2 (51)	$1\frac{7}{8}$ (48)	$1\frac{1}{2}$ (38)	$1\frac{3}{8}$ (35)
27 (690)	$3\frac{1}{2}$ (89)	$3\frac{1}{4}$ (83)	$2\frac{3}{4}$ (70)	$2\frac{1}{2}$ (64)	$2\frac{1}{4}$ (57)	$2\frac{1}{8}$ (54)	$1\frac{11}{16}$ (43)	$1\frac{9}{16}$ (40)
30 (760)	$3\frac{5}{8}$ (92)	$3\frac{3}{8}$ (86)	3 (76)	$2\frac{3}{4}$ (70)	$2\frac{1}{2}$ (64)	$2\frac{3}{8}$ (60)	$1\frac{7}{8}$ (48)	$1\frac{3}{4}$ (44)
33 (840)	$3\frac{3}{4}$ (95)	$3\frac{1}{2}$ (89)	$3\frac{1}{4}$ (83)	3 (76)	$2\frac{5}{8}$ (67)	$2\frac{1}{2}$ (64)	2 (51)	$1\frac{3}{4}$ (44)
36 (915)	4 (102)	$3\frac{3}{4}$ (95)	$3\frac{1}{2}$ (89)	$3\frac{1}{4}$ (83)	$2\frac{3}{4}$ (70)	$2\frac{5}{8}$ (67)	$2\frac{1}{16}$ (52)	$1\frac{7}{8}$ (48)
39 (990)	$4\frac{1}{8}$ (105)	$3\frac{7}{8}$ (98)	$3\frac{3}{4}$ (95)	$3\frac{3}{8}$ (86)	...	...	$2\frac{3}{4}$ (70)	$2\frac{5}{8}$ (67)
42 (1070)	$4\frac{1}{8}$ (105)	$3\frac{7}{8}$ (98)	4 (102)	$3\frac{1}{2}$ (89)	...	...	$2\frac{3}{4}$ (70)	$2\frac{5}{8}$ (67)

* The minimums for inside diameter of socket and depth of socket may be waived where such dimensions are conducive to the proper application of the joint.

† Plain-end pipe shall conform to the dimensions in this table, except those dimensions pertaining to sockets.

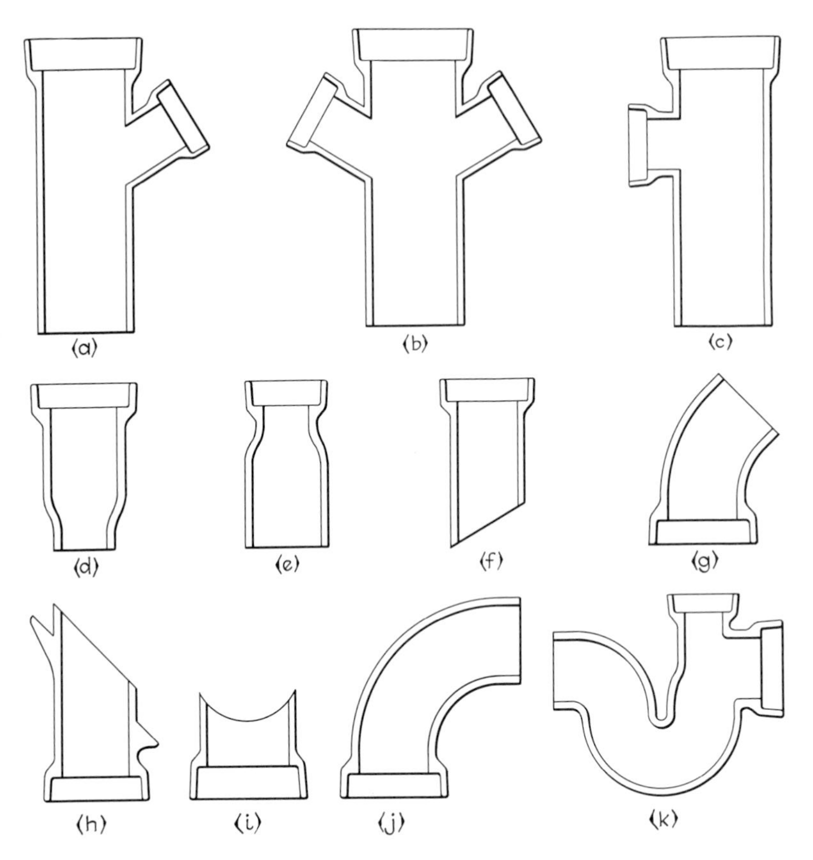

FIGURE 14-1
Sections of bell-and-spigot fittings for vitrified clay pipe. (*a*) Wye; (*b*) double wye; (*c*) tee; (*d*) reducer; (*e*) increaser; (*f*) slant; (*g*) $\frac{1}{8}$ bend; (*h*) wye saddle; (*i*) tee saddle; (*j*) $\frac{1}{4}$ bend; (*k*) running trap.

TABLE 14-3
Physical and dimensional requirements for nonreinforced concrete pipe (*reprinted by permission of the American Society for Testing and Materials, copyright 1977*)

	Class 1		Class 2		Class 3	
Internal diameter, mm (in)	Minimum thickness of wall, mm	Minimum strength, kN/linear m, three-edge bearing	Minimum thickness of wall, mm	Minimum strength, kN/linear m, three-edge bearing	Minimum thickness of wall, mm	Minimum strength, kN/linear m, three-edge bearing
100 (4)	15.9	21.9	19.0	29.2	22.2	35.0
150 (6)	15.9	21.9	19.0	29.2	25.4	35.0
200 (8)	19.0	21.9	22.2	29.2	28.6	35.0
250 (10)	22.2	23.3	25.4	29.2	31.8	35.0
310 (12)	25.4	26.3	34.9	32.8	44.5	37.9
380 (15)	31.8	29.2	41.3	37.9	47.6	42.2
460 (18)	38.1	32.1	50.8	43.8	57.2	48.1
530 (21)	44.5	35.0	57.2	48.1	69.9	56.2
610 (24)	54.0	37.9	76.2	52.5	95.3	64.2

A —Typical cross sections of joints with mortar or mastic packing

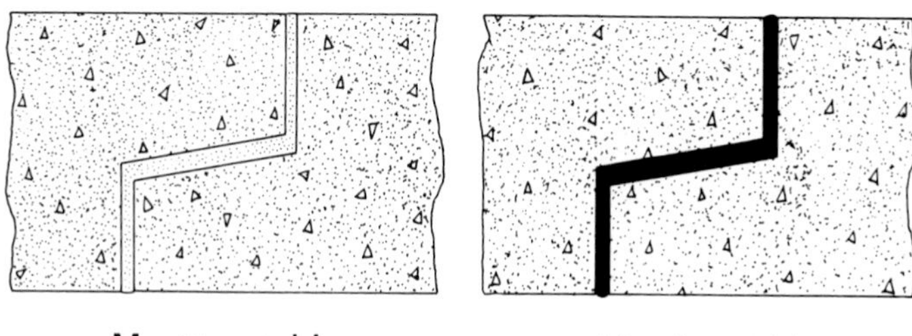

B —Typical cross sections of basic compression type rubber gasket joints

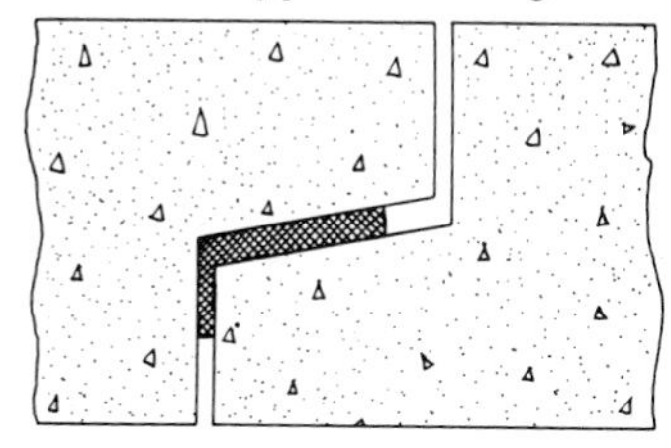

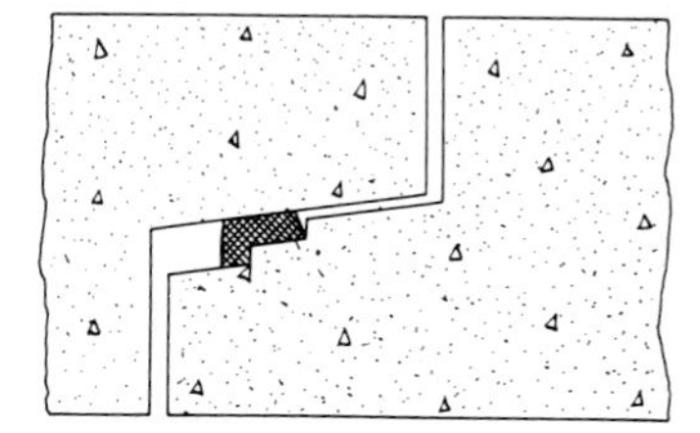

C —Typical cross sections of opposing shoulder type joint with O-ring gasket

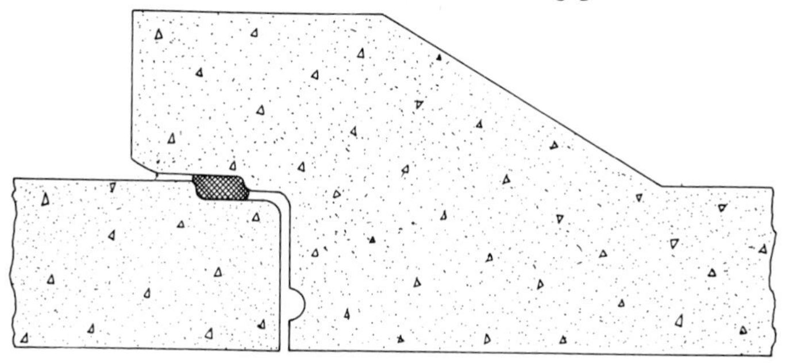

D —Typical cross section of spigot groove type joint with O-ring gasket

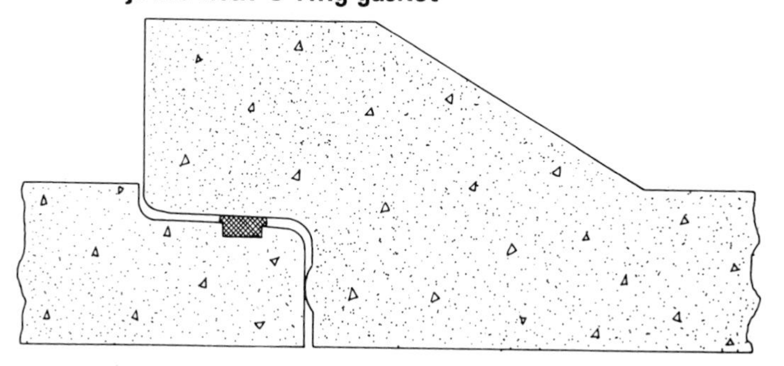

FIGURE 14-2
Typical concrete pipe joints. (*Courtesy American Concrete Pipe Association.*)

TABLE 14-4
Dimensions of reinforced concrete pipe*

Internal diameter, mm (in)	Wall thickness, mm (in)		
	Wall A	Wall B	Wall C
310 (12)	44 (1¾)	51 (2)	
380 (15)	47 (1⅞)	57 (2¼)	
460 (18)	51 (2)	63 (2½)	
530 (21)	57 (2¼)	70 (2¾)	
610 (24)	63 (2½)	76 (3)	95 (3¾)
690 (27)	66 (2⅝)	83 (3¼)	101 (4)
760 (30)	70 (2¾)	89 (3½)	108 (4¼)
840 (33)	73 (2⅞)	95 (3¾)	114 (4½)
910 (36)	76 (3)	101 (4)	120 (4¾)
1070 (42)	89 (3½)	114 (4½)	130 (5¼)
1220 (48)	101 (4)	127 (5)	146 (5¾)
1370 (54)	114 (4½)	140 (5½)	159 (6¼)
1520 (60)	127 (5)	152 (6)	171 (6¾)
1680 (66)	140 (5½)	165 (6½)	184 (7¼)
1830 (72)	152 (6)	178 (7)	197 (7¾)
1980 (78)	165 (6½)	190 (7½)	209 (8¼)
2130 (84)	178 (7)	203 (8)	222 (8¾)
2290 (90)	190 (7½)	216 (8½)	235 (9¼)
2440 (96)	203 (8)	229 (9)	248 (9¾)
2590 (102)	216 (8½)	241 (9½)	260 (10¼)
2740 (108)	229 (9)	254 (10)	273 (10¾)
2900 (114)	241 (9½)		
3050 (120)	254 (10)		
3200 (126)	267 (10½)		
3350 (132)	279 (11)		
3500 (138)	292 (11½)		
3650 (144)	305 (12)		
3800 (150)	318 (12½)		
3960 (156)	330 (13)		
4110 (162)	343 (13½)		
4270 (168)	356 (14)		
4420 (174)	368 (14½)		
4570 (180)	381 (15)		

* Not all sizes are available in all classes.

sizes up to 760 mm (30 in) and tongue-and-groove above that size. Joints are made by mortaring the cleaned and wetted tongue and groove before assembly or by mechanical or O-ring joints similar to those shown in Fig. 14-3.

Large concrete pipe lines are often assembled without fittings, although anything which might be necessary is available on special order. Curves are made by deflecting the joints and pouring concrete to fill the opening, while connections are made either at manholes (Chap. 15) or by cutting the pipe as described above.

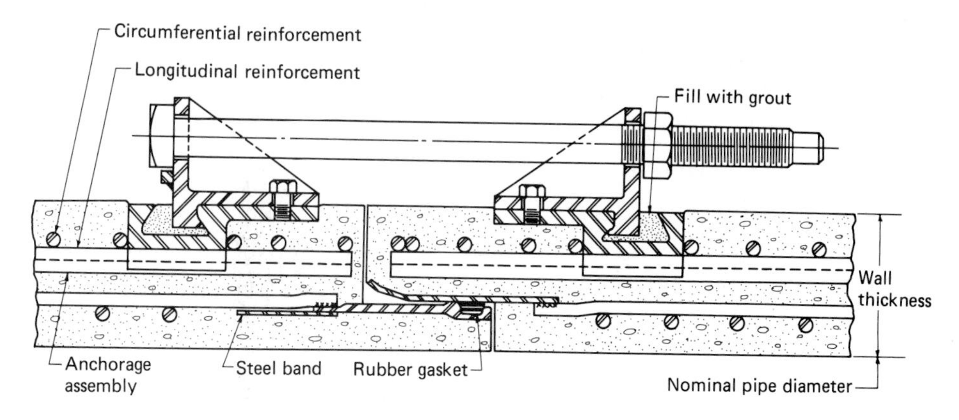

(a) Subaqueous joint for concrete pipe outfall

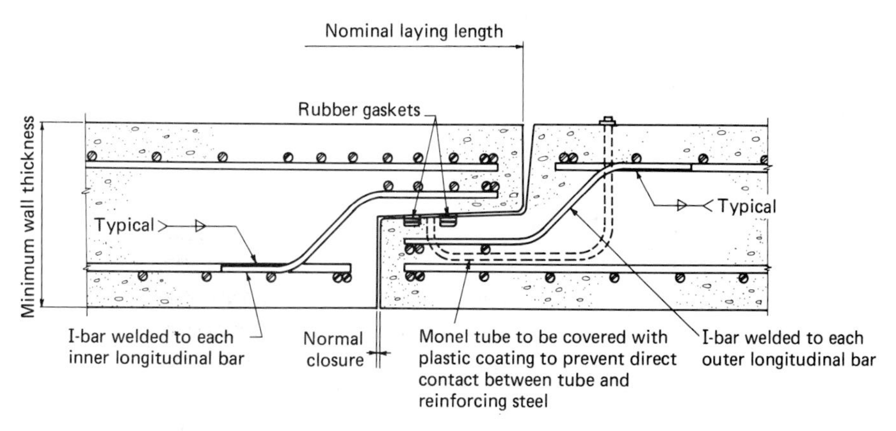

(b) Double rubber gasket joint for large-diameter concrete pipe outfall

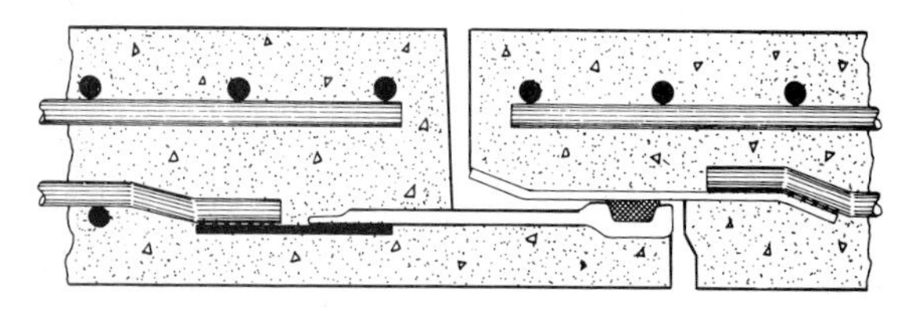

(c) Typical cross section of steel end ring joint with spigot groove and O-ring gasket

FIGURE 14-3
Reinforced-concrete pipe joints. (*Courtesy Portland Cement Association and American Concrete Pipe Association.*)

Asbestos cement pipe is manufactured in the sizes and classes shown in Table 14-5. Joints are made in the same manner as with asbestos cement water mains, with a sleeve and rubber gaskets that slip over the ends of adjoining sections.

TABLE 14-5
Minimum crushing loads for asbestos cement pipe (*reprinted by permission of the American Society for Testing and Materials, copyright 1977*)

Nominal size in (mm)	Crushing strength per linear foot, lb (kN/m)		
	Class 100	Class 150	Class 200
4 (102)	4,100 (59.8)	5,400 (78.8)	8,700 (126.9)
6 (152)	4,000 (58.4)	5,400 (78.8)	9,000 (131.3)
8 (203)	4,000 (58.4)	5,500 (80.2)	9,300 (135.8)
10 (254)	4,400 (64.2)	7,000 (102.1)	11,000 (160.5)
12 (304)	5,200 (75.8)	7,600 (110.8)	11,800 (172.3)
14 (356)	5,200 (75.8)	8,600 (125.5)	13,500 (197.1)
16 (406)	5,800 (84.6)	9,200 (134.2)	15,400 (224.8)
18 (457)	6,500 (94.8)	10,100 (147.4)	17,400 (254.0)
20 (508)	7,100 (103.6)	10,900 (159.0)	19,400 (283.2)
24 (610)	8,100 (118.2)	12,700 (185.3)	22,600 (329.9)
30 (762)	9.700 (141.5)	15,900 (231.9)	28,400 (414.6)
36 (914)	11,200 (163.4)	19,600 (285.9)	33,800 (493.5)

14-2 Strength and Bedding of Sewers

Since sewers are not ordinarily pressurized, are often more deeply buried than water mains, and are normally made of brittle, rather weak materials, the effect of soil and other external loads is quite important. The static load produced on buried pipe may be calculated from Marston's equation:

$$W = CwB^2 \tag{14-1}$$

where W = load on the pipe per unit length
w = weight of fill material per unit volume
B = width of trench just below top of pipe
C = coefficient that depends on the depth of trench, character of construction, and fill material

For ordinary trench construction, C may be calculated from

$$C = \frac{1 - e^{-2K\mu' H/B}}{2K\mu'} \tag{14-2}$$

where H = depth of fill above pipe
B = width of trench just below top of pipe
K = ratio of active lateral pressure to vertical pressure
μ' = coefficient of sliding friction between fill material and sides of trench

The product $K\mu'$ ranges from 0.1 to 0.16 for most soils, as shown in Table 14-6.

TABLE 14-6
Value of the product $K\mu'$

Soil type	Maximum value of $K\mu'$
Cohesionless granular material	0.192
Sand and gravel	0.165
Saturated top soil	0.150
Clay	0.130
Saturated clay	0.110

Graphical solutions of Eqs. (14-1) and (14-2) and of similar equations for other construction conditions are presented in Refs. 3 and 4. Tabular listings of allowable loadings or depths of fill for the bedding conditions shown in Figures 14-4 and 14-5 are also available from manufacturer's literature.[3–5]

The weights of materials commonly used for backfill are presented in Table 14-7. Figure 14-4 presents standard bedding conditions for clay pipe and load factors provided by each condition. The load factor is the ratio of the strength of the pipe when installed in the manner shown to that obtained in a standardized three-point bearing test. The values obtained in the standardized test are tabulated in Table 14-1.

Example 14-1 A 610-mm (24-in) clay sewer is to be placed in an ordinary trench 3.66 m (12 ft) deep and 1.22 m (4 ft) wide which will be filled with wet clay weighing 1920 kg/m³ (120 lb/ft³). Determine the load on the pipe and the type of bedding required if the installation is to have a factor of safety of 1.5.

Solution
From Eq. (14-2),

$$C = \frac{1 - e^{-0.110(2)(3.66/1.22)}}{2(0.110)} = 2.20$$

From Eq. (14-1),

$$W = 2.20(1920)(1.22)^2 = 6290 \text{ kg/m } (4220 \text{ lb/ft})$$

TABLE 14-7
Unit weight of backfill material

Material	Unit weight	
	kg/m³	lb/ft³
Dry sand	1600	100
Ordinary sand	1840	115
Wet sand	1920	120
Damp clay	1920	120
Saturated clay	2080	130
Saturated topsoil	1840	115
Sand and damp topsoil	1600	100

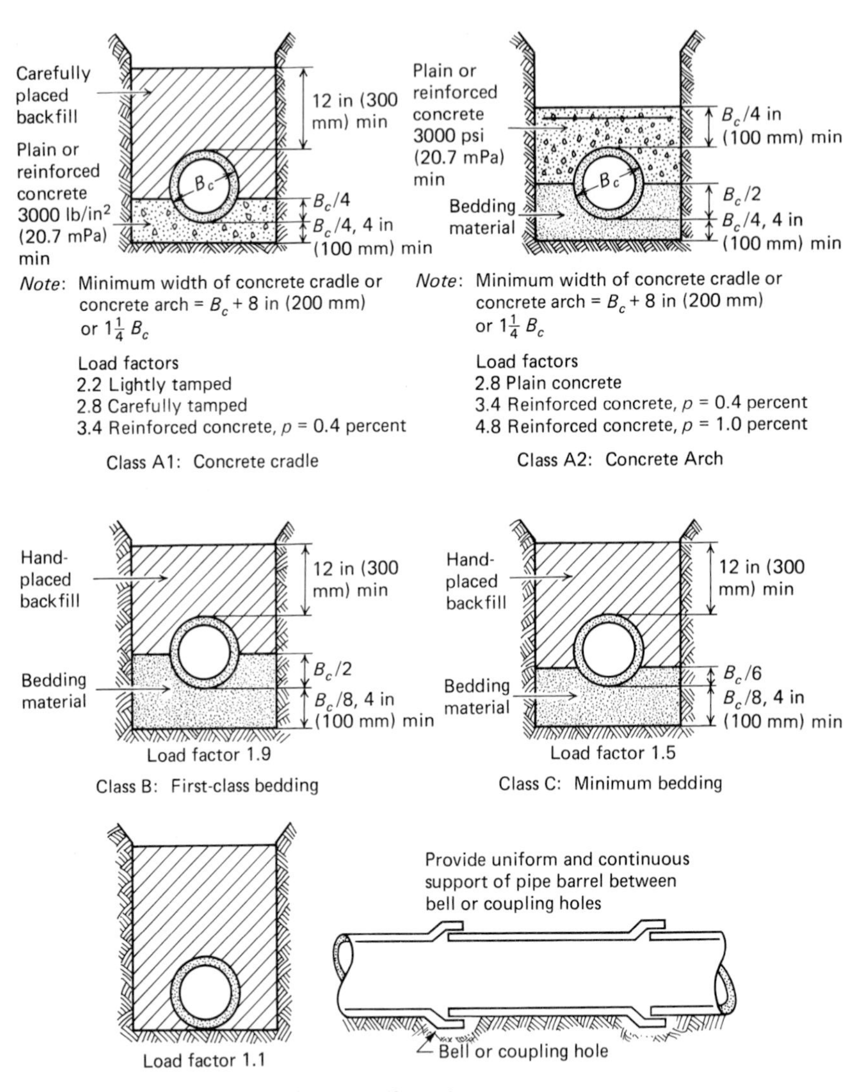

FIGURE 14-4
Methods of bedding clay pipe and load factors applicable to strength. (*Reprinted by permission of the American Society for Testing and Materials, Copyright 1977.*)

From Table 14-1, standard-strength pipe and extra-strength pipe have crushing strengths, respectively, of 3870 and 6550 kg/m. To apply the factor of safety, one may either increase the applied load by a factor of 1.5 or reduce the crushing strength by a factor of 1.5. In either case, the ratio of (load × safety factor)/strength becomes 2.44 for the standard-strength pipe and 1.4 for extra-strength pipe. Considering Fig. 14-4, one can see that only concrete encasement (class A) will provide a load factor of 2.44, while minimum bedding (class C) will provide a load factor of 1.4.

Although extra-strength pipe is somewhat more expensive than standard-strength pipe, the additional cost of concrete bedding would be much greater than the difference in pipe cost. Hence, the most economical selection would be extra-strength pipe with class C bedding. When the bedding is the same for the two strengths, standard-strength pipe is selected. The choice between class B with standard pipe and class C with extra-strength pipe may be properly left to the contractor.

Provision of proper bedding is very important in developing the strength of the pipe, assuring it is laid to the proper grade, and preventing subsequent settlement. In unfavorable soil conditions (wet clays, organic soils, etc.) bedding is particularly important. In some cases it may be necessary to leave sheeting in place, construct a floor, or even support the pipe on piles. The Class D bedding illustrated at the bottom of Fig. 14-4 is not considered satisfactory and should not be permitted in new construction.

In addition to the loads imposed by backfill, superficial loads on the soil produced by buildings, stockpiled materials, vehicles, and similar sources may reach buried sewers. The proportion of such loads reaching the pipe may be estimated from Tables 14-8 and 14-9. In Table 14-8 "long" loads are those longer than the width of the trench. "Short" loads, in Table 14-9, are those applied over lengths shorter than the trench width or perpendicular to the trench. The maximum values in Table 14-9 are for a load length equal to the trench width. The minima are for a length equal to one-tenth the width.

TABLE 14-8
Proportion of "long" superficial loads reaching pipe in trenches

Ratio of depth to width	Sand and damp topsoil	Saturated topsoil	Damp yellow clay	Saturated yellow clay
0.0	1.00	1.00	1.00	1.00
0.5	0.85	0.86	0.88	0.89
1.0	0.72	0.75	0.77	0.80
1.5	0.61	0.64	0.67	0.72
2.0	0.52	0.55	0.59	0.64
2.5	0.44	0.48	0.52	0.57
3.0	0.37	0.41	0.45	0.51
4.0	0.27	0.31	0.35	0.41
5.0	0.19	0.23	0.27	0.33
6.0	0.14	0.17	0.20	0.26
8.0	0.07	0.09	0.12	0.17
10.0	0.04	0.05	0.07	0.11

TABLE 14-9
Proportion of "short" superficial loads reaching pipe in trenches

Ratio of depth to width	Sand and damp topsoil		Saturated topsoil		Damp clay		Saturated clay	
	Max.	Min.	Max.	Min.	Max.	Min.	Max.	Min.
0.0	1.00	1.00	1.00	1.00	1.00	1.00	1.00	1.00
0.5	0.77	0.12	0.78	0.13	0.79	0.13	0.81	0.13
1.0	0.59	0.02	0.61	0.02	0.63	0.02	0.66	0.02
1.5	0.46		0.48		0.51		0.54	
2.0	0.35		0.38		0.40		0.44	
2.5	0.27		0.29		0.32		0.35	
3.0	0.21		0.23		0.25		0.29	
4.0	0.12		0.14		0.16		0.19	
5.0	0.07		0.09		0.10		0.13	
6.0	0.04		0.05		0.06		0.08	
8.0	0.02		0.02		0.03		0.04	
10.0	0.01		0.01		0.01		0.02	

Example 14-2 A concrete structure 0.91 m wide with a weight of 1340 kg/m crosses a trench 1.22 m wide in damp clay. The structure bears on the soil 1.83 m above the top of the pipe. Find the load transmitted to the pipe.

Solution
The load applied by the structure is

$$F = 1340 \times 1.22 = 1635 \text{ kg}$$

The pressure applied to the soil above the pipe is

$$P = \frac{1635}{0.91} = 1795 \text{ kg/m}$$

The ratio of depth to width is 1.83/1.22 = 1.5. From Table 14-9, the maximum proportion of the load reaching the pipe will be 0.51. Therefore the load reaching the pipe will be

$$P = 1795 \times 0.51 = 915 \text{ kg/m}.$$

This load must be added to the soil load calculated from the Marston formula before the type of pipe and bedding are selected.

Loads on concrete pipe are calculated in the same manner as for clay pipe. Bedding classes and corresponding load factors for concrete pipe are illustrated in Fig. 14-5. Allowable loads, based on a safety factor of 1.5, are presented in Table 14-10 for the different bedding classes. These values may be calculated from the strengths given in Table 14-3, the various load factors, and a value of 1.5 for the safety factor. It should be noted that the bedding classes for concrete pipe, while similar, are not identical to those for clay pipe.

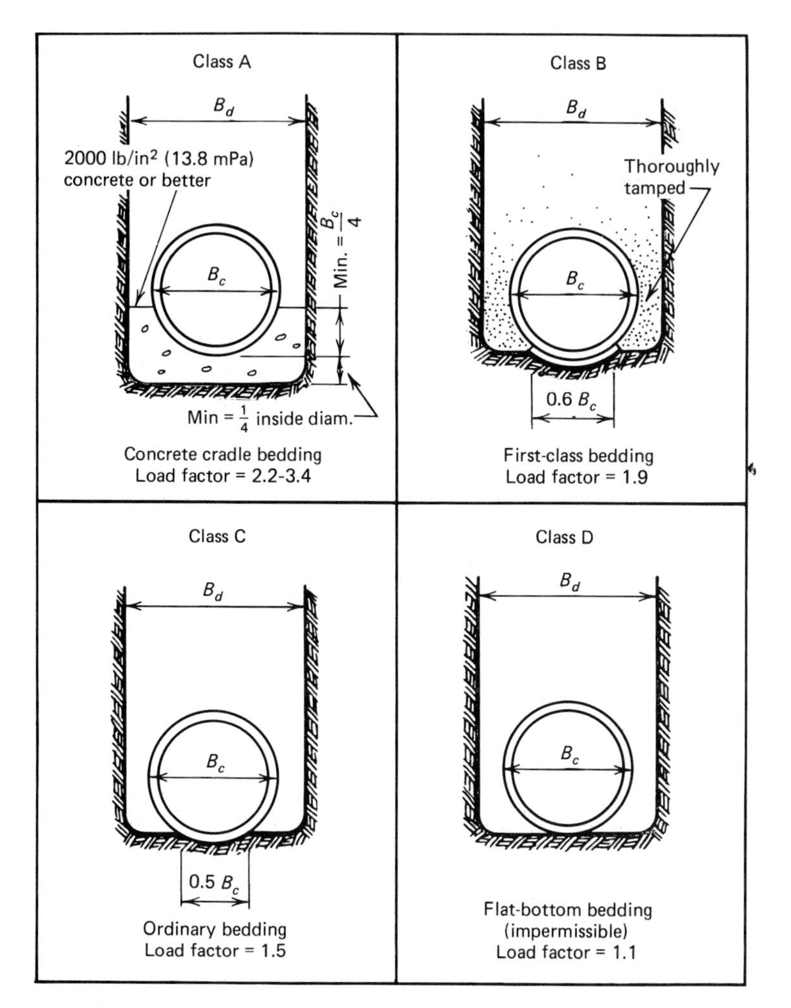

FIGURE 14-5
Methods of bedding concrete pipe and load factors applicable to strength.[6] (*Courtesy Portland Cement Association.*)

Reinforced-concrete pipe is made in five classes with two wall thicknesses in class I and three wall thicknesses in the other four classes. Strength is based on either the load which produces a 0.25-mm (0.1-in) crack or the ultimate load. The cracking load and ultimate load for the different classes are presented in Table 14-11. Allowable design loads based on cracking and a factor of safety of 1.0 are presented in Table 14-12. These loads do not necessarily have to be reduced by a safety factor since, as may be noted from Table 14-11, in most cases the ratio of ultimate to cracking load is approximately 1.5.

Loads on asbestos cement pipe are calculated as for other rigid pipe. The strengths listed in Table 14-5 are ultimate strengths in the three-point bearing

TABLE 14-10
Supporting strength of concrete pipe* per linear foot of pipe in thousands of pounds (kips)

ASTM spec. no.	Standard strength concrete sewer pipe, C 14, safety factor = 1.5				Extra strength concrete sewer pipe, C 14, safety factor = 1.5			
Bedding class	**D**	**C**	**B**	**A**	**D**	**C**	**B**	**A**
Load factor	**1.1**	**1.5**	**1.9**	**3.0**	**1.1**	**1.5**	**1.9**	**3.0**
Internal diameter of pipe, in								
6	0.8	1.1	1.4	2.2	1.5	2.0	2.5	4.0
8	0.9	1.3	1.6	2.6	1.5	2.0	2.5	4.0
10	1.0	1.4	1.8	2.8	1.5	2.0	2.5	4.0
12	1.1	1.5	1.9	3.0	1.6	2.2	2.8	4.5
15	1.2	1.7	2.2	3.5	2.0	2.8	3.5	5.5
18	1.4	2.0	2.5	4.0	2.4	3.3	4.2	6.6
21	1.6	2.2	2.8	4.4	2.8	3.8	4.9	7.8
24	1.7	2.4	3.0	4.8	2.9	4.0	5.1	8.0

* Supporting strengths are for concrete pipe meeting ASTM specifications (three-edge bearing test) and include safety and bedding factors as indicated (kips/ft × 14.6 = kN/m, in × 25.4 = mm).

test and an appropriate safety factor should be applied before selecting the pipe class and type of bedding to be used.

14-3 Other Fabricated Sewers

Plastic truss pipe consists of an extruded shell with integral diagonal stiffeners between the inner and outer membranes (Fig. 14-6). The space between the inner and outer surfaces is filled with lightweight concrete, which increases the pipe stiffness. The sections are joined by solvent welding of the parent

TABLE 14-11
Design loads for reinforced-concrete pipe (*reprinted by permission of the American Society for Testing and Materials*)

	Design load			
	To produce a 0.25-mm crack		Ultimate	
Class	N/m per mm dia.	lb/ft per ft dia.	N/m per mm dia.	lb/ft per ft dia.
I	38.3	800	57.4	1200
II	47.9	1000	71.8	1500
III	64.6	1350	95.8	2000
IV	95.8	2000	144.0	3000
V	144.0	3000	180.0	3750

TABLE 14-12
Supporting strength of concrete pipe* per linear foot of pipe in thousands of pounds (kips)

Reinforced concrete culvert, storm drain, and sewer pipe C76
Safety factor = 1.0. Based on 0.01-in (0.25-mm) crack

ASTM spec no.	Class I				Class II				Class III				Class IV				Class V			
Bedding class	D	C	B	A	D	C	B	A	D	C	B	A	D	C	B	A	D	C	B	A
Load factor	1.1	1.5	1.9	3.0	1.1	1.5	1.9	3.0	1.1	1.5	1.9	3.0	1.1	1.5	1.9	3.0	1.1	1.5	1.9	3.0
Internal diameter of pipe, in																				
12					1.1	1.5	1.9	3.0	1.5	2.0	2.6	4.0	2.2	3.0	3.8	6.0	3.3	4.5	5.7	9.0
15					1.4	1.9	2.4	3.8	1.8	2.5	3.2	5.0	2.8	3.7	4.8	7.5	4.1	5.6	7.1	11.3
18					1.6	2.2	2.8	4.5	2.2	3.0	3.8	6.1	3.3	4.5	5.7	9.0	5.0	6.8	8.6	13.5
21					1.9	2.6	3.3	5.2	2.6	3.5	4.5	7.1	3.9	5.3	6.7	10.5	5.8	7.9	10.0	15.8
24					2.2	3.0	3.8	6.0	3.0	4.0	5.1	8.1	4.4	6.0	7.6	12.0	6.6	9.0	11.4	18.0
27					2.5	3.4	4.3	6.8	3.3	4.6	5.8	9.1	4.9	6.7	8.5	13.5	7.4	10.1	12.9	20.2
30					2.7	3.7	4.7	7.5	3.7	5.1	6.4	10.1	5.5	7.5	9.5	15.0	8.2	11.2	14.2	22.5
33					3.0	4.1	5.2	8.2	4.0	5.6	7.0	11.1	6.0	8.2	10.4	16.5	9.0	12.4	15.7	25.0
36					3.3	4.5	5.7	9.0	4.4	6.1	7.7	12.2	6.6	9.0	11.4	18.0	9.9	13.5	17.1	27.0
42					3.8	5.2	6.6	10.5	5.2	7.1	8.9	14.2	7.7	10.5	13.3	21.0	11.5	15.7	20.0	31.5
48					4.4	6.0	7.5	12.0	6.0	8.1	10.2	16.2	8.8	12.0	15.2	24.0	13.2	18.0	22.8	36.0
54					4.9	6.7	8.5	13.5	6.7	9.1	11.5	18.2	9.9	13.5	17.1	27.0	14.8	20.2	25.7	40.5
60	4.4	6.0	7.6	12.0	5.5	7.5	9.5	15.0	7.4	10.1	12.8	20.2	11.0	15.0	19.0	30.0	16.5	22.5	28.5	45.0
66	4.8	6.6	8.3	13.2	6.0	8.2	10.4	16.5	8.1	11.1	14.1	22.3	12.1	16.5	21.0	33.0	18.1	24.8	31.3	49.5
72	5.3	7.2	9.1	14.4	6.6	9.0	11.4	18.0	8.9	12.1	15.4	24.3	13.2	18.0	22.8	36.0	19.8	27.0	34.2	54.0
78	5.7	7.8	9.9	15.6	7.1	9.7	12.3	19.5	9.6	13.2	16.7	26.3	14.3	19.5	24.7	39.0				
84	6.1	8.4	10.6	16.8	7.7	10.5	13.3	21.0	10.4	14.2	18.0	28.4	15.4	21.0	26.6	42.0				
90	6.6	9.0	11.4	18.0	8.2	11.2	14.2	22.5	11.1	15.2	19.2	30.4								
96	7.0	9.6	12.2	19.2	8.3	12.0	15.2	24.0	11.9	16.2	20.5	32.4								
102	7.5	10.2	12.9	20.4	9.3	12.7	16.1	25.5	12.6	17.2	21.8	34.4								
108	7.9	10.8	13.7	21.6	9.9	13.5	17.1	27.0	13.4	18.2	23.1	37.0								

* Supporting strengths shown in table are for concrete pipe meeting ASTM specifications (three-edge bearing test) and include safety and bedding factors as indicated (kips/ft × 14.6 = kN/m, in × 25.4 = mm).

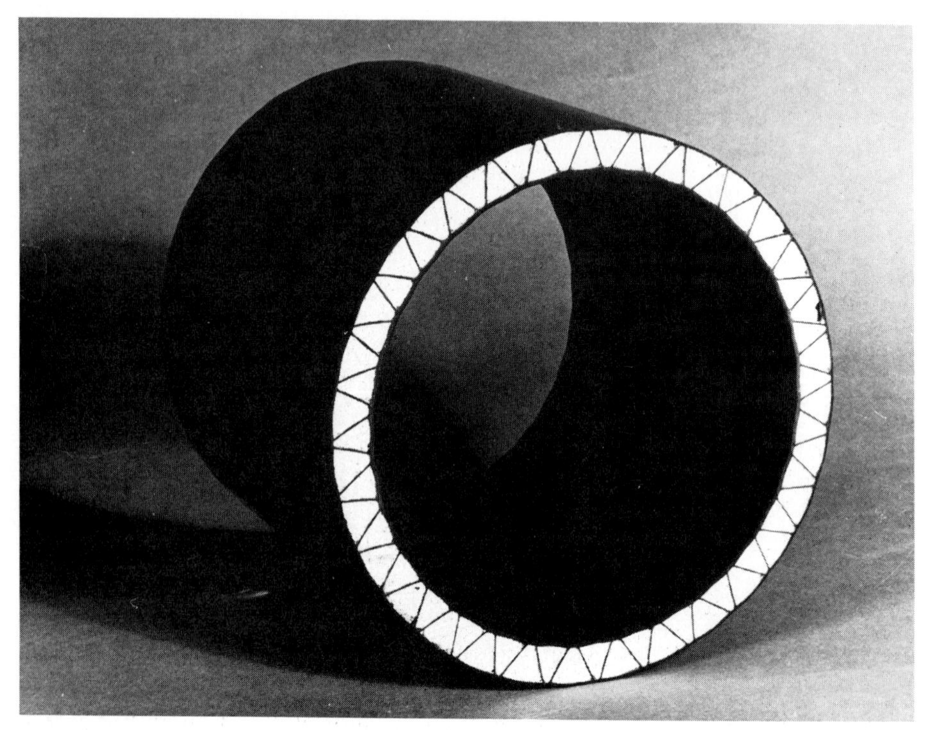

FIGURE 14-6
Plastic truss pipe.

acrylonitrile-butadiene-styrene (ABS) material which yields, in effect, a continuous pipe. Miscellaneous fittings are available to permit connection of household sewers of clay, solid plastic, asbestos cement, impregnated fiber, etc. This pipe is available only in sizes of 200 to 380 mm (8 to 15 in); hence, its application is limited to laterals and submains.

The strength of truss pipe is not based on the three-point bearing test used to evaluate rigid pipes. When subjected to external vertical loads, this pipe will deform, developing lateral support which permits it to act as an arch. Truss pipe can be subjected to rather large differential settlements without failure, and is thus particularly useful in areas with poor soil conditions. Its durability in routine service appears to be satisfactory. The initial installations of this material, which were made in 1965, were still in service in 1989.

Solid-wall plastic pipe is manufactured of polyvinyl chloride (PVC) in diameters from 100 to 380 mm (4 to 15 in). The smaller sizes are commonly used for household plumbing, but may also be applied in vacuum and pressurized collection systems. The small range of sizes, as with truss pipe, limits the application of solid-wall plastic pipe in gravity sewer systems to laterals and submains. Like truss pipe, the cemented joints are watertight.

Corrugated metal pipe is sometimes used for storm sewers, although its primary application is in highway drainage. The pipe may be galvanized or provided with other protective coatings (such as asphalt) to increase its life. Corrugated pipe is available in a large variety of cross sections and wall thicknesses. Like truss pipe, it derives a large part of its strength from its ability to deform and develop lateral support. Structural design procedures may be found in the manufacturer's literature.[7]

Iron pipe, as noted earlier, may be used in conveying sewage in circumstances in which other, less expensive materials, are unsuitable. Such applications include force mains, small outfall lines, inverted siphons, and lines installed under very high external loads. Lines within sewage treatment plants are usually constructed of iron. Joints in iron sewer lines are made the same way as in iron water lines (Chap. 6).

14-4 Infiltration and Sewer Joints

All water which enters a sewer is likely to remain in it and pass through whatever pumping and treatment units are incorporated in the system. Since both pumping and treatment are expensive, it is often more economical to exclude extraneous flow than to handle it once it has entered.

The cost of treating wastewater depends on the size of the plant and the actual processes employed. Present costs (1989) may be as much as $0.25/m^3 ($1.00/1000 gal). The cost of treating infiltration depends on the rate of infiltration (commonly expressed in terms of flow per unit diameter per unit length), the diameter of the sewer, its length, and the treatment cost per unit volume. The annual cost of treating infiltration may approach $4000 per kilometer of sewer under unfavorable circumstances. Reduction of operating costs by that amount could justify a present investment of as much as $40,000/km ($60,000/mi) or more.

Pumping costs depend on the total flow and the head against which the flow is pumped. For typical pumping conditions, the cost of transporting infiltration might be as much as $600/km per year. Removal of this cost could justify a present investment of up to $6000/km ($10,000/mi). The costs presented here are probable maxima, but indicate the savings which are possible through careful design, construction, and maintenance of a sewer system.

The most common source of infiltration is poor joints—particularly those where household sewers join common sewers. These connections are normally made by building contractors as a part of house construction and are seldom made properly unless city inspection is required. Use of pipe with gasketed joints, provision of wyes and tees for household connections, and insistence, on the part of the city, that all connections be made at wyes or tees will reduce infiltration substantially.

14-5 Built-in-Place Sewers

Except in very large communities, sanitary sewage can generally be conveyed by conduits which do not exceed the range of sizes available in clay pipe. Concrete sewers, although they are not always suitable for sanitary sewage, extend the range of sizes further and are normally adequate for most storm drainage applications. For large storm sewers, corrugated metal pipe or pipe arches may be used in many applications. There remain, however, some circumstances in which flows, soil or subsurface conditions, hydraulic considerations, or other factors may dictate construction of cast-in-place sewers.

Large concrete sewers may be analyzed as closed rings or fixed arches by the techniques of structural design.[8] For small diameters or spans, empirical designs are more likely to be used, since theoretical calculations lead to thicknesses too small for ordinary construction techniques. Reinforced concrete arches are often constructed with the thickness of the crown equal to one-twelfth the span, with a minimum of 125 mm (5 in). The thickness of the invert is 25 mm (1 in) greater than the crown, and the haunches are 2 to 3 times the crown thickness. The sewer must be designed to conform to the bearing capacity of the foundation material, which may require placement of a subbase of crushed stone or gravel or, in some cases, piles.

The shape of the sewer depends on hydraulic considerations, construction conditions, and available space. The lower surface is generally curved to concentrate low flows and maintain self-cleansing velocities. The bottom should be placed as soon as excavation and preparation of the foundation are completed. The remainder of the structure may be placed in two or more lifts with waterstops at the construction joints. When a constant sewer section is maintained for some distance, collapsible steel forms may be used to form the arch. A number of cast-in-place sewer sections are shown in Fig. 14-7.

14-6 Corrosion of Sewers

Organic material is likely to accumulate in sanitary sewers as a result of deposition at low flow velocities and coagulation of grease at the junction of the water surface and the pipe. This accumulated material will be slowly degraded by the bacteria in the sewage and this degradation, under the conditions prevalent in many sewers, may be accompanied by the biological reduction of sulfates present in the flow.

The anaerobic oxidation of complex organics involves, as an intermediate step, the production of short-chain volatile organic acids. These acids are water-soluble and may depress the pH in the sewer. The combination of sulfate reduction and low pH can cause the release of hydrogen sulfide into the air space of the sewer, where it may redissolve in condensed moisture accumulated at the crown (Fig. 14-8). In that location, whatever oxygen is available can be used by the bacterium *Thiobacillus* in the oxidation of hydrogen sulfide to sulfuric acid.

In sewers made of acid-soluble materials such as concrete, iron, or steel, this acid formation may lead to destruction of the crown and failure of the sewer. The problem is aggravated by flat sewer grades, which produce low velocities and long detention times, and by warm temperatures. Extensive damage has occurred to concrete sewers in the Gulf coast region of the United States, where both conditions exist.

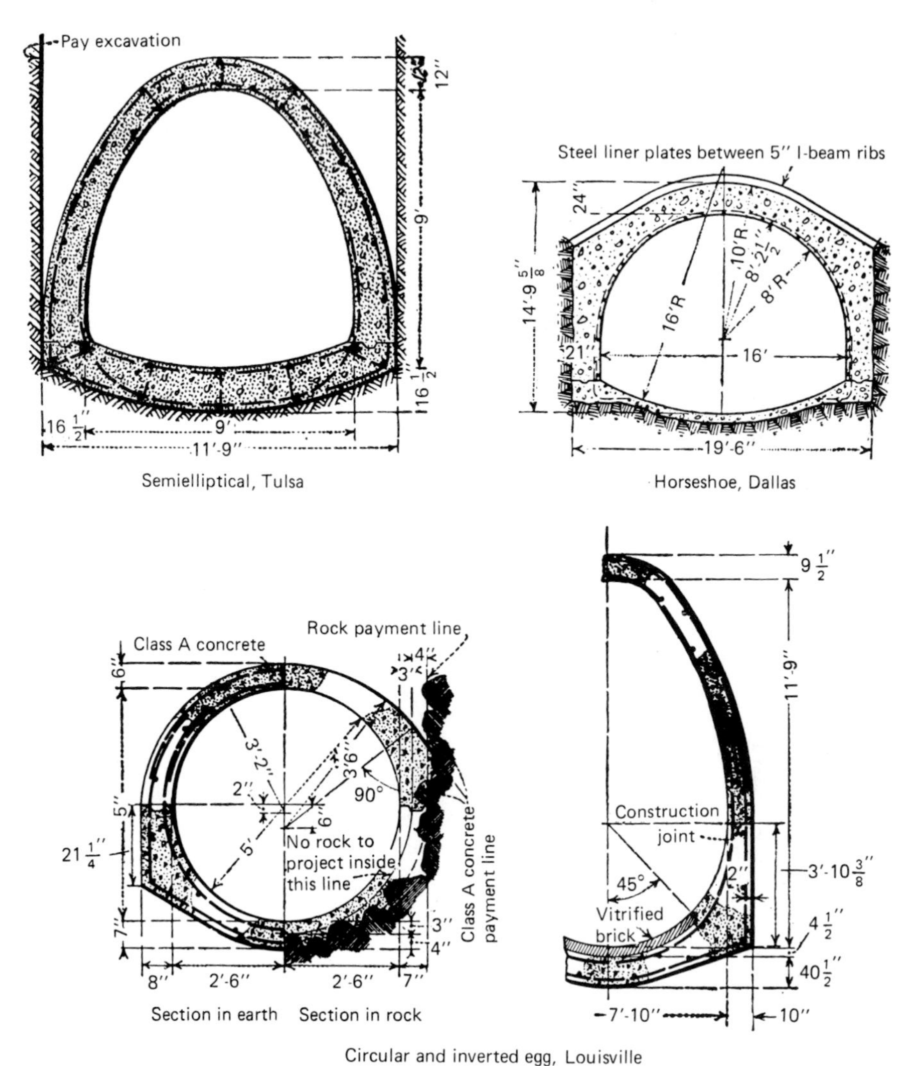

FIGURE 14-7
Typical built-in place sewer sections. Circular section in earth is suitable for soft material.

Sewer corrosion has been combatted by chlorination, forced ventilation, and lining with inert materials. Chlorination halts biological activity—at least temporarily. Forced ventilation reduces crown condensation, strips H_2S from the atmosphere of the sewer, and may provide sufficient oxygen to prevent sulfate reduction and production of organic acids.

Sewers flowing full and outfall lines carrying sewage which has been treated to secondary standards do not provide the conditions necessary for sewer corrosion. Ordinary sewer lines are customarily built of vitrified clay, since this is the only material which has been demonstrated, through long service, to be resistant to corrosion. In new construction, particularly where

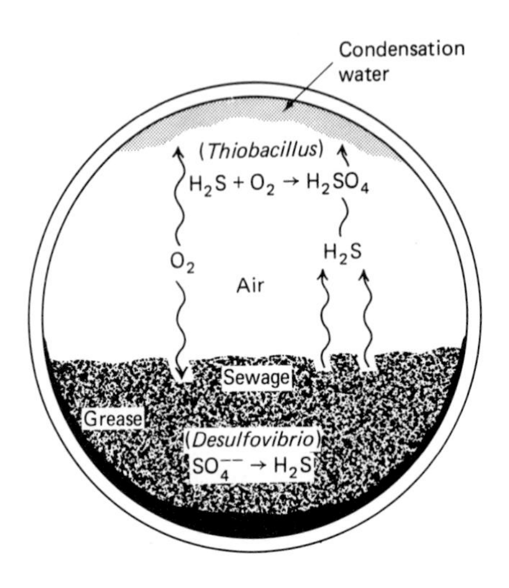

FIGURE 14-8
Schematic diagram of sewer corrosion.

foundation conditions are poor or the groundwater table is high, solid-wall plastic or plastic truss pipe may be expected to give satisfactory service.

When sanitary sewers larger than 1070 mm (42 in) are required, concrete is commonly used. If the conditions of the installation are conducive to corrosion, the sewer may be cast with an integral lining or be lined in place with plastic, clay tile, or asphaltic compounds.

PROBLEMS

14-1. A 250-mm clay pipe is to be placed in a trench 5 m deep to the top of the pipe. The backfill material is clay weighing 1920 kg/m³. If the trench width at the top of the pipe is 1.5 pipe diameters plus 300 mm, determine the load per unit length on the pipe. Using a factor of safety of 1.5, determine the type of bedding required for standard and extra-strength pipe. Which would you select? Why?

14-2. A 200-mm clay sewer is to be buried 3 m deep (to the top of the pipe) in saturated topsoil weighing 1840 kg/m³ in a trench with width defined as in Prob. 14-1. A building foundation 300 mm wide crosses the pipeline at right angles, applying a load of 3000 kg/m which bears 2.5 m above the pipe. Determine the load on the sewer under the foundation and remote from the foundation. Using a factor of safety of 1.5, select a suitable bedding and pipe class for each condition.

14-3. A precast reinforced-concrete sewer 1220 mm in diameter is buried under 5 m of saturated clay cover in a trench 2 m wide. Consider the safe load to be that which produces a 0.25-mm crack modified by a safety factor of 1.25. Determine what types of bedding and pipe classes are suitable. Which would you select? Why?

REFERENCES

1. "Standard Specification for Compression Joints for Vitrified Clay Bell-and-Spigot Pipe," C-425, American Society for Testing and Materials, Philadelphia.
2. "Standard Specification for Compression Couplings for Vitrified Clay Plain End Pipe," C-594, American Society for Testing and Materials, Philadelphia.

3. *Clay Pipe Engineering Manual*, National Clay Pipe Institute, Washington, D.C., 1974.
4. *Concrete Pipe Design Manual*, American Concrete Pipe Association, Vienna, Va., 1987.
5. "Standard Specification for Extra Strength and Standard Strength Clay Pipe and Perforated Clay Pipe," C-700, American Society for Testing and Materials, Philadelphia.
6. *Design and Construction of Concrete Sewers*, Portland Cement Association, Chicago, 1968.
7. *Modern Sewer Design*, American Iron and Steel Institute, Washington, D.C., 1980.
8. *Analysis of Arches, Rigid Frames and Sewer Sections*, Portland Cement Association, Chicago, 1974.

CHAPTER 15

SEWER APPURTENANCES

15-1 Manholes

Manholes provide access to sewers for inspection and cleaning and are located at changes in direction, changes in pipe size, substantial changes in grade, and at intervals of 90 to 150 m (300 to 500 ft) in straight lines. Sewers larger than 1.5 m (5 ft) in diameter can be entered readily and thus need fewer manholes.

The design of manholes is fairly well standardized in most cities. A typical brick manhole, illustrated in Fig. 15-1, has a cast iron frame and cover with a 500 to 600 mm (20 to 24 in) opening. The frame rests on brickwork which is corbeled as shown to form a cylinder from 1 to 1.25 m (40 to 48 in) in diameter which extends downward to the lowest sewer. The walls are typically 200 mm (8 in) thick for depths up to 4 m (12 ft) and increase by 100 mm (4 in) for each additional 2 m (6 ft) of depth. The interior of brick manholes is often plastered with portland cement or mortar.

The bottom of the manhole is normally concrete, sloping toward an open channel which is an extension of the lowest sewer. The open channel is sometimes lined with half-round or split sections of sewer pipe. The channel should be sufficiently well-defined and deep enough to prevent sewage from spreading over the bottom of the manhole. Changes in direction in the lower sewer are made by deflecting the open channel as shown in Fig. 15-2, which represents a cast-in-place reinforced-concrete manhole.

Figure 15-3 illustrates a precast reinforced-concrete manhole which, with the exception of the tapered section, is fabricated of tongue-and-groove concrete pipe. This figure also shows a common technique of matching the grades of sewers without requiring all the lines to be placed at the same elevation. Such a structure is called a drop manhole and is usually provided when the difference in elevation between the high and low sewer exceeds 0.6 m (2 ft). The sewage falls through the vertical line, but the sewer itself is open for inspection and cleaning through the horizontal extension. When large sewage

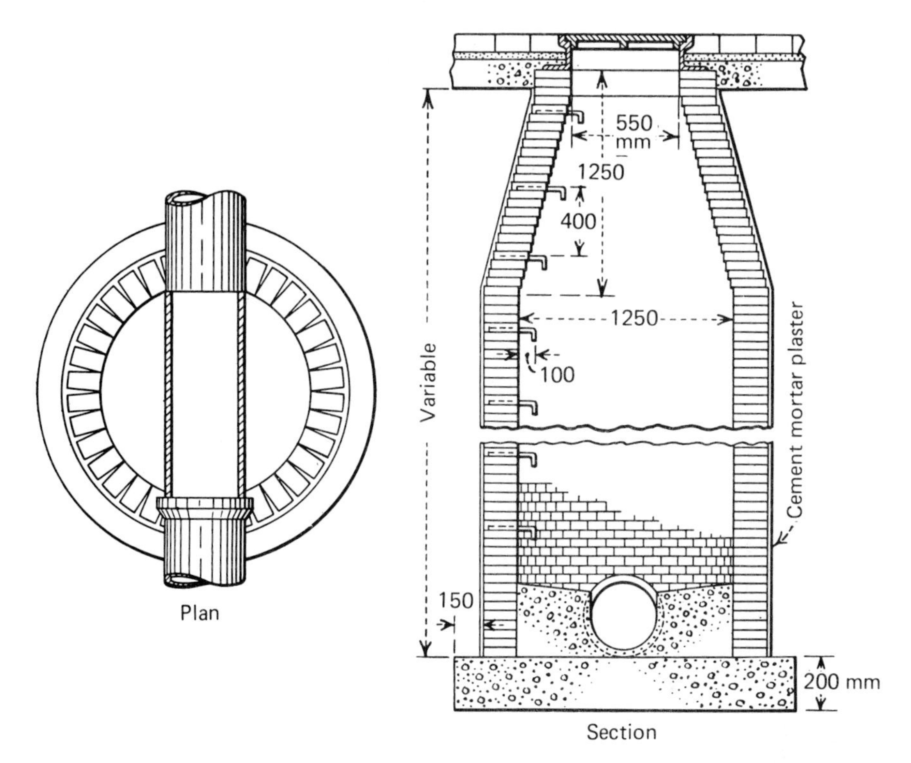

FIGURE 15-1
Brick manhole.

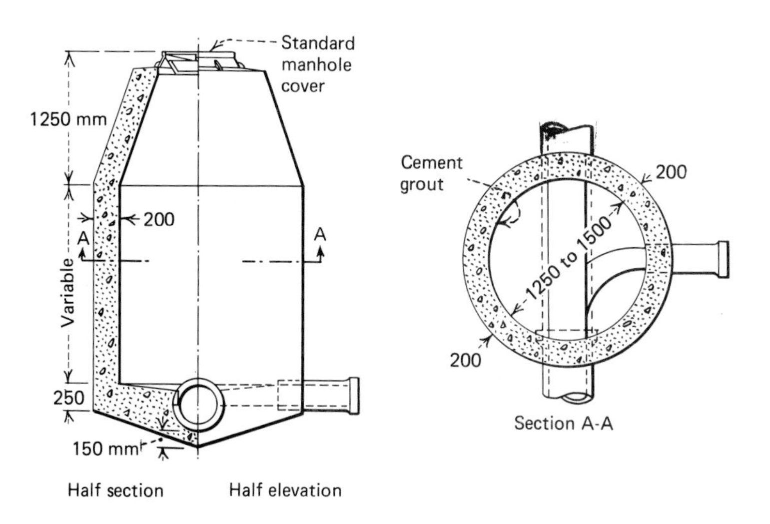

FIGURE 15-2
Concrete manhole with junction of branch sewer.

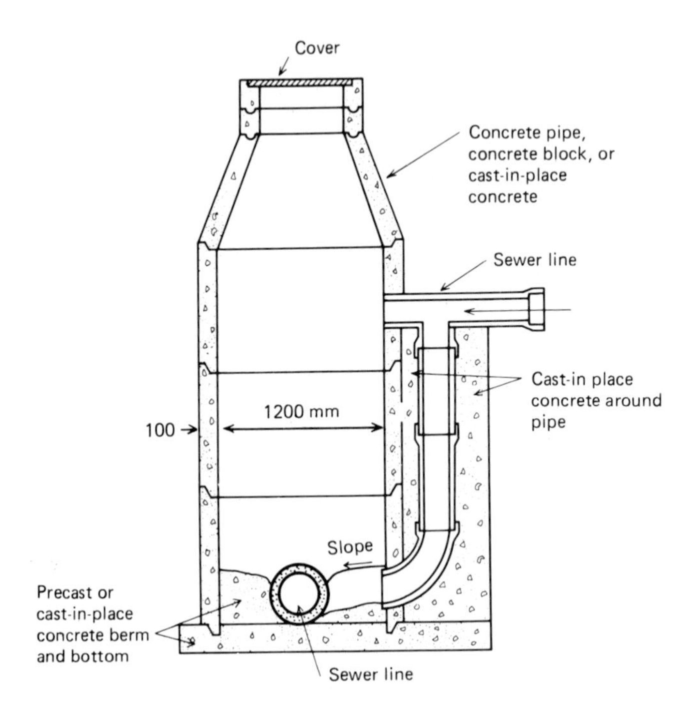

FIGURE 15-3
Precast concrete manhole with drop inlet.

flows must fall long distances in order to reach a lower sewer, the fall is generally interrupted by staggered horizontal plates within the shaft or by a steplike arrangement. These devices prevent excessive kinetic energy from damaging the bottom of the structure. Manholes for large sewers may be constructed as shown in Fig. 15-4 or as a separate structure connected to the sewer by a short tunnel.

Manhole covers and frames are manufactured in several standard weights for different traffic conditions. The heaviest covers and frames weigh about 340 kg (750 lb), those intended for city streets about 245 kg (540 lb), and the lightest about 70 kg (150 lb). Light and heavy frames are illustrated in Fig. 15-5. Covers are generally cast with a raised pattern on their surface to make them less slippery when wet. Openings through the cover should not be allowed, since these contribute to infiltration during rainfall events.

Manholes may be provided with metal rungs inserted in the walls, as shown in Figs. 15-1 and 15-4. Such rungs and the manholes themselves are subject to corrosion and may present a danger to workers in older sewers. It may be safer to lower a ladder into the manholes than to depend on such rungs.

Fiber-glass-reinforced plastic manhole structures have been manufactured and used since about 1975. These are, of course, much lighter than concrete and thus much easier to install. Fiber glass is also more resistant than

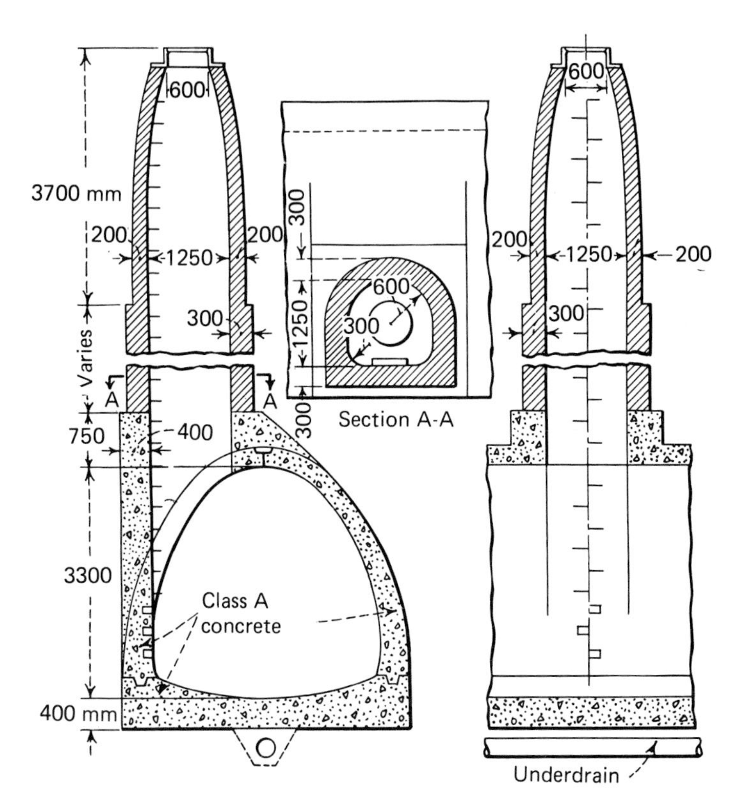

FIGURE 15-4
Manhole access to large sewer.

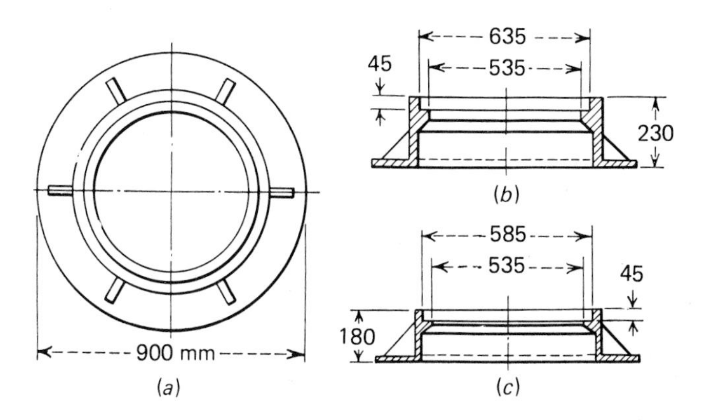

FIGURE 15-5
Manhole frames. (*a*) Plan; (*b*) section of heavy frame; (*c*) section of light frame.

concrete to the mechanism of sewer corrosion (Chap. 14). The drawbacks to fiber glass lie in its uncertain design life and its susceptibility to structural failure if it is improperly installed. Very careful placing and compaction of backfill is required to ensure structural stability.

Cleanouts are sometimes used to permit cleaning of small sewers, particularly at the upper end of laterals. A standard cleanout used in Dallas is illustrated in Fig. 15-6. A light cast iron cover provides access to a line of pipe leading to the sewer. The connection to the sewer is made through a special wye with its side outlet making an angle of 27° with the main rather than the standard 60°. When the sewer is fairly deep, as in the illustration, a $\frac{1}{16}$ bend is installed above the wye to bring the cleanout to the surface at a 45° angle. The sewer may be rodded or flushed with water through such structures and small television cameras can be inserted to permit inspection of the condition of the pipe and joints.

15-2 Inlets

Inlets are structures through which storm water enters the sewers. Their design and location require consideration of how far water will be permitted to extend into the street under various conditions. The permissible depth of water in the gutter in most cities is limited to 150mm (6 in) on residential streets and to that depth which will leave two lanes clear of standing water on arterials and one lane on major streets. On curved streets the gutter depth must be decreased to

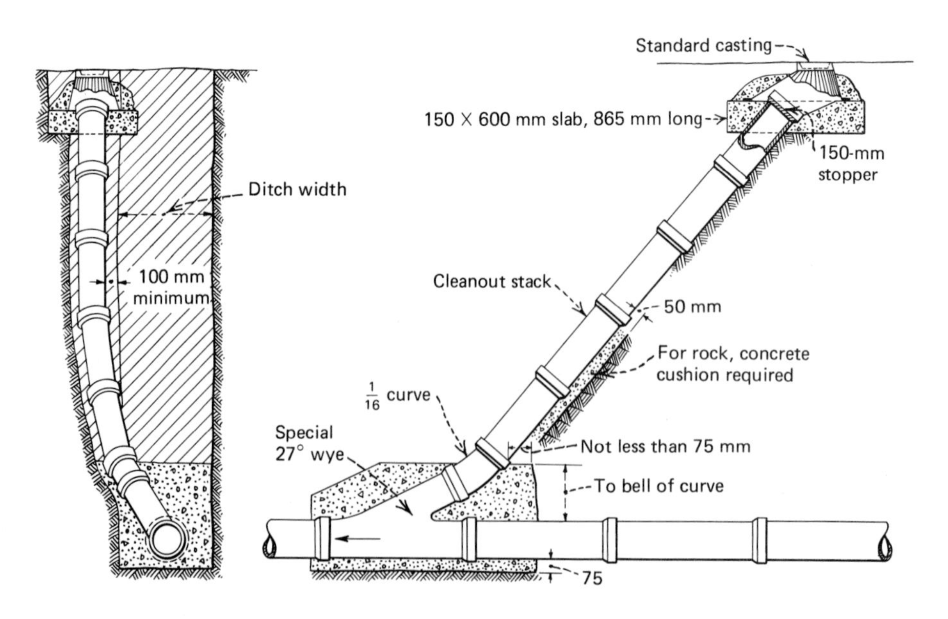

FIGURE 15-6
Standard sewer cleanout.

prevent the flow from jumping the curb at driveways or other openings. These criteria are applied to the design storm condition which is used to size the sewers. During high-intensity storms with recurrence intervals greater than that of the design storm, the streets are expected to be flooded.

Flow in streets is commonly calculated by using Manning's equation, modified for a triangular gutter cross section:

$$Q = K\frac{z}{n}s^{1/2}y^{8/3} \tag{15-1}$$

where Q = gutter flow
z = reciprocal of the cross slope of the gutter
n = roughness coefficient
s = gutter slope
y = depth at the curb
K = constant depending on units and equal to 0.38 (m^3/s, m) or 0.56 (ft^3/s, ft)

If Manning's equation can be used to describe the flow condition in the street (see Chap. 3), then from the flow and the street cross section and slope one may calculate the depth at the curb. The width to which the water will spread is equal to zy, which can be compared to the criteria above.

Example 15-1 A street has a slope of 1 percent, n = 0.018, a cross slope of 4 percent, and a curb height of 150 mm. The street width is 10 m and 3.5 m must be kept clear at the design condition. What is the maximum flow which can be carried by the gutter?

Solution

The gutter flow will be limited either by the spread (3.25 m) or the curb height. For a cross slope of 4 percent, the spread limits the curb depth to 3250 × 0.04 or 130 mm.

$$Q = 0.38\frac{25}{0.018}(0.01)^{1/2}(0.130)^{8/3} = 0.23 \text{ m}^3/\text{s}$$

The first inlet on the street in the example above would be located at the point where the flow reached 0.23 m^3/min. The location of subsequent inlets depends upon their design, since normally a portion of the gutter flow is permitted to pass by. In business areas where there is heavy pedestrian traffic, inlets may be located as shown in Fig. 15-7 to keep the crosswalks relatively dry. Inlets may be classified according to location as inlets in sumps, inlets on grade, and inlets on grade with gutter depression. They may further be classified according to design as curb opening, grate, and combination inlets. Inlets in sumps are those which are located at low points in the street system where water which is not removed by the inlet will pond rather than pass by. Curb inlets in sumps have a capacity equal to

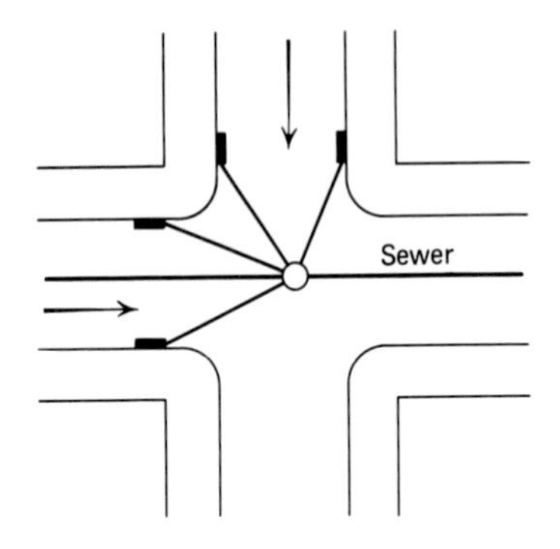

FIGURE 15-7
Street intersection showing inlets and branch lines from inlets to main sewer. Arrows show direction of surface flow.

$$Q = Ky^{3/2}L \tag{15-2}$$

where y = depth at the gutter
L = length of curb opening
K = constant dependent on units and equal to 1.66(m^3/s, m) or 3.0(ft^3/s, ft)

The flow calculated from Eq. (15-2) is often reduced by 10 percent as an allowance for clogging.

Grate inlets in sumps have a capacity equal to

$$Q = KAy^{1/2} \tag{15-3}$$

where y = depth at curb
A = open area of grate
K = constant dependent on units and equal to 2.96 (m^3/s, m) or 5.37 (ft^3/s, ft)

The flow given by Eq. (15-3) is often reduced by 25 percent or more to account for clogging, which is more likely on grates. Inlets in sumps must be sized to accommodate the entire flow which will reach them under design conditions, since no flow will pass by to other inlets.

Inlets on grade are usually designed to permit between 5 and 15 percent of the upstream flow to pass the structure. This results in a less expensive design than would result if the street were dewatered completely at each inlet. The percentage of the gutter flow intercepted by inlets which are shorter than those required for total dewatering may be determined from Fig. 15-8. The flow intercepted per unit length by a *curb inlet on grade* is given by

$$\frac{Q}{L} = \frac{K}{y}[(a + y)^{5/2} - a^{5/2}] \tag{15-4}$$

where Q/L = flow intercepted per unit length
y = depth at curb above normal gutter grade
a = depression of gutter at inlet below its normal level elsewhere
K = constant dependent on units and equal to 0.39 (m^3/s, m) or 0.70 (ft^3/s, ft)

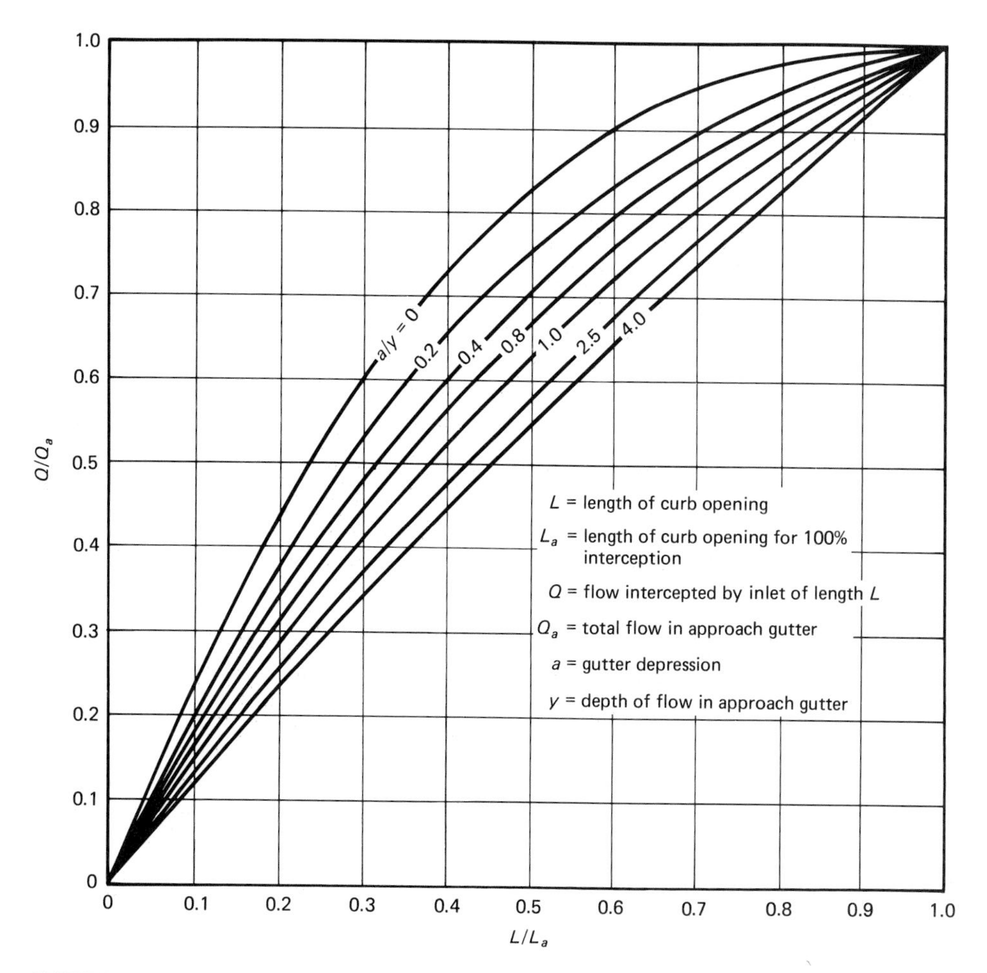

FIGURE 15-8
Ratio of intercepted flow to total flow for inlets on grade.

The capacity of *grate inlets on grade* is given by Eq. (15-3). If the gutter is depressed, the value of y should be replaced by $(y + a)$. The calculated capacity may be reduced by up to 25 percent to allow for clogging. A *combined inlet on grade* has a capacity equal to the sum of the flows given by Eqs. (15-3) and (15-4).

Example 15-2 A gutter with $z = 20$, $n = 0.015$, and a slope of 1 percent carries a flow of 0.25 m³/s. For a curb depression of 60 mm, find the required inlet length to intercept the entire flow and the capacity of a 3-m-long curb inlet.

Solution
The depth of flow in the gutter is given by Eq. (15-1):

$$y = \left[\frac{0.25(0.015)}{20(0.38)(0.01)^{0.5}}\right]^{3/8} = 0.137 \text{ m}$$

Substituting this value of y in Eq. (15-4) gives

$$\frac{Q}{L} = \frac{0.39}{0.137}[(0.197)^{5/2} - (0.06)^{5/2}] = 0.0465 \text{ m}^3/\text{s per meter}$$

The inlet length for complete interception is

$$L = \frac{0.25}{0.0465} = 5.38 \text{ m}$$

If a 3-m inlet is used, Fig. 15-8 will yield the fraction of the flow intercepted.

$$\frac{L}{L_a} = \frac{3}{5.38} = 0.56$$

$$\frac{a}{y} = \frac{0.06}{0.137} = 0.44$$

From the figure, $Q/Q_a = 0.75$; therefore

$$Q = 0.75(0.25) = 0.19 \text{ m}^3/\text{s}$$

The example above illustrates the importance of bypassing a portion of the flow. One may observe that an inlet which is 56 percent of the length required for total interception will remove 75 percent of the flow. Bypassing thus uses the inlet structure more efficiently.

Storm sewer inlets permit the gutter flow to leave the street and enter a drop structure which is connected to the subsurface drainage system. The subsurface portion of the inlet sometimes is designed with the outlet located above the bottom of the structure as shown in Fig. 15-9. Such designs are called *catch basins* and are intended to permit heavy debris to settle out before it enters the sewers. In modern practice, it is more common to design the sewers

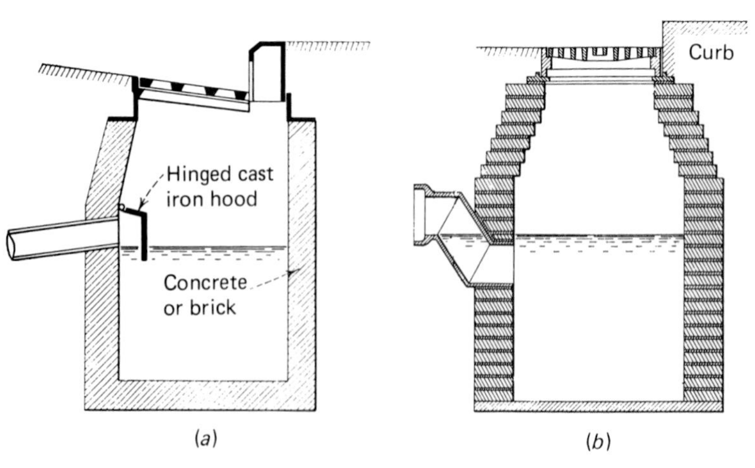

FIGURE 15-9
Catch basins. (*a*) Combined grate and curb inlet; (*b*) grate inlet.

with grades which are adequate to maintain self-cleansing velocities and to employ simple drop inlets.

15-3 Inverted Siphons

An inverted siphon is a section of sewer which is dropped below the hydraulic grade line in order to avoid an obstacle such as a railway or highway cut, a subway, or a stream. Such sewers will flow full and will be under some pressure, hence they must be designed to resist low internal pressures as well as external loads. It is also important that the velocity be kept relatively high (at least 0.9 m/s) to prevent deposition of solids in locations which would be very difficult or impossible to clean.

Since sewage flow is subject to large variations, a single pipe will not serve adequately in this application. If it is small enough to maintain a velocity of 0.9 m/s at minimum flow, the velocity at peak flow will produce very high head losses and may actually damage the pipe. Inverted siphons normally include multiple pipes and an entrance structure designed to divide the flow among them so that the velocity in those pipes in use will be adequate to prevent deposition of solids.

Figure 15-10 illustrates an inverted siphon designed to avoid an obstacle. The three pipes are designed to carry, respectively, the minimum flow, the difference between minimum flow and average flow, and the difference between average and maximum flow. The difference in elevation from one end to the other is dictated by maintaining self-cleansing velocity (0.9 m/s) in the smallest sewer. At the same differential head, the velocities will be greater in the larger sewers.

The inlet structure has two side flow weirs which direct low flows to the central pipe. As the depth in the structure increases with increasing flow, the

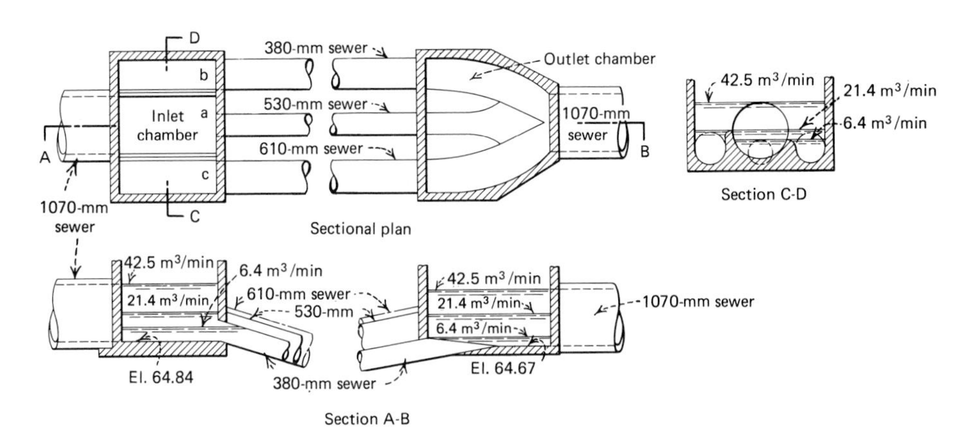

FIGURE 15-10
Inverted siphon.

excess flow will spill over the lower weir and enter the second pipe. Further increases in flow and depth will cause flow to be diverted to the third pipe. The hydraulic design of such structures is based on the principles detailed in Chap. 3. Inlet and outlet structures for inverted siphons should be installed in manholes or other access should be provided for maintenance and cleaning.

15-4 Sewer Outlets and Outfalls

Stormwater and treated wastewater may be discharged to surface drainage or to bodies of water such as lakes, estuaries, or the ocean. Outlets to small streams are similar to the outlets of highway culverts, consisting of a simple concrete headwall and apron to prevent erosion. Some wastewater treatment plants are located at elevations which might be flooded. Present regulations require that sewage treatment works be protected against a 100-year flood, which may require levees around low-lying installations and pumping of the treated flow when stream levels are high. Gravity discharge lines in such circumstances must be protected by flap gates or other automatically closed valves which will prevent the stream flow from backing up into the plant.

Sewers discharging into large bodies of water are usually extended beyond the banks into fairly deep water where dispersion and diffusion will aid in mixing the discharge with the surrounding water. The outfall lines are constructed of either iron or reinforced concrete and may be placed from barges or joined by divers. Iron is generally preferred for outfalls 610 mm (24 in) in diameter or less. In bodies of water which are sufficiently large to permit heavy wave action, the outfall may be protected by being placed in a dredged trench or by being supported on pile bents. Subsurface discharges normally employ multiple outlets to aid in distribution and dilution of the wastewater.

15-5 Alternative Sewer Systems

In order to facilitate cleaning and inspection, gravity sanitary sewers are generally at least 200 mm (8 in) in diameter. To maintain self-cleansing velocities at low flow, the sewer grade must be steep; hence, gravity sewers installed in areas of low population density may be very expensive.

Alternative collection systems have been developed that may be useful anywhere, but which are particularly valuable where the sources of flow are widely scattered and the total flow is low.

Vacuum systems such as that shown in Fig. 15-11 utilize 50 to 150 mm (2 to 6 in) plastic pipe which is buried deep enough to prevent freezing and in which a vacuum is maintained. Each source is connected to the vacuum main through a special entry valve which opens automatically when a sufficient volume of sewage has accumulated in the gravity household sewer. The valve recloses automatically after a timed interval sufficient to permit the accumulated waste to be drawn into the main. The vacuum is maintained by pumps at a central collecting tank to which a number of mains may be connected. The

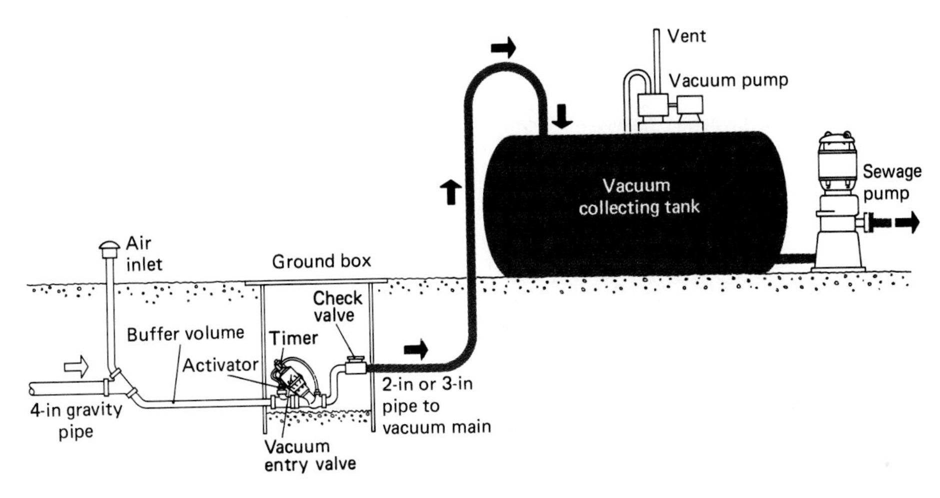

FIGURE 15-11
Vacuum collection system. (*Courtesy Colt Industries.*)

accumulated waste in the tank is periodically pumped to treatment or to another part of the collection system. Reference 1 presents a detailed description of one manufacturer's recommendations and a computer-aided design procedure based on those recommendations.

Pressurized systems have also been used to avoid the deep sewers and expensive construction required for gravity systems in areas of low population density.[2] These systems depend on *grinder pumps* (Fig. 15-12), which reduce the size of the solids to levels which will not plug small lines. The pumps are usually submersible and are installed in small wet wells which may serve from one to five residences. Some users have encountered difficulty with accumulations of floating solids in the wet wells which, in time, may be great enough to fill the sump and cause the pump to run dry, overheat, and fail. Regular maintenance and inspection can correct this problem. Detailed design procedures for pressurized collection systems will be found in Ref. 3.

15-6 Pumping of Sewage

There are many communities in which it is possible to convey all the sewage to a central treatment location or point of discharge in only a gravity system. In other areas with flat terrain, more than one drainage area, low-lying sections, or similar complications, pumping may be required. Pumping may also be required at or within sewage treatment plants, in the basements of buildings which are below the grade of the sewer, and to discharge treated wastewater to streams which are above the elevation of the treatment plant.

Pumping of *storm water* requires moving very large flows against relatively low heads. Large-diameter axial-flow pumps are most suitable for this service. Small axial-flow pumps may operate at speeds as high as 1200 r/min,

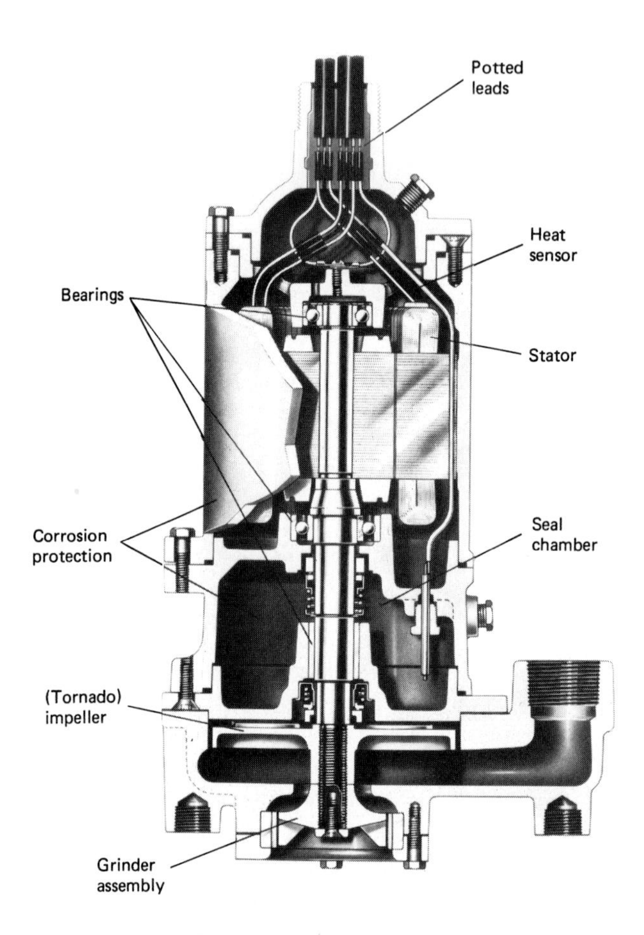

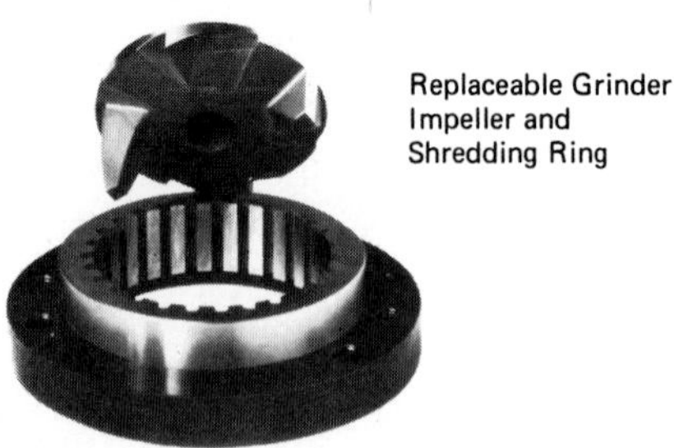

FIGURE 15-12
Grinder pump. (*Courtesy F. E. Myers Co., Division of McNeil Corporation.*)

but large pumps such as those used in New Orleans, which are up to 4.25 m (14 ft) in diameter, operate at speeds less than 100 r/min. A pump of this size and type may deliver a flow of 43 m^3/s (1500 ft^3/s) against a head of 3 to 4 m (10 to 12 ft).

Pumping of untreated sanitary sewage requires special designs, since sewage often contains large solids. *Nonclog* pumps have impellers which are usually closed and have, at most, two or three vanes. The clearance between the vanes is sufficiently large that anything which will clear the pump suction will pass through the pump. A bladeless impeller, sometimes used as a fish pump, has also been applied to this service. For a specified capacity, bladeless impellers are larger and less efficient than vaned designs.

Manufacturers of sewage pumps specify the sphere size which the pump will pass. A pump with a 100-mm (4-in) discharge might pass a 75-mm (3-in) sphere, while a 250-mm (10-in) discharge would pass a 200-mm (8-in) sphere. Pump suctions are usually larger than the discharge by about 25 percent. The smallest discharge size commonly used for sanitary sewage is 75 mm (3 in).

Sewage *pumping stations* within the collection system include a *wet well* which serves to equalize the incoming flow, which is always variable. Although pumps that can operate at variable speed are available, their cost and the complexity of their control systems generally make them an expensive alternative. Ordinary constant-speed pumps with standard motors should not be turned on and off too frequently since this can cause them to overheat. In small pumping stations there may be only two pumps, each of which must be able to deliver the maximum anticipated flow. Lower flows are allowed to accumulate in the wet well until a sufficient volume has been accumulated to run the pump for about 2 min. The wet well may also be sized to ensure that the pump will not start more often than once in about 5 minutes. The specific values of running time and cycle time depend upon the characteristics of the motor used and must be obtained from the manufacturers.

Example 15-3 A small subdivision produces an average wastewater flow of 120,000 L/day. The minimum flow is estimated to be 15,000 L/day and the maximum 420,000 L/day. Using a 2-min running time and a 5-min cycle time, determine the design capacity of each of two pumps and the required wet well volume.

Solution

Each pump must be able to deliver the peak flow, or 420,000 L/day. The pump running time is the working volume of the wet well divided by the net discharge, which is the pumping rate minus the inflow:

$$t_r = \frac{V}{Q_{\text{out}} - Q_{\text{in}}}$$

The filling time, with the pump off, is

$$t_f = \frac{V}{Q_{\text{in}}}$$

The total cycle time is therefore

$$t_c = t_r + t_f = \frac{V}{Q_{\text{out}} - Q_{\text{in}}} + \frac{V}{Q_{\text{in}}}$$

To assure a 2-min running time,

$$V = \frac{2}{1440}(420{,}000 - 15{,}000) = 562.5 \text{ L}$$

To assure a 5-min cycle time, it is first necessary to consider the circumstances under which the cycle time will be shortest. This can be shown to occur when Q_{in} is $\frac{1}{2}Q_{out}$ (Prob. 15-4). Therefore,

$$5 = \frac{V}{Q_{out} - 0.5Q_{out}} + \frac{V}{0.5Q_{out}}$$

$$V = \frac{5}{1440} \times \frac{0.5 \times 420{,}000}{2} = 365 \text{ L}$$

The required working volume is dictated by the running time and will be about 600 L.

A minimum depth, variable with the pump inlet features,[4] must be maintained over the pump suction. At an intake velocity of 0.6 m/s (2 ft/s) a submergence of about 300 mm (1 ft) is required. It is common to provide some freeboard above the anticipated maximum water level, typically about 600 mm (2 ft). A wet pit suitable for the example above is illustrated in Fig. 15-13.

Larger pump stations may employ pumps of different capacities to permit a closer match of influent and effluent flow. The different pumps are controlled by floats with mercury switches or other level switches to come on at different elevations in the wet well. The size of the wet well is kept as small as is compatible with minimum running time and cycle time criteria in order to reduce the amount of anaerobic decay which occurs in the sump. The air space in the sump must be vented, and anaerobic decomposition produces unpleasant odors which may create nuisance conditions in the neighborhood.

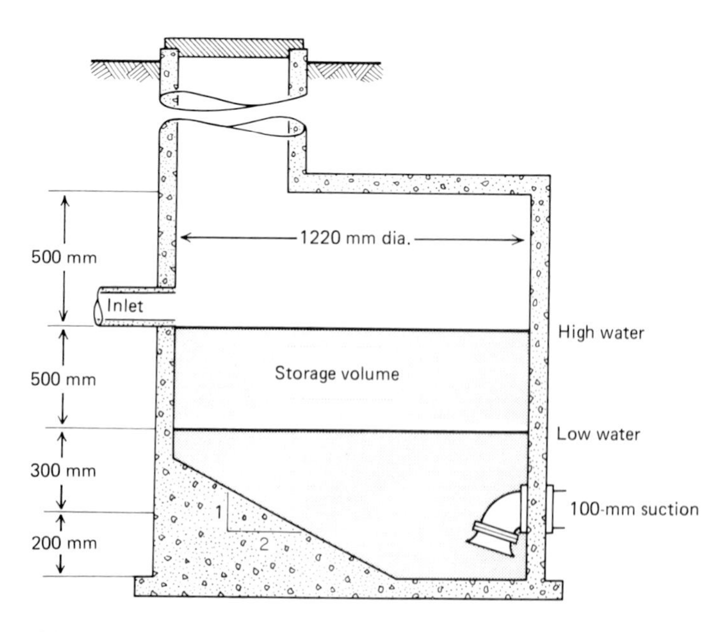

FIGURE 15-13
Wet pit details.

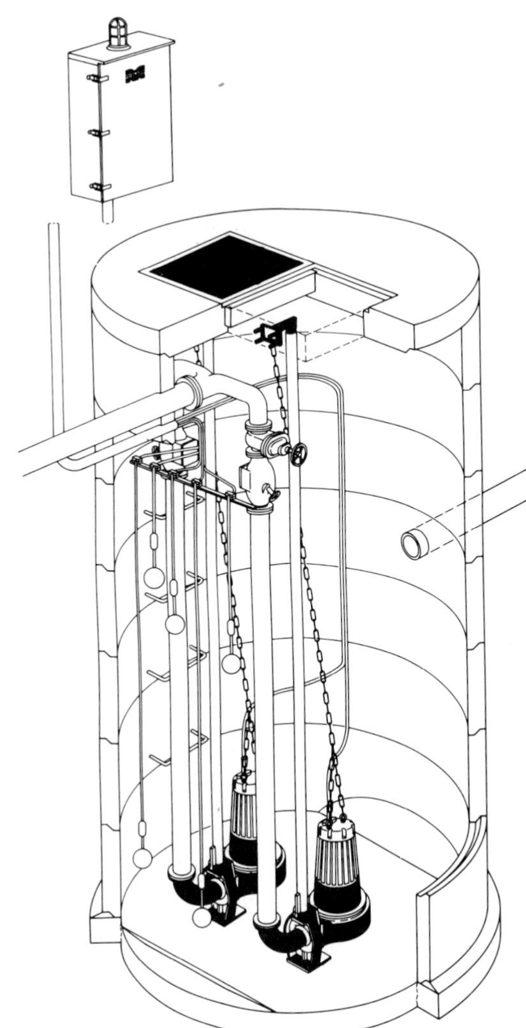

FIGURE 15-14
Submersible sewage pump installation. (*Courtesy LFE Corp., Fluids Control Division*)

The pumps themselves may be submersible designs which are submerged in the wet well (Fig. 15-14) or may be located in an adjacent dry well (Fig. 15-15). The dry well design has the advantage of permitting easier access to the pumps, but is substantially more expensive. The detailed mechanical design of pumping stations is beyond the scope of this text. For most small applications, premanufactured stations made of steel, fiber glass, or concrete are the most economical alternative.

15-7 Miscellaneous Appurtenances

Combined sewers may incorporate regulators and specially formed junctions. *Regulators* are devices which are used to divert water from one sewer to another and which may be employed to divert storm water around a waste-

FIGURE 15-15
Wet pit–dry pit sewage pumping station. (*Courtesy Marolf, Inc.*)

water treatment plant. Designs used in the past either bypassed all the flow once a preset value was reached or bypassed all flow in excess of a preset value. The level of flow at which diversion was begun was approximately 3 times the maximum sanitary sewage flow. This permitted the first storm water—which may be heavily contaminated—to be treated and avoided bypassing any sanitary sewage until the level of dilution in the sewer (and presumably the receiving stream) was fairly high. Most modern sewer systems are separate systems which are carefully designed to exclude any storm drainage from the sanitary sewers, hence regulators are likely to be used only in the few combined systems which still exist.

Junctions in large storm sewers are not made in manholes but, rather, by joining the lines in a gradual fashion which will minimize head losses. Two such junctions are illustrated in Fig. 15-16. The bellmouth junction employs a single arch over both sewers from their first contact, with the height and span gradually being reduced to match the common outlet. In the flat-top junction, both sewers terminate where their walls join and a flat-top transition section carries the combined flow to the outlet. Similar structures are employed for sewers of other cross sections.

PROBLEMS

15-1. A suburban street receives storm water runoff at a rate of 5 L/(m · sec) (on each side). The street has a slope of 3 percent and a cross slope of 3 percent; $n = 0.015$. If the street is 10 m wide and can be totally covered by water (depth = 0 at the midpoint), what is the maximum distance between inlets?

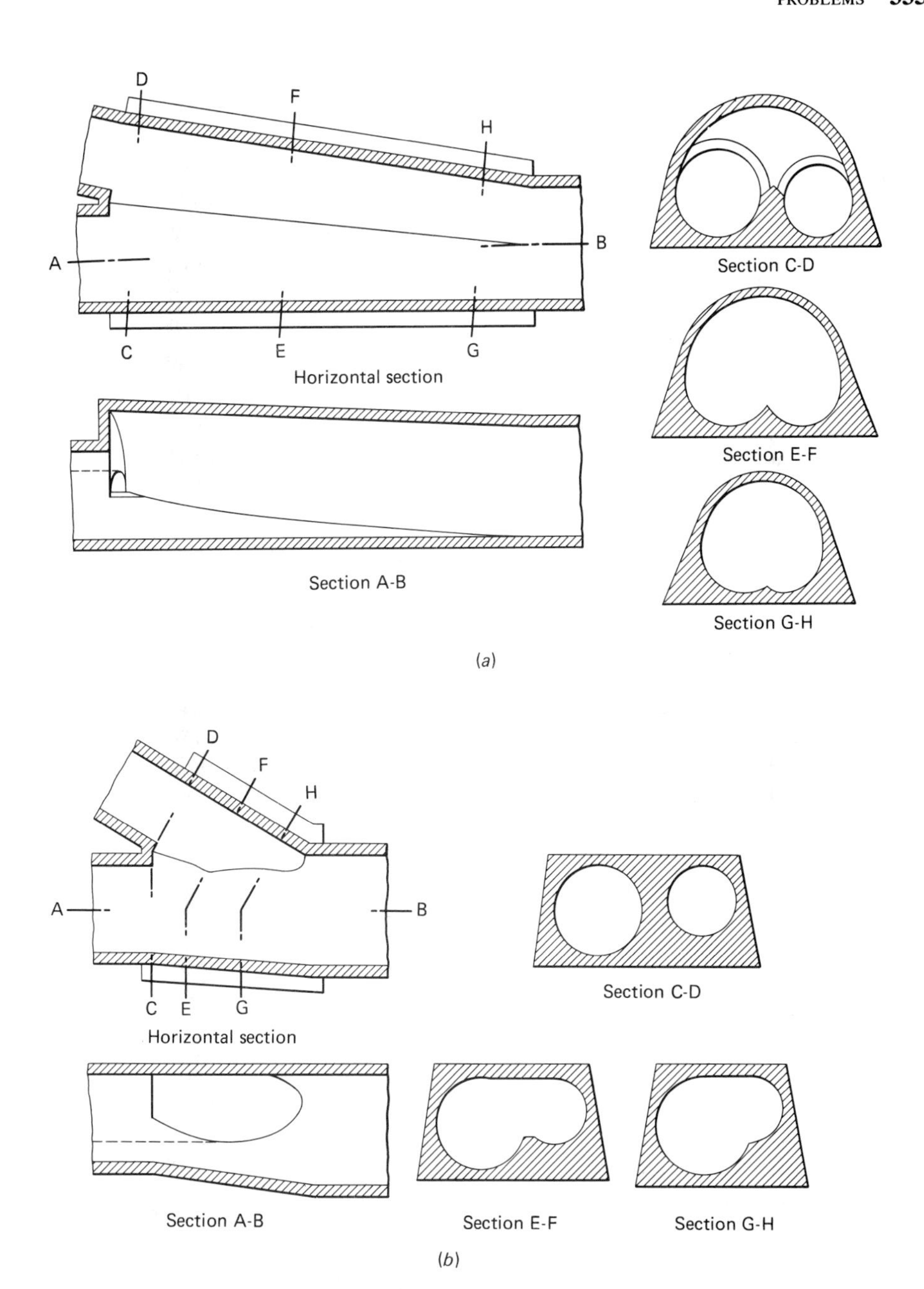

FIGURE 15-16
Large sewer junctions. (*a*) Bellmouth junction; (*b*) flat-top junction.

15-2. If the street of Prob. 15-1 is drained by a curb inlet 3 m long with $a = 50$ mm, determine the percentage of the full street flow removed by each inlet and the spacing of inlets other than the first.

15-3. A combined inlet in a sump has a length of 2 m on the curb portion and an open area of 1.5 m^2 on the grate. How much flow can this inlet accommodate if the water depth is limited to 250 mm?

15-4. Show that the minimum cycle time for a pump station occurs when the inflow is equal to one-half the discharge. (*Hint*: differentiate the equation for t_c with respect to Q_{in} and solve for Q_{in}.)

15-5. A small sewage pumping station with two identical pumps is to be selected for an average flow of 450,000 L/day. The minimum flow rate is one-quarter the average and the maximum 175 percent of average. Determine the working volume required, the pump capacity, and the cycle time at minimum and average flow. The minimum cycle time is 5 min and the minimum running time 1.5 min.

REFERENCES

1. Mark A. Pacheco, Donald D. Gray, and Michael E. Barber, "Computer Aided Design of Vacuum Sanitary Sewer Systems" Technical Report No. CE-HSE-82-2, Purdue University, Lafayette, Ind., 1982.
2. I. G. Carcich et al., "Pressure Sewer Demonstration," *Journal Environmental Engineering Division, American Society of Civil Engineers*, **100**:EE1:25, 1974.
3. Michael E. Barber and Donald D. Gray, "Computer Aided Design of Pressure Sanitary Sewer Systems," Technical Report No. CE-HSE-82-3, Purdue University, Lafayette, Ind., 1982.
4. Tyler G. Hicks and Theodore W. Edwards, *Pump Application Engineering*, McGraw-Hill, New York, 1971.

CHAPTER 16

DESIGN OF SEWER SYSTEMS

16-1 Preliminary Investigations

Preliminary investigations provide a basis for cost estimates which are used to evaluate the feasibility of a project and to justify bond issues, assessments against property, or other methods of fund raising.

Fairly detailed maps are available for most communities. Towns which do not maintain official maps may have been mapped by assessors, insurance companies, or public utilities that will normally permit their maps to be copied. If no maps are available, aerial strip photography is probably the least expensive method of obtaining a map with the necessary detail.

Preliminary designs are based on estimated flows, approximate ground contours, the location of the streets or sewer easements, and the location or locations to which the sewage is to be taken. These preliminary designs will permit estimation of the quantity of pipe of various sizes, the quantity of excavation, the quantity of pavement repair, and the various appurtenances which will be required.

Cost estimates are made for the alternatives identified as being physically practicable and environmentally acceptable. Costs should be based, when possible, on bid tabulations for recent similar construction. When such information is not available, one may use national averages corrected to correspond to the local cost index.

16-2 Detailed Design Requirements

Before final lines and grades are established for a sewer system, an *underground survey* should be conducted to establish the location of existing sewers; water and gas lines; electrical, telephone, and television wires; tunnels; foundations; and other construction which might present obstacles to the proposed design. Many city engineering departments maintain maps showing all underground structures. When such city maps are not available, the designer must compile the information from the various utility companies.

The presence of rock or other difficult subsurface conditions in the construction area will have a significant effect on costs, hence soil borings or soundings are desirable. Soundings may be made by driving a sharpened steel rod into the ground until rock is struck. For depths in excess of 5 to 6 m (15 to 20 ft), borings are necessary. The number of soundings or borings should be sufficient to establish the location of the rock surface throughout the area of the project, and will thus depend on the geological character of the area.

Preparation of construction drawings requires knowledge of street pavement types, the location of all underground structures, the location and basement elevations of all buildings (elevations are usually estimated for residences), the profiles of all streets in which sewers are to be placed, and the elevations of the maximum water surface and invert of all streams, culverts, and ditches. Permanent benchmarks should be established during the survey for use during construction. A detailed map must be prepared, on which the information listed above, together with ground contours, elevations of street intersections, and any abrupt changes in street grade are noted. The scale of the map is typically 1:500 to 1:1000 with a contour interval of about 300 mm (1 ft).

A tentative layout of the proposed system is made by locating lines along the streets or utility easements with arrows showing the direction of flow—normally in the direction of the ground slope. The result will be a main sewer leaving the area at its lowest point with submains and laterals radiating to outlying areas and following the natural slope of the ground to the extent possible. Ridges within the area served may require construction of systems with separate discharges or pumping across the high area. In flat terrain, all sewers may be sloped to a common point from which the collected flow is pumped.

Location of sewers in backlot utility easements will minimize damage to pavement but makes access somewhat more difficult. Generally water, gas, and sewer lines are placed within the street right of way. It is desirable that water lines and sanitary sewers be separated, preferably by the width of the street. On very wide streets, sewers may be placed on both sides to reduce the length of the household sewers.

The vertical location is limited by the need to provide minimum cover and service to basement sanitary facilities and the desirability of minimum excavation. In northern states, 3 m (10 ft) of cover may be required to prevent freezing. In southern states, minimum cover is dictated by traffic loads and ranges upward from 0.75 m (2.5 ft), depending on the pipe size and the anticipated loads.

Manholes are located at sewer intersections, abrupt changes in horizontal direction or slope, changes in size, and at regular intervals along straight runs. Manhole spacing generally does not exceed 100 m (300 ft) and should never be greater than 150 m (500 ft) except in sewers which can be walked through. Each manhole is numbered and the numbers of the manholes also serve to identify the sewers, which run from manhole to manhole.

The area tributary to each manhole is sketched on the map, on the basis of ground contours and the location of lots and buildings. Every manhole need not have a tributary area. For storm sewers a similar procedure is used except the lines are considered to run from inlet to inlet or from intersection to intersection of the streets.

A vertical profile is prepared for each sewer line at a horizontal scale of 1:500 to 1:1000 and a vertical scale about 10 times greater. The profile shows the ground or street surface, tentative manhole locations, elevation of important subsurface strata such as rock, locations of borings, all underground structures, basement elevations, and cross streets. A plan of the line and relevant other structures is usually shown on the same sheet.

The profile assists in the design and is used as the basis of construction drawings. When the design is completed, it may be presented as shown in Fig. 16-1, with line sizes and slopes and elevations at changes in size or grade. The values listed at each manhole represent the ground elevation, entering invert elevation, leaving invert elevation, and the cut to the leaving invert. A tabulation of pipe sizes, lengths, depths, and the number and depth of manholes may also be presented on the drawing. Manhole (MH) 19 of Fig. 16-1 is a drop manhole used to maintain minimum cover at the low point without using a steep grade which would produce a deep cut at manhole 20.

16-3 Design Principles

The principles of hydraulic design are discussed in Chap. 3. Sewers present some special problems since they are normally closed, but are seldom designed to flow full. Manning's equation may be used to determine the required size of

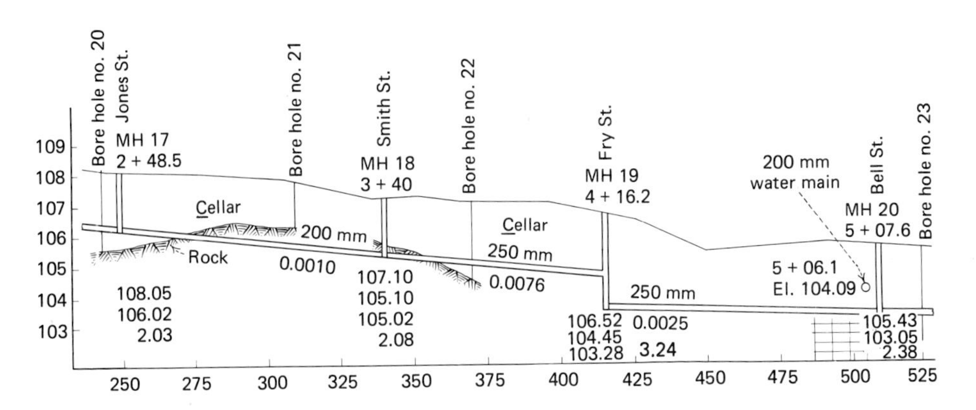

FIGURE 16-1
Profile of a sanitary sewer.

individual conduits, but cannot be applied directly since the hydraulic radius and area of flow are not simple functions of depth in standard sewer sections.

Velocities in sewers are selected with the goal of keeping the solids in the sewage in suspension or at least in traction. Sanitary sewers should be sized to provide a velocity of at least 0.6 m/s (2 ft/s), which is adequate to keep grit in traction. Some regulatory agencies specify minimum slopes for sewers of various diameters. These slopes are those which are calculated to give a velocity of 0.6 m/s when the sewers are full. Since the sewers are commonly not full and the hydraulic radius is thus different from that of a full sewer, the actual velocity will differ from 0.6 m/s, generally being less. In flat terrain the designer may be tempted to use larger pipes since the "minimum" slope is less. This is not good practice, since a large sewer carrying a low flow will have a velocity far less than that corresponding to full flow.

Storm sewer velocities are normally higher than those in sanitary sewers because of the relatively coarse solids which they must convey. Velocities flowing full are generally kept between 0.75 and 2.5 m/s (2.5 to 8 ft/s). The maximum velocity is limited to reduce the potential for abrasive damage to the sewers. Large storm sewers may be lined with vitrified tile blocks to reduce wear. Storm sewer design is based on full flow. Although the sewer may not flow full at the design condition, there will be larger runoff events which will cause the sewer to flow full and aid in scouring out accumulated debris. If ground contours do not permit maintenance of self-cleansing velocities in storm sewers, the appurtenances should be designed to facilitate manual or mechanical cleaning.

In most sewer design procedures, the hydraulic grade line in sewers which are flowing partially full is assumed to parallel the sewer invert. Additional energy losses will occur at changes in direction or pipe size. Sewers larger than 915 mm (36 in) are sometimes curved by deflecting the joints to provide a gradual change in direction. The losses produced by this may be calculated using the procedures of Chap. 3, but are often accounted for more simply by increasing the value of n in Manning's equation by 25 to 40 percent. Changes in direction in smaller sewers should always be made at a manhole. The loss resulting from this change is commonly assumed to be about 30 mm (0.1 ft) and is provided for by dropping the invert by that amount from one side to the other of the manhole.

Losses resulting from increases in pipe size may be provided for by matching the crowns or the 0.8 depth points of the smaller and larger sewer. The drop in the invert provided by matching the crowns will always exceed 30 mm, thus losses due to change in direction may be neglected if the sewer size increases at the manhole. Dropping the invert of the lower sewer by matching the crowns also ensures that the smaller sewer will not be caused to flow full by backwater from the larger unless the larger is also full.

The assumption that the hydraulic grade line parallels the invert is reasonable if the sewer is fairly long and is not surcharged. If the sewer is surcharged, its slope is irrelevant.

16-4 Sanitary Sewer Design

The full flow capacity of circular pipes may be calculated directly from Manning's equation. The nomograms of Figs. 16-2 through 16-4 permit a graphical solution of this problem, which is sometimes quicker than other computational techniques if only a few conduits are involved. From a knowledge of any two parameters, the other two may be found by placing a straightedge across the diagram and connecting the known values. The other two are then read from the intersection with their scales. The typical problem involves selecting a size and slope adequate to carry a given flow at some minimum velocity. The slope, of course, must be considered in relation to the natural slope of the ground.

Some engineers design sanitary sewers to flow half full when at design flow. This is not unreasonable for laterals and submains, since it provides a reasonable factor of safety against peak flows; however, it is not justified for mains or outfall sewers. The diagrams of Fig. 16-2 through 16-4 may also be used to select sewers which will flow half-full. The design flow must be increased by a factor of 2.44 in this procedure. A sewer selected to be full at 2.44 times the design flow will be half-full at the actual flow.

In the typical design application, with a known flow, a desired minimum velocity, and certain restrictions on slope presented by terrain and subsurface conditions, there will usually be a number of possible combinations of pipe size and slope which will serve. A small pipe on a steeper slope can carry the same flow as a larger line on a flat slope. Which is the best choice is not always clear, although as a rule it is cheaper to use a larger pipe if this reduces the excavation required. This may not always occur, however, since the drop in elevation of the invert occasioned by the change in size may be greater than the difference in drop between the two slopes.

Once a pipe diameter and slope have been tentatively selected, one must determine the actual depth and velocity which will exist at design flow. This may be calculated for circular sections since the relationship among depth, area, and hydraulic radius is mathematically definable, although not in a convenient fashion. For example, the area of flow of a partially full circular pipe is given by

$$A_p = r^2 \cos^{-1}\frac{r-y}{r} - (r-y)(2ry - y^2)^{0.5} \qquad (16\text{-}1)$$

in which r is the radius and y the depth of flow. Similarly, the wetted perimeter is given by

$$P_p = 2r \cos^{-1}\frac{r-y}{r} \qquad (16\text{-}2)$$

Consideration of Manning's equation leads to the conclusion that the ratio of partial flow to full flow is

$$\frac{Q_p}{Q_f} = \frac{n_f A_p R_p^{2/3}}{n_p A_f R_f^{2/3}} \qquad (16\text{-}3)$$

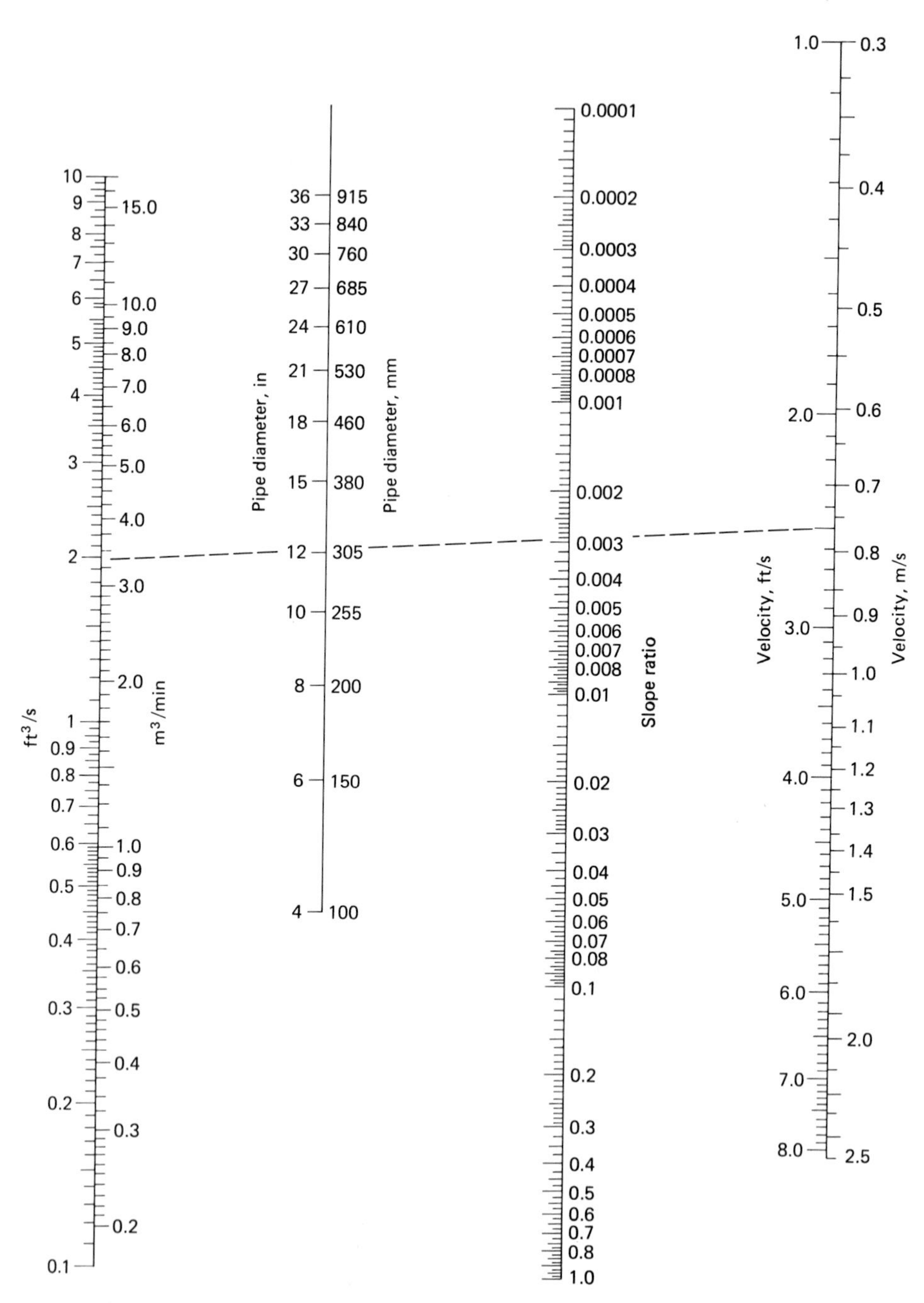

FIGURE 16-2
Nomogram for solution of Manning's equation for circular pipes flowing full ($n = 0.013$).

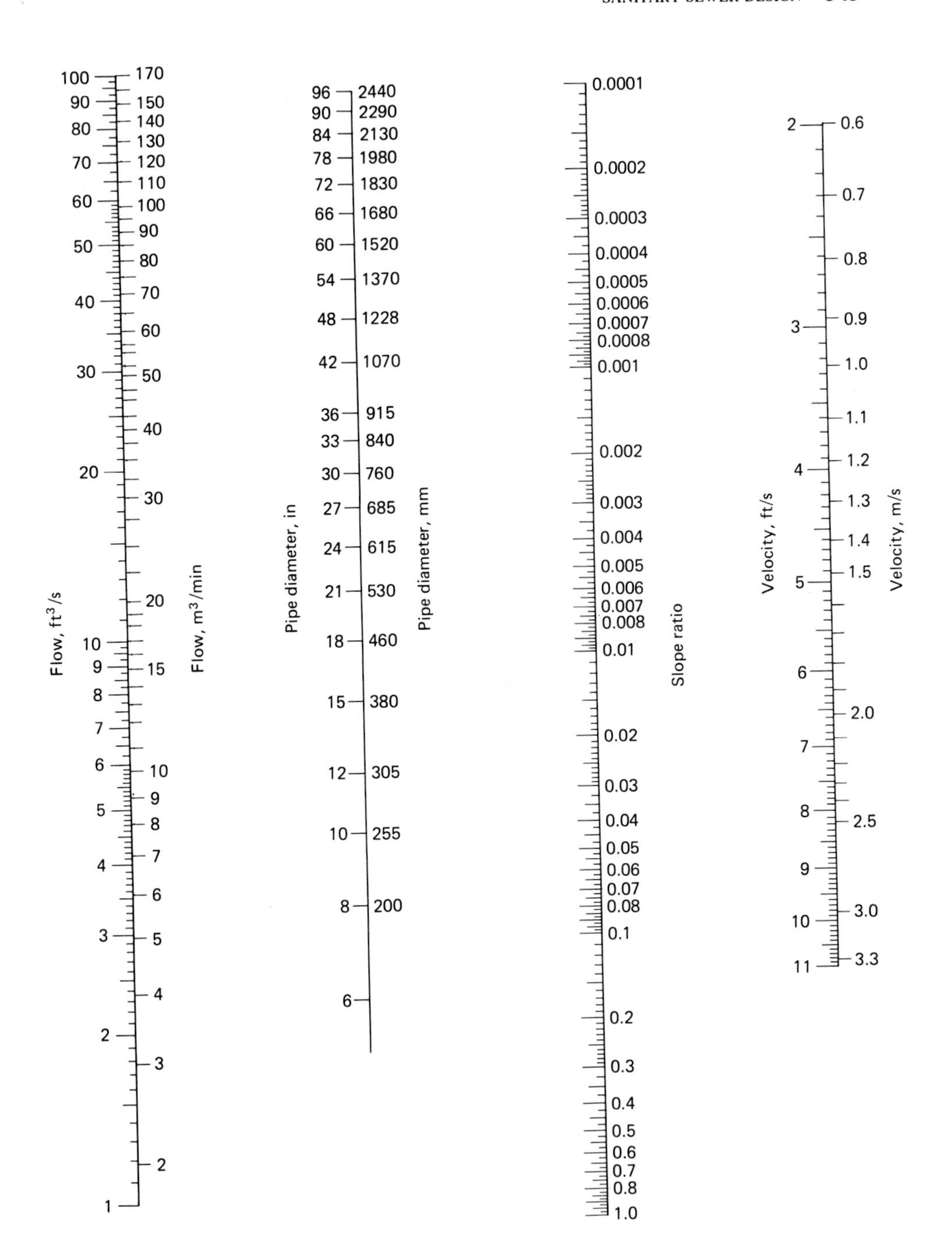

FIGURE 16-3
Nomogram for solution of Manning's equation for circular pipes flowing full ($n = 0.013$).

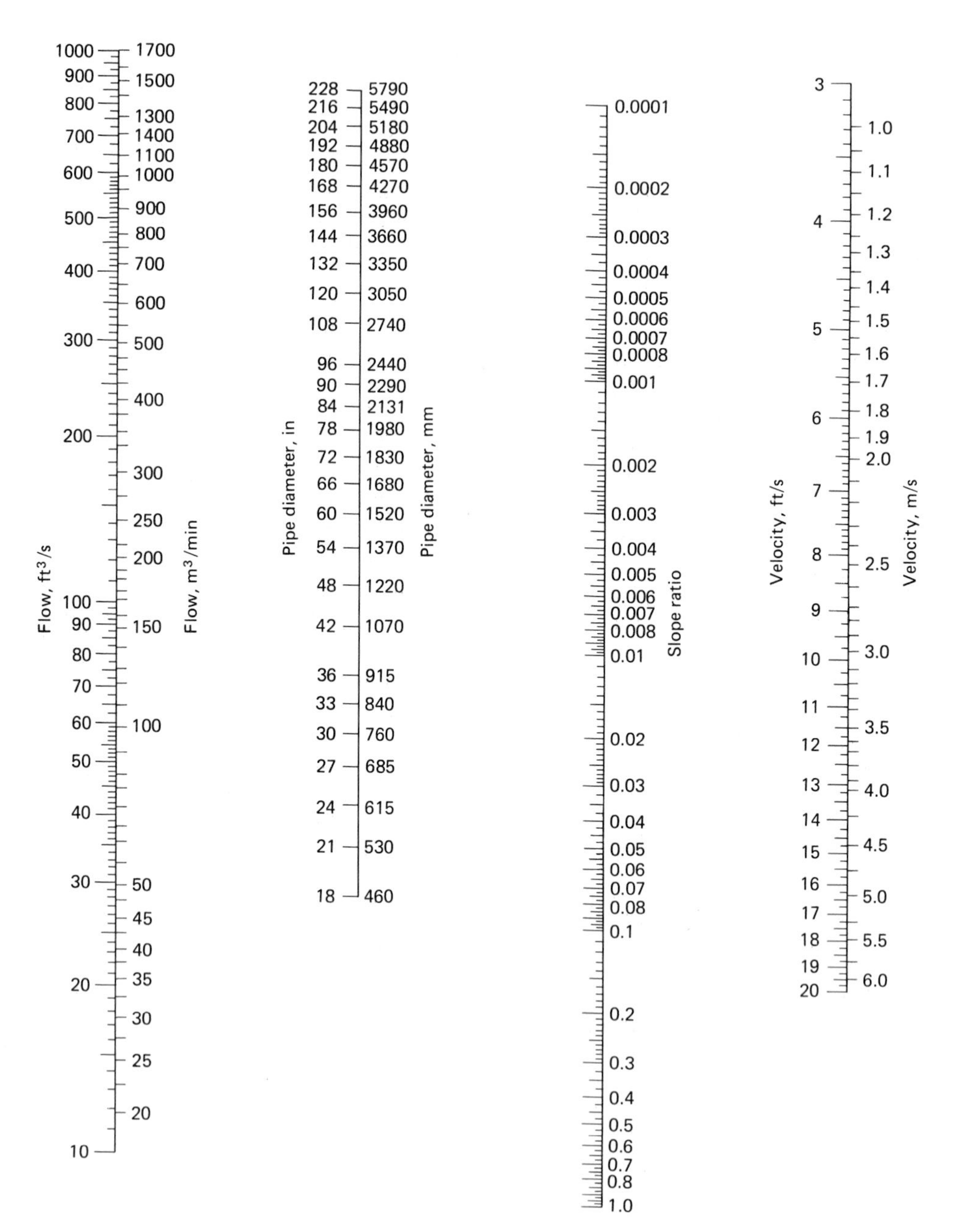

FIGURE 16-4
Nomogram for solution of Manning's equation for circular pipes flowing full (n = 0.013).

and the ratio of partially full velocity to full velocity is

$$\frac{V_p}{V_f} = \frac{n_f R_p^{2/3}}{n_p R_f^{2/3}} \tag{16-4}$$

in both of which R is the hydraulic radius, A/P.

For the typical situation, Q_p/Q_f is known and it is desired to find y and V_p/V_f. The solution to this problem is not straightforward mathematically, but can be obtained readily from a *partial flow diagram* such as Figs. 16-5 and 16-6. The first of these shows the variation of R and A with depth for a circular conduit and the second the variation of Q and V with depth. The latter includes the effect of an observed change in the effective value of n with depth.

Example 16-1 A 915-mm (36-in) sewer is installed on a slope of 0.001. The sewer is 100 m long and runs from a manhole at which its invert is at 98.750 m to a river discharge at which its invert is at 98.650 m. The pipe is to carry a flow of 0.28 m³/s (10 ft³). What is the depth of the water at the upstream manhole when the downstream water surface is at 98 m? At 100 m?

Solution

In the first case the discharge is free and the pipe will not be surcharged. The full-flow capacity of a 36-in sewer on the given slope is approximately 0.63 m³/s (22.2 ft³/s). This may be determined by calculation from Manning's equation or from the nomographs. At a flow of 0.28 m³/s, this pipe will be about 52 percent full (from

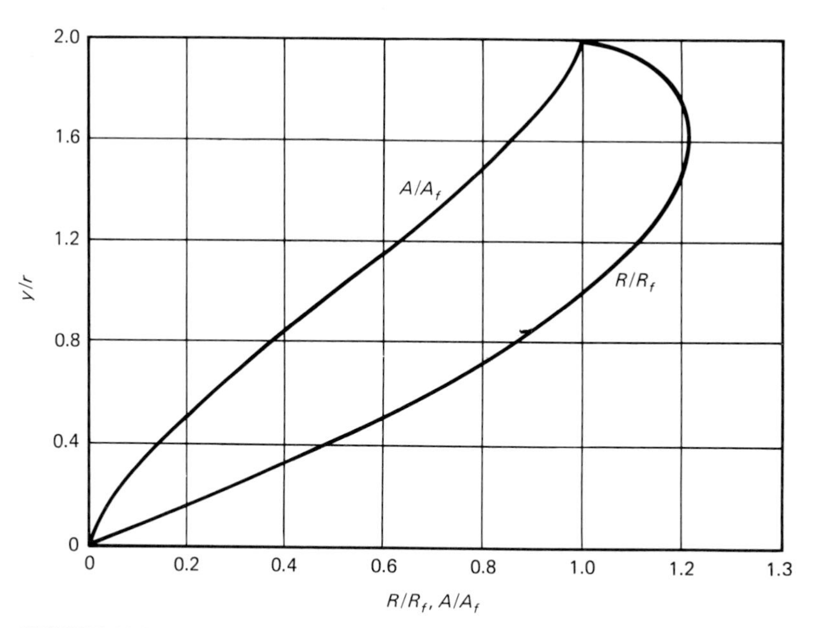

FIGURE 16-5
Variation of R and A with depth in circular pipes.

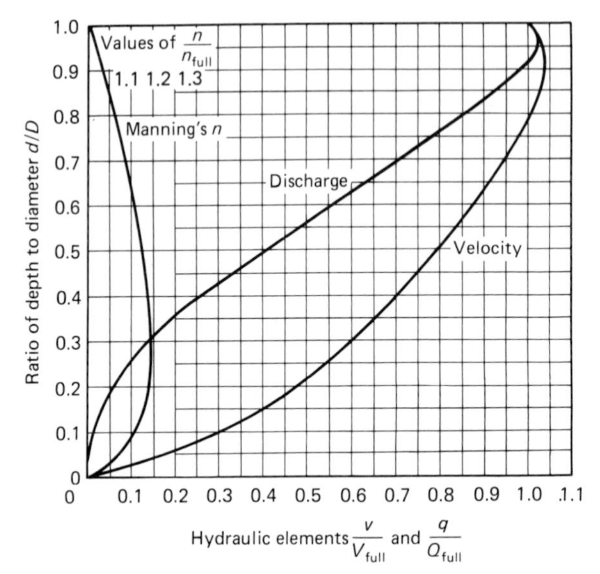

FIGURE 16-6
Variation of flow and velocity with depth in circular pipes.

Fig. 16-6). The depth at the upstream end will thus be 98.750 + 0.52(0.915) = 99.225 m.

In the second case, the pipe is full throughout its length (the water surface at the lower end is above the crown at the upper end). The head loss in a 36-in pipe flowing full at a flow of 0.28 m^3/s is 0.00023 m/m (from Manning's equation or the nomographs). There will be an additional entrance and exit loss equal to about 0.014 m (Chap. 3), hence the water surface at the upstream end will be at 100.00 + 0.00023 × 100 + 0.014 = 100.037 m.

One may observe, as was noted above, that in the second case the slope of the pipe does not enter the calculation, other than indirectly in determining that the pipe is full throughout its length.

Example 16-2 A 915-mm (36-in) sewer is laid on a slope of 0.003. What will the depth of flow and velocity be when the flow is 8.5 m^3/min?

Solution

From Manning's equation or the nomograms, the flow when the sewer is full will be 62 m^3/min and the velocity 1.57 m/s. The ratio of actual to full flow is 8.5/62 = 0.14. From Fig. 16-6, the ratio of depth to diameter is 0.3, and, from this, the ratio of actual velocity to velocity full is 0.60. The depth of flow is thus 0.3 × 915 = 275 mm and the velocity 0.6 × 1.57 = 0.94 m/s.

It must be noted that partial flow diagrams give only approximate results, particularly for high velocities. The discrepancies between computed and actual flow conditions may be caused by wave formation and other factors which are not well defined for small channels. In practical designs it is therefore common practice to apply a factor of safety in order to assure that the sewers do not flow full.

Example 16-3 An engineering handbook lists "minimum" sewer slopes as follows:

Sewer diameter, in	Minimum slope
6	0.0043
8	0.0033
10	0.0025
12	0.0019
15	0.0014
18	0.0011
21	0.00092
24	0.00077

In an area with a ground slope of 0.0015, a sanitary sewer is required to carry a flow of 0.05 m^3/min (0.29 ft^3/s). What sewer size and slope should be used?

Solution

The conclusion invited by the table is that one should select a diameter of 15 in and place the sewer on a slope equal to the ground slope. Investigation of this solution using the nomograms and Fig. 16-6 leads to:

$$Q_f = 4.2 \text{ m}^3/\text{min} \qquad V_f = 0.62 \text{ m/s} \qquad \text{(Fig. 16-2)}$$

$$\frac{Q_p}{Q_f} = \frac{0.5}{4.2} = 0.12$$

$$\frac{y}{D} = 0.28 \qquad \frac{V_p}{V_f} = 0.58 \qquad \text{(Fig. 16-6)}$$

Therefore,

$$V_p = 0.58(0.62) = 0.36 \text{ m/s}$$

This velocity is substantially less than the desirable minimum of 0.6 m/s. A smaller sewer on a steeper slope should be used. Possible solutions include the following:

Sewer dia., in	Slope	Q_f	V_f	q/Q	d/D	v/V	v
8	0.0055	1.5	0.80	0.33	0.45	0.75	0.60
10	0.0052	2.7	0.90	0.19	0.35	0.65	0.59

If the crown elevation at the upper end is fixed regardless of the sewer size (this is usually the case) and the sewer is 100 m long, the change in elevation from the crown at the upstream end to the invert at the downstream end will be 0.753 m for the 8-in line and 0.774 m for the 10-in line. The 8-in line will therefore require less excavation even though it is on a steeper slope.

The application of the general principles discussed above will be illustrated by a small example. The details of the design may vary from location to location depending on state recommendations or regulations with respect to flow per capita, allowance for infiltration, manhole spacing, minimum cover, etc.

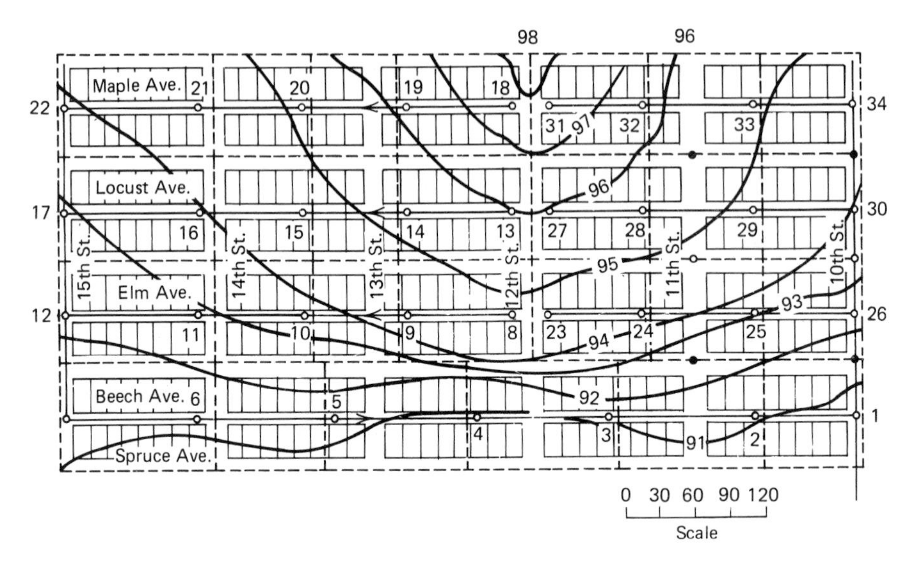

FIGURE 16-7
Portion of a sanitary sewer system.

Figure 16-7 shows a portion of a city for which sewers are to be designed. Sewers for the area north of Maple Avenue have already been designed and a sewer leading from that area flows south on 15th Street to manhole 22. The illustrative design covers the area west of 12th Street and that portion east of 12th Street and south of Beech Avenue. The main sewer will flow south on 10th Street from manhole 1. The area east of 12th Street and north of Beech Avenue is not a part of this design.

Following the ground slope, sewers are located on the map in the streets and alleys in a manner which provides service to all of the buildings. Arrows on the map show the assumed direction of flow. Manholes (which are numbered) are located at intersections of the sewer lines, changes in direction, and at intervals of not more than 90 m along the upper ends of the laterals and 120 m elsewhere. The area tributary to each line is shown by the dashed lines on the drawing. These boundaries are determined by field inspection of the location of buildings and lot lines. Some of the lines have no tributary area, although they would receive some infiltration if this were considered separately from the sewage flow.

It is assumed that the maximum population density in the area will be 10,000 persons/km^2. It is also assumed that the maximum rate of sewage flow, including infiltration, will be 1500 L/capita per day (see Chap. 3). This value might be appropriately reduced for submains and mains since the ratio of peak to average flow will decrease as the average flow increases. The calculation of the design flows in the various lines is tabulated in Table 16-1. Line 0 is the sewer flowing south on 15th Street and terminating at manhole 22. It has already been designed and its tributary population and flow contribution are

TABLE 16-1

Flow calculation for a sanitary sewer system

Line no. (1)	On street (2)	From man-hole (3)	To man-hole (4)	Length of line, m (5)	Increment of area, m^2 (6)	Increment of population (7)	Total tributary population (8)	Sewage flow, L/day (9)	Sewage flow, m^3/min (10)
0	15th	...	22	...	...	...	5725	8,587,500	5.96
1	Alley between Maple and Locust	18	19	90	10,000	100	100	150,000	0.10
2	Alley between Maple and Locust	19	20	90	7,000	70	170	255,000	0.18
3	Alley between Maple and Locust	20	21	90	7,000	70	240	360,000	0.25
4	Alley between Maple and Locust	21	22	120	12,000	120	360	540,000	0.38
5	15th	22	17	87	...	...	6085	9,127,500	6.34
6	Alley between Locust and Elm	13	14	90	10,000	100	100	150,000	0.10
7	Alley between Locust and Elm	14	15	90	7,000	70	170	255,000	0.18
8	Alley between Locust and Elm	15	16	90	7,000	70	240	360,000	0.25
9	Alley between Locust and Elm	16	17	120	12,000	120	360	540,000	0.38
10	15th	17	12	87	...	...	6445	9,667,500	6.71
11	Alley between Elm and Beech	8	9	90	10,000	100	100	150,000	0.10
12	Alley between Elm and Beech	9	10	90	7,000	70	170	255,000	0.18
13	Alley between Elm and Beech	10	11	90	7,000	70	240	360,000	0.25
14	Alley between Elm and Beech	11	12	120	12,000	120	360	540,000	0.38
15	15th	12	7	87	...	...	6805	10,207,500	7.09
16	Alley between Beech and Spruce	7	6	120	12,000	120	6925	10,387,000	7.21
17	Alley between Beech and Spruce	6	5	120	9,000	90	7015	10,522,500	7.31
18	Alley between Beech and Spruce	5	4	120	12,000	120	7135	10,702,500	7.43
19	Alley between Beech and Spruce	4	3	120	11,000	110	7245	10,867,500	7.55
20	Alley between Beech and Spruce	3	2	120	11,000	110	7355	11,032,500	7.66
21	Alley between Beech and Spruce	2	1	90	7,000	70	7425	11,137,500	7.73

entered in the table. If infiltration were separately considered, additional columns would be inserted in the table. Flows are calculated by estimating the tributary area (column 6), multiplying by the assumed maximum population density to obtain the incremental population (column 7), totaling the tributary population for each line (column 8) and multiplying by the flow per capita to obtain the design flow (columns 9 and 10).

In selecting appropriate sewer sizes and slopes, a minimum cover of 2 m is assumed and a minimum sewer diameter of 200 mm (8 in) and a minimum velocity of 0.6 m/s are used. If a sewer changes in direction, a drop of 30 mm is provided in the manhole. At changes in size, the crowns of the upstream and downstream sewers are matched. Drop manholes are used only if the invert of the branch line is more than 0.6 m above the elevation which would be obtained by matching the crowns. The design proceeds from the upstream end of the system as shown in the tabular calculations of Table 16-2. In the actual design,

TABLE 16-2
Design of a sanitary sewer system

						Ground elevations	
Line no. (1)	To street (2)	From manhole (3)	To manhole (4)	Length of line, m (5)	Sewage flow, m³/min (6)	Upper manhole (7)	Lower manhole (8)
0	15th	...	22	...	5.96	...	93.69
1	Alley between Maple and Locust	18	19	90	0.10	97.74	96.40
2	Alley between Maple and Locust	19	20	90	0.18	96.40	95.27
3	Alley between Maple and Locust	20	21	90	0.25	95.27	93.93
4	Alley between Maple and Locust	21	22	120	0.38	93.93	93.69
5	15th	22	17	87	6.34	93.69	92.99
6	Alley between Locust and Elm	13	14	90	0.10	96.04	95.37
7	Alley between Locust and Elm	14	15	90	0.18	95.37	94.57
8	Alley between Locust and Elm	15	16	90	0.25	94.57	93.81
9	Alley between Locust and Elm	16	17	120	0.38	93.81	92.99
10	15th	17	12	87	6.71	92.99	92.32
11	Alley between Elm and Beech	8	9	90	0.10	94.85	94.30
12	Alley between Elm and Beech	9	10	90	0.18	94.30	93.48
13	Alley between Elm and Beech	10	11	90	0.25	93.48	92.90
14	Alley between Elm and Beech	11	12	120	0.38	92.90	92.32
15	15th	12	7	87	7.09	93.32	91.92
16	Alley between Beech and Spruce	7	6	120	7.21	91.92	91.74
17	Alley between Beech and Spruce	6	5	120	7.31	91.74	91.71
18	Alley between Beech and Spruce	5	4	120	7.43	91.71	91.40
19	Alley between Beech and Spruce	4	3	120	7.55	91.40	91.43
20	Alley between Beech and Spruce	3	2	120	7.66	91.43	91.40
21	Alley between Beech and Spruce	2	1	90	7.73	91.40	90.61

a profile like that of Figure 16-1 would be available for each street and the sewers would be sketched on those drawings.

The design begins with line 1. The ground elevations at the upper and lower ends are noted from columns 7 and 8 and the invert at the upper end is set at 97.74 − 0.2 − 2.0 = 95.54, which provides for the minimum cover and the diameter of the sewer. A tentative elevation at the lower end, obtained in the same manner, is 94.20 m. These elevations are entered in columns 17 and 18. The slope of the sewer is calculated to be 0.0146 which, for a 200-mm pipe, gives a full flow of 2.45 m^3/min and a full velocity of 1.28 m^3/s. Upon consideration of the partial flow diagram of Fig. 16-6, it is found that the actual velocity at the design flow of 0.1 m^3/min is only 0.44 m/s on this slope. The slope is therefore increased as shown in the table to provide a velocity of at least 0.6 m/s at design flow. It may be necessary to try a number of slopes before a satisfactory velocity is obtained. The slope is entered in column 10, the drop calculated as the product of columns 10 and 5, and the value of column 18 obtained as the difference between columns 17 and 11. Note that the invert

Dia. of pipe, mm (9)	Grade of sewer (10)	Fall of sewer, m (11)	Velocity flowing full, m/s (12)	Capacity flowing full, m^3/min (13)	Q/Q_{full} (14)	V/V_{full} (15)	V, m/s (16)	Invert elevations: Upper manhole (17)	Invert elevations: Lower manhole (18)
305	...	...	...	...	...	...	...	...	91.23
200	0.0180	1.62	1.42	2.72	0.04	0.44	0.62	95.54	93.92
200	0.0130	1.17	1.20	2.30	0.08	0.53	0.64	93.92	92.75
200	0.0113	1.02	1.10	2.10	0.12	0.58	0.64	92.75	91.73
200	0.0070	0.84	0.89	1.70	0.22	0.68	0.61	91.73	90.89
380	0.0040	0.35	1.04	6.95	0.91	1.02	1.06	90.71	90.36
200	0.0180	1.62	1.42	2.72	0.04	0.44	0.62	93.84	92.22
200	0.0130	1.17	1.20	2.30	0.08	0.53	0.64	92.22	91.05
200	0.0113	1.02	1.10	2.10	0.12	0.58	0.64	91.05	90.03
200	0.0070	0.84	0.89	1.70	0.22	0.68	0.61	90.03	89.14
460	0.0015	0.13	0.71	6.97	0.96	1.03	0.73	88.88	88.75
200	0.0180	1.62	1.42	2.72	0.04	0.44	0.62	92.65	91.03
200	0.0130	1.17	1.20	2.30	0.08	0.53	0.64	91.03	89.86
200	0.0113	1.02	1.10	2.10	0.12	0.58	0.64	89.86	88.84
200	0.0070	0.84	0.89	1.70	0.22	0.68	0.61	88.84	88.00
530	0.00092	0.08	0.62	8.18	0.87	1.02	0.63	87.67	87.59
530	0.00092	0.11	0.62	8.18	0.88	1.02	0.63	87.56	87.45
530	0.00092	0.11	0.62	8.18	0.89	1.02	0.63	87.45	87.34
530	0.00092	0.11	0.62	8.18	0.91	1.02	0.63	87.34	87.23
530	0.00092	0.11	0.62	8.18	0.92	1.02	0.63	87.23	87.12
530	0.00092	0.11	0.62	8.18	0.94	1.03	0.64	87.12	87.01
530	0.00092	0.08	0.62	8.18	0.94	1.03	0.64	87.01	86.93

elevation at the upper end of a line is the same as that at the lower end of the upstream line except where there is a change in size or direction.

Manhole 22 is at the lower end of lines 4 and 0 and the upper end of line 5. Above the manhole line 0 is 305 mm in diameter. Trial indicates that a 305-mm line would require a slope steeper than the ground to carry the flow of 6.34 m^3/min. Since the invert of line 4 is already deeper than required by minimum cover, it is preferable to use a larger sewer on a lower slope. A 380-mm line is required. Since a change in size occurs, the crowns of the sewers are matched. The crown of line 4 is at 91.09, that of line 0 at 91.54; therefore the crown of line 5 must be at 91.09, giving an invert elevation of 90.71. The drop provided by matching crowns (0.18 m) exceeds that required by the change in direction. Manholes 17 and 12 are handled in a similar fashion. At manhole 7 there is a change in direction with no change in pipe size. The invert is thus dropped 30 mm across the manhole.

The procedure outlined in Tables 16-1 and 16-2 can be readily programmed to permit computer assisted design. Pacheco and Gray[1] have developed a program which, subject to certain constraints, will design a system that minimizes excavation and one which minimizes pipe size. The program does not ensure that the velocity at design flow is at least 0.6 m/s, but does offer a rapid method for preliminary design. The program listing could be modified to permit inclusion of partial flow calculations.

SWMM[2] and others of the more sophisticated storm-water simulation models include provisions for modeling sanitary sewage flow, including its diurnal variations. Although not as economical as the model of Pacheco and Gray, SWMM's design option can be used to determine preliminary sizes for sanitary sewers by omitting the RUNOFF (stormwater flow generation) block and using FILTH and INFIL (sanitary sewage flow and infiltration) within the TRANSPORT block.

In addition to the models discussed above, there are a large number of other systems with widely varying capabilities which are commercially available. Except for very small systems or for small additions or modifications to existing systems, it is unlikely to be economical to use manual calculation techniques in the design of sanitary sewers.

16-5 Storm Sewer Design

In Chap. 13 it is shown that the flow which storm sewers must be designed to carry is very much dependent on the design of the sewers themselves. Systems which minimize surface ponding and storage will have higher peak flows, will require larger structures, and will increase the potential for downstream flooding. The flow at any point in a storm-water collection system is a function not only of rainfall and abstractions but also of the design of the system upstream.

Design concepts for storm sewers have changed considerably in recent years. In the past, sewers were frequently designed to drain local areas rapidly, with little thought being given to the cumulative effect on an urban area. The results of such development have included flooding because of increased runoff and loss of groundwater recharge because of both increased imperviousness and decreased detention time. It is now considered desirable, and in some cases essential, to prevent the peak discharge from developed areas from increasing above that which would have occurred prior to development. It is generally not possible to maintain the total discharge at the same level as existed prior to development, but peak flows can definitely be reduced.

Storm drainage systems should be considered to consist of two components, sometimes called the *minor system* and the *major system*. The *minor system* consists of ditches, canals, and sewers designed to accommodate a storm of moderately short recurrence interval, perhaps 2 to 5 years, depending on the character of the area. The purpose of this system is to prevent flooding of roadways and adjoining areas by moderate storms which occur relatively frequently. The *major system* is the path followed by the flow during runoff events which exceed the capacity of the minor system. The components of the major system include the streets and adjoining land, drainage rights of way for surface flow, and natural drainage channels which are preserved during development. The major system should be adequate to carry flows resulting from storms of long recurrence interval—at least 25 years.

The design for a modern drainage system should always incorporate techniques for reducing the quantity of runoff. This requires legal action in the form of building and zoning regulations as well as proper engineering design. The following methods can reduce runoff and thereby reduce both the cost of flood protection and the likelihood of flood damage:

1. Discharge of roof drains to grassed surfaces rather than pavement provides increased detention and increased opportunity for infiltration. This reduces both the total quantity of runoff and the magnitude of the peak flow.
2. Contour grading, which maintains natural drainage patterns, will produce more circuitous flow paths than those which exist in gridlike systems. This will reduce the peak flow, and, to the extent the flow channels are pervious, reduce the total discharge.
3. Porous pavements of asphalt, interlocking paving blocks, or gravel will permit some infiltration. The rougher surface of these materials will also reduce the velocity of flow and thereby both reduce the peak flow and provide additional opportunity for infiltration.
4. Grassed ditches used in place of curbed streets or sewer pipes will reduce the peak flow and provide a reduction in total discharge as well.
5. Surface detention can be increased by designed infiltration basins—detention ponds which may either be normally dry or normally part full, and by deliberate ponding on roadways, parking lots, and flat roofs. The latter can be achieved by reducing the size of minor drainage structures.
6. Subsurface disposal using perforated storm sewers, infiltration trenches, or dry wells will reduce both peak flows and total discharge.

Since the minor system is intended to accommodate storms of relatively low intensity, a great deal of precision is not required. Subarea flows can be determined using the rational method, synthetic hydrographs, or the simpler simulation techniques such as ILLUDAS (see Chap. 13). The application of the rational method to a section of a city is illustrated below. The only difference between the rational method and the other techniques lies in the manner of determining the design flows for the individual sewers.

Figure 16-8 shows the section that is to be provided with storm sewers. The area is residential, and it is assumed that the runoff coefficient will be 0.4 at the time of maximum development. In residential areas it is appropriate to use high-frequency rainfall events, thus a 2-year intensity-duration curve will be used. This curve can be fitted by the equation $i = 1590/(t + 17)$. The minimum sewer size is 305 mm, the minimum velocity is 0.75 m/s with the sewer flowing full, and the minimum cover above the crown is 1.5 m. The location of all existing underground structures must be known in order to avoid conflicts. In this example it is assumed that no conflicts exist.

As with sanitary sewers, the first step is to establish a tentative location for the sewers, in this case in the streets, following the natural slope of the ground. Since no household connections are required (or permitted), the sewers need not abut the individual lots. The drainage areas tributary to each inlet are then sketched on the map, based on ground countours and a field inspection in which the grading of lots and location of roof drains is noted. Corner lots on the drawing are shown as draining in part to each abutting street. The areas are measured on the map and the resulting values in hectares are noted within the dashed lines defining each area.

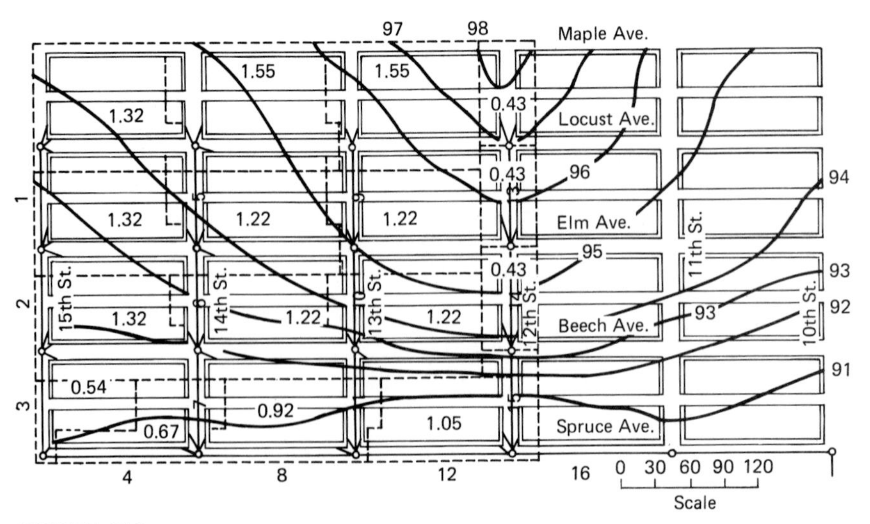

FIGURE 16-8
Portion of a storm sewer system.

Laterals run south on 12th, 13th, 14th, and 15th streets to a main located along Spruce Avenue. Inlets are located at street intersections to serve the areas shown. Since 12th Street is nearly at the crest of the watershed, it may not be necessary to provide a sewer—gutter details at the intersections could be designed to divert the flow to the cross streets. It is also a matter of judgment whether to extend the laterals to Maple Avenue or to permit the flow to be carried in the gutters to Locust Avenue. In this case a sewer is provided on 12th Street and the laterals are not extended to Maple Avenue. All lines have been numbered on the drawing for identification in the tabular computations presented in Table 16-3.

The first four columns in the table show the line numbers and their locations. The area directly tributary to each line is shown in column 5, the runoff coefficient in column 6 (constant over the area in this example, but normally variable), and the product *Ca* in column 7. The total area tributary to each line (directly and indirectly) is summed in column 8.

In order to calculate the flow in the rational method, the design intensity must be determined and this, as noted in Chapter 13, depends upon the time of concentration. The time of concentration for the directly contributing area may be calculated as shown in Chap. 13. In this example it is assumed to be 10 min except for line 13. This line serves a very small area for which the time of concentration is reduced to 6 min. The times of concentration for each line are tabulated in column 9. At the beginning of the design these values are known only for those lines which have only directly contributing areas (1, 5, 9, and 13).

Beginning at the upstream end of the most remote lateral, using the time of concentration of 10 min, the rainfall intensity is determined to be 96 mm/h. The product of this value and column 8 yields the design flow in column 11. Once the flow is known, selection of an appropriate conduit can begin. In storm sewers it is particularly desirable to prevent the flow line from falling too far below the ground surface, hence the initial slope tried should be that which will provide minimum cover. A pipe may then be selected from the nomograms of Figs. 16-2 through 16-4 or from Manning's equation. If the velocity is too high or too low, this initial selection may be modified.

The values tabulated in columns 18 and 19 are not derivable from the figure, but are obtained from survey data. A trial slope is based on the difference between the ground elevations at the upstream and downstream ends and the length of the line (column 15). A pipe is selected which will carry at least the flow of column 11 on the trial slope. The velocity and flow capacity of this line are then determined and entered in columns 14 and 17. Assuming the velocity of flow is satisfactory, the time of flow in the line can then be calculated (column 16). This is added to the time of concentration to find the inlet time for the next sewer downstream. The invert elevations of columns 20 and 21 result from subtracting the cover and pipe diameter from the ground elevation.

In this design, as in the sanitary sewer design, crowns are matched at changes in diameter, and inverts are dropped by 30 mm at changes in direction.

TABLE 16-3
Design of a storm sewer system

Line no. (1)	Location (2)	From street (3)	To street (4)	Increment of area, 10^4 m² (5)	C (6)	Equivalent area Ca 100 percent, 10^4 m² (7)	Total area ΣCa, 10^4 m² (8)	Time of concentration, min (9)	mm/h (10)
1	15th St.	Locust	Elm	1.32	0.40	0.53	0.53	10.0	96
2	15th St.	Elm	Beech	1.32	0.40	0.53	1.06	11.0	93
3	15th St.	Beech	Spruce	1.32	0.40	0.53	1.59	11.9	90
4	Spruce Ave.	15th	14th	0.54	0.40	0.22	1.81	12.7	87
5	14th St.	Locust	Elm	1.55	0.40	0.62	0.62	10.0	96
6	14th St.	Elm	Beech	12.2	0.40	0.49	1.11	11.0	93
7	14th St.	Beech	Spruce	1.22	0.40	0.49	1.60	11.7	90
8	Spruce Ave.	14th	13th	0.67	0.40	0.27	3.68	15.5	79
9	13th St.	Locust	Elm	1.55	0.40	0.62	0.62	10.0	96
10	13th St.	Elm	Beech	1.22	0.40	0.49	1.11	11.0	93
11	13th St.	Beech	Spruce	1.22	0.40	0.49	1.60	11.5	91
12	Spruce Ave.	13th	12th	0.92	0.40	0.37	5.65	17.3	76
13	12th St.	Locust	Elm	0.43	0.40	0.17	0.17	6.0	113
14	12th St.	Elm	Beech	0.43	0.40	0.17	0.34	6.8	109
15	12th St.	Beech	Spruce	0.43	0.40	0.17	0.51	7.5	106
16	Spruce Ave.	12th	11th	1.05	0.40	0.42	6.58	19.6	71

The time of concentration for lines which have more than one upstream contributor is the longest of those possible. The time of concentration for line 8, for example, is that of line 4 (which is longer than that of line 7) plus the time of flow in line 4, or 12.7 + 2.8 = 15.5 min. On Spruce Avenue the grade has been increased above that of the ground to minimize increases in pipe size. This is economical in this case, since the extra excavation occasioned by the greater slope is less than that which would be required by an increase in pipe diameter.

16-6 Major System Design

The major system, together with the minor system, is expected to be able to prevent flooding of houses and other valuable property resulting from storms of relatively low frequency. The design storm normally will have a recurrence interval of at least 25 years. The major system may also incorporate features intended to reduce the peak flow from the entire development, such as controlled-discharge detention ponds.

Flow calculations for the major system must be based on sound hydrologic and hydraulic principles—the rational method will not serve. The SCS technique, the unit hydrograph method, or continuous simulation models, all of which are discussed in Chap. 13, may be used for this level of design and

Q m³/min (11)	Grade (12)	Dia. of pipe, mm (13)	Velocity flowing full, m/s (14)	Length of line, m (15)	Time of flow, min (16)	Capacity of sewer, m³/min (17)	Ground elevations Upper end (18)	Ground elevations Lower end (19)	Invert elevations Upper end (20)	Invert elevations Lower end (21)
8.5	0.0077	380	1.4	87	1.0	9.3	93.29	92.62	91.41	90.74
16.4	0.0070	530	1.7	87	0.9	22.6	92.62	92.01	90.59	89.98
23.9	0.0074	610	1.9	87	0.8	32.3	92.01	91.37	89.90	89.26
26.3	0.0011	840	0.9	140	2.8	28.9	91.37	91.22	89.03	88.88
9.9	0.0052	460	1.3	87	1.0	12.7	94.05	93.60	92.09	91.64
17.2	0.0147	460	2.2	87	0.7	22.1	93.60	92.32	91.64	90.36
24.0	0.0126	530	2.2	87	0.6	28.9	92.32	91.22	90.29	89.19
48.5	0.0020	915	1.3	140	1.8	51.0	91.22	91.10	88.81	88.53
9.9	0.0084	380	1.5	87	1.0	10.2	95.46	94.73	93.58	92.85
17.2	0.0243	460	2.8	87	0.5	27.2	94.73	92.62	92.77	90.66
24.3	0.0193	460	2.5	87	0.6	25.0	92.62	91.10	90.66	88.98
71.6	0.0009	1220	1.0	140	2.3	72.4	91.10	91.07	88.22	88.09
3.2	0.0168	305	1.8	87	0.8	7.8	96.92	95.46	95.12	93.66
6.2	0.0256	305	2.2	87	0.7	9.5	95.46	93.23	93.66	91.43
9.0	0.0248	305	2.3	87	0.6	9.9	93.23	91.07	91.43	89.27
77.9	0.0011	1220	1.2	140	2.0	79.9	91.10	91.19	88.35	88.20

evaluation. Whatever method be used, the engineer must be certain that it has been properly calibrated for the area being studied.

Storm-water runoff in excess of that which can be carried by the minor system will flow in the streets and over the surface of the ground. The hydraulic capacity of the streets may be estimated from Manning's equation. Figure 16-9 presents the capacity of several street cross sections as a function of their slope.

The peak flow resulting from a 25-year storm on the area of Fig. 16-8 will be estimated here using the SCS technique. The total area is 164,500 m² (40.6 acres). The slope of the tract is approximately 1.5 percent. The curve number is taken as 75 (corresponding, more or less, to a runoff coefficient of 0.4). A 25-year return period storm would have a total precipitation of about 150 mm (6 in) in the area corresponding to the 2-year intensity-duration relationship used above (the midwestern United States).

From Table 13-6, the runoff depth is 82 mm. From Fig. 13-6, the peak discharge on a 1 percent slope is 2.15×10^{-1} m³/s per millimeter of runoff. This must be adjusted by a factor of 1.13 obtained from Table 13-7. The peak discharge is therefore

$$0.215 \times 1.13 \times 82 = 19.9 \text{ m}^3/\text{s}$$

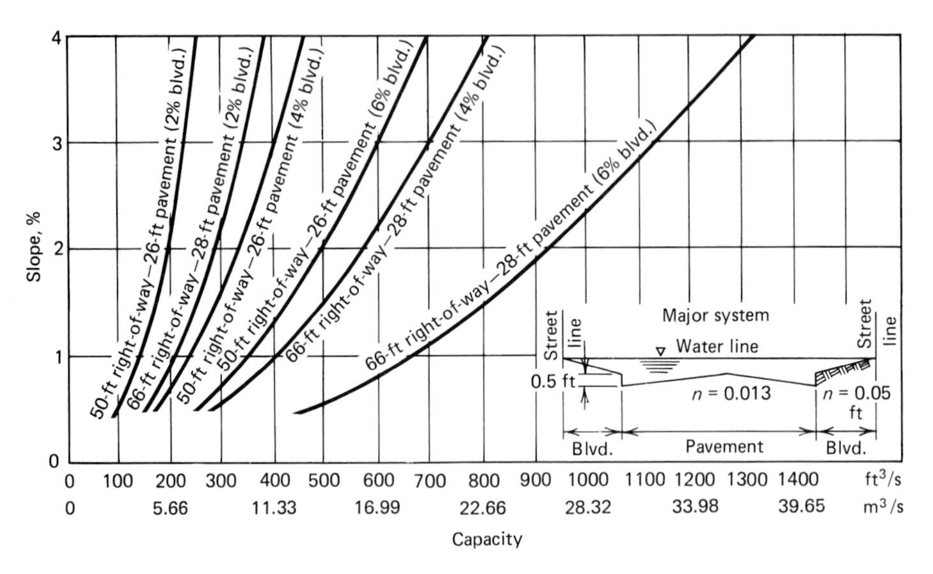

FIGURE 16-9
Hydraulic capacity of roadways.[3] (*Reprinted from Modern Sewer Design, Copyright 1980, American Iron and Steel Institute. Used with permission of the Institute.)*

The peak capacity of the minor system, from above, is 79.9 m^3/min or 1.33 m^3/s. The ratio of capacity of the major to minor system is thus about 15 to 1.

If the ratio applied above were applied to the individual streets (this is not precisely correct), the flows at their lower ends would range from 2.3 m^3/s on 12th Street to 6.1 m^3/s on 13th Street. The slopes of these streets at their lower ends range from 1 to 3 percent. From Fig. 16-9 it appears that these streets should be able to accomodate these flows. Spruce Avenue, however, is another matter. Spruce Avenue is nearly level, and the flow at its intersection with 12th Street will be about 20 m^3/s. If flooding is to be prevented in this area, a right of way would need to be obtained through which this flow could be channeled.

As noted above, the peak discharge from an urban area can be reduced by a number of methods, many of which must be applied throughout the entire drainage basin if they are to be effective. *Controlled-discharge detention ponds* offer a technique for reducing the peak flow at the lower end of the system in addition to any reductions which may have been provided upstream.

Controlled-discharge detention ponds are small reservoirs with outlet structures. Detailed design should be done with an actual discharge hydrograph, which, as noted in Chap. 13, can be obtained from a number of the more sophisticated urban hydrology analysis methods. Reservoir routing procedures are, in fact, included in some urban models. HEC-1[4] and TR-20[5] are particularly useful in design of these structures.

Reference 6 presents a generalized method which is useful for preliminary estimates but which should not be used for final design. Figures 16-10 and

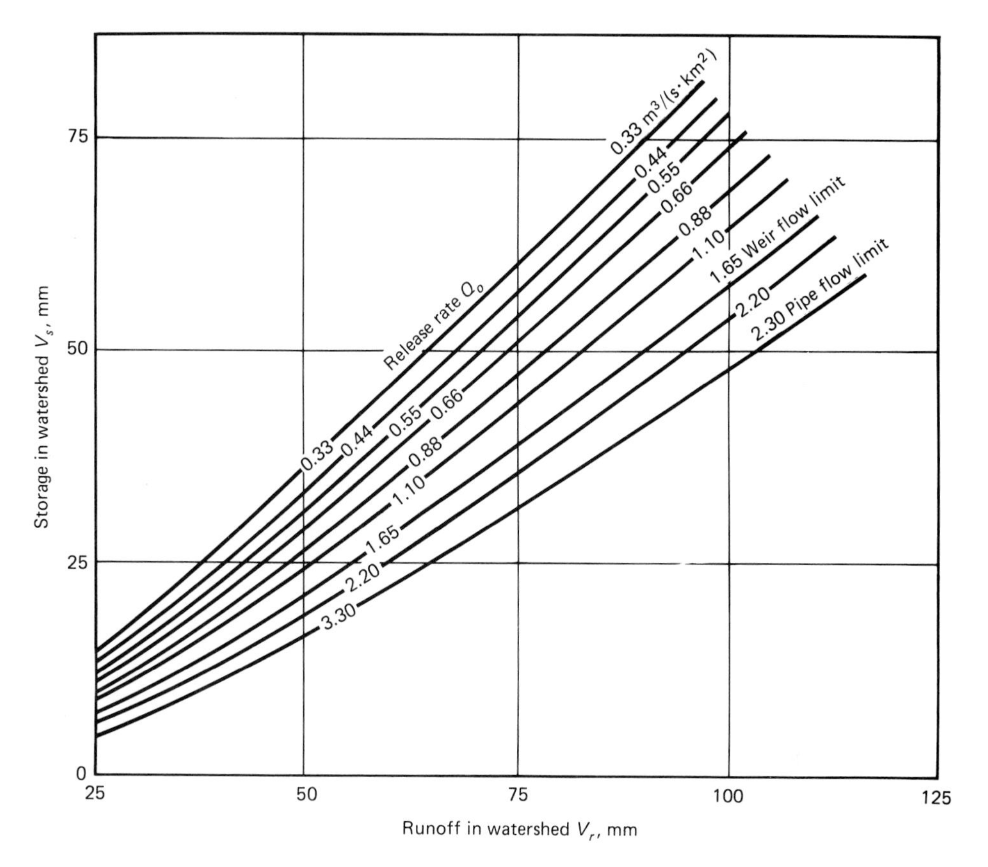

FIGURE 16-10
Approximate single-stage routing for weir flow structures up to 1.65-m³/(s · km²) release rate and pipe flow structures up to 3.30 m³/(s · km²). (Type II distribution, 24-lh rainfall.)

16-11, which are adapted from Ref. 6, relate the volume of inflow and storage volume to the peak release rate.

In order to use these figures, one must first determine the volume of runoff and the peak rate of flow from the watershed, and establish the desired peak rate of outflow from the structure. The latter might be done by using the SCS technique to estimate the peak flow existing prior to development.

Example 16-4 Using the data of the SCS method example in Chap. 13, determine the required detention volume to reduce the peak discharge of the developed area to that which existed before development.

Solution
From Chap. 13, the peak discharge from the undeveloped area is 8.6 m³/s and that following development is 19.0 m³/s. The total volumes of runoff are 73,560 m³ and 86,760 m³ respectively. The desired discharge per unit area is 8.6/1.2 or

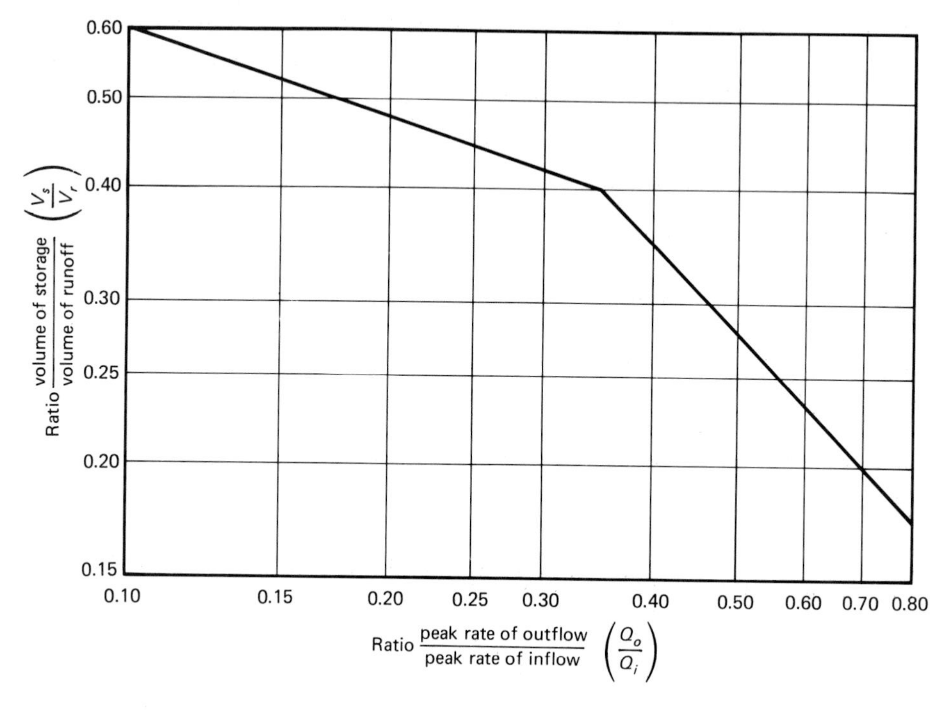

FIGURE 16-11
Approximate single-stage routing for weir flow structures over 1.65 m³/ (s · km²) and pipe flow structures over 3.30 m³/(s · km²). (Type II distribution, 24-lh rainfall.)

7.17 m³/(s · km²). Figure 16-11 must be used, since the discharge exceeds 3.3 m³/(s · km²).

$$\frac{Q_o}{Q_i} = \frac{8.6}{19} = 0.45$$

From Fig. 16-11,

$$\frac{V_s}{V_r} = 0.31$$

Hence the required storage volume is

$$V_s = 0.31 \times 86{,}760 = 26{,}900 \text{ m}^3 \text{ (21.8 acre-ft)}$$

The outlet and the detention basin must, of course, be designed so that the peak discharge of 8.6 m³/s coincides with a storage volume of 26,900 m³.

PROBLEMS

16-1. Design the sewers required to serve the area east of 12th Street and north of Beech Avenue in Figure 16-7. Assume that the line from the north of 10th Street ending at manhole 34 is 305 mm in diameter and serves a population of 5300 with a total sewage flow of 5.52 m^3/min. The ground elevation at this manhole is 94.27 and its invert elevation is 92.13. The line south from manhole 26 must enter manhole 1 no lower than elevation 88.31. Assumptions are to be the same as in the example. Assume ground surface elevations from the contours of the map. Make tabular calculations as in the example.

16-2. Using the same assumptions as in the storm sewer design example, lay out a system of storm sewers to serve the balance of the district. Design the line on Spruce Avenue from 11th to 12th Streets and find the total storm water flow from the district. Assume the ground elevation at 10th Street and Spruce Avenue to be 90.5 m.

16-3. A developer proposes to install a 5200 m^3 detention reservoir at the outlet of a 30-ha subdivision. The present peak discharge from the design storm is calculated to be 5.1 m^3/s. The discharge from the developed area will total 26,230 m^3 and the peak flow will be 10.2 m^3/s. Will the proposed reservoir be adequate to reduce the discharge to 5.1 m^3/s?

REFERENCES

1. Mark A. Pacheco and Donald D. Gray, "Computer Aided Design of Gravity Sanitary Sewerage Systems," Technical Report CE-HSE-82-1, Purdue University, Lafayette, Ind., 1982.
2. *Stormwater Management Model Users Manual Version III*, U.S. Environmental Protection Agency, Cincinnati, 1982.
3. *Modern Sewer Design*, American Iron and Steel Institute, Washington, D.C., 1980.
4. "HEC-1 Flood Hydrograph Package Users Manual," 723-X6-L2010, U.S. Army Corps of Engineers, 1981.
5. "Computer Program for Project Formulation—Hydrology (1982 Version)," Technical Release No. 20, U.S. Department of Agriculture, 1983.
6. "Urban Hydrology for Small Watersheds," Technical Release No. 55, U.S. Department of Agriculture, 1975.

CHAPTER 17

SEWER CONSTRUCTION AND MAINTENANCE

17-1 Maintenance of Line and Grade

In order to ensure that a sewerage system will function as intended, it must be constructed in accordance with the plans and specifications. It is particularly important that the line and grade of each sewer be carefully established and maintained so that self-cleansing velocities will be obtained.

Prior to construction, the contractor should establish an offset line where it will not be disturbed or covered, then measure from the offset line to lay out the trench on the ground. When the trench has been brought close to its final grade, batter boards are placed across it at intervals of 10 to 15 m (30 to 50 ft) as shown in Fig. 17-1. The centerline of the sewer is established on the batter boards by measuring from the offset line and nailing an upright cleat so that one edge is on the centerline. The elevation of each cleat is then established and a mark is made thereon at an elevation which is an identical distance above the finished grade of the sewer at each batter board. A nail is driven into each cleat at the grade mark and a string is run from nail to nail. This string is at the slope of the sewer and directly above its centerline. Line is established by dropping a plumb bob from the string, and grade is checked with an ell-shaped gauge marked at the distance equal to the vertical displacement between the string and the invert of the sewer. When the gauge is installed in the sewer as shown, the mark should match the string. Grade is checked in this manner on each length of pipe.

A more modern technique of maintaining line and grade is illustrated in Fig. 17-2. In this method, a laser is mounted at one manhole and directed toward the next, with the beam coinciding with the centerline of the sewer. A target is placed over the end of each section and the line and grade are adjusted until the beam is in the center of the target. Accuracy of line and grade can be

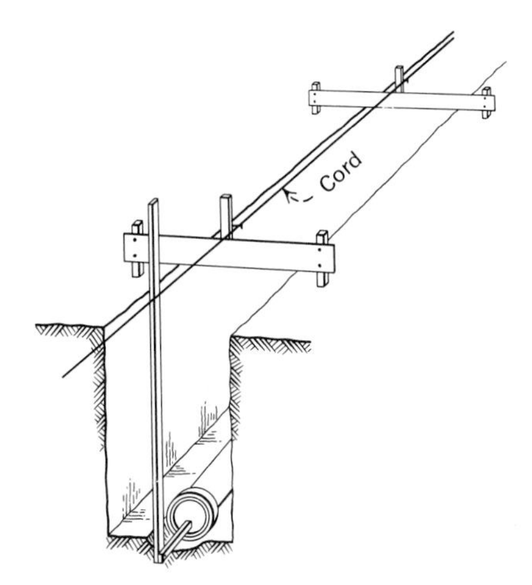

FIGURE 17-1
Establishment of line and grade of a sewer.

held to within 0.01 percent over a distance of up to 300 m (1000 ft) by this procedure.

17-2 Excavation Techniques

The type of excavation which the contractor will encounter should be indicated on the drawings, and may be divided into several classes. *Solid rock* includes solid rock in its original bed, or in well-defined ledges removable only by blasting, and boulders over 0.25 m^3 (8 ft^3) in volume. *Hardpan* includes mate-

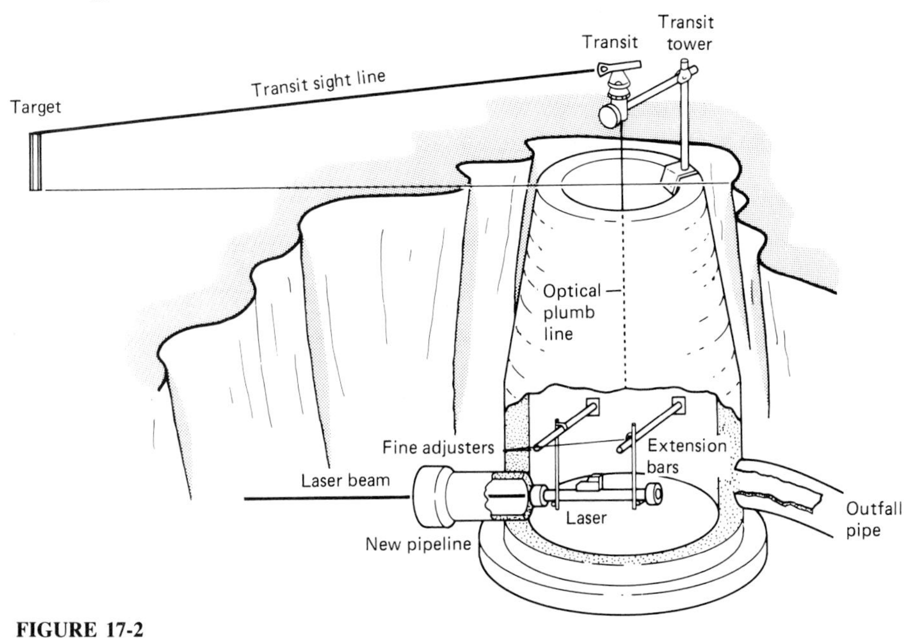

FIGURE 17-2
Establishment of line and grade with a laser.

rials such as disintegrated limestone, shale, soapstone, slate, fire-clay, cemented gravel, and boulders less than 0.25 m^3 in volume. This type of material can be excavated by hand or machinery with minor difficulty. Where *quicksand* is known to occur, it should also be included as a separate item.

Excavation for sewers should be done with mechanical equipment wherever possible. Specialized equipment for excavating trenches is available which employs continuous chain drives carrying buckets that cut the soil, bring it to the surface, and discharge it to a conveyor which carries it to the side of the trench. With side cutters on the buckets, such equipment can excavate trenches up to 1.3 m (4 ft) wide and 9 to 10 m (30 to 33 ft) deep in a single pass. In shallow trenches with favorable soil conditions, the trench may be dug at a rate of up to 10 m/min (30 ft/min). If the contractor chooses to do so, trenches can also be excavated with standard construction equipment such as backhoes, clamshells, or draglines.

The trench should be excavated a minimum of 200 mm (8 in) below the final grade so that suitable bedding material can be placed beneath the pipe (Chap. 14). When bell-and-spigot pipe is used, hand excavation of the bedding material is required at each bell. Hand excavation is also required in the vicinity of other subsurface utilities in order to ensure that they are not damaged. Except for such locations, machinery should be employed.

Excavation in rock should be carried below the bottom of the pipe to a depth equal to one-fourth the pipe diameter or 100 mm (4 in), whichever is greater. The space between the rock and the pipe is then filled with appropriate bedding material. Small amounts of rock may be excavated by drilling and hammering with either hand tools or machinery similar to pavement breakers. When explosives are used, the technique should conform to modern safety codes—including the use of mats to contain debris. Once the rock has been broken, it can be removed with ordinary construction equipment.

17-3 Sheeting and Bracing

Trenches in unstable materials require sheeting and bracing to prevent caving or collapse of the sidewalls. Recommendations with regard to supporting open excavations should be based on geotechnical evaluation of the site conditions. The danger to workers and adjacent structures and the extra cost occasioned by soil failures justify whatever expense may be necessary to prevent such occurrences. Trenches more than 1.5 m (5 ft) deep and 2.5 m (8 ft) in length should be held by shoring or bracing or sloped to the angle of repose of the soil. Trenches which are excavated by sloping the sides are dug vertically at the bottom to reduce the soil load on the pipe.

Sheeting includes the support materials in contact with the walls of the excavation. *Bracing* refers to crosspieces extending from one side to the other. *Rangers* are structural members which transfer the load from sheeting to braces.

Stay bracing consists of vertical boards placed against opposite walls of the trench and supported by two cross braces. The spacing between vertical members depends upon the soil. The vertical members should be at least 50 × 100 mm (2 × 4 in) and the cross braces at least 100 × 100 mm (4 × 4 in). This type of support is not very dependable and should be used only in shallow trenches in fairly stable materials.

Poling boards are short pieces of board placed against the walls of the trench and supported by rangers and cross bracing as shown in Fig. 17-3. The system may be constructed of random lengths of timber and is not continuous over the wall. Poling boards are used in materials which will stand without support at depths of 1 to 1.5 m (3 to 5 ft). The sheeting can thus be installed after the trench has been partially excavated and need not be driven ahead of the excavation. Sheeting in this method is usually 50 mm (2 in) thick, the rangers 100 × 150 or 100 × 200 mm (4 × 6 or 4 × 8 in), and the braces 150 × 150 or 200 × 200 mm (6 × 6 or 8 × 8 in), depending on the width of the trench.

Box sheeting employs horizontal sheeting and vertical rangers as shown in Fig. 17-4. In relatively stable material, there may be gaps between the boards. The size requirements for sheeting, rangers, and braces are similar to those for poling boards. Box sheeting is best suited to unconsolidated soils which are excavated in stages equal to the height of an individual board which is then placed and temporarily braced. When three or four boards have been installed, the rangers and permanent bracing are put in position. This method may also be used for the upper 1 to 1.5 m (3 to 5 ft) of an excavation which employs vertical sheeting at lower levels.

Vertical sheeting is the strongest and most elaborate method employed in trench construction and is used in soft soils and in those where groundwater may be encountered. The trench is first excavated as far as possible without the danger of bank failure. The first set of horizontal rangers is then placed about 300 mm (1 ft) below the ground surface against three planks which bear against the squared and trimmed sides of the trench at the ends and midpoint of the ranger. Rangers on opposite sides of the trench are crossbraced and the

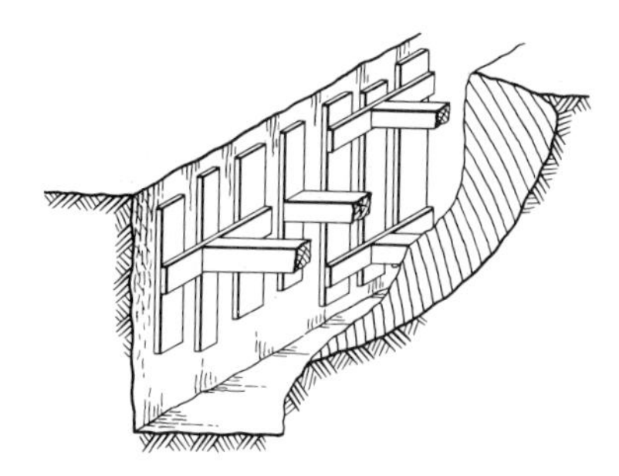

FIGURE 17-3
Poling boards.

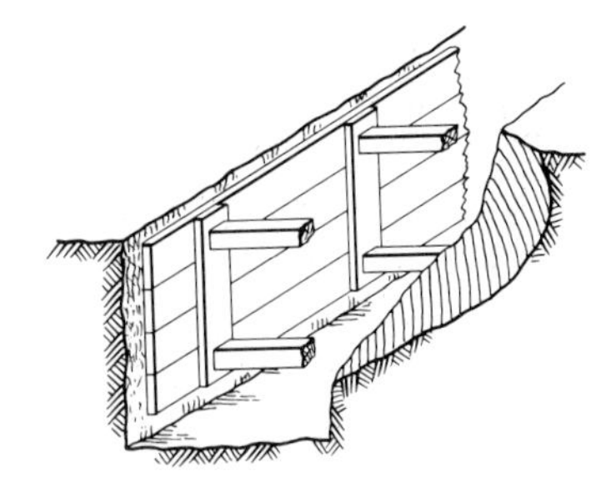

FIGURE 17-4
Box sheeting.

sheeting is then driven vertically between the rangers and the wall of the trench.

As the sheeting is driven downward, the excavation proceeds with additional rangers and cross braces being placed at intervals of about 1 m (3 ft). A second tier of sheeting is begun after the lowest rangers and braces for the upper tier are in place. The cross braces of the upper tier prevent driving the sheeting that is immediately below them, hence these openings must be closed by boards nailed in place. The sheeting boards are usually 50 × 250 mm (2 × 10 in). Rangers range from 150 × 150 or 150 × 200 mm (6 × 6 or 6 × 8 in) at the top to 200 × 250 mm (8 × 10 in) at the bottom of a 9-m- (30-ft-) deep trench. Cross braces range from 100 × 150 to 100 × 200 mm (4 × 6 to 4 × 8 in).

It is desirable to use separate cross braces at the end of each ranger (Fig. 17-5) so that the individual sections are structurally independent. The sheeting is beveled at the end in two directions (Fig. 17-6) so that it will fit snugly against adjoining pieces and the walls of the trench. Individual sheets are capped with malleable iron to protect them during driving. Braces are tightened by wedges driven between their ends and the rangers, but they should also be nailed together.

If the trench is expected to extend below the groundwater table, sheet piling may be used instead of planks. A variety of types of sheet piling are illustrated in Fig. 17-7. Steel sheet piling is stronger than wooden types and can be pulled and reused more often. Sheet piling also requires longitudinal and cross bracing. Its major advantage is its relative watertightness.

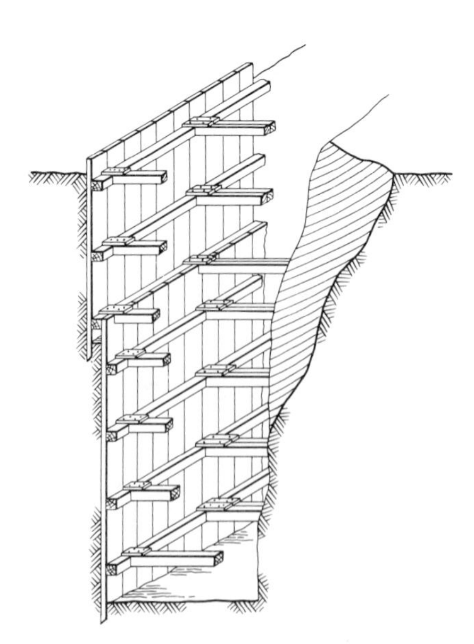

FIGURE 17-5
Vertical sheeting.

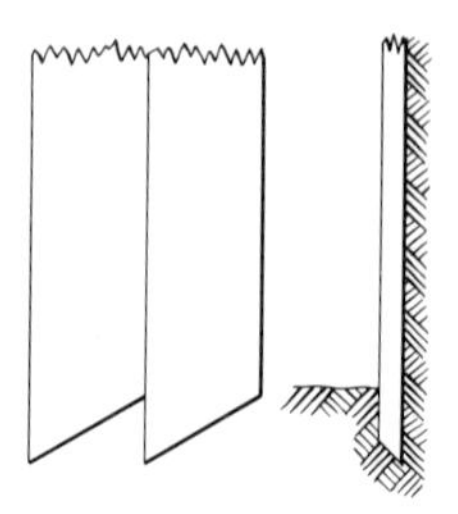

FIGURE 17-6
Beveling of vertical sheeting.

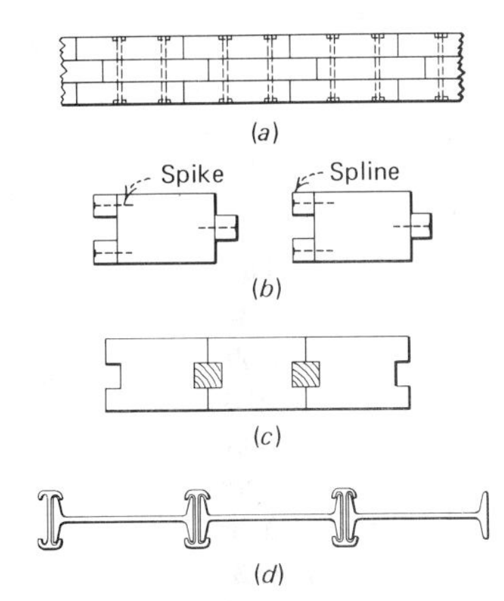

FIGURE 17-7
Sheet piling. (*a*) Wakefield (timber); (*b*) tongue-and-groove (timber); (*c*) groove-and-spline (timber); (*d*) steel.

Skeleton or *open sheeting* consists of rangers and cross braces placed as in vertical sheeting, but with sheeting planks placed only at the ends and midpoints of the rangers. It may be used as a preliminary step before installation of more elaborate protection. It is readily converted to vertical sheeting by driving the missing boards.

Sheeting and bracing are usually recovered before the trench is closed. As backfilling proceeds, the braces are removed and the sheeting can then be pulled either by machinery or by hand. The soil load on the pipe depends on the development of friction and arching against the walls of the excavation. If the trench is completely filled before the sheeting is pulled, the entire weight of the column of soil may momentarily bear on the pipe, causing its failure. In particularly difficult soil conditions, it may be necessary to leave the sheeting in place when the trench is filled.

17-4 Dewatering of Excavations

When excavations extend below the groundwater table, water will flow into the opening. If the subsurface strata are sufficiently permeable, the velocity of flow may be sufficiently great to fluidize the soil, creating a quick condition. This can undermine the sheeting and cause failure of the trench.

Quick conditions in the trench can be prevented by lowering the groundwater table with well points—a line of small driven wells parallel to the trench, which are pumped at a rate large enough to depress the groundwater table below the bottom of the opening. Well points may be placed along either or both sides of the trench, about 2 m (6 ft) from the centerline, 1 m (3 ft) apart, and extending well below the bottom of the trench. The individual wells are attached in series to a header which is connected to a common pump.

When water is encountered unexpectedly, the trench bottom may be stabilized temporarily with gravel, rock, or rubble. Material of this sort will cause arching between the grains of the soil and will prevent its fluidization. The flow of water will not be halted by this technique.

If the flow is not so great as to cause fluidization, it may be removed by letting it run along the trench bottom to a sump, from which it is then pumped. In the construction of large sewers, an open-jointed tile underdrain may be placed in the bottom of the trench below the location of the major structure to provide more thorough drainage and a dry trench bottom for construction. The drain is left in place when the sewer is completed but should be plugged so that it does not result in permanent drainage of the soil.

17-5 Pipe Laying and Jointing

Before the pipe is lowered into the trench, the grade of the bedding material should be checked with levels or a device of the sort illustrated in Fig. 17-1. The grade of the sewer should be held to within 10 mm ($\frac{1}{2}$ in) of that shown on the plans.

The pipe should be inspected to ensure that it is sound and that the ends are undamaged. If the sections are to be joined by a ring, as is the case with plain-end clay pipe, some plastic pipe, and asbestos cement pipe, the ring may be installed on one end before the assembly is lowered into the trench. The technique used to pick up and lower the pipe should be selected to ensure that neither the pipe nor its gasketing material is damaged. Slings or rolling hitches are generally suitable.

The pipe sections are placed on line and grade in the bottom of the dewatered trench and are pressed together with a hand lever or a winch. The gasketing material should be lubricated according to the manufacturer's instructions before the sections are joined. It may be necessary to support a portion of the weight of the section being placed by the sling and to place some backfill on and around each section before the joining force is released. In the case of plastic pipe that is solvent-welded, placement of additional sections must be delayed until the last joint has set up.

Wye or tee sections which have been provided in the sewer for future household connections should be plugged and mortared shut.

Joints in storm sewers may be made as illustrated in Chap. 14, with asphaltic material, polymeric gaskets, or mortar.

The trench should be filled as soon as the pipe has been placed and the installation inspected. When class A (concrete) bedding is required, the backfill is delayed until the concrete has set up sufficiently to support its weight. Draining of the trench must be continued until the backfill is completed.

Fill material must be free of brush, debris, frozen material, and large rocks. No rock should be placed within 900 mm (3 ft) of the top of the pipe nor within 400 mm (15 in) of the ground surface. The fill must be carefully placed in layers not more than 150 mm (6 in) thick and be tamped under, around and over the pipe to a height of 600 mm (2 ft) above the crown. Until this level has been reached, the earth should be placed very carefully. Backfilling with a bulldozer on a bare pipe will almost certainly displace it and may break it. Fill beneath streets or other construction must be placed carefully, at optimum moisture content, with appropriate compaction until the ground surface is reached. In easements where the surface will be unused for traffic or other load-bearing purposes, the backfill above a level 600 mm (2 ft) above the crown need not be compacted.

Installation of sewers below highways, railroads, airfields, etc. may be accomplished without interference with surface traffic by *jacking and boring*. In this technique, pipes are driven by hydraulic jacks mounted in a pit at one side of the obstacle. A cutting ring may be installed on the first section and lubrication fittings may be placed on subsequent sections. A cutter head operating ahead of the pipe opens the hole into which the pipe is jacked, while an auger operating on the same shaft as the cutter draws the excavated material through the pipe to the jacking pit. Jacking is generally done on a slight upgrade to keep the cutting face and pipe dry. The pipe should be kept in motion once the operation is begun. If the pipe remains stationary for any time, particularly under surfaces subject to vibratory loads, the soil may consolidate around the pipe, freezing it in position. It may then be necessary to jack and bore from the other side to meet the frozen pipe.

17-6 Maintenance of Sewers

Maintenance of sewer systems depends not only on proper design and construction and the availability of a competent work force, but also on protection of the system against damaging materials which may be discharged by the public.

Most communities have enacted *sewer ordinances* which regulate the materials that may be discharged to the sewers. These ordinances may also require that connections of household sewers to the common sewers be made by city crews or be inspected by the city in order to ensure that the work is properly done. The materials excluded include steam; corrosive, flammable, and explosive liquids; gases and vapors; garbage; and dead animals. Wastes from kitchen sinks and floor drains in restaurants, hotels, and boarding houses and from packing houses, creameries, bakeries, cleaners, laundries, garages, and other industries may be required to pass through grease traps or grit separation facilities before they enter the sewers.

Sewer maintenance requires certain specialized equipment for clearing plugged or partially plugged lines. Sanitary sewers may be stopped by roots which enter through small leaks or by deposits of grease. Storm sewers are more commonly blocked by large objects and by sand and other sediments.

Roots are removed in sewers up to 380 mm (15 in) in diameter by flexible rods driving an augerlike cutter (Fig. 17-8). The auger may be rotated by hand or by a machine as it is advanced into the sewer. In larger sewers, a cutting drag is pulled through by a cable and winch. The problem with roots can be prevented by eliminating leakage, since the roots follow the water into the

FIGURE 17-8
Sewer cleaning. (*Courtesy Flexible Rod Equipment Co.*)

sewer. Particularly troublesome sections are sometimes replaced by iron pipe, but newer sewer joint designs are much less likely to leak than those used in the past.

Grease is the most common cause of blockage in household sewers, and larger sewers may have their capacity substantially reduced by deposits of grease on the walls. House sewers may be cleaned with rotating tools mounted on the end of a flexible tape which is driven by hand or a small electric motor. Common sewers are cleaned of grease deposits by cutting tools similar to that in Fig. 17-8, generally followed by brushing with a wire brush.

Sand and grit may be removed by buckets or scoops pulled through by a cable and winch. If the deposits are not too extensive, they may be removable by devices similar to the turbine cleaner of Fig. 17-9. This tool has a water-powered rotating cutter which is drawn through the sewer by a cable while it flushes the deposits from the line.

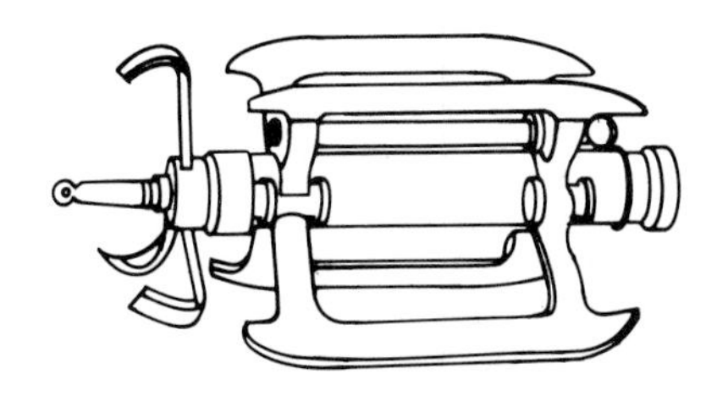

FIGURE 17-9
Turbine sewer cleaner. (*Courtesy Turbine Sewer Machine Company.*)

Routine cleaning of some sewer lines may be necessary if their slopes or flows are particularly low. *Flushing* with water taken from fire hydrants is sometimes sufficient to remove deposits of grit, but is seldom adequate to remove grease and will have no effect on roots. Flushing must be done cautiously, since the flow introduced from a hydrant may be sufficient to surcharge the sewer and cause water to back up into basement drains or household plumbing fixtures.

A soft rubber ball inflated to a size slightly less than the diameter of the sewer aids in removing grit and grease deposits. The ball adjusts itself to irregularities in the pipe while water held behind it escapes around its edges at high velocity, flushing away the deposits. Roots up to 6 mm ($\frac{1}{4}$ in) may be broken off by the ball. The operation is most effective when the head behind the ball is 0.3 to 1.2 m (1 to 4 ft), but can be satisfactory at heads as low as 50 mm (2 in) in circumstances where higher heads would cause flow to back up into adjoining property.

Material dislodged in sewer cleaning operations should be removed at the next manhole to prevent it from forming another blockage in a downstream line. The debris can be removed at a manhole by inserting an ell section in the outlet with the open end turned upward. The manhole itself will then serve as a trap and retain the solids.

Routine inspection of sewers is quite helpful in avoiding severe blockages. Sewers on flat grades or with a history of problems may be inspected every 3 months, while those with no known problems are checked once or twice a year. Inspections are made visually, from manhole to manhole with a bright light being placed in the manhole toward which the inspector is looking. A mirror on a pole lowered into the manhole will often permit the inspection to be made from street level. Sewers are also inspected, at times, by passing small television cameras through them. This permits close-up inspection of joints and detection of any breakage or other damage as well as location of blockages.

17-7 Sewer Repairs and Connections

Damage to sewers caused by corrosion, surface loads, or settlement should be repaired without delay. Openings in sewers may permit soil to enter and thereby undermine roadways or buildings. Damaged sections should be located as closely as possible by inspection and rodding from manholes. The excavation can then be reasonably small.

A broken section of clay or concrete pipe can be chipped out of the line while the flow is diverted by pumping from manhole to manhole. A replacement section is easily installed if the pipe is of the plain-end type. Bell-and-spigot sections require that the upper section of the bell of the section downstream be carefully chipped off. The replacement section also has half its bell removed and is rotated once in place, so the gap is at the top. The gaps in the bells are then filled with mortar.

Sections of sanitary sewer which have deteriorated badly and which are permitting excessive infiltration may be repaired in place by slip lining from manhole to manhole with plastic pipe made specifically for this purpose. Slip lining cuts off the household connections, which must be remade, but avoids excavation of the entire length of the sewer.

New connections to existing sewers should never be made by breaking the sewer and simply inserting the new line. Wyes or tees should be inserted or, as a minimum, wye or tee saddles should be used. All new connections should be either made by city crews or be inspected by the city to ensure that the work is done properly.

17-8 Sewer Gases

Explosions in sewers are not uncommon and workers have been killed by breathing toxic gases produced by biological activity or industrial discharges.

The most common cause of explosions in sewers is gasoline which has leaked from corroded subsurface storage tanks or been discharged deliberately by garages. Additional sources of explosive gases are industrial chemicals such as calcium carbide, which may react with the water, and biologically produced hydrogen sulfide and methane.

Biological activity in the sewers may reduce the oxygen content of the atmosphere and this, by itself, or coupled with the presence of hydrogen sulfide, may kill unprotected workers. The gas content of manholes should be tested before work crews enter them and the workers below ground should wear a harness connected to an overhead hoist which will permit them to be pulled from the sewer if they should lose consciousness. At least one person should remain above ground to help those below if they should encounter difficulties. Hose masks or gas masks should be provided and used if it is known that the atmosphere is dangerous. It is sometimes possible to flush toxic concentrations of gas from the sewers by forced ventilation with portable blowers.

CHAPTER 18

CHARACTERISTICS OF WASTEWATER

18-1 Variability of Wastewater and Wastewater Analyses

Wastewater is not constant in character from place to place nor from time to time. Further, the techniques commonly used in its sampling and analysis are subject to substantial error. The combination of inherent variability and experimental error produce considerable uncertainty regarding the actual characteristics in any given situation, and extensive testing programs may be necessary to determine the actual nature of the waste. The "typical" values which are presented here and in other texts are national averages which should never be assumed to represent the waste of a particular community.

As sewage passes through the collection system, solid organic material will tend to be solubilized by microbial action and some solids may be removed by sedimentation—or at least be carried in traction along the bottom of the sewer, while grease and oils will tend to move to the upper surface and, perhaps, to be deposited along the walls of the pipes. Materials deposited in the bottom or along the sides of the pipes may be removed by later, higher flows which produce greater velocities.

Sampling a flow which is likely to be stratified requires that aliquots be taken across the depth in proportion to the velocity profile and area if a representative sample is to be obtained. Analyzing a flow which is variable in time requires that a large number of samples be taken in order to define the range of concentrations to be expected. Wastewater characterization studies are seldom sufficiently detailed to establish the variability with much certainty, hence engineers must be very cautious in designing systems intended to treat domestic wastes.

The various tests used in characterizing water and wastewater are described with regard to their precision and accuracy.[1] *Precision* refers to the reproducibility of an analytic technique when it is repeated on a homogeneous sample under controlled conditions, without regard to whether the measured values correspond to the actual value. Precision is measurable by the standard deviation of the test results. *Accuracy* refers to the correspondence between the measured value and the actual value. *Relative error* is the difference between the actual and measured value as a percentage of the actual value. One method may be precise (i.e., reproducible) but inaccurate, with all measurements closely grouped about the wrong value, while another may be accurate but imprecise, with all measurements widely scattered about the correct value. The accuracy and precision of the various test methods are presented below, but it must be recognized that these values refer only to the laboratory determination. Additional uncertainty is introduced by the inherent variability of wastewater and the difficulty of obtaining truly representative samples.

18-2 Physical Characteristics

Sewage is over 99.9 percent water, but the remaining material has very significant effects upon the nature of the mixture. Fresh domestic sewage has a slightly soapy or oily odor, is cloudy, and contains recognizable solids, often of considerable size. As the waste ages, its character changes as a result of biological and chemical phenomena. Stale sewage has a pronounced odor of hydrogen sulfide, is dark gray, and contains smaller but occasionally recognizable solids.

The change from fresh to stale requires 2 to 6 h at a temperature of 20°C, with the time depending primarily on the concentration of organic matter. The concentration varies with per capita water use, infiltration, and the quantity of industrial waste which enters the collection system. The quantity of domestic waste produced per person is relatively invariant on a dry solids basis, but the quantity of carriage water is not.

18-3 Solids Determinations

The solids in sewage may be either suspended or dissolved. As noted above, biological action will tend to dissolve some of the suspended organic material as it passes through the sewer. *Total solids* includes both suspended and dissolved species and is determined by evaporating a sample of known weight or volume at 103 to 105°C and weighing the residue. Results are expressed in milligrams per liter or percent (1 percent is equal to 1 gram of solids per 100 grams of sewage). Assuming the specific gravity of the mixture is 1 (which is nearly always justifiable), this is equivalent to 10 grams per liter or 10,000 mg/L. In testing in a single laboratory, the standard deviation of the differences between analyses for total solids on duplicate samples was 6 mg/L.[1]

Suspended solids and *dissolved solids* determinations require filtration of the sample. The filtration is made through a membrane filter similar to those used in bacteriological analysis of water (Chap. 9). If suspended solids are to be determined, the filter is dried and preweighed, a measured volume of sample is drawn through it, and it is dried and reweighed. The increase in weight divided by the volume of the sample yields the concentration of suspended solids. A measured volume of the filtrate may be evaporated to dryness and the residue weighed to determine the dissolved solids mass. This, divided by the sample volume, will give the dissolved solids concentration. If any of the two concentrations are measured, the third may be calculated as either the sum or difference of the others. In studies by two analysts using four sets of 10 determinations, the standard deviation of this test was found to range from 5.2 mg/L at 15 mg/L to 13 mg/L at 1707 mg/L.[1]

Volatile solids are those solids ignitable at 550°C. The concentration of total volatile solids is considered to be a rough measure of organic content, while suspended volatile solids, in some instances, is considered a measure of the concentration of biological solids such as bacteria and protozoa. Volatile solids may be measured for the total sample (total volatile solids), the suspended fraction (suspended volatile solids), or the filtrate (dissolved volatile solids). The determination is made by ignition of the residue from the total solids test in a muffle furnace. For volatile suspended solids determinations, the filter is made either of glass (which will undergo only a slight weight loss, which is corrected by a blank) or cellulose acetate (which will leave no ash). The volatile fraction is determined from the difference between the weight of the residue after drying and that after ignition. The residue following ignition is called *nonvolatile solids* or *ash* and is a rough measure of the mineral content of the wastewater. The standard deviation of this determination was 11 mg/L at 170 mg/L total volatile solids in studies conducted in three laboratories on four samples and 10 replicates.[1]

Samples used for solids determinations on wastewater are typically small (10 to 100 mL) and, as noted above, are not likely to be truly representative of the total flow.

18-4 Chemical Characteristics

Wastewater contains both inorganic and organic chemicals. The *inorganic constituents* are present in the carriage water and increase because of water use. Ordinary sewage treatment is not directed toward altering the concentration of inorganic contaminants, although the concentrations of phosphorus and nitrogen are sometimes significant in biological treatment processes. Tertiary treatment, which may be required in some cases to maintain water quality, removes inorganic contaminants by techniques similar to those employed in water treatment.

Nitrogen may be present in wastewater in both inorganic and organic forms and in both reduced and oxidized states. In untreated wastewater it is chiefly present as ammonia or as a constituent of protein (organic nitrogen). Its concentration is determined by a digestion-distillation process followed by colorimetric, titrimetric, or electrophoretic analysis. The relative standard deviation for ammonia determinations ranges from less than 10 to nearly 70 percent.[1] Organic nitrogen determinations have relative standard deviations ranging from about 40 to over 100 percent.[1]

Phosphorus is primarily present in wastewater in the form of phosphates—the salts of phosphoric acid. It may be complexed in organic matter, hence a preliminary digestion step precedes analysis by a variety of colorimetric techniques. The preferred method had a relative standard deviation of about 2 to 4 percent in eight laboratories.[1]

The *alkalinity* of wastewater is important, since it provides a buffer against acids produced by bacterial action in anaerobic or nitrifying systems. As sewage ages, its pH tends to drop because of the production of organic acids by bacterial metabolism. When the wastewater is treated or undergoes natural stabilization, these acids are oxidized to carbon dioxide and water and the pH will rise. The organic acids produced by biological activity are *weak acids*, similar in character to carbonic acid, and can produce an apparent increase in alkalinity. No general conclusion can be reached about the accuracy of alkalinity determinations in wastewater because of the potential variability of its constituents. In samples containing only carbonate and bicarbonate, a standard deviation of less than 10 percent can be expected.[1]

The *organic constituents* of wastewater may be divided into carbohydrates, proteins, and fats, but separate analyses for these constituents are seldom made. Rather, the total quantity of organic material is measured by determining the quantity of some oxidizing agent which is required to convert it to CO_2, H_2O, and other oxidized end products. The techniques presently in use, which are discussed in detail below, include *biochemical oxygen demand* (BOD), *chemical oxygen demand* (COD) and *total organic carbon* (TOC). These methods do not measure the same thing, hence they are not directly comparable.

18-5 Biochemical Oxygen Demand

Bacteria placed in contact with organic material will utilize it as a food source. In this utilization, the organic material will eventually be oxidized to stable end products such as CO_2 and water. The amount of oxygen used in this process is called the *biochemical oxygen demand* and is considered to be a measure of the organic content of the waste. It also represents, to some extent, the amount of oxygen which would be required to stabilize the waste in a natural environment such as a stream or lake.

The BOD determination has been standardized[1] and measures the amount of oxygen utilized by microorganisms in the stabilization of wastewater during 5 days at 20°C. For ordinary domestic sewage the 5-day value, or BOD_5, represents approximately two-thirds of the demand which would be exerted if all the biologically oxidizable material were, in fact, oxidized.

In conducting the test on domestic wastewater, it can be assumed that a suitable bacterial inoculum will be present. If relatively clean industrial wastes are to be analyzed, an inoculum may need to be added.[1,2] A suitable inoculum may ordinarily be obtained by culturing bacteria taken from domestic wastewater upon the waste to be tested.

The solubility of oxygen in water is quite limited (App. 1), hence nearly all BOD measurements require that the sample be diluted. The dilution water is carefully manufactured and contains a mixture of salts providing all the trace nutrients necessary for biological activity plus a phosphate buffer to maintain a neutral pH. The water is aerated to saturate it with oxygen before mixing with the sewage sample.

The exertion of BOD is considered to be a first-order reaction describable by

$$\frac{dy}{dt} = -K_1 y \tag{18-1}$$

in which y is the BOD (or concentration of organic matter) remaining at time t and K_1 is a constant. Integrating, and setting L equal to the BOD at $t = 0$, gives

$$y = Le^{-K_1 t} \tag{18-2}$$

The BOD exerted at any time t is, of course, the difference between that initially present and that remaining, whence

$$\text{BOD}_t = L - y = L(1 - e^{-K_1 t}) \tag{18-3}$$

Although a relation of the form of Eq. (18-3) cannot be linearized, Thomas[3] noted the similarity between the first terms for the series expansions of $1 - e^{-K_1 t}$ and $K_1 t[1 + K_1 t/6]^{-3}$ and developed the approximate formula

$$\text{BOD} = LK_1 t\left(1 + \frac{K_1 t}{6}\right)^{-3} \tag{18-4}$$

which can be linearized as

$$\left(\frac{t}{\text{BOD}}\right)^{1/3} = (K_1 L)^{-1/3} + \frac{K_1^{2/3}}{6L^{1/3}}\, t \tag{18-5}$$

By plotting $(t/\text{BOD})^{1/3}$ versus t, a straight line may be obtained with intercept $(K_1 L)^{-1/3}$ at $t = 0$ and slope $K_1^{2/3}/6L^{1/3}$

Example 18-1 The following data are to be used to determine the value of K_1 and L for the waste tested:

Time, days	$\frac{1}{2}$	1	2	3	4	5	7	10	15
BOD, mg/L	5	20	90	160	200	220	260	285	320

Solution

The data are first plotted as shown in Fig. 18-1 to determine if the oxygen uptake began immediately or if a lag was evident. A lag may result from a lack of acclimation of the microbial inoculum to the waste. In this case it appears that the time of beginning which best fits the data is $t_0 = 0.8$ days. The times at which the BOD data were taken are thus corrected by this amount:

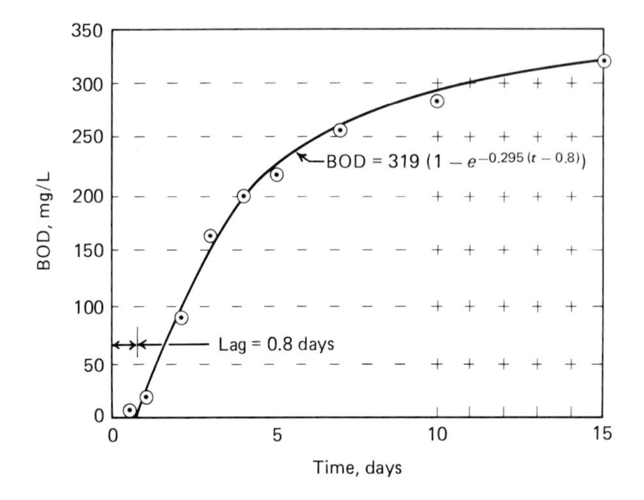

FIGURE 18-1
BOD exerted versus time.

Time, days	Corrected time, days	BOD, mg/L	$(t/\text{BOD})^{1/3}$
0.5	.....	5	
1.0	0.2	20	0.22
2.0	1.2	90	0.24
3.0	2.2	160	0.24
4.0	3.2	200	0.25
5.0	4.2	220	0.27
7.0	6.2	260	0.29
10.0	9.2	285	0.32
15.0	14.2	320	0.35

$(t/\text{BOD})^{1/3}$ is plotted versus t in Fig. 18-2. From the figure, $(K_1L)^{-1/3} = 0.22$ and $K_1^{2/3}/6L^{1/3} = 0.0108$. Solving simultaneously gives $K_1 = 0.295$/day and $L = 319$ mg/L. The derived BOD equation is thus BOD $= 319[1 - e^{-0.295(t-0.8)}]$, which is superposed on Fig. 18-1.

Not all experimental data will fit the theoretical curve as well as that of the example. The data which is best is generally that which falls between 2 and 10 days.

The BOD_5 determination is effected by incubating the diluted sample at 20°C for 5 days. The dissolved oxygen in the diluted sample is measured at the beginning and the end of the test period, and the BOD is equal to the change in dissolved oxygen multiplied by the dilution factor of the sample. Since the saturation value of oxygen in water is less than 10 mg/L at 20°C and the BOD of "typical" domestic wastewater is about 200 mg/L, the dilution factor may range from 10:1 to 100:1. A standard 250 mL BOD bottle will thus contain only 2.5 to 25 mL of the waste, which is then considered to represent the heterogeneous and variable flow. The potential for sampling error is obvious, and the

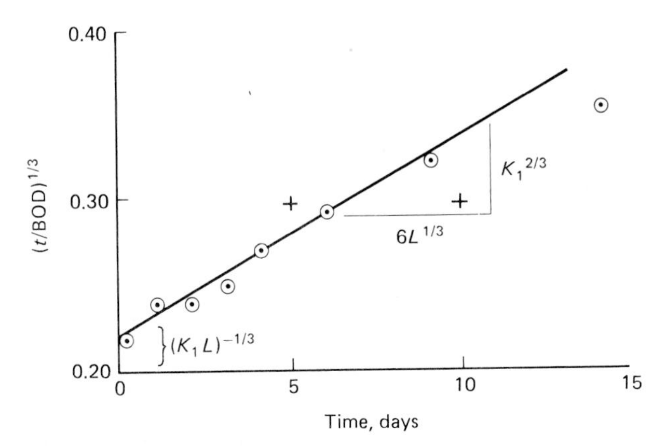

FIGURE 18-2
Graphical determination of K_1 and L.

test itself, since it depends on a living system, is also quite imprecise. It is, however, important, since it is the best measure available of what may occur in the living systems in wastewater treatment plants and natural waters. The engineer must remember, however, that the BOD test is quite inexact, and treatment systems must be designed to be able to accommodate substantial variations in waste strength about those which may be measured in short-term studies. In a series of interlaboratory studies on prepared primary standards of known BOD, the mean measurement was about two-thirds of the actual value and the standard deviation was about 12 percent.[1]

The constant K_1 varies with temperature in accordance with

$$K_{1(t)} = K_{1(20)}[1.047^{(T-20)}] \tag{18-6}$$

in which $K_{1(20)}$ is the value determined at 20°C in the BOD test and T is the actual temperature in degrees celsius. Equation (18-6) can be used to determine the rate of oxygen demand exertion at temperatures other than 20°C. This is important in assessing the effect of waste discharges on natural waters and in the design of wastewater treatment plants.

18-6 Chemical Oxygen Demand

The BOD test, while the best available representation of what will occur in a natural water, requires a minimum of 5 days, and thus is not useful in control of processes in sewage treatment plants, which have retention times on the order of hours. The COD test involves an acid oxidation of the waste by potassium dichromate. A measured volume of standardized dichromate is added to a measured volume of the waste, the sample is acidified with concentrated sulfuric acid, and the mixture is boiled (while connected to a reflux condenser) for 2 h. The sample is then cooled and the remaining dichromate measured by titration with ferrous ammonium sulfate. No clear correlation exists between BOD and COD in general, but at specific plants it may be possible to establish a relationship between the two values. COD values are typically higher than BODs, since the strong oxidizing agent will oxidize materials which are only slowly (if at all) biodegradable. At COD levels of about 200 mg/L, measurements at different laboratories produced standard deviations of 10 to 20 mg/L, depending on the technique used.[1]

18-7 Total Organic Carbon

The total organic carbon test involves acidification of the sample to convert all the inorganic carbon to CO_2, which is then stripped. The sample is then injected into a furnace, where it is oxidized in the presence of a catalyst. The CO_2 produced is measured by infrared analysis and converted instrumentally to the original organic carbon content. The test is rapid, accurate, and correlates moderately well with BOD. The major obstacles to its wider use are the cost of the equipment and the skill necessary in its operation. The precision of the test is limited primarily by the difficulty of sampling wastes containing particulate matter. On clear or filtered samples, precision is within about 1 to 2 percent or 1 to 2 mg/L, whichever is greater.[1]

18-8 Microbiology of Sewage and Sewage Treatment

By its nature, domestic wastewater contains enormous quantities of microorganisms. Depending on its age and the quantity of dilution water, bacterial counts in raw sewage may be expected to range from 500,000 to 5,000,000 per mL. Viruses, protozoa, worms, and other forms are also found, but their presence is seldom important enough to require measurement.

Bacteria are single-celled plants which metabolize soluble food and reproduce by binary fission. They are capable of solubilizing food particles outside their cell wall by means of extracellular enzymes, and hence can remove soluble, colloidal, and solid organic matter from wastewater.

In the presence of adequate food and a suitable environment (temperature, pH etc.), bacteria will reproduce as illustrated in Fig. 18-3, which shows number of organisms versus time. The end of log growth and beginning of declining growth indicates the point at which the available food supply has been largely depleted and food becomes the limiting factor in further growth. Engineers seldom actually count bacteria, but infer their number from the concentration of volatile suspended solids. Figure 18-4 presents a plot of bacterial mass rather than bacterial number versus time. It may be noted that this curve exhibits only three rather than seven phases and that the beginning of growth is immediate.

Although the log growth phase coincides with the maximum rate of substrate (waste) removal, this is not the optimum zone of operation for waste

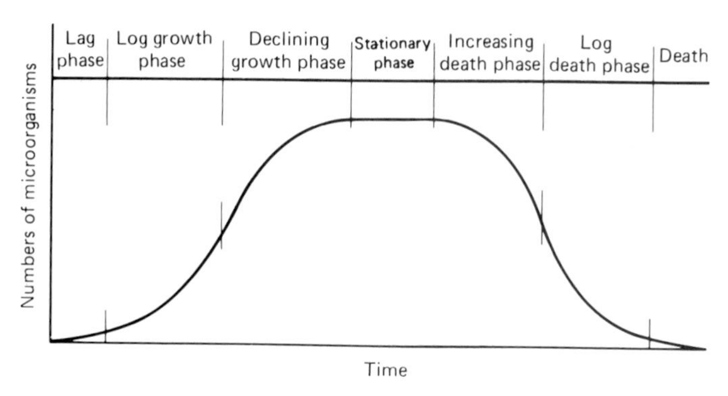

FIGURE 18-3
Growth pattern based on number of organisms. (*From Microbiology for Sanitary Engineers by Ross E. McKinney, Copyright 1962, McGraw-Hill Book Company. Used with permission of McGraw-Hill Book Company.*)

treatment systems. In order to maintain log growth, the food must be in ample supply, but a low concentration is desired in the treated waste. Further, the maximum rate of utilization requires that other growth factors, such as oxygen, be supplied at a maximum rate and this may be difficult to do. Finally, bacteria in the log growth phase of development have a great deal of energy available, have limited accumulation of waste products, and hence are likely to be motile, dispersed, and difficult to remove by sedimentation.

The declining growth phase is generally used for biological treatment systems since the problems listed above can be avoided. Some systems such as extended aeration (Chap. 22) and sludge digestion (Chap. 23) are operated in the endogenous phase of Fig. 18-4.

Anaerobic bacteria oxidize organic matter utilizing electron acceptors other than oxygen. In carrying out their metabolic processes they produce CO_2, H_2O, H_2S, CH_4, NH_3, N_2, reduced organics, and more bacteria. A large

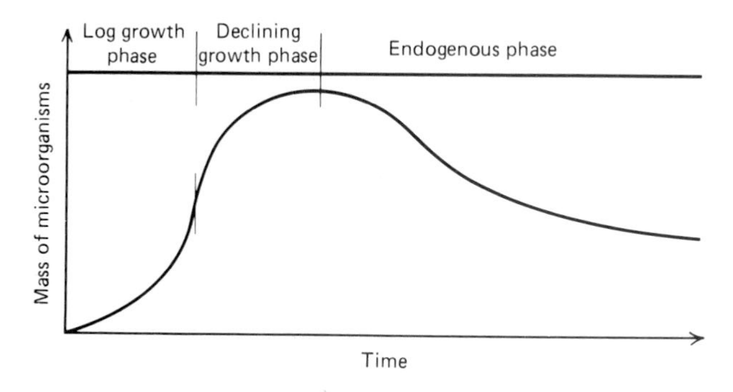

FIGURE 18-4
Growth pattern based on mass of organisms. (*From Microbiology for Sanitary Engineers by Ross E. McKinney, Copyright 1962, McGraw-Hill Book Company. Used with permission of McGraw-Hill Book Company.*)

part of the available energy appears in the form of end products, hence cell production is low and the by-products, such as methane, may be utilized as an energy source (Fig. 18-5).

The end products of anaerobic fermentation are likely to be odorous, and intermediates, such as the volatile organic acids, may be toxic to the methane-forming bacteria, thus promoting upset of the process. The production of a stable effluent is unlikely, since wastes do not usually contain sufficient electron acceptors to permit complete oxidation. Anaerobic processes are, however, useful in reducing the strength of concentrated wastes prior to other treatment and may offer substantial economy in some cases.

Aerobic bacteria utilize free oxygen as an electron acceptor. The end products of aerobic activity are CO_2, H_2O, SO_4^{-2}, NO_3^-, NH_3, and more bacteria. The bulk of the available energy is converted into either cell mass or heat, yielding a stable effluent which will not undergo further decomposition. The energy division is shown schematically in Fig. 18-6. The oxygen required may be furnished naturally from the atmosphere or mechanically by bubble aeration, thin-film aeration, or droplet aeration. The oxidation of ammonia to nitrate may or may not occur, depending on the retention time of the biological cell mass, the oxygen available, the temperature, and other factors discussed in Chap. 24.

Most of the bacteria encountered in wastewater and wastewater treatment are not *strict* or *obligate* aerobes or anaerobes, but rather can function in both aerobic and anaerobic environments. Such microorganisms are called *facultative* and are the major contributor to stabilization of wastewater. There are certain specific reactions, such as reduction of CO_2 to CH_4 and oxidation of NH_3 to NO_3^-, which are effected only by a limited population of obligate anaerobes or obligate aerobes. When such reactions are important, the engineer must ensure that the proper conditions are maintained—introduction of oxygen into an anaerobic process, for example, may kill all the methane-forming bacteria.

Other microorganisms of occasional importance in wastewater treatment include algae, protozoa, fungi, and rotifers.

Algae are photosynthetic microorganisms which can produce oxygen and organic cell mass from inorganic chemicals. They are not significant in most waste treatment processes, but play a role in oxidation ponds (Chap. 22) in which a symbiotic relationship exists between them and the saprophytic bacteria which oxidize the organic material in the waste.

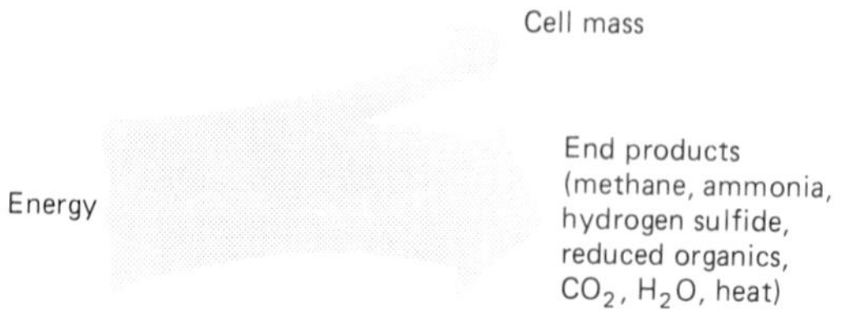

Cell mass

Energy

End products (methane, ammonia, hydrogen sulfide, reduced organics, CO_2, H_2O, heat)

FIGURE 18-5
Energy conversion in anaerobic processes.

FIGURE 18-6
Energy conversion in aerobic processes.

Protozoa are single-celled *animals* which reproduce by binary fission. There are many different species with differing shape, size, motility, and substrate. Protozoa may be aerobic, anaerobic, or facultative. Many species can utilize soluble food; however, the concentration must be far higher than that found in domestic sewage, hence the major food source of the protozoa is the bacteria. By reducing the number of bacteria the protozoa alter the food/mass ratio, thus stimulating further bacterial growth and further waste stabilization.

Fungi are multicellular nonphotosynthetic plants. Most fungi are aerobic, but anaerobic species are known. Because of their different cellular composition, fungi tend to predominate over bacteria in wastes which are deficient in nitrogen or low in pH. Because of their relatively large filamentous shape, fungi tend to settle poorly and are thus difficult to remove by sedimentation. Their presence is thus undesirable in waste treatment processes.

Rotifers are the simplest multicellular animal. They feed on bacteria and small protozoa, thus further stabilizing the waste. Since they require a relatively high dissolved oxygen content, their presence is a good indication of the relative stability of a treated waste. The relative predominance of microorganisms in waste stabilization is illustrated in Fig. 18-7.

18-9 Sampling

The difficulty of obtaining a representative sample of a heterogeneous and variable material such as wastewater has been mentioned above. There are essentially three fundamental methods: grab sampling, composite sampling, and continuous sampling.

A *grab sample* is simply a portion of the flow removed in a manner which will enhance the probability that it is representative of the flow at the instant it is taken. Grab samples may be taken from the discharge of a pump, be manually dipped from the flow, or be automatically dipped or siphoned from the stream. Grab samples are required to establish the variability of the waste with respect to time.

A *composite sample* is a mixture of grab samples taken over a period of time, with the volume of individual samples usually being proportional to the flow at the time the sample is taken. Composite samples may be obtained manually or automatically, either on a timed basis or on reaching a specified

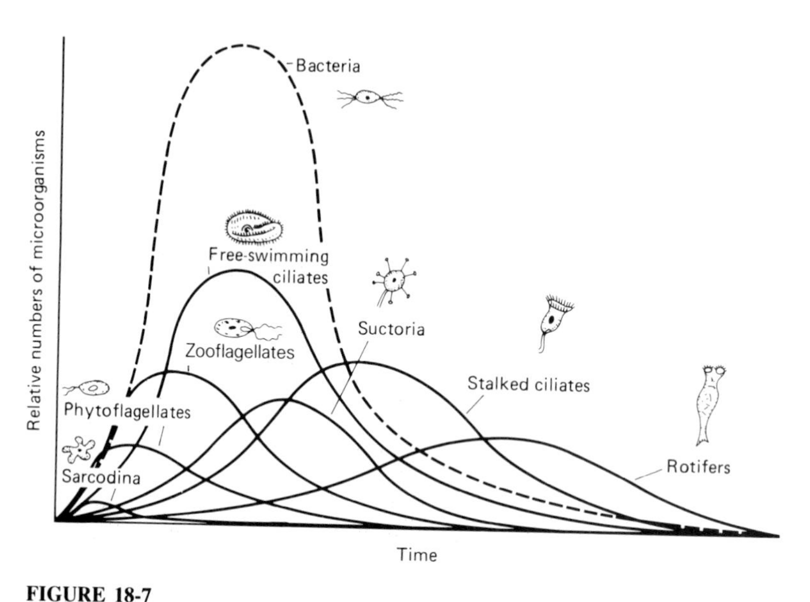

FIGURE 18-7
Relative growth of microorganisms in stabilization of wastes. Individual curves are not to the same scale. (*From Microbiology for Sanitary Engineers by Ross E. McKinney, Copyright 1962, McGraw-Hill Book Company. Used with permission of McGraw-Hill Book Company.*)

total flow. Composite samples are most useful for analyses of average characteristics such as daily waste loads.

A *continuous sample* represents diversion of a small fraction of the total flow over some period of time. Continuous samplers are usually not flow-proportional. Rather, they extract the sample at a constant rate. Continuous samplers are most suitable for instrumental measurements which can be performed virtually instantaneously, such as temperature, dissolved oxygen, pH, etc.

Representative samples must sample the entire cross section of the flow, since, as has been noted, sewage contains multiple phases which tend to be segregated to some extent. A scoop sampler which cuts across the entire stream is the best method of obtaining such a specimen. Since it may not be possible to install such a sampler in some locations, an alternative design involves pumping a portion of the flow to a chamber in which the sampler can be located.

The use of tube samplers should be avoided, since such designs do not provide representative samples unless the velocity within the tube is identical to that in the stream. If the velocities are different (typically greater in the sampler), substantial differences in solids concentrations must be expected. A tube sampler, by its nature, can sample only a point in the flow; hence, even if the isokinetic condition is achieved, the sample will not be representative of a segregated flow.

Samples must be suitably preserved until they can be analyzed. For most purposes, domestic sewage samples can be preserved satisfactorily by storage at 4°C. Freezing alters the character of the solids and thus should be avoided.

Because of the inherent variability of wastewater, the difficulty in obtaining representative samples, and the difficulties in analyzing the samples (even assuming they are representative), an extended period of study is desirable before the design of new wastewater treatment facilities is undertaken. If a special sampling program is necessary, it should cover at least 2 weeks and should be continued as long as may be necessary to ensure that it includes data from at least 3 wet days and 3 dry days. The definition of a "wet" day for this purpose is arbitrary, but has been taken to mean any 24-h period during which at least 25 mm (1 in) of rain falls.

18-10 Typical Characteristics

The variability of characteristics and inaccuracies of sampling and testing notwithstanding, there is some value in presenting concentrations of contaminants found in domestic wastewater. Table 18-1 shows the values which might be expected in domestic sewage, depending on the degree of dilution provided by infiltration and per capita water consumption. The values shown are *average* concentrations over an extended period of time. Instantaneous values must be expected to fluctuate widely about these figures. The *measured* concentrations of BOD in domestic wastewater normally fit a log normal distribution better than an arithmetic normal distribution. This means that the mean value will be more than the median and that extreme high values are more likely than extreme low values.

The per capita contribution of both suspended solids and organic matter (BOD) to domestic sewage is relatively invariant on a daily basis. In the United States, the per capita contributions of BOD and suspended solids are approximately 80 and 90 g/day respectively. The variations which are observed in concentration are largely a result of variations in flow and the other factors discussed above. For the design of some wastewater treatment processes in

TABLE 18-1
Typical domestic sewage characteristics, mg/L

Parameter	Weak	Medium	Strong
Total suspended solids	100	200	350
Volatile suspended solids	75	135	210
BOD	100	200	400
COD	175	300	600
TOC	100	200	400
Ammonia-N	5	10	20
Organic-N	8	20	40
PO_4-P	7	10	20

communities which have no significant industrial wastes, it is better to consider mass loadings (which will be relatively constant) rather than to focus on concentrations (which are inherently variable).

PROBLEMS

18-1. The 5-day BOD of a waste is 190 mg/L. Determine the ultimate oxygen demand L assuming that $K_1 = 0.25$/day.

18-2. A BOD determination yields the following data. Determine the value of K_1 and L for this waste.

Time, days	1	2	4	5	6	7	9	11	13
BOD, mg/L	4	12	19	21	23	27	30	35	35

18-3. The following BOD_5 values are obtained in daily composites collected over a 12-day period. Determine the mean, median, and standard deviation of these values. What is the BOD_5 of this waste?

Day	1	2	3	4	5	6	7	8	9	10	11	12
BOD	210	180	160	260	300	205	170	180	200	260	190	200

18-4. If the waste of Prob. 18-1 is discharged to a stream at an average temperature of 30°C, what fraction of the BOD would be exerted in 5 days? How long would be required for the same degree of stabilization if the temperature were 10°C?

18-5. A grab sample of treated wastewater is collected by dipping a bottle in the outfall of the treatment plant. A sample of the influent is collected in a similar fashion. BOD analyses of the two samples yield values of 33 mg/L and 188 mg/L respectively. The plant, supposedly, was designed to produce an effluent BOD of 30 mg/L or less when the influent BOD is 200 mg/L. Can any conclusion be drawn concerning the adequacy of the plant? Explain. Are the two samples equally likely to be representative of the flows which they sample?

REFERENCES

1. *Standard Methods for the Examination of Water and Wastewater*, 16th ed., American Public Health Association, New York, 1985.
2. Ross E. McKinney, *Microbiology for Sanitary Engineers*, McGraw-Hill, New York, 1962.
3. H. A. Thomas, Jr., "Graphical Determination of BOD Curve Constants," *Water and Sewage Works*, **97**:123, 1950.

CHAPTER 19

SEWAGE DISPOSAL

19-1 Disposal Techniques

The liquid wastes from industrial and domestic sources must eventually be disposed of in some manner, whether by reuse, by discharge to surface waters, by injection or percolation to groundwaters, or by evaporation to the atmosphere. In nearly all cases the water must first be treated to remove the bulk of the contaminants, either as a matter of engineering necessity or to meet the requirements of state and federal environmental regulations.

In order to determine the degree of treatment which will be required, it is necessary to consider the effects of the various pollutants on the environment to which they will be discharged, as well as any statutory or regulatory requirements which may have been established.

19-2 Effects of Stream Discharge

In natural streams there is a balance between plant and animal life, with considerable interdependence among the various life forms. Waters of good quality are characterized by multiplicity of species with no dominance. Organic matter which enters the stream is metabolized by bacteria and converted into ammonia, nitrates, sulfates, carbon dioxide, etc., which are used, in turn, by plants and algae to produce carbohydrates and oxygen. The plant life is fed upon by microscopic animals (protozoa, rotifers, etc.) which serve as a food source for crustacea, insects, worms, and fish. Some of the animals feed on the wastes of others, thus assisting bacterial degradation.

The introduction of excessive quantities of pollutants can upset this natural balance in a variety of ways. Changes in pH or the concentration of some organic and inorganic species may be toxic to specific life forms. Excessive quantities of organic material may cause rapid bacterial growth and depletion of the dissolved oxygen resources of the stream. Polluted waters are typically characterized by very large numbers of relatively few species and the absence of higher forms.

As the concentration of pollutants is reduced by dilution, precipitation, aeration, bacterial oxidation, or other natural processes, the normal cycle and distribution of life forms will tend to be reestablished. Water quality standards are based on the maintenance of minimum dissolved oxygen concentrations, nontoxic concentrations of specific chemical species, and a near-neutral pH. When a healthy environment is maintained in a stream, its natural assimilative capacity can be used to assist in waste treatment without adversely affecting downstream users.

The *self-purification* of natural waters results from a variety of physical, chemical, and biological phenomena.

Dilution greatly reduces the impact of all contaminants and is the only mechanism by which the concentration of some chemical species is naturally reduced. In addition, dilution of a contaminated flow with relatively clean water will improve the biological environment and enhance the natural stabilization processes.

Currents assist in dispersion of the waste in the receiving water, thus reducing the likelihood of locally high concentrations of pollutants. Lack of current, as in eddies and backwaters of streams, may encourage sedimentation of solids, formation of sludge banks, and production of odors. Sedimentation removes pollutants from the water which is passing by, thereby improving its quality, but creates unfavorable conditions in the locations where the contaminants accumulate. Deposits may also be scoured by later, higher, velocities and the resuspended pollutants may then cause water quality problems downstream. High velocity improves the transfer of oxygen from the atmosphere and enhances the stripping of volatile contaminants from the flow. It also carries the pollutants more rapidly and may therefore cause their effect to be more widely distributed.

Sedimentation results from differences in density between solid pollutants and the water which carries them. If the velocity of the flow is not sufficiently great to scour the solids from the bottom, thus resuspending them, sedimentation can result in improved water quality downstream. The environment at the locations where the solids accumulate will, however, usually be adversely affected and, as noted above, higher flows at other seasons may resuspend the material which has been removed.

Bottom deposits and non-point-source runoff provide diffuse sources of contaminants which can cause water quality degradation. The materials in bottom deposits, whatever their source, can release soluble contaminants to the water above as they decay. Non-point-source runoff produced by rainfall on urban or agricultural land can contribute significant pollutant loads to surface waters, and should be included in an evaluation of anticipated water quality.

Sunlight acts as a disinfectant and stimulates the growth of algae. The algae produce oxygen during the day (sometimes creating supersaturated dissolved oxygen levels), but use oxygen at night. Waters which contain heavy algal growths may thus have high dissolved oxygen levels in daylight hours and be anaerobic at night.

Temperature affects the solubility of oxygen in water, the rate of bacterial activity, and the rate at which gases are transferred to and from the water. With regard to dissolved oxygen levels, the critical condition is generally in warm weather when bacterial utilization rates are high, the saturation concentration is reduced, and lower flows limit the effects of dilution.

19-3 Water Quality Modeling

In order to assess the effect of pollutants on the receiving water body, it is necessary that the physical, chemical, and biological processes be described by mathematical equations which can be solved to yield a prediction of the resulting water quality. Empirical methods such as the multiple correlation technique of Churchill and Buckingham[1] have been used in the past, but most modern models are deterministic. Although the models are based on theoretical equations, the actual processes are often so complicated and interrelated that the equations describe them rather poorly, and, in order to obtain reasonable results, it is necessary to calibrate the models for each specific application.

In the most general case, one might consider the transport of pollutants in three dimensions. While there are three-dimensional water-quality models, their complexity is beyond the scope of this text, and they are seldom used in practice. The discussion here will be limited to transport in one dimension, which yields a satisfactory representation for most streams. Reservoir and estuarine models are frequently two-dimensional, with the second dimension being either vertical or horizontal, although some one-dimensional reservoir models (with vertical segmentation) are used.[2]

The general equation for transport by one-dimensional continuous flow is

$$\frac{\partial C}{\partial t} = \frac{1}{A(x, t)}\frac{\partial}{\partial x}\left[E(x, t)A(x, t)\frac{\partial C}{\partial x}\right] - \frac{1}{A(x, t)}\frac{\partial}{\partial x}[Q(x, t)C] - S(C, x, t) \quad (19\text{-}1)$$

where C = concentration
A = cross-sectional area of flow
E = longitudinal dispersion coefficient
x = distance
t = time
Q = flow

S includes all sources and sinks of the constituent under consideration. If longitudinal dispersion is neglected ($E = 0$), Eq. (19-1) reduces to

$$\frac{\partial C}{\partial t} = -\frac{1}{A(x, t)}\frac{\partial}{\partial x}[Q(x, t)C] - S(C, x, t) \quad (19\text{-}2)$$

Dispersion can be neglected without introducing any error, provided the situation being modeled is at a steady state. Equation (19-2) can be applied to the transport of any constituent of the flow. If we consider the concentration of *dissolved oxygen*, the term $S(C, x, t)$ includes the utilization of oxygen to satisfy biological oxygen demand, reaeration from the atmosphere, and any other sources or sinks (such as benthal utilization or algal production).

The rate of oxygen depletion by biological oxygen demand is equal to the rate of BOD exertion, or

$$\frac{dC}{dt} = \frac{dL}{dt} = -K_1 L(x, t) \tag{19-3}$$

For gases of low to moderate solubility, such as oxygen, the rate of change of concentration due to reaeration is given by

$$\frac{dC}{dt} = K_L \frac{A}{V}(C_s - C) \tag{19-4}$$

where K_L = mass-transfer coefficient
A = surface area over which transfer occurs
V = liquid volume
C_s = saturation concentration of the gas

The product $K_L(A/V)$ may be considered a volumetric mass-transfer coefficient. Then

$$\frac{dC}{dt} = K_2(C_s - C) \tag{19-5}$$

For dissolved oxygen, then, in a reach of a stream for which flow and area are constant, Eq. (19-2) may be written as

$$\frac{\partial C}{\partial t} = -\frac{Q}{A}\frac{\partial C}{\partial x} + K_2(C_s - C) - K_1 L(x, t) - S_R(x, t) \tag{19-6}$$

If $(C_s - C)$ is replaced by D, the oxygen deficit, then

$$\frac{\partial D}{\partial t} = -\frac{Q}{A}\frac{\partial D}{\partial x} - K_2 D + K_1 L(x, t) + S_R(x, t) \tag{19-7}$$

Under steady-state conditions $\partial D/\partial t = 0$ at any point in the stream, hence

$$U\frac{dD}{dx} = -K_2 D + K_1 L(x) + S_R(x) \tag{19-8}$$

where U = velocity of flow (Q/A)
D = oxygen deficit
$L(x)$ = BOD remaining

$S_R(x)$ includes all other sources and sinks of oxygen.

In a similar fashion, Eq. (19-2) may be applied to the transport of other constituents of the flow. In the case of BOD, for example, the term $S(C, x, t)$ includes the removal of BOD by biological action,

$$\frac{dL}{dt} = -K_1 L(x, t) \tag{19-9}$$

the removal of BOD by sedimentation or adsorption,

$$\frac{dL}{dt} = -K_3 L(x, t) \tag{19-10}$$

and addition of BOD by resuspension of sediments or from surface runoff,

$$\frac{dL}{dt} = L_a(x, t) \tag{19-11}$$

For BOD, then,

$$\frac{\partial L}{\partial t} = -\frac{Q}{A}\frac{\partial L}{\partial x} - K_1 L(x, t) - K_3 L(x, t) + L_a(x, t) \tag{19-12}$$

which, for steady-state conditions, $\partial L/\partial t = 0$, yields

$$U\frac{dL}{dx} = -(K_1 + K_3)L(x) + L_a(x) \tag{19-13}$$

For the condition in which K_1, K_3, and L_a are constant, Eq. (19-13) may be integrated to

$$L(x) = L_0 e^{-(K_1+K_3)x/U} + \frac{L_a}{K_1 + K_3}(1 - e^{-(K_1+K_3)x/U}) \tag{19-14}$$

in which L_0 is the ultimate BOD in the stream at $x = 0$, that is, at the beginning of the reach. Substituting Eq. (19-14) in Eq. (19-8) and integrating with K_1, K_2, K_3, U, L_a, and S_R constant, one obtains

$$D(x) = D_0 e^{-K_2 x/U} + \frac{K_1}{K_2 - (K_1 + K_3)}\left(L_0 - \frac{L_a}{K_1 + K_3}\right)(e^{-(K_1+K_3)x/U} - e^{-K_2 x/U}) + \left(\frac{S_R}{K_2} + \frac{K_1 L_a}{K_2(K_1 + K_3)}\right)(1 - e^{-K_2 x/U}) \tag{19-15}$$

D_0 is the deficit at $x = 0$ and the other terms are as defined above. Equation (19-14) permits calculation of the BOD remaining at any point in the stream, while Eq. (19-15) yields the dissolved oxygen deficit at any point. The variation of these factors with distance is shown in Fig. 19-1.

The equations for transport of BOD and dissolved oxygen are seen to be linked. Other potential constituents of the stream may be independent of each other or linked in even more complex fashions. The original partial differential equations, in some cases, may not be simplifiable by the assumptions of steady-state behavior and uniformity of rate constants and thus may not be directly integrable. In such circumstances, the solution may require the application of numerical methods to a finite-element model of the water body.[2,3,4]

Equation (19-15) is sometimes called the *oxygen sag equation*. Although the entire dissolved oxygen profile of a stream is often of interest, water quality analyses frequently focus on the point of minimum dissolved oxygen or maximum deficit. The *maximum* or *critical deficit*, labeled D_c in Fig. 19-1, occurs at the inflection point of the oxygen sag curve and may thus be found directly by taking the derivative of Eq. (19-15) with respect to x, setting the resulting expression equal to zero, and solving for x_c. Thus,

$$x_c = \frac{U}{K_2 - (K_1 + K_3)} \ln\left\{\frac{K_2}{K_1 + K_3} + \frac{K_2 - (K_1 + K_3)}{(K_1 + K_3)L_0 - L_a}\left[\frac{L_a}{K_1 + K_3} - \frac{K_2 D_0 - S_R}{K_1}\right]\right\} \tag{19-16}$$

The value of x_c may then be substituted for x in Eq. (19-15) to find the critical deficit, D_c.

19-4 Application of Analysis Techniquès

In order to apply any mathematical model, it is necessary to evaluate the various constants in the equations and establish whatever initial and boundary conditions may be required to obtain a solution. Determining the boundary conditions in non-steady-state models is particularly difficult if there are multiple tributaries[5] and may greatly complicate evaluation of the rate constants. In general, once the required boundary conditions have been determined, a number of sets of water quality data are collected for the stream. The mathematical model is then run and the results are compared to one of the data sets. The rate constants (BOD, reaeration, dispersion, etc.) are then adjusted to provide the best possible match between the measured and predicted values. After this *calibration*, the model is rerun for another set of conditions and the predicted values are compared to the second data set. Subsequent adjustment and verification may require additional data until a satisfactory calibration is

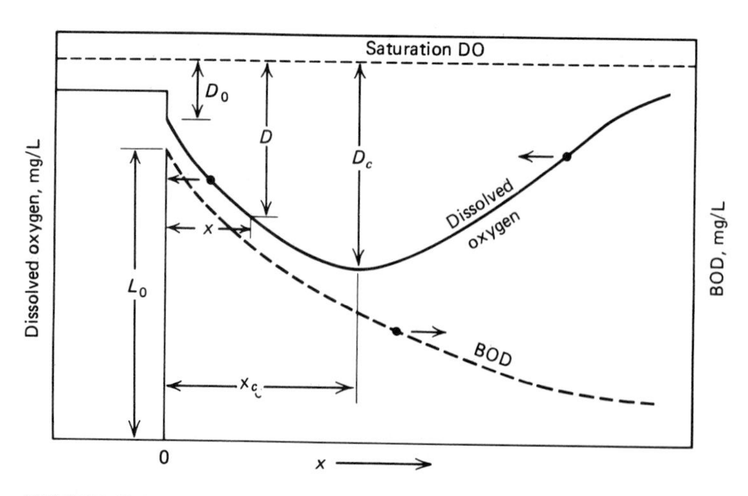

FIGURE 19-1
Oxygen sag and BOD removal in a stream.

obtained. The model then may be used to assess the response of the stream to other waste loads, flows etc. In calibrating a water quality model, one must recognize the inaccuracies inherent in sampling, laboratory analyses, and the theoretical models themselves. A perfect match between field data and model prediction should not be expected.

In some circumstances, specific studies may be conducted to determine particular rate constants, particularly dispersion and reaeration coefficients.[6,7] These values may then be taken as established, thus limiting the number of variables in the calibration process.

K_1, the deoxygenation constant, is usually measured in the laboratory as detailed in Chap. 18. The laboratory measurement is made at 20°C and must be corrected for the actual stream temperature by using Eq. (18-6). K_2, the reaeration constant, has been found to be a function of stream turbulence[8] and, in some circumstances, may be satisfactorily calculated from

$$K_2 = \left(\frac{D_m U}{H^3}\right)^{1/2} \tag{19-17}$$

where U = average stream velocity
H = average depth
D_m = molecular diffusion coefficient = 2.037×10^{-5} cm²/s at 20°C

K_2 varies with temperature in accord with

$$K_{2(T)} = K_{2(20°C)}(1.025)^{T-20} \tag{19-18}$$

As noted above, K_2 may also be measured directly by a variety of gas tracer techniques. One such procedure[7] involves the simultaneous release of a homogeneous mixture of three tracers: tritium in the form of tritiated water, dissolved krypton-85, and a fluorescent dye. The last indicates when to sample for the other two tracers and provides a measure of flow time between sampling stations. The tritiated water provides a measure of the total dispersion of the tracer mixture and may be used to estimate E, the dispersion coefficient. Since the tritium is in the form of water molecules, it is not absorbed or otherwise lost in any significant amount, hence the decrease in its concentration between stations is attributable to the total dispersion in the stream.

Because of the simultaneous release of the three tracers, the dissolved krypton-85 is dispersed to the same extent as the tritiated water, but is, in addition, lost to the atmosphere as a result of turbulent mixing and surface replacement. The rate of krypton-85 loss has been shown to be directly related to the rate of reaeration, and K_2 may be calculated from the concentrations C of tritium and krypton-85 measured at two stations and the time of flow between them:

$$K_2 = \frac{\ln\left[(C_{Kr_a}/C_{Tr_a})/(C_{Kr_b}/C_{Tr_b})\right]}{0.83t} \tag{19-19}$$

where the subscripts a and b refer to the two stations and t is the time of flow.

K_3 and L_a are measures of independent physical properties which are often considered negligible. Values of these parameters may be obtained from field data as follows. If the BOD decreases along a reach and the BOD at some point x is less than predicted by $L(x) = L_0 e^{-K_1 x/U}$, then K_3 is positive and exceeds L_a. L_a is assumed to be zero, and K_3 is calculated from

$$L(x) = L_0 e^{-(K_1 + K_3)x/U} \tag{19-20}$$

If the BOD increases with x or decreases less than predicted by $L(x) = L_0 e^{-K_1 x/U}$, then L_a is positive and exceeds K_3. K_3 is assumed to be zero and L_a is determined from

$$L(x) = L_0 e^{-K_1 x/U} + \frac{L_a}{K_1}(1 - e^{-K_1 x/U}) \tag{19-21}$$

S_R is a measure of oxygen sources and sinks other than reaeration and BOD exertion. It is often neglected, but may be determined from field measurements of dissolved oxygen once L, K_1, K_2, K_3, and L_a are known.

There are a large number of water quality models which are available for prediction of the effects of waste discharges. Many of these models are in the public domain and are available at little or no cost.[9] The EPA's Center for Water Quality Modeling schedules regular training sessions in the use of established and newly developed models.

In simple one-dimensional steady-state cases where the differential equations may be directly integrated, direct solutions to water quality problems may be obtained. If K_3, L_a, and S_R are neglected, Eq. (19-15) and (19-16) may be reduced to the classic Streeter-Phelps formulation:[10]

$$D(x) = D_0 e^{-K_2 x/U} + \frac{K_1}{K_2 - K_1} L_0 (e^{-K_1 x/U} - e^{-K_2 x/U}) \tag{19-22}$$

and

$$x_c = \frac{U}{K_2 - K_1} \ln \frac{K_2}{K_1}\left[1 - \frac{K_2 - K_1}{K_1 L_0}(D_0)\right] \tag{19-23}$$

In these equations, $t = x/U$ is often substituted to yield

$$D(t) = D_0 e^{-K_2 t} + \frac{K_1}{K_2 - K_1} L_0 (e^{-K_1 t} - e^{-K_2 t}) \tag{19-24}$$

and

$$t_c = \frac{1}{K_2 - K_1} \ln \frac{K_2}{K_1}\left[1 - \frac{K_2 - K_1}{K_1 L_0}(D_0)\right] \tag{19-25}$$

Example 19-1 A community discharges a waste flow of 1000 m³/day to a small stream. The 7-day, 10-year low flow in the stream is 5.74 m³/min. The maximum stream temperature is 30°C and this coincides with low flow. At this condition, the stream dissolved oxygen concentration is 6.1 mg/L and the BOD_5 is 5 mg/L above the sewage outfall. The values of K_1 and K_2 at 20°C are 0.23/day and 0.46/day respectively. K_3, L_a, and S_R are neglected. Determine the effluent BOD_5 from the sewage treatment plant which will not deplete the dissolved oxygen concentration to less than 4 mg/L under this condition.

Solution

In the absence of values for K_3, L_a, and S_R, Eqs. (19-24) and (19-25) may be used. The solution requires assumption of a value of L_0 and calculation of t_c and D_c with repeated trials until a value equal to that specified is obtained. This procedure is easily programmed.

From App. 1, the saturation value of dissolved oxygen at 30° is found to be 7.63 mg/L. The waste, since a fairly high degree of treatment is expected to be required, is assumed to contain at least 2.0 mg/L of dissolved oxygen.

The dissolved oxygen concentration in the stream following addition of the waste may be calculated from a simple mass balance:

$$DO(Q_s + Q_w) = 6.1Q_s + 2.0Q_w$$

$$DO = \frac{6.1(8265.6) + 2.0(1000)}{8265.6 + 1000}$$

$$= 5.66 \text{ mg/L}$$

The initial deficit, D_0, is then

$$D_0 = C_s - C = 7.63 - 5.66 = 1.97 \text{ mg/L}$$

At $T = 30°C$, K_1 and K_2 are

$$K_{1(30)} = 0.23(1.047)^{10} = 0.364/\text{day}$$

$$K_{2(30)} = 0.46(1.025)^{10} = 0.589/\text{day}$$

Whence

$$D_c = 1.97e^{-0.589t_c} + 1.618L_0(e^{-0.364t_c} - e^{-0.589t_c})$$

$$t_c = 4.444 \ln 1.618\left(1 - \frac{1.218}{L_0}\right)$$

The maximum allowable value of D_c if a minimum DO concentration of 4 mg/L is to be maintained is 7.63 − 4 = 3.63 mg/L. Values of L_0 are assumed, and t_c and D_c are calculated:

L_0	t_c	D_c
15	1.76	4.88
10	1.56	3.50
11	1.62	3.77
10.5	1.59	3.64

The maximum allowable ultimate BOD in the stream is thus approximately 10.5 mg/L. In terms of BOD_5, this equals

$$\text{BOD}_5 = L_0(1 - e^{-K_1(5)})$$

$$\text{BOD}_5 = 10.5(1 - e^{-0.23(5)}) = 7.18 \text{ mg/L}$$

Note that the calculation is made with $K_{1(20)}$, not $K_{1(30)}$. The BOD_5 of the waste is then calculated from another mass balance:

$$7.18(Q_w + Q_s) = 5Q_s + \text{BOD}_w(Q_w)$$

$$\text{BOD}_w = \frac{7.18(9265.6) - 5(8265.6)}{1000}$$

$$\text{BOD}_w = 25.2 \text{ mg/L}$$

In this case the water quality standard (minimum DO = 4 mg/L) governs rather than the secondary effluent design standard of $\text{BOD}_5 \leq 30$ mg/L. The treatment plant for this community must be designed and operated to yield an average effluent BOD_5 of approximately 25 mg/L. This can be dependably achieved with some, but not all, secondary-type treatment systems.

19-5 Lake and Ocean Discharge

Self-purification phenomena in lakes, estuaries, and the ocean are similar to those which operate in streams. Since currents are normally less pronounced in larger bodies of water, there is more opportunity for sedimentation to occur in the immediate vicinity of the discharge. The cyclic reversal of flow provided by tidal action in estuaries may also result in long flow-through times for contaminants. The decomposition which occurs in deep water may be slowed by low temperatures and lack of dissolved oxygen, which is supplied only by diffusion from the surface. It is not uncommon, even in the absence of sewage discharges, to find that the upper layers in ponds and lakes contain ample dissolved oxygen and support clean-water plankton and fishes, while the lower levels show the characteristics of polluted waters—low dissolved oxygen, anaerobic decomposition, and production of odors. Shallow lakes present favorable conditions for quick self-purification, having a large water surface relative to their volume and ample opportunity for algal growth, reaeration, and mixing by wind-driven currents.

The saturation concentration of dissolved oxygen in water tends to decrease with increasing salt content (App. 1). In seawater the saturation concentration is approximately 80 percent of that in freshwater.

The density of saline waters is greater than that of freshwater, hence sewage may tend to spread, without mixing, over the surface. When mixing is limited by large differences in density, dilution will be commensurately less. The lesser dilution coupled with the lower availability of oxygen in saline waters may lead to nuisance conditions which would not occur in a freshwater impoundment.

Two- and three-dimensional water quality models which employ finite-element segmentation techniques and provide numerical solutions to the partial differential equations have been developed. An example of such models is

WASP3,[4] which can be applied in one, two, or three dimensions and is available from the EPA in a personal computer (PC) version. The theory and development of such models is beyond the scope of this text, but their application is reasonably straightforward, provided the data necessary for their calibration is available. The results of any simulation technique, however sophisticated it may be, are only as good as the data used in its calibration.

19-6 Submarine Outfalls

Cities located along coastlines may elect to discharge their wastewater to the sea. The degree of treatment required by the EPA may be less for ocean discharges than for those to rivers, lakes, and estuaries. The actual NPDES (National Pollutional Discharge Elimination System) permit limits depend on the specific circumstances and are based on a careful assessment of the potential environmental impact of various levels of treatment. Where it has been shown that the dilution is very great, levels of treatment less than secondary have been approved. As a minimum, the treatment should include removal of all solids readily separable by sedimentation, since these might float in salt water and be returned to the coast by tidal and wind-driven currents.

Submarine outfalls are expensive. First costs are high, and pumping is nearly always required. The prevention of adverse impacts on coastal activities may require outfalls several kilometers in length. The outfalls are constructed of reinforced concrete, iron, or steel. Iron pipe may be provided with a cement mortar lining. Steel pipe used for this purpose is nearly always coated with either mortar or asphaltic materials and may be provided with cathodic protection as well. Joints in the pipe must have substantial mechanical strength and be resistant to biological and chemical attack. Ball-and-socket joints are often used for iron pipe, while steel pipe is usually welded. The pipe may be placed in trenches on bottoms of soft rock, sand, or gravel. On unstable bottoms, piling is necessary to secure the pipe against damage from wave action. Outfalls may employ single outlets or a variety of diffuser structures. A typical diffuser consists of a large number of small ports distributed over a large length of the pipe, perhaps as much as a third of its total length. The ports may be simple openings or may be fitted with tees to discharge the sewage in two directions.

19-7 Land Disposal and Treatment

Wastewater may be discharged to the land either for disposal or for treatment prior to discharge to surface waters. Although plants, microscopic forms in the upper layers of the soil, and the soil matrix itself have the ability to treat ordinary domestic wastes and many industrial wastes to a very high degree, ordinarily some treatment is provided before land application. The reasons for this include reduction in stress on the soil system, minimization of nuisance conditions, and the need to store the wastewater for extended periods of time when local conditions are unfavorable for disposal. In northern states, land disposal may be physically impossible or be prohibited by regulatory agencies during much of the winter, and, in any area, rainy periods may saturate the soil without any supplemental addition of wastewater. Required storage volumes vary with climate and soil characteristics and may range from as little as one week's to as much as four months' flow.

Land disposal may be broadly classified into *slow rate*, *rapid infiltration*, *overland flow*, *wetland*, and *subsurface* techniques.[11,12] Slow rate, rapid infiltration, and subsurface techniques depend on moving the water downward through the soil and are thus limited by infiltration and percolation capacity. Percolation capacity is a function of the soil characteristics while infiltration depends upon the degree of clogging at the point of application. If clogging is minimized, percolation will limit the rate at which liquid can be applied. A comparison of the features of the various systems is presented in Table 19-1, and the required characteristics of the site are summarized in Table 19-2.

Slow rate systems have also been called *irrigation systems*, since the application techniques are normally identical to those employed in agricultural irrigation. Surface runoff of the applied waste is usually not permitted; all the flow must either percolate to groundwater or be returned to the atmosphere by evapotranspiration. The potential benefits of slow rate systems include treatment of the wastewater, production of crops with a cash value, and conservation of water, nitrogen, and phosphorus. Potential problems include transmission of disease by spread of droplets through the air or by surface contamination of crops, increased nitrate concentration in the groundwater below the site, and changes in the natural vegetation because of increased moisture content in the root zone. Crops which respond well to slow rate systems include nearly all grains and grasses. Corn and hay grasses remove significant quantities of nutrients from the waste and are potentially salable. Corn requires annual plowing and planting, while grasses need only be harvested. Additionally, the grasses have a fully established root system at the start of the season and can provide immediate nutrient uptake. Forested areas may be adversely affected by wastewater irrigation, probably through saturation of the root zone; however, successful silvicultural applications have been widely reported.[11]

Rapid infiltration may be used either for waste disposal, groundwater recharge, or both. The water is applied at relatively high rates and percolates, either vertically or horizontally, away from the application zone. The application sites are typically large basins underlain by sand and soils of high permeability. The bottom of the basin may be covered by grasses such as bermuda or reed canary, which can tolerate both wet and dry conditions. Liquid is applied for periods of about 2 weeks followed by 1 to 3 weeks of drying, depending upon climate, season, soil characteristics, and other factors.[12] The percolate may be recovered from beneath the site by drains or wells and is typically of high quality—comparable to that provided by advanced waste treatment systems. Problems encountered in rapid infiltration systems are generally associated with errors in estimating the infiltration/percolation capac-

TABLE 19-1
Comparison of design features for land treatment processes (*after Ref. 11*)

	Principal processes			Other processes	
Feature	**Slow rate**	**Rapid infiltration**	**Overland flow**	**Wetlands**	**Subsurface**
Application techniques	Sprinkler or surface*	Usually surface	Sprinkler or surface	Sprinkler or surface	Subsurface piping
Annual application rate, ft	2 to 20	20 to 560	10 to 70	4 to 100	8 to 87
Field area required, acres†	56 to 560	2 to 56	16 to 110	11 to 280	13 to 140
Typical weekly application rate, in	0.5 to 4	4 to 120	2.5 to 6‡ 6 to 16¶	1 to 25	2 to 20
Minimum preapplication treatment provided in United States	Primary sedimentation§	Primary sedimentation	Screening and grit removal	Primary sedimentation	Primary sedimentation
Disposition of applied wastewater	Evapotranspiration and percolation	Mainly percolation	Surface runoff and evapotranspiration with some percolation	Evaportranspiration, percolation, and runoff	Percolation with some evapotranspiration
Need for vegetation	Required	Optional	Required	Required	Optional

* Includes ridge-and-furrow and border strip.

† Field area in acres not including buffer area, roads, or ditches for 1 Mgal/day (43.8 L/s) flow.

‡ Range for application of screened wastewater.

¶ Range for application of lagoon and secondary effluent.

§ Depends on the use of the effluent and the type of crop.

1 in = 2.54 cm; 1 ft = 0.305 m; 1 acre = 0.405 ha

TABLE 19-2
Comparison of site characteristics for land treatment processes (*after Ref. 11*)

	Principal processes			Other processes	
Characteristics	**Slow rate**	**Rapid infiltration**	**Overland flow**	**Wetlands**	**Subsurface**
Slope	Less than 20% on cultivated land; less than 40% on noncultivated land	Not critical; excessive slopes require much earthwork	Finish slopes 2 to 8%	Usually less than 5%	Not critical
Soil permeability	Moderately slow to moderately rapid	Rapid (sands, loamy sands)	Slow (clays, silts, and soils with impermeable barriers)	Slow to moderate	Slow to rapid
Depth to groundwater	2 to 3 ft (minimum)	10 ft (lesser depths are acceptable where underdrainage is provided)	Not critical	Not critical	Not critical
Climatic restrictions	Storage often needed for cold weather and precipitation	None (possibly modify operation in cold weather)	Storage often needed for cold weather	Storage may be needed for cold weather	None

1 ft = 0.305 m

ity of the site. Such errors may result from inadequate subsurface exploration, failure to consider hydraulic interactions among individual basins, reduction in permeability by construction activity, and the presence of excessive suspended solids in the wastewater. Before design of a rapid infiltration system, the soil profile must be investigated to considerable depth (30 m or more) and the range of groundwater table variation be established. During construction, care must be taken to ensure that the surface soils are not compacted and that fines are not accumulated at the final grade.

Overland flow is not a true disposal system, since most of the flow must be collected after it has passed over the soil. The process consists of applying wastewater along the upper edges of sloping, grass-covered fields and collecting it at the bottom in ditches which intercept the surface flow. This method is typically applied in areas with soils of low permeability, although it can also be used effectively under other conditions.[12] Some percolation and evapotranspiration occurs and some nutrients are removed by the plant growth. Effluent quality is reasonably good—somewhat better than is normally attainable in secondary treatment systems. Advantages of overland flow systems include reduced requirements for pretreatment and storage. Disadvantages include somewhat reduced effluent quality, potential contamination of rainwater on the treatment site, and the need for more extensive site preparation.

Wetlands, either natural or artificial, have a substantial capacity for wastewater renovation. Organic material is oxidized by bacteria in suspension or on the surface of aquatic plants, while nutrients and many heavy metals and other contaminants may be taken up by the plants themselves. The aquatic plants must be harvested if the removal which they provide is to be taken advantage of since their constituents would otherwise be released to the water when they die. Even then, the bulk of the plant mass is below the ground so most of the nutrient material will still be recycled. Wetland systems are similar to overland flow systems in that they are primarily treatment processes with little disposal of flow. The effluent quality, like that in overland flow, is typically slightly better than that obtained in secondary systems.[11]

Subsurface application is an alternative which is widely used in small-scale systems employing septic tanks (Chap. 15) in areas of permeable soils. The method has been extended to areas with unfavorable soil conditions by construction of earthen mounds containing permeable materials which provide a locus for bacterial stabilization of organic material and plant uptake of nutrients. If the mound is underlain by impermeable soils, the product water must be collected by underdrains. If the subsurface consists of fractured rock, the effluent may be allowed to flow to the groundwater if no adverse impacts will result from this discharge.

The major activity in soil systems occurs in the upper 300 mm (12 in) of the soil mantle. Adsorption of phosphates and heavy metals may occur at deeper levels as the capacity of the upper layer is exhausted. Readily *biodegradable compounds* are oxidized in the upper few millimeters of the soil. From 0.45 to 1 kg/m^2 of organic material is required each year for general soil equilibrium, and such levels are seldom approached in land treatment systems. *Other organic compounds* such as pesticides, cellulose, polysaccharides, and humic materials which may be present in wastewater are adsorbed to the soil matrix and are slowly degraded.

Nitrogen may be present as organic nitrogen, ammonia, nitrate, or a combination of two or more species. *Nitrate* can percolate to the groundwater if it is not removed by plant uptake or reduced to nitrogen by bacterial action. *Ammonia* may be adsorbed on the soil or be fixed in clays. It is removed by plants and oxidized to nitrate by soil bacteria in an aerobic environment. The oxidation is slow, and the delay in passage through the soil increases the likelihood of plant utilization. *Organic nitrogen* is released as ammonia on the oxidation of the organic material of which it is part.

Phosphorus is utilized by plants as a nutrient and is fixed by adsorption and by exchange reactions with compounds in the soil which contain aluminum or iron. The adsorptive capacity of fine-textured soils is high (up to 2.25 kg/m^2) and offers a useful site life of as much as 100 years under favorable conditions. Overland runoff and wetland systems do not provide much phosphorus removal, since little interaction with the soil is involved.

Inorganic compounds which may be harmful include heavy metals and monovalent cations. *Heavy metals* are removed by adsorption on soil particles. While no clear limit to capacity is evident, the adsorptive capacity is thought to exceed the tolerance level of plants, hence the first effect would be on the crop, not on the groundwater. It should be noted, however, that the agricultural use of land may be destroyed by heavy metal poisoning. Under acid conditions, heavy metals may be leached from the soil.

Monovalent ions tend to exchange for divalent ions in the soil matrix. In some clays, such exchange leads to swelling and loss of permeability. Additionally, high salinity decreases the ease of water utilization by plants. Water which has an ionic composition suitable for general agricultural use will not cause swelling of clay materials.

19-8 Total Retention Systems

In some climates it is possible to dispose of wastewater by evaporation. Most so-called total retention systems, however, discharge a portion of the flow to the soil.

The most common evaporation system is an oxidation pond with no outlet. The design criteria and biological processes of oxidation ponds are discussed in Chap. 22; in this article it is sufficient to know that the liquid in the pond is more or less equivalent in quality to that obtained after secondary treatment.

Average annual precipitation and average annual pan evaporation for different areas of the United States are presented in Chap. 4. Broadly speaking, if the annual precipitation exceeds 70 percent of the pan evaporation, an evaporation system is not feasible. The area of the United States most suitable for such systems covers the Great Plains from the 100th meridian to the Rocky Mountains. In some states, so-called total retention system design criteria permit percolation of a small quantity of water (perhaps 3 mm/day) through the bottom of the impoundment. This extends the practical range of the system to the upper Missouri River valley.

Total retention systems are designed on the basis of a mass balance. Since the total inflow must equal the water lost, in the long term,

$$A(i_a) + Q_w = (e_a + p_a)A \tag{19-26}$$

where i_a is the annual precipitation, Q_w is the annual waste flow, e_a is the annual evaporation (0.7 × pan evaporation) and p_a is the allowable annual percolation.

Example 19-2 Determine the area required for a total retention system for a location in eastern Nebraska for which the annual precipitation is 600 mm/year and the average pan evaporation is 1200 mm/year. The waste flow is 190 m^3/day, and the pond may be constructed to permit 3 mm of loss per day by percolation.

Solution
Converting all values to commensurate units gives i_a = 0.6 m/year, Q_w = 69,350 m^3/year, e_a = 0.7(1.2) = 0.84 m/year, p_a = 1.1 m/year.

$$A(0.6) + 69{,}350 = (0.84 + 1.10)A$$

$$A = 51{,}750 \text{ m}^2$$

An equivalent pond designed for treatment alone (see Chap. 22) would have an area of approximately 20,000 m^2. In an actual case the designer would need to consider extreme as well as average values. Oxidation ponds are constructed so that their area varies significantly with depth, hence some variability in flow, rainfall, and evaporation can be accommodated.

19-9 Selection of a Disposal System

There is no single system which is best suited to the disposal of all wastewaters. The engineer must investigate each system which is physically practicable to determine the cheapest technique that is environmentally and socially acceptable.

Stream disposal is the commonest technique in the United States and is generally cheapest, provided water quality standards do not require advanced treatment.

Land disposal is often socially and politically desirable, and may be economical in water-poor areas where suitable land is available and stream standards are restrictive. Land systems may also require somewhat less skill in operation, which can be a significant factor in ensuring protection of the environment. Viewed simply as a disposal technique in areas where substantial storage is required, land disposal is generally quite expensive in comparison to discharge to surface waters.

Evaporation is practicable only in limited areas, and in those areas the water might be more profitably used to recharge groundwater or irrigate crops. The designer in such a case must weigh the cost of the more expensive land disposal system against the value of the benefit obtained.

PROBLEMS

19-1. In Example 19-1, Art. 19-4, assume that the waste flow is 2000 m³/day. What effluent BOD_5 must be provided by the treatment plant?

19-2. A stream has an average depth of 2 m, a velocity of 0.5 m/s, a flow of 30 m³/s, and a temperature of 30°C. Determine the value of the reaeration coefficient.

19-3. In a field evaluation of K_2, the concentrations of tritium (Tr) and krypton-85 (Kr) are recorded at the time when the peak concentration of a fluorescent dye reaches the sampling point. Data obtained on three different dates is presented below:

		Point A (km 100.6)			Point B (km 108.2)		
Date	**Q, m³/s**	**Time**	**Kr**	**Tr**	**Time**	**Kr**	**Tr**
6/15	10.5	10:00	10.2	18.1	16:10	7.6	16.9
6/18	12.1	09:55	9.8	17.4	15:25	6.8	14.7
6/21	13.3	10:10	9.5	16.9	15:15	6.2	13.3

Determine the value of K_2 from Eq. (19-19) for each date. Does K_2 vary with flow or velocity in a reasonable manner? Do the results seem useful? Explain.

19-4. A small stream is sampled just below a wastewater discharge and at various stations downstream. From the following data, estimate the values of K_1 and K_2 assuming K_3, L_a, and S_R are negligible.
Date: July 5, 1990, average velocity 0.1 m/s, temperature 23°C

Distance, km	0	5	12	25	40
DO, mg/L	7.0	5.7	5.0	5.0	5.7
BOD, mg/L	9	. . .	. . .	. . .	3

Date: July 11, 1990, average velocity 0.08 m/s, temperature 24.5°C

Distance, km	0	5	12	25	40
DO, mg/L	6.8	5.1	4.3	4.8	5.5
BOD, mg/L	10	. . .	. . .	. . .	3

19-5. In Prob. 19-4 assume that K_1 and K_2 have been independently determined to be 0.2/day and 0.6/day at 20°C. What values would you use for K_3, L_a, and/or S_R?

19-6. Determine the land area required for a slow rate land disposal system for a community with a design population of 100,000 persons. The average daily flow is 50,000 m³/day, the application rate is 50 mm/week during the period March 1

through November 30, and all flow is stored during December through February. Assume the storage basins are inside the site, have a useful depth of 3 m, and do not contribute any net water loss (or gain) due to seepage, evaporation, or rainfall. Find the storage basin area, the irrigation area, and add a buffer zone 50 m wide around the entire periphery. Assume the area is square.

19-7. In the Example 19-2, Art. 19-8, determine the required pond area if no percolation is permitted. Does this appear to be a practical system? Why?

REFERENCES

1. M. A. Churchill and R. A. Buckingham, "Statistical Method for Analysis of Stream Purification Capacity," *Sewage and Industrial Wastes*, **28**:517, 1956.
2. "CE-QUAL-R1: A Numerical One-Dimensional Model of Reservoir Water Quality; User's Manual," IR E-82-1, U.S. Army Engineer Waterways Experiment Station, Vicksburg, Miss., 1982.
3. "Computer Program Documentation for the Enhanced Stream Water Quality Model QUAL2E," EPA/600/3-85/065, Environmental Protection Agency, Washington, D.C., 1985.
4. "WASP3, A Hydrodynamic and Water Quality Model—Model Theory, User's Manual, and Programmer's Guide," EPA/600/3-86/034, Environmental Protection Agency, Washington, D.C., 1986.
5. Harvey E. Jobson, "Simulating Unsteady Transport of Nitrogen, Biochemical Oxygen Demand and Dissolved Oxygen in the Chattahoochee River Downstream from Atlanta, Georgia," Water-Supply Paper 2264, U.S. Geological Survey, Alexandria, Va., 1985.
6. H. E. Jobson and R. E. Rathburn, "Use of the Routing Procedure to Study Dye and Gas Transport in the West Fork Trinity River, Texas," Water-Supply Paper 2252, U.S. Geological Survey, Alexandria, Va., 1984.
7. E. C. Tsivoglou, "Symposium on Direct Tracer Measurement of the Reaeration Capacity of Streams and Estuaries," Water Pollution Series Report No. 16050, U.S. Environmental Protection Agency, 1972.
8. D. O'Connor and W. Dobbins, "The Mechanism of Reaeration in Natural Streams," *Journal of Hydraulics Division, Proceedings of American Society of Civil Engineers*, **101**:H11:1315, 1975.
9. "Water Quality Exposure and Risk Modeling," EPA/M-86/018, U.S. Environmental Protection Agency, Washington, D.C., 1986.
10. Harold W. Streeter and Earl Phelps, Bulletin 146, U.S. Public Health Service, 1925.
11. *Process Design Manual for Land Treatment of Municipal Wastewater*, EPA 625/1-81-013, Environmental Protection Agency, Washington, D.C., 1981.
12. *Process Design Manual for Land Treatment of Municipal Wastewater Supplement on Rapid Infiltration and Overland Flow*, EPA 626/1-81-013a, Environmental Protection Agency, Washington, D.C., 1984.

CHAPTER 20

PRELIMINARY TREATMENT SYSTEMS

20-1 Classification of Treatment Systems

Wastewater treatment has been separated into *preliminary*, *primary*, *secondary*, and *advanced* systems, a division which, while somewhat arbitrary, is fairly well established. *Preliminary* systems include measurement and regulation of the incoming flow and removal of large floating solids, grit, and perhaps grease. The quality of the wastewater is not substantially improved by primary systems. Rather, the operation of subsequent processes is enhanced through measurement and control of the flow and by removal of materials which might interfere with mechanical, chemical, or biological treatment.

Flow measurement is discussed in detail in Chap. 3. Devices used to measure wastewater flow include a variety of flumes, magnetic meters, and sonic meters. Which device is best depends on the particular circumstances and whether the flow is to be measured in a closed or open conduit. Instruments which can be fouled by solids or grease should be avoided.

20-2 Racks and Coarse Screens

The racks and screens used in preliminary treatment are intended to remove only fairly large suspended solids. Openings in these screens are typically 25 mm (1 in) or more. In small plants, manually cleaned racks similar to that shown in Fig. 20-1 may be used either alone or in parallel with a channel containing a mechanically cleaned screen or a comminutor. In the latter case, the manually cleaned system serves as a backup to be used in the event the mechanical system fails. As the rack is plugged by solids, the operator periodically rakes the accumulation to the screening platform where some drainage occurs. Mechanically cleaned screens such as that shown in Fig. 20-2 are

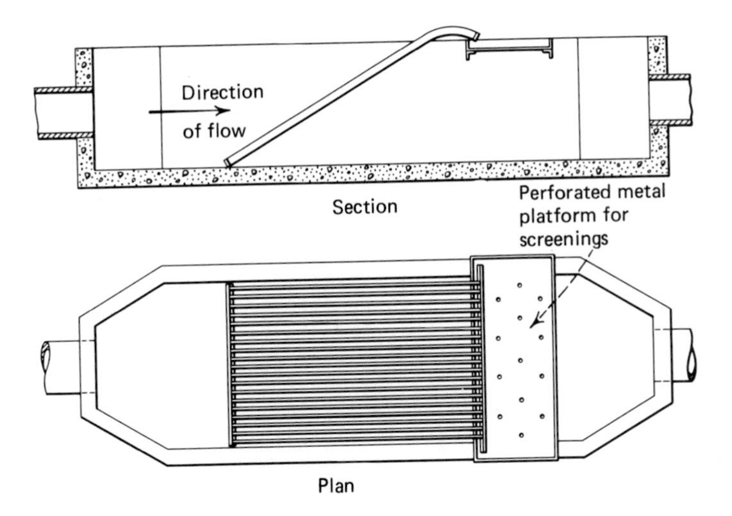

FIGURE 20-1
Manually cleaned bar screen.

commonly used in modern plants. Multiple units are provided so that mechanical failure or maintenance will not cause hydraulic overload. In very small plants, where the smallest available unit can handle the entire flow, a manually cleaned screen may be used as a backup.

Mechanical screens may be cleaned continuously, on a timed basis, or on the development of a specified head loss. The screenings are conveyed out of the flow path and are dropped on a platform like that of Fig. 20-1 or are deposited in a container. In either case, the screenings must be periodically removed, since much of the material is putrescible, is impregnated with fecal material, produces odors, and attracts insects. The quantity of screenings depends on the age of the sewage and the size of the screen opening. Fresh sewage and fine screens will tend to maximize production. The *volume* of screenings reportedly ranges from 1.3×10^{-6} to 3.7×10^{-5} m^3 per cubic meter of flow, with an average value of about 1.5×10^{-5} m^3/m^3. Solids-handling systems should be sized to accommodate the largest anticipated volume.

The screenings are sometimes fed directly to a grinder, which reduces their size so that they will not interfere with subsequent treatment. The ground screenings are then returned to the flow or combined with primary sludge (Chap. 21), a process that is equivalent to comminution (Art. 20-3). Other methods of managing this material include anaerobic digestion (Chap. 23), incineration, and burial. Before incineration it is desirable to dewater the screenings as much as possible, since they normally are about 80 percent water by weight. Incineration is usually carried out in conjunction with management of other sewage solids (Chap. 23) because the quantity of screenings by itself is not adequate to justify such a process. Following digestion, dewatering, or

FIGURE 20-2
Mechanically cleaned bar screen *(Courtesy Envirex, a Rexnord Company.)*

incineration, the screenings will still leave a solid residue which must be properly managed. Disposal in a sanitary landfill is the normal management technique and this can be used directly as well, without any prior treatment other than draining of free water.

20-3 Comminutors

Instead of removing large suspended solids, comminutors reduce them in size so they will not interfere with other systems. The chopped or ground solids are then removed from the flow in subsequent sedimentation processes.

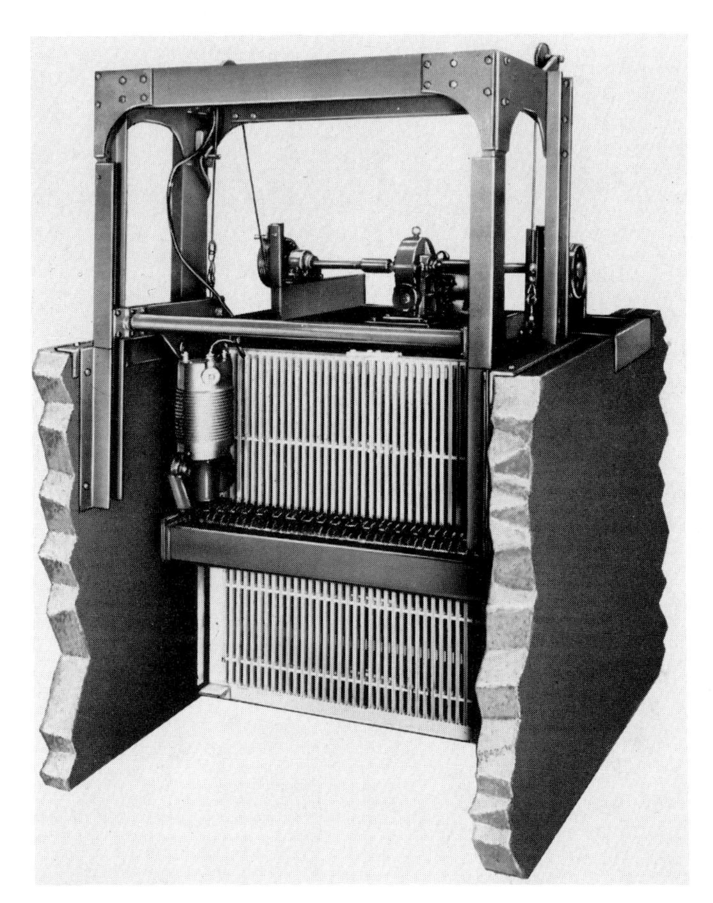

FIGURE 20-3
Barminutor. *(Courtesy FMC Corp.)*

Comminutors consist of a fixed screen and a moving cutter. The rack of Figure 20-3 is a comminuting device, but a more typical system employs a curved screen and a rotating or oscillating cutting blade (Fig. 20-4). Some rotating comminutor designs have a tendency to draw flexible materials through the screen rather than chop them. This can produce nuisance conditions in unskimmed clarifiers, trickling filters, and aeration basins.

Selection of comminutors is based on the flow rate. In small plants, a single unit rated for the peak flow may be used in parallel with a manually cleaned bar screen. In larger facilities multiple identical units are used, sized so that the remaining machines can handle the peak flow with one or two out of service. Head loss across comminutors depends on the screen details and the flow. Normal values are on the order of 50 to 100 mm (2 to 4 in) at design flow. Values for specific devices can be obtained from the manufacturers. When possible, comminutors should be protected against excessive wear by installing grit-removal processes upstream.

FIGURE 20-4
Comminutor. *(Courtesy Worthington Pump Corporation.)*

20-4 Grit Removal

A portion of the suspended solids in municipal sewage consists of inert inorganic material such as sand, metal fragments, eggshells, etc. This *grit* is not benefited by secondary treatment or sludge processing techniques and can block conduits and promote excessive wear of mechanical equipment.

Grit-removal devices rely on the difference in specific gravity between organic and inorganic solids to effect their separation. In *gravity separators*, the principles developed in Chap. 9 are applied. All particles are assumed to settle in accord with Newton's law:

$$v_s = \left[\frac{4g(\rho_s - \rho)d}{3C_D\rho}\right]^{1/2} \tag{20-1}$$

and to be scoured at a velocity

$$v_h = \left[\frac{8\beta(s - 1)gd}{f}\right]^{1/2} \tag{20-2}$$

To ensure removal of grit while permitting the organic matter which may settle to be resuspended by scour, the horizontal velocity should be close to but less than the scour velocity of the grit.

Example 20-1 A suspension contains particles of grit with a diameter of 0.2 mm and specific gravity of 2.65. For particles of this size, $C_D \approx 10, f \approx 0.03$, and $\beta \approx 0.06$. The suspension also contains organic solids of the same size for which the specific gravity is 1.10, and f and β are unchanged. Determine the settling velocity of the grit and the scour velocity of both grit and organic material. Select a horizontal velocity for the basin.

Solution
Substitution in Eqs. (20-1) and (20-2) yields for the grit:

$$v_s = 2.1 \text{ cm/s}$$

$$v_h = 23 \text{ cm/s}$$

and for the organic material:

$$v_h = 5.6 \text{ cm/s}$$

Thus, if the basin is designed to have a surface overflow rate of 2.1 cm/s and a horizontal velocity greater than 5.6 cm/s and less than 23 cm/s, the grit will be removed without removing organic material. To assure that the grit is reasonably clean, the horizontal velocity should be close to the scour velocity of the grit.

The horizontal velocity, as illustrated above, is very important to the proper function of gravity grit separators. The velocity can be held constant, regardless of the flow, by proper combination of basin cross section and the control device. For a constant velocity, the basin cross section must be proportioned so that

$$v_h = C = \frac{Q}{\int_0^y x \, dy} \tag{20-3}$$

In the control, in general,

$$Q = K'y^n \tag{20-4}$$

Equating (20-3) and (20-4),

$$K'y^n = C \int_0^y x \, dy \tag{20-5}$$

which, when differentiated, yields

$$y^{n-1} = Kx \tag{20-6}$$

Thus the condition of constant velocity is maintained, provided the width of the basin varies so that $y^{n-1} = Kx$, where n is the discharge coefficient of the control section.

If the control section is rectangular in cross section (like a Parshall flume), n will be approximately 1.5, thus

$$y = Cx^2 \tag{20-7}$$

and the channel cross section must be parabolic, as shown in Fig. 20-5. With a proportional flow weir (Chap. 3), $n = 1$ and $y = C$. The channel cross section in this case is rectangular, which somewhat simplifies construction. The actual proportions of the channel and weir must be selected together to provide the necessary conditions for grit removal.

Example 20-2 Design a grit-removal system consisting of four identical channels for a plant which has a peak flow of 80,000 m³/day, an average flow of 50,000 m³/day and a minimum flow of 20,000 m³/day. Use parabolic channels and assume a minimum of three basins will be in operation at any time. The design velocity is 0.25 m/s.

Solution

The peak flow per channel will be 26,667 m³/day, the normal maximum 20,000 m³/day, the average 12,500 m³/day, and the minimum 5000 m³/day. For a parabolic channel,

$$A = \frac{2}{3} WD$$

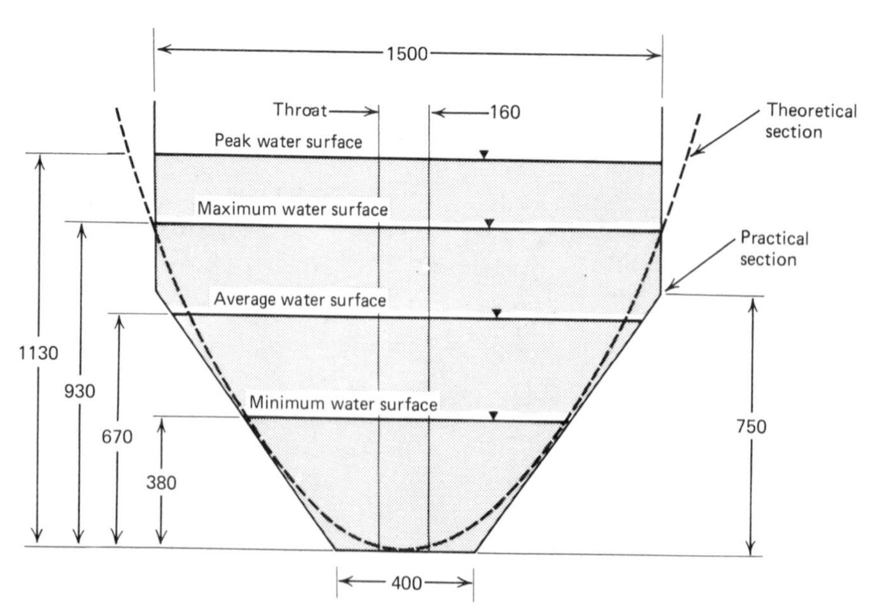

FIGURE 20-5
Parabolic grit chamber cross section.

Therefore,

$$A_{peak} = 1.23 \text{ m}^2$$

$$A_{max} = 0.93 \text{ m}^2$$

$$A_{avg} = 0.58 \text{ m}^2$$

$$A_{min} = 0.23 \text{ m}^2$$

The channel, in principle, can have any appropriate combination of width and depth. A number of depths might be tested before finding a combination which is neither unduly shallow and wide nor too deep and narrow. For a width of 1.5 m at maximum depth, $D_{max} = 0.93$ m.

The total energy head in the flume flow at Q_{max} is

$$\frac{v_h^2}{2g} + D = 0.93 \text{ m}$$

The control section will produce critical depth (see Chap. 3); thus, in the control, $v_h = v_c$ and $d_c = v_c^2/g$.

The total energy head in the control is $v_c^2/g + v_c^2/2g$. If we assume the head loss in the control is 10 percent of the velocity head, then

$$D = \frac{v_c^2}{g} + \frac{v_c^2}{2g} + \frac{0.1v_c^2}{2g} = \frac{3.1v_c^2}{2g}$$

at maximum flow, with $D = 0.93$ m, $v_c = 2.42$ m/s, and $d_c = 0.60$ m, and the width of the control section is $w = Q/(v_c \cdot d_c) = 0.16$ m.

For the other flow conditions, with the width of the control known, the value of d_c can be directly calculated, and from that value the upstream depth in the grit basin is calculated. The width at that depth is then found from the equation of the parabola. The results of these calculations are shown in the table below and in Fig. 20-5.

Q	d_c	w	D	W
5,000	0.24	0.16	0.38	0.91
12,500	0.43	0.16	0.67	1.30
20,000	0.60	0.16	0.93	1.50
26,666	0.73	0.16	1.13	1.63*

* Theoretical value. The actual constructed width is as shown in Fig. 20-5.

The length of the basin depends on the ratio of the settling velocity of the particles to the horizontal velocity of the flow, $v_h/v_s = L/D$, as shown in Chap. 9. Thus $L = D(v_h/v_s)$. For the peak depth of 1.13 m and a surface overflow rate of 21 mm/s, this yields a length of about 13.5 m.

The example above is not the only feasible solution to the given problem. A variety of different controls and cross sections could serve equally well. Variations in the cross section will change the depth and thus the length, and there is no reason why four basins must be used. The theoretical cross section shown in Fig. 20-5 would not actually be built, since it would be very difficult to form. Rather, a simplified section which approximates the parabolic shape would be selected. A possible section is shown on Fig. 20-5, but other approximations could also be used.

Example 20-3 Design a set of rectangular grit basins with proportional flow weirs to serve in the plant described above. Use five basins and assume a minimum of four will always be in use. Make the peak depth equal to the width.

Solution
At peak flow, $Q_{peak} = 20{,}000\ m^3/day$. For a horizontal velocity of 0.25 m/s, $A = 0.93\ m$ and $w = d = 0.96$ m. We might choose to make the basin 1 m wide by 1 m deep. This would provide a freeboard of 0.07 m (3 in) at peak flow. The weir must be shaped so that

$$Q = 8.18 \times 10^{-6} w y^{1.5}$$

(from Chap. 3). At the various flows which might occur:

Q, m³/day	y, mm	w, mm
20,000	926	60.3
16,000	741	67.3
10,000	463	85.2
4,000	185	135.0

The length, as above, would be approximately 12 times the maximum depth, or about 12 m. A typical rectangular grit basin is illustrated in Fig. 20-6. The head loss through a proportional flow weir is approximately equal to the head upstream since any submergence of the weir will alter its characteristics.

Aerated grit chambers control the separation of organic and inorganic solids by producing a rolling motion through introducing diffused air on one side of the basin (Fig. 20-7). The solids are carried to the bottom of the basin by

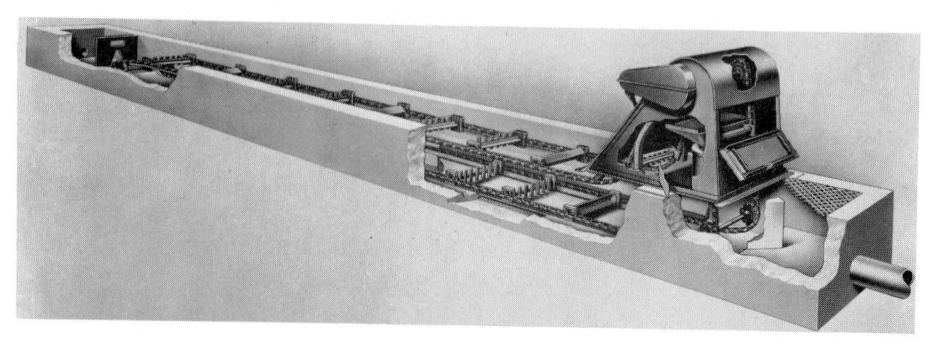

FIGURE 20-6
Rectangular gravity grit chamber with continuous chain-driven removal. *(Courtesy Envirex, a Rexnord Company.)*

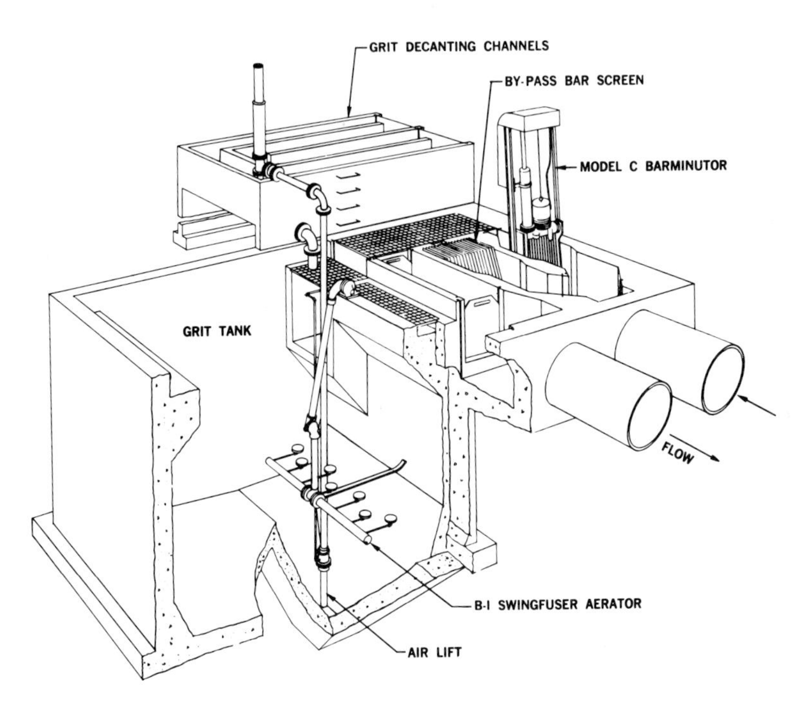

FIGURE 20-7
Aerated grit chamber. *(Courtesy FMC Corporation.)*

the rolling motion of the water, and the velocity of the flow is maintained at a level which is sufficient to carry the organic materials back up on the other side while leaving the inorganics at the bottom. The air flow is adjusted to achieve the separation desired. Detention times are usually less than 3 min with air flows on the order of 0.3 to 0.5 m^3/min per meter of tank length (3 to 5 ft^3/min per foot). The depth is 70 to 100 percent of the width and ranges from 3 to 5 m (10 to 16 ft). The ratio of length to width should be at least 2.5:1. Head loss through aerated grit chambers is negligible.

Constant-level grit chambers consist of shallow, normally circular sedimentation tanks with a rapidly rotating scraping mechanism which helps in separating grit and the organic solids which also are removed. These basins do not give a clean separation of grit and organics, and subsequent washing of the solids is necessary. Although some older plants contain such systems there is no reason to continue building them, since better processes are available at equal cost.

Grit from properly designed and operated gravity or aerated grit basins is low in organic content and relatively innocuous. If equipment malfunctions occur, it may contain up to 50 percent organic material and be putrescible. Clean grit may be used for fill material. If contaminated, it must be disposed of in sanitary landfills or other approved solid waste facilities.

The *quantity of grit* varies with the condition of the sewer system, the quantity of storm water and the proportion of industrial waste. Recorded quantities range from as little as 2.5×10^{-6} to 1.8×10^{-4} m^3 per cubic meter of wastewater.[1] An average value is about 6×10^{-5} m^3/m^3, but considerable variation should be anticipated in designing grit-handling facilities. The grit has the physical characteristics of saturated sand—it is heavy, moderately cohesive, abrasive, and difficult to move. Mechanical equipment used to move grit includes clamshell buckets, augers, and chain-mounted buckets. Mechanical failures are common in grit-handling equipment, and very heavy duty features should be specified to minimize maintenance and repair.

20-5 Grease Removal

Excessive quantities of grease may plug trickling filters or coat biological floc in activated sludge processes. Grease is removed to a degree by surface skimming devices in primary sedimentation tanks, but communities with particularly high concentrations of grease or treatment plants which omit primary clarification may need other processes.

Skimming tanks employ baffled subsurface entrance and exit structures which permit floating material to be retained. Retention times are 15 min or less, and continuous mechanical skimming is usually employed. The horizontal velocity of the water is kept in the range of 50 to 250 mm/s (0.2 to 0.8 ft/s) in order to prevent deposition of organic particles on the bottom.

Grease is sometimes also removed by *flotation processes* similar to those employed in sludge thickening (Chap. 23). The accumulation of small air bubbles at the surface of grease particles will decrease their density and increase their effective diameter. From a consideration of Stoke's law (Chap. 9), it is evident that the increase in size and decrease in density will result in a more rapid separation. Flotation processes include *aeration*, which is not very effective; *injection* of air to the pressurized flow which is then depressurized, releasing the dissolved air; and *vacuum flotation* in which the atmospheric pressure is reduced to draw dissolved gases from solution. Design procedures for flotation systems are presented elsewhere.[2] Most flotation systems are provided as package plants, but selection of such equipment should always be based on pilot plant studies to ensure that it will be effective.

20-6 Preaeration

Aeration of sewage before any other treatment is provided can have a number of desirable effects. First, aeration will strip volatile compounds (which are typically odorous) and increase the dissolved oxygen content of the flow. Both of these effects will tend to reduce odor production. Aeration, through the mixing which it provides, may also improve grease removal slightly, offer additional opportunity for flocculation of suspended solids, and help to make the flow uniform in character as it enters subsequent processes. Aeration, as has been noted earlier, can also be employed as a means of grit removal.

Preaeration generally is done in expanded grit basins which have a retention time of about 30 min. Aeration rates range from 0.01 to 0.05 m^3 of air per cubic meter of waste. Basin cross-sectional dimensions are similar to those given above for aerated grit basins.

20-7 Equalization

The principal purpose of equalization is to dampen the fluctuations in wastewater flow which occur on a daily and longer-term basis. Delivering the flow to treatment processes at a constant or near-constant rate would be expected to optimize their operation and also permit them to be smaller, since the maximum flow rate would be less. In fact, these anticipated results are achieved when the flow is equalized.

In order to hold untreated wastewater in a reservoir without creating nuisance conditions, it is necessary that the contents be mixed. The mixing provides aeration and tends to equalize concentration by dispersing strong flows in the contained volume. It must be noted that optimum equalization of both flow and concentration are not achievable in a single process. To equalize flows, the equalization basin must, at times, be empty. To equalize concentrations the basin should always be full. Nonetheless, a basin which equalizes flows will also produce some reduction in peak concentrations.

A mixed basin containing wastewater, dissolved oxygen, and bacteria will also provide a degree of biological treatment. This matter and its effect upon subsequent processes is discussed in Chap. 22. Here our attention is directed simply to the equalization of flow.

Design of an equalization basin is identical, in principle, to the design of reservoirs in water supply. In both cases the goal is the conversion of a variable supply (or demand) rate to a uniform flow. The mass diagram analysis presented earlier can be applied to problems of this sort to yield the storage volume required for equalization. Figure 20-8 presents the results of such an analysis. Lines A and B are drawn parallel to the average flow line and tangent to the inflow mass diagram. The vertical distance between the lines represents the storage required for equalization. The basin would be empty at about 8 a.m. and full at about 5 p.m. The volume required (750 m^3) represents approximately 4 h flow at the average rate, hence it is comparable in size, as will be seen, to some biological treatment processes.

PROBLEMS

20-1. Design a single grit basin for a peak flow of 4000 m^3/day and a minimum flow of 500 m^3/day. Use a rectangular channel with a proportional-flow weir designed for a horizontal velocity of 250 mm/s. Find the combination of width, depth, and

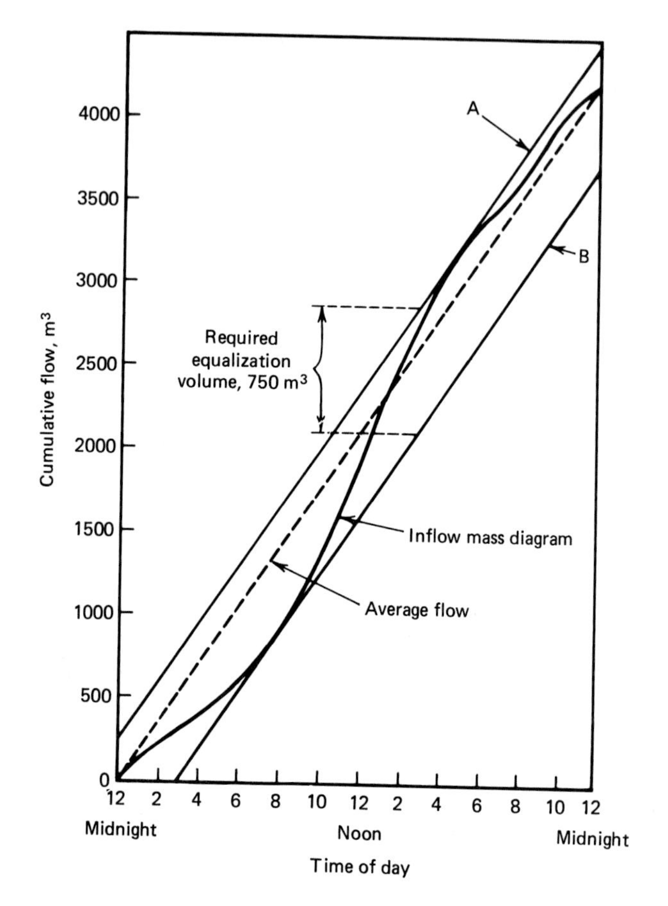

FIGURE 20-8
Equalization basin mass diagram.

length that appears most suitable to you. Why did you pick these particular dimensions? Is it likely that there is an optimum configuration?

20-2. Rework Prob. 20-1 for a parabolic channel. Again, pick those dimensions that you consider most suitable and discuss the reasons for your choice. Is it likely that there is an optimum configuration for this type of basin?

20-3. Rework Prob. 20-1 for an aerated grit chamber. Determine appropriate dimensions and the air flow rate required. Compare the hydraulic detention time of the basins of Probs. 20-1 and 20-2 to those of this design. There is a common perception that aerated grit basins are "small"—presumably in comparison to other designs. Does this perception seem to be justified?

20-4. Rework the Example 20-2, Art. 20-4, using three channels and limiting the maximum width to 2 m. Compare the resulting design to that of the example. Which appears better? Why?

20-5. Estimate the average, maximum, and minimum daily volumes of grit which would be produced in a plant designed for an average wastewater flow of 30,000 m^3/day.

Consider anticipated variations in flow (Chap. 2) as well as variations in concentration.

20-6. Estimate the average, maximum, and minimum quantities of screenings which would be expected in the plant of Prob. 20-5. As in that problem, consider variations in flow as well as in concentration.

REFERENCES

1. *Wastewater Treatment Plant Design*, Manual of Practice No. 36, American Society of Civil Engineers, New York, 1977.
2. Metcalf and Eddy, Inc., *Wastewater Engineering: Treatment, Disposal, Reuse*, 2d. ed., McGraw-Hill, New York, 1979.

CHAPTER 21

PRIMARY TREATMENT SYSTEMS

21-1 Purpose of Primary Treatment

What are now called primary treatment processes were originally designed to remove suspended solids in wastewater prior to its discharge, since these were the most obvious source of pollution. Additional treatment methods developed to remove dissolved and colloidal solids were called *secondary* processes, and, by analogy, the suspended solids removal processes were *primary*. Primary treatment in present-day practice usually involves a simple sedimentation process, although fine screens are sometimes used for the same purpose and chemicals are sometimes added to assist in removal of finely divided or colloidal solids or to precipitate phosphorus.

21-2 Plain Sedimentation

The theory of sedimentation of discrete, flocculent, and hindered suspensions is developed in Chap. 9. Sedimentation processes in wastewater treatment include all three types. Grit removal in gravity basins (Chap. 20) and plain sedimentation approximate discrete settling processes; secondary settling following activated sludge processes is usually flocculent, and waste sludge thickeners exhibit hindered settling, as may the lower levels of other clarification processes.

Primary clarifiers are usually designed to remove particles with settling rates of 0.3 to 0.7 mm/s. Plants operated within that design range normally provide suspended solids removals from 30 to 60 percent, depending in part on the original concentration and the age of the sewage. Figure 21-1 presents reported solids removals in existing sewage treatment plants as a function of their average surface overflow rate. Since the solids are largely organic, their removal also results in a reduction in biological oxygen demand. BOD removal is often considered to vary with surface overflow rate as shown in Fig. 21-2.

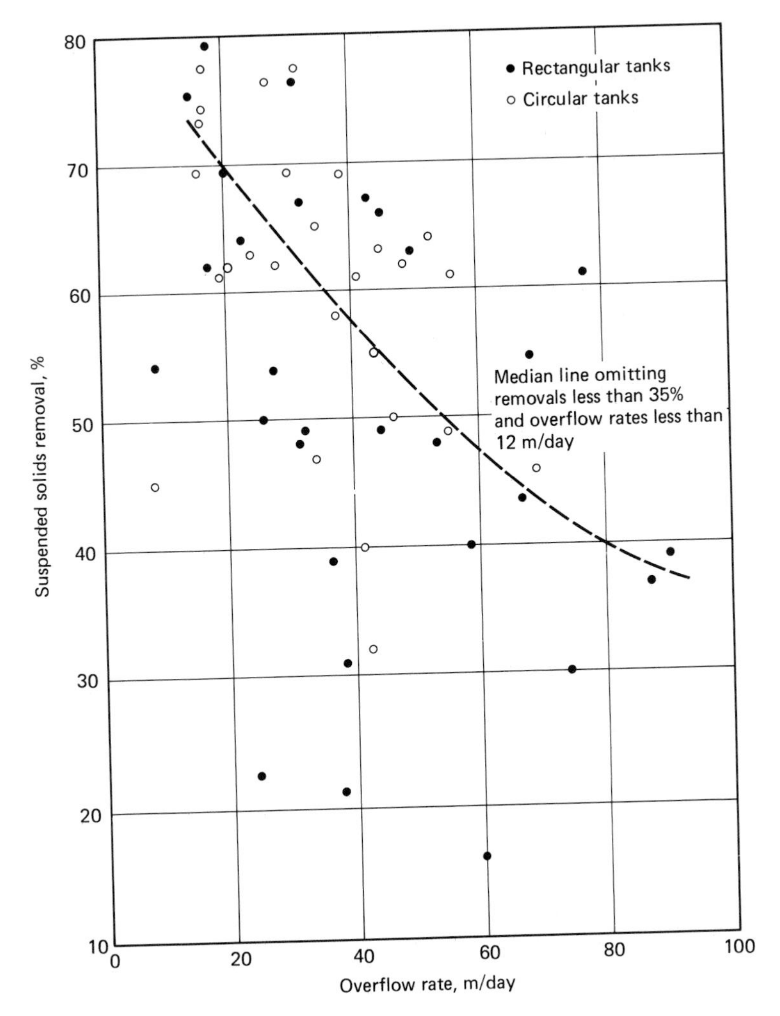

FIGURE 21-1
Suspended solids removal in primary clarifiers. (*From Sewage Treatment Plant Design. Copyright 1959 by the American Society of Civil Engineers and the Water Pollution Control Federation.*[1])

Actual removals vary considerably with the characteristics and age of the waste.

Retention times in primary clarifiers are generally short, from 1 to 2 h at peak flow. Combining this criterion with a surface overflow rate of 0.3 to 0.7 mm/s yields a depth of 1 to 5 m (3 to 16 ft). Practical basins are seldom less than 2 m (6 ft) or more than 5 m (16 ft) in depth. Recent research indicates that great depth in primary clarifiers is not helpful and that shallower basins function best, subject to limitations imposed by the need to install mechanical equipment for sludge removal.[3]

Multiple basins should be provided in all but the smallest plants. Primary

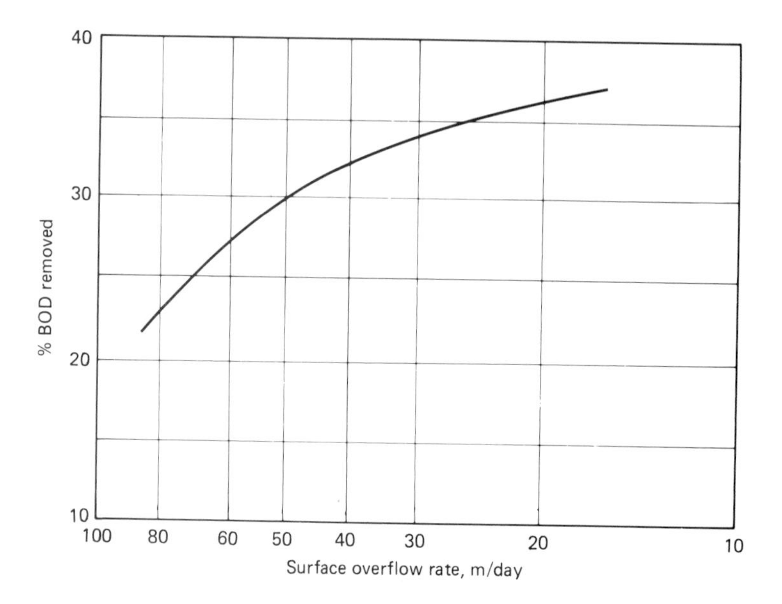

FIGURE 21-2
BOD removal in primary clarifiers. (*Modified from Recommended Standards for Sewage Works, 1971 ed.*[2])

clarifiers may be bypassed to secondary processes, but this is likely to overload the secondary and is certain to produce a poorer overall effluent quality. Hydraulic overload of primary clarifiers designed in the usual range produces only modest losses in efficiency, provided the channels and launders are capable of transporting the flow.

Example 21-1 Design a primary clarification system for a community with an average wastewater flow of 7500 m^3/day, a maximum of 18,000 m^3/day and a minimum of 4000 m^3/day. Use at least two basins and a surface overflow rate of 30 m/day at average flow. Determine the efficiency of BOD removal at all flow conditions with all basins working and with one basin out of service.

Solution
At a surface overflow rate of 30 m/day, the required total area is 7500/30 = 250 m^2. This might be divided into two, three, or more basins. Evaluating a three-basin system, the area with one basin out of service will be 167 m^2. The calculations are

Flow, m^3/day	No. of basins	Area, m^2	SOR	Efficiency, %
18,000	3	250	72	25
18,000	2	167	108	< 20
7,500	3	250	30	34
7,500	2	167	45	31
4,000	3	250	16	37
4,000	2	167	24	35

The efficiencies are obtained from Fig. 21-2. It may be observed that, at any flow, the loss of one-third the capacity produces a reduction in efficiency of 6 percent or less. In a larger plant, which would contain more basins and would exhibit less difference between minimum and maximum flow, the variation would be less.

The depth of the basins would be dictated by the criterion of a 1-hr detention time at peak flow. For two basins in service, the required volume is 18,000/24 = 750 m^3 and the required depth is 750/167 = 4.5 m.

Primary clarifiers are constructed in rectangular, square, and circular configurations. The choice of tank shape is generally based on site and economic considerations.

Rectangular clarifiers permit the use of common-wall construction in plants containing multiple basins, offer a reduced likelihood of short circuiting, and allow optimum use of land area. Skimming and sludge removal systems are somewhat more complicated than those in circular basins. Some engineers believe rectangular basins to be more efficient than circular designs, but data such as that of Fig. 21-1 do not support the theory.

Inlets in rectangular clarifiers are placed at one end and are intended to minimize the effect of density and velocity currents. Typical designs consist of small pipe with upturned ells, perforated baffles, multiple ports discharging against baffles, single pipes turned back to discharge against the headwall, simple weirs, submerged weirs sloping upward to a horizontal baffle, etc. The inlet structure should be designed so it does not trap either floating or settleable solids. This may require careful consideration of velocities under all anticipated flow conditions and provision of special features for scum removal.

Baffles are provided 600 to 900 mm (2 to 3 ft) in front of the inlet and are submerged 450 to 600 mm (18 to 24 in) with 50 mm (2 in) water depth above the baffle so that floating solids are not trapped. *Scum baffles* are installed ahead of the outlet structure and extend 150 to 300 mm (6 to 12 in) below the water surface. Rotating slotted scum pipes are provided in some designs in an attempt to incorporate a scum baffle and a scum removal channel in a single structure. These devices are not the best technique for scum removal and are completely unsatisfactory in the absence of positive mechanical skimming.

Outlets in rectangular basins consist of weirs located toward the discharge end of the basin. *Weir overflow rates* range from 120 to 370 m^3/day per meter of weir length [10,000 to 30,000 gal/(ft · day)]. The upper value is used at peak flow conditions. For the flow and clarifiers of the example above, the required weir length would be 18,000/370 = 49 m. This weir length would have to be developed in two basins, thus the length per basin would be 24.5 m. A clarifier with an area of 83 m^2 would have a width considerably less than this. If the length-to-width ratio were 3:1, for example, the width would be 5.3 m. To provide a length of 24.5 m, multiple launders would be required.

Optimum tank proportions have not been established, but existing plants typically have length-to-width ratios in the range of 4:1 with minimum depths of 2 m (6 ft). The average depth of existing tanks is about 3.5 m (12 ft). The floor is usually sloped gently (about 1:24 or less) toward the sludge hopper to facilitate draining the basin. Scum and sludge removal equipment may be chain-driven or supported on bridges or floats. The velocity of the mechanism is usually less than 1 m/min (3 ft/min). The sludge is moved along the bottom of the basin to a sump from which it may be pumped or be drawn by gravity. The floating material is moved along the surface by a blade which extends both above and below the water level. At the end of the basin the skimmer pushes the scum into some sort of collector. The best systems include a floating beach, which consists of a ramp extending below the water surface and rising above it to form the edge of a collection trough. The skimming blade rides up on the beach, pushing the floating solids up the ramp and into the trough from which they are then removed by augers, pumps, or gravity flow. Positive systems are very desirable for moving both scum and settled solids. Relying on a steeply sloped bottom to convey the solids or on the bulk flow of the water to move the scum invites operational problems.

Circular and square basins are commonly used in small plants which contain only a few units and in expansions of existing plants where common-wall construction techniques may not be applicable. Circular tanks have a minimum perimeter for a given area and thus may be more economical in such circumstances. Additionally, in small diameters they carry a substantial portion of the hydrostatic load in hoop tension, which may permit somewhat lighter construction.

Inlets in circular or square tanks are usually at the center, and the flow is directed radially outward through a baffle which has a diameter of 10 to 20 percent that of the basin and which extends 1 to 2 m (3 to 6 ft) below the water surface. There is some tendency for the incoming flow to short-circuit and reach the outlet without passing through the total volume of the basin, particularly the lower portion. *Peripheral flow* designs (Chap. 9) are less likely to have such problems, since the influent enters around the entire periphery, is directed downward, flows toward the center, and then returns through the upper portion of the basin to peripheral weirs.

Outlet weirs typically extend around the periphery of the tank with baffles extending 200 to 300 mm (8 to 12 in) below the water surface. The weir length may be increased by supporting the launder on brackets some distance from the wall, thus providing a weir on each side of the channel. Serpentine weirs may also be used to increase the length. The design shown in Fig. 21-3 provides an effective length 2.5 times that of an ordinary weir. The weir pans shown extend 1.2 m (4 ft) from the effluent trough to which they are connected. Radial weirs should not be used in circular sewage clarifiers, since they make mechanical skimming impossible. It is normally possible to provide adequate weir length in circular clarifiers without radial weirs.

Example 21-2 Calculate the maximum-radius clarifier for which a single peripheral weir surface will provide adequate length. Assume SOR = 30 m/day and WOR = 370 m^2/day.

FIGURE 21-3
Serpentine weir pan. (*Courtesy Leopold Co., Division of Sybron Corporation*).

Solution
The area required is $Q/30$, and the weir length (perimeter) is $Q/370$. Thus,

$$\pi r^2 = \frac{Q}{30}$$

and

$$2\pi r = \frac{Q}{370}$$

whence

$$r = 24.6 \text{ m } (81 \text{ ft})$$

For larger clarifiers, additional length would need to be developed by using a serpentine weir or a launder supported on brackets. Few clarifiers exceed 50 m in diameter.

Tank proportions are dictated by flow. Side depths range from 2 to 4 m (6 to 13 ft). Floors slope toward the center on grades of 1:12 to 1:24, with the lower slopes in larger tanks. Slopes may have to be selected to match the sludge-removal device which is to be installed; these, however, are usually made to order so the designer should be able to specify any slope that is desired. The sludge-removal devices rotate at peripheral speeds of 1.5 to 2.5 m/min (5 to 8 ft/min). Scum is swept outward to a small floating beach with a hopper, while sludge is windrowed to a central pit. The accumulated solids are removed from these collection points by equipment similar to that used in rectangular basins.

Large circular clarifiers which use rotating scum collection mechanisms may be ineffective at times because of movement of the scum layer by wind. If the floating beach is on the upwind side of the basin, the accumulated floating material may be pushed away each time the skimmer approaches the beach. Some rotating skimming mechanisms have multiple suction ports along the sweep mechanism. Such designs are preferable in large-diameter basins.

21-3 Chemical Coagulation

Addition of the common metallic coagulants and polymeric coagulants discussed in Chap. 9 will increase the removal of suspended solids in primary clarifiers. If the dosage of metallic coagulants is sufficiently large (Chap. 24) significant quantities of phosphorus may also be precipitated.

The justification for chemical addition is based upon special conditions. Historically, it has been used when seasonal loads required special treatment to avoid plant expansion, when room for expansion was lacking, or when a degree of treatment intermediate between primary and secondary was desired. The last condition is no longer likely in the United States, since the EPA, in general, requires that all municipal plants provide the equivalent of secondary treatment.

The principles of chemical coagulation are developed in Chap. 9. When this technique is applied to wastewater in modern practice, the major goal is usually phosphate precipitation. Suspended solids removals in such processes range from 70 to 90 percent.[4] Similar solids removals were obtained in the past with somewhat lower chemical dosages (100 mg/L or less).[1] BOD removals can be as high as 70 percent on fresh sewage, but decrease as the sewage ages and organic material is solubilized by microbial action. The same basic design criteria govern chemical coagulation processes in both water and wastewater treatment. The chemicals must be flash-mixed with the flow, flocculation is required, and surface overflow rates are similar to those in primary sedimentation.

21-4 Fine Screens

Fine screens are available in the form of rotating drums (Fig. 21-4) and as fixed surfaces (Fig. 21-5). The manner of operation of these devices is somewhat different, since in the rotating screens a portion of the treated water is used to clean the surface of accumulated solids, while the static screen depends pri-

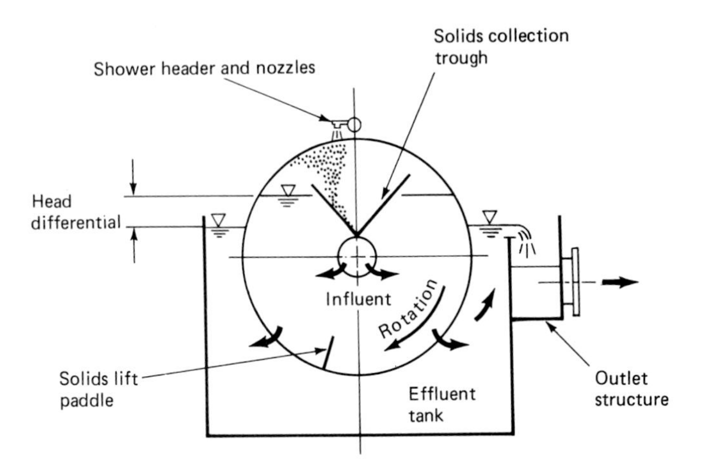

FIGURE 21-4
Rotating fine screen.

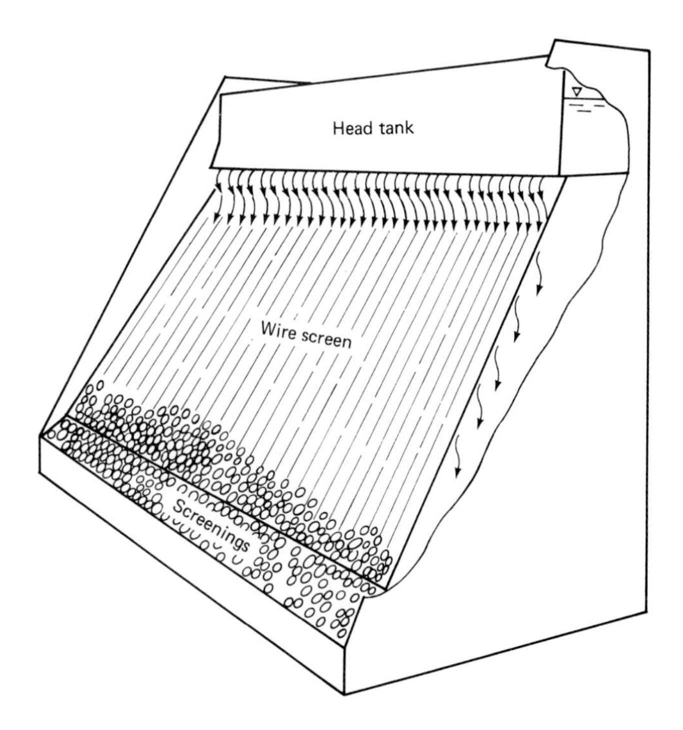

FIGURE 21-5
Static fine screen.

marily on gravity and the incoming flow to carry the screenings to the recovery area.

Hydraulic loading rates on sewage screens are in the range of 0.4 m/min (1.3 ft/min), hence screens can be far smaller than primary clarifiers. The screen openings range from as little as 0.8 mm to about 2.3 mm (0.03 to 0.10 in). Suspended solids removals on the finer screens sometimes approach those obtained in primary clarifiers, but pilot tests should be conducted before depending on these devices for such efficiency. Fine screens are sometimes blinded by grease and finely divided solids, which creates substantial operational problems. Passage of the sewage through the screen provides a degree of aeration, which may be beneficial.[5]

Screens can serve the purpose of removing fairly large solids which might create nuisance conditions in biological processes, but it is unwise to depend on them for high removals of solids or BOD or for removal of grease. The selection of screens in place of primary clarifiers will reduce the cost of that phase of treatment, but may create additional costs for downstream processes.

21-5 Estimation of Solids Quantity

The amount of sludge produced in a primary clarifier is a direct function of its efficiency. Figure 21-1 permits estimation of suspended solids removal as a function of surface overflow rate. In general, removals of 50 to 60 percent can be expected for ordinary domestic sewage. Thus a wastewater containing 250 mg/L of suspended solids and having a flow of 18,000 m^3/day might be expected to produce 2250 kg of solids per day. The moisture content of primary solids ranges from 93 to 98 percent, with an average value of 95 percent. The total mass of wet solids would therefore be 45,000 kg/day or about 2.5 kg/m^3 of flow (0.02 lb/gal).

Coagulation sludges naturally contain more solids because of the precipitation of metallic hydroxides and the increased efficiency of the process. The sludge *volume*, however, may not be increased and may even be less because of the inclusion of less water in the mass. Moisture content of chemically precipitated sludges may be as low as 90 percent.

If waste biological solids are recycled through the primary clarifiers prior to being wasted, the mass of solids removed will be increased and the moisture content will be altered as well. Waste trickling filter solids may yield a denser sludge, while waste activated sludge will increase the moisture content.

Scum removed averages about 7.5×10^{-6} m^3 per cubic meter of flow. It is important that the scum removal facilities be capable of handling peak loads of up to 4.5×10^{-5} m^3/m^3 for brief periods.[4]

Both the primary solids and the scum have a tendency to plug pipelines, and positively acting conveyance equipment should be used. Sludge, under a positive head of several meters, will flow through pipelines, and such heads are typically available at the bottom of primary clarifiers. Scum, on the other hand, is usually deposited in rather shallow sumps and its viscosity may prevent it from flowing through the gravity drainage structures often provided. Water jets are sometimes added to thin the scum so it will flow, but this is counterproductive in the sense that it remixes the scum with water from which it was just separated.

21-6 Other Primary Processes

A number of manufacturers provide proprietary equipment which incorporates a flocculation or gentle-mixing zone prior to sedimentation. The mixing may be effected mechanically or by aeration, and the settled solids may be returned to the mixing zone to increase contact opportunity. Suspended solids and BOD removals are reportedly increased by about 10 percent over that obtainable in plain sedimentation. These systems are not widely used.

Imhoff tanks incorporate solids separation and digestion (Chap. 23) in a single unit. Modern versions of these processes employ mixing and heating in the digestion zone, but the advantage of these units over separate sedimentation and digestion systems is questionable. Imhoff tanks were widely used in the past, particularly in small plants designed to provide only primary treatment. Old Imhoff tanks have been converted to other uses in recent years. Design criteria are presented in Ref. 1, but the use of this process is not recommended.

PROBLEMS

21-1. A raw wastewater containing 250 mg/L suspended solids and 200 mg/L BOD passes through a clarifier with a surface area of 500 m^2. The flow ranges from 10,000 to 30,000 m^3/day. Estimate the effluent BOD and suspended solids at maximum and minimum flow and the maximum and minimum rates of sludge production. Assume the sludge is 95 percent water.

21-2. Design a primary clarifier system for a community with a design population of 50,000 persons. Estimate the average, maximum, and minimum flows and assume that each person contributes 85 g/day of both BOD and suspended solids regardless of the flow. Use at least four basins with a minimum detention time of 1 h and a weir overflow rate of 370 m^2/day at peak flow with one basin out of service. Estimate the effluent quality and the quantity of sludge produced under different conditions as in Example 21-1, Sec. 21-2. How much does the quantity of sludge vary? Is the manner of variation surprising?

21-3. Using the sludge production figures from Prob. 21-2, determine the maximum and minimum quantities of sludge to be expected per day *per clarifier*. The values will vary with the number of clarifiers selected in Prob. 21-2. Using these figures determine the required flow capacity of a sludge pump for each clarifier. The pump is to run, at most, 15 min/h and must maintain a velocity of at least 0.75 m/s (2.5 ft/s) in the sludge pipe, which cannot be less than 150 mm (6 in) in diameter.

21-4. Compare the quantities (both mass and volume) of sludge to be expected in simple sedimentation and chemical coagulation of a sewage flow of 50,000 m^3/day which contains 225 mg/L suspended solids. The sedimentation process has an efficiency of 55 percent and the chemical process 80 percent. In addition, the chemical process produces 150 mg/L of metallic hydroxides which are also in the sludge. The moisture content from the sedimentation process is 95 percent, while that from the chemical process is 90 percent.

REFERENCES

1. *Sewage Treatment Plant Design*, Manual of Practice 36, American Society of Civil Engineers, New York, 1959.
2. Great Lakes—Upper Mississippi River Board of State Sanitary Engineers, *Recommended Standards for Sewage Works*, Health Education Service, Albany, NY, 1971.
3. David W. Ostendorf and Bradley C. Botkin, "Sediment Diffusion in Primary Shallow Rectangular Basins," *Journal of Environmental Engineering Division, American Society of Civil Engineers*, **113**:597, 1988.
4. *Process Design Manual for Suspended Solids Removal*, U.S. Environmental Protection Agency, 1975.
5. *Wastewater Treatment Plant Design*, Manual of Practice 36, American Society of Civil Engineers, New York, 1977.

CHAPTER 22

SECONDARY TREATMENT SYSTEMS

22-1 Purpose of Secondary Treatment

Secondary treatment systems are intended to remove the soluble and colloidal organic matter which remains after primary treatment. While removal of this material can be effected by physicochemical means, secondary treatment is usually understood to imply a *biological process*.

It is noted in Chaps. 18 and 19 that wastewater, in addition to containing organic matter, also carries a large number of microorganisms which are able to stabilize the waste in a natural purification process. Biological treatment consists of application of a controlled natural process in which microorganisms remove soluble and colloidal organic material from the waste and are, in turn, removed themselves.

In order to carry out this natural process in a reasonable time, it is necessary that a very large number of microorganisms be available in a relatively small container. Biological treatment systems are designed to maintain a large active mass of bacteria within the system confines. While the basic principles remain the same in all biological processes, the techniques used in their application may vary widely. A useful classification divides these systems into *attached (film) growth* or *suspended growth processes*, although there are techniques in which both types are incorporated.

22-2 Attached Growth Biological Processes

Attached growth processes utilize a solid medium on which bacterial solids are accumulated in order to maintain a high population. The area available for such growth is an important design parameter, and a number of processes have been developed which attempt to maximize area as well as other rate-limiting factors. Surface growth processes include intermittent sand filters, trickling filters, rotating biological contactors, fluidized beds, and a variety of similar systems.

Intermittent sand filters are no longer used by large communities because of the extensive area required. They may, however, still have applications in rural areas. The advantages of intermittent sand filtration include low head loss, simple operation, a satisfactory effluent, and limited sludge production.

Operation consists of intermittent application of wastewater (normally following primary clarification) to the sand surface. Solids are trapped in the sand while bacterial growth developed on the surface of the grains adsorbs the soluble and colloidal organic matter. Between dosing cycles, air penetrates the bed to permit biological oxidation of the bulk of the accumulated organics.

The sand usually used has an effective size of 0.2 to 0.5 mm and a uniformity coefficient of 2 to 5, although unsieved sand is sometimes used.[1] The depth of the bed ranges from 460 to 760 mm (18 to 30 in), with deeper beds yielding a somewhat better effluent. The sand is underlain by approximately 300 mm (12 in) of graded gravel ranging from 6 to 50 mm ($\frac{1}{4}$ to 2 in) in which perforated pipe or unjointed drain tile are placed in order to collect the treated waste. Typical filter bed details are shown in Figs. 22-1 and 22-2.

Hydraulic loading rates for settled sewage range from 70 to 235 mm/day [2 to 6 gal/(ft^2 · day)] with the flow applied once or twice a day. BOD reduction may reach 95 percent in treating ordinary domestic sewage on ripened filters,

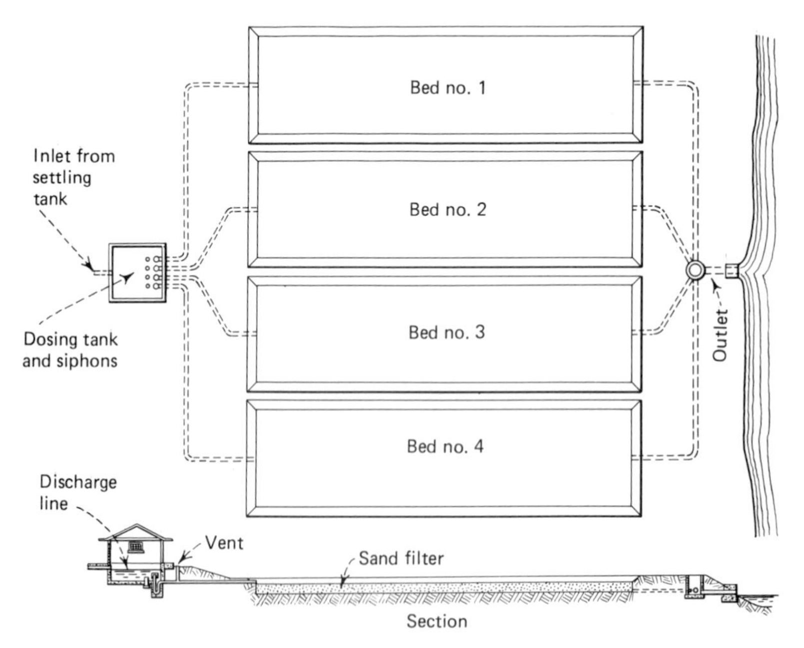

FIGURE 22-1
Plan view of intermittent sand filter plant.

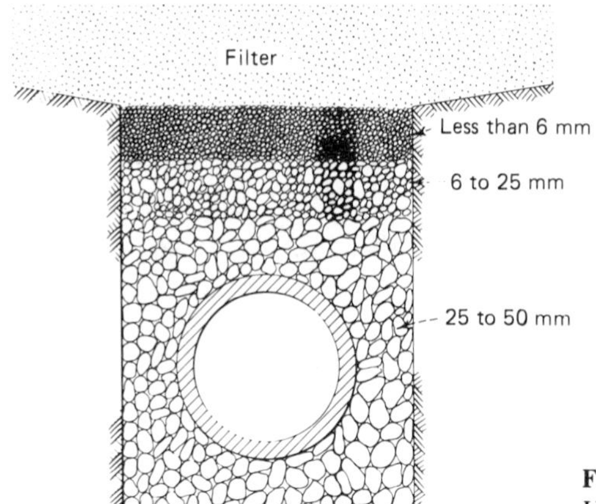

FIGURE 22-2
Underdrain section of intermittent sand filter.

but is lower when operation is first begun. Eventually the filter surface will become clogged with accumulated solids to the point that the hydraulic loading can no longer be maintained. When this occurs, the upper 50 to 75 mm (2 to 3 in) of the sand is removed and replaced with clean material. The dirty sand may be disposed of by land spreading or filling. It cannot be cleaned economically. The duration of filter runs is typically several months, but depends on temperature, mode of operation, sand gradation, and influent BOD and suspended solids.

Trickling filters utilize a relatively porous bacterial support medium such as rock or formed plastic shapes. Bacterial growth occurs on the surface while oxygen is provided by diffusion through the void spaces. The wastewater is applied to the filter in either an intermittent or continuous fashion and percolates through the medium, flowing over the bacterial growth in a thin film.

The process can be represented as shown in Fig. 22-3. Nutrients and oxygen are transferred to the fixed water layer, and waste products are transferred to the moving layer, primarily by diffusion. As the bacteria on the filter surface metabolize the waste and reproduce, they will gradually cause an increase in the depth of the slime layer. With thickening of the biological layer, the bacteria in the interior layers find themselves in a nutrient-limited situation, since the organic matter and oxygen are utilized near the surface. Eventually these interior cells die and lyse, breaking the contact between the slime layer and the support medium. When sufficient cells have lysed, the slime layer will slough off and be carried from the filter by the waste flow. The solids in the filter effluent are removed from the flow in a secondary clarifier.

Although trickling filters have been used for a great many years, their operation is still not readily described mathematically. The process rate is affected by mass transfer of oxygen to the liquid from the air and from the liquid to the bacterial slime; by transfer of biodegradable material from the liquid to the slime; and by the rate of utilization of the organic matter by the bacteria.[2]

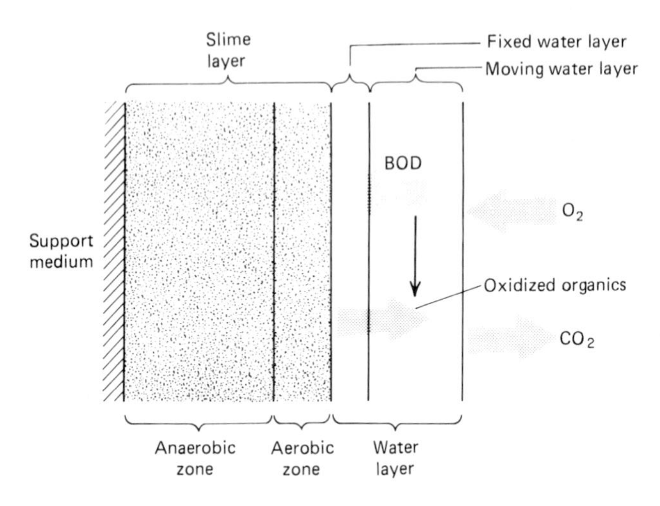

FIGURE 22-3
Schematic diagram of attached growth process.

Mathematical models have been suggested which are empirical[3,4] or based on a postulated reaction,[5,6] a simplified analogy,[7] or more complicated hypotheses.[2,8,9] The theoretical models are generally impractical for design purposes, although they may aid considerably in understanding the interplay of process variables.[10,11,12] In particular, it may be concluded that the Velz and NRC equations (see below), which are based on an implicit assumption that oxygen is the rate-limiting factor,[13] are not invalid on that account provided the waste is not excessively dilute.[11,13] It has been concluded that oxygen is, in fact, the limiting variable for wastes with a soluble BOD greater than 40 mg/L[11]—which includes most wastes of interest.

Trickling filters are described by their hydraulic or organic loading rate and by the medium provided to support the bacterial growth. The terms *low rate* or *conventional* refer to trickling filters in which the wastewater passes through the filter only once. The organic loading rate on these units ranges from 0.3 to 1.5 kg BOD per day per cubic meter of bulk filter volume and the hydraulic loading rate from about 2 to 4 m/day [45 to 90 gal/($ft^2 \cdot$ day)] based on the plan area. Either the hydraulic or organic loading rate may govern, depending on the characteristics of the waste. Low-rate filters were widely used in the past and may still be encountered in older plants. A common feature of these systems is a *dosing chamber*, which is used to correct problems of poor distribution and inadequate velocity associated with low flows. These chambers hold the incoming flow until a fairly large volume has been accumulated. The contents are then released quickly, providing a relatively high flow rate, adequate to turn rotary distributors or to provide equal flow to fixed nozzles.

High-rate trickling filters employ recirculation of wastewater around the filter and are operated at higher organic loading rates as well. The recirculation provides a constant minimum flow, hence dosing chambers are not needed. The organic load on high-rate filters ranges from 1.5 to 18.7 kg BOD per day per cubic meter of bulk volume. The hydraulic loading rate, including recirculation, ranges from 10 to 30 m/day [230 to 690 gal/($ft^2 \cdot$ day)] based on the plan area. Some newer designs use even higher hydraulic loading rates, sometimes exceeding 90 m/day, particularly on some types of plastic media.

Roughing filters may be employed as a pretreatment device to reduce the strength of particularly concentrated wastewaters. Loadings exceed those on ordinary high-rate filters and have no arbitrary upper limit. The effluent of such a process requires further treatment.

Standard *rock* filters utilize crushed stone, slag, and gravel. The rock ranges from 60 to 90 mm ($2\frac{1}{4}$ to $3\frac{1}{2}$ in) and should be composed of traprock, granite, or slag and be free of sand and clay.[14]

Plastic media are made of either interlocking sheets or molded or extruded shapes. The former are manufactured in a variety of configurations, some permitting the water to simply run over vertical surfaces, others creating a zigzag path which provides a longer detention time and a longer flow path within the same volume. The latter medium, called *crossflow*, has been reported to be more effective than others by some investigators,[15] but this may not be true at all loading conditions. Vertical media have been reported to provide better results at high loadings.[16]

Media made of formed sheets are carefully placed within the basin to ensure uniform flow (Fig. 22-4). Molded or extruded shapes are placed in the

FIGURE 22-4
Sheet-plastic trickling filter. (*Courtesy Imperial Chemical Industries, Ltd.*)

FIGURE 22-5
Random dumped plastic trickling filter medium. (*Courtesy Koch Engineering Co. Inc.*)

same manner as rock and are commonly called *random* media (Fig. 22-5). Plastic media are often cheaper than rock in areas where suitable rock does not occur naturally, are significantly lighter, and appear to provide somewhat higher efficiency, particularly at high loadings.[16]

Wood, particularly redwood, has been used as a substitute for rock or plastic media in trickling filters. Small slats, approximately 6 by 25 mm ($\frac{1}{4}$ by 1 in), are fabricated into flats which are stacked one upon another to fill the basin. Such media are sometimes called *horizontal* and are incorporated in some proprietary processes which combine both suspended and fixed growth treatment systems. Considered alone, horizontal media are reportedly the least efficient of all those considered here.[16]

The important characteristics of filter media are considered to be the *specific surface* (the area per unit volume) and the percent *void space*. The surface area, of course, is directly related to the available active biological population, while the void space is important in conveying both the waste and the oxygen required for its stabilization. Typical characteristics of various media are presented in Table 22-1.

Recirculation is generally practiced in modern trickling filter plants since it provides a more uniform hydraulic and organic load, increases the mass of biological solids in the system, continuously reseeds the filter with sloughed bacteria, dilutes the influent with better-quality water, and thins the biological slime layer. It should be noted that recirculation may not increase efficiency under all circumstances,[13,17] particularly with relatively dilute wastes. This is

TABLE 22-1
Physical properties of trickling filter media

Medium	Size, mm	Unit weight, kg/m³	Specific surface, m³/m³	Void space, %
Plastic sheet	600 × 600 × 1200	32–96	82–115	94–97
Redwood	1200 × 1200 × 500	165	46	76
Granite	25–75	1440	62	46
Granite	100	1440	47	60
Slag	50–75	1090	67	49

usually not reflected in empirical or simplified formulas for trickling filter performance.

Techniques of recirculation vary widely, with at least 14 different configurations in use (Fig. 22-6).[18] The procedure used reportedly has no effect on the efficiency of the process, hence the choice of configuration should be based on other considerations. Flow patterns which introduce additional flow into either the primary or secondary clarifiers may require that those units be larger, hence patterns such as those in Fig. 22-6*c*, *d*, *k*, and *m* may be preferable. Recirculation *rates* range from 50 to 1000 percent of the wastewater flow, with usual rates being 50 to 300 percent.

Rotating biological contactors (RBCs) consist of basins containing large plastic disks mounted on rotating shafts similar to those shown in Fig. 22-7. The wastewater passes through the basin as the disks slowly rotate, exposing the biological growth which develops on their surface alternately to the wastewater and to the oxygen in the air. To the biological film, the process is substantially the same as that in a trickling filter with a rotating distributor.

The disks are 3 to 4 m (10 to 12 ft) in diameter and 10 mm (0.4 in) thick and are placed 30 to 40 mm on centers (1 to 1.5 in) along a shaft of variable length. The shaft rotates at 1 to 2 r/min and may be driven by electric motors, by diffused air, or by a combination of the two. This process, developed in Europe, has been in use in the United States since the late 1960s. A number of treatment plants have reported problems associated with shaft imbalance and/or overload which have caused failures of mechanical equipment and resulting failure of the process. The difficulties are reportedly worse at low temperatures.[19] Additional difficulties have been reported with maintenance of aerobic conditions in the basins, particularly at the head of the plant. These problems are lessened, but not always eliminated, when air drive is used. Design procedures, which are discussed below, are still largely empirical, as are those for trickling filters.

Fluidized beds take advantage of the very high ratio of surface area to volume provided by small media such as sand. Utilizing fine media presents a potential problem in that the small openings may plug, as in intermittent sand filters. This problem can be avoided by directing the flow upward through the

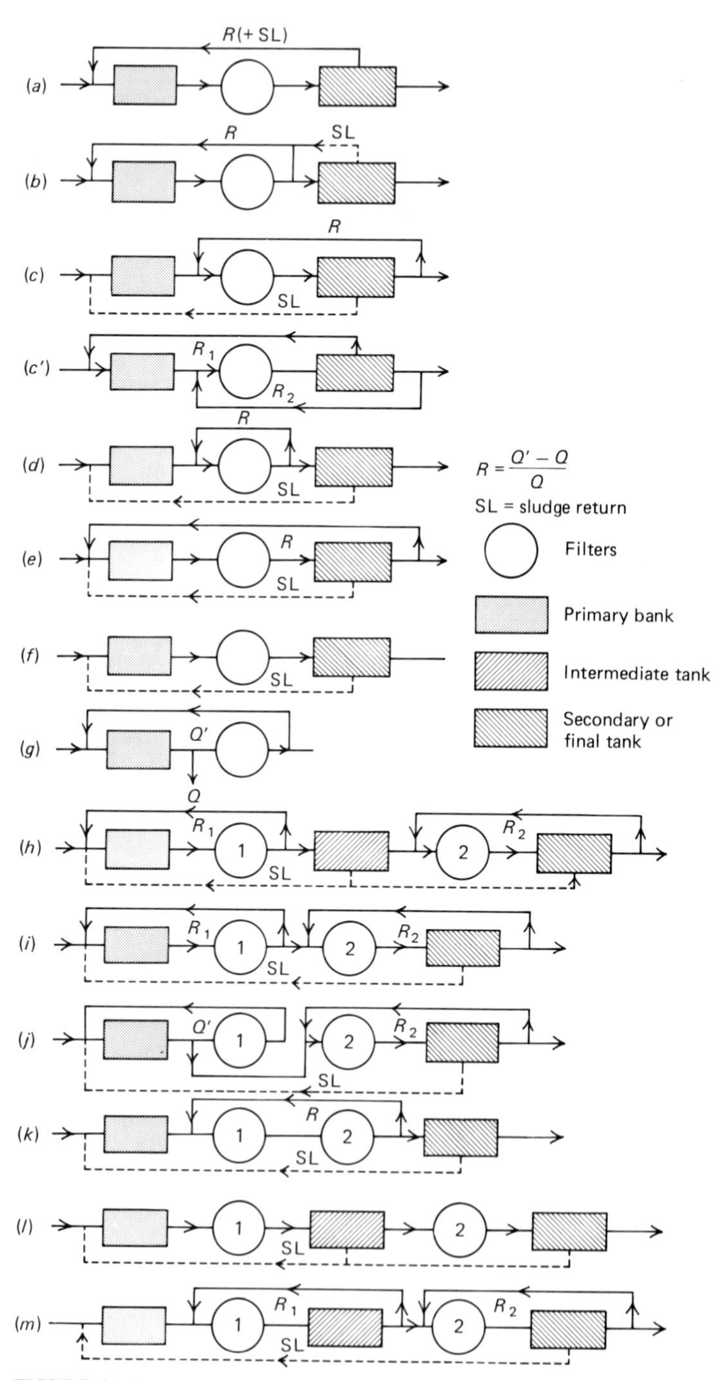

FIGURE 22-6
Flow diagrams of one- and two-stage trickling filter plants. (*From Wastewater Treatment Plant Design,*[18] *Copyright 1977 by the American Society of Civil Engineers and the Water Pollution Control Federation. Used with Permission of the Water Pollution Control Federation.*)

FIGURE 22-7
Rotating-disk assemblies. (*Courtesy Biosystems Divison, Autotrol Corporation.*)

bed at a velocity sufficient to cause modest expansion (fluidization). The required upward velocity depends on the density and size of the medium and will change somewhat as biological growth on the surface alters both the effective diameter and density of the particles. At a constant velocity, a coated bed will expand further than a clean one, but it is possible to maintain the expansion within reasonable limits under most circumstances. The very high surface area permits maintenance of higher concentrations of active biological solids than can be achieved in other processes, hence fluidized beds are potentially smaller than other systems. The high treatment rates theoretically possible may not be achieved because of limitations on the ability to transfer oxygen; nevertheless, this process offers some definite advantages over other biological systems.

22-3 Design of Trickling Filters

As noted above, theoretical formulas for the performance of trickling filters are not useful for the design of practical systems, although they are helpful in describing the interplay of process variables. Design is presently based on empirical and semiempirical formulas which give satisfactory results, provided they are not applied in situations different than those from which they were developed. All the empirical formulas discussed here include the effect of secondary clarifiers, which must be designed as described below.

The *Velz formula*[5] takes the form

$$C_e = \left(\frac{C_i + rC_e}{1 + r}\right)e^{-KD} \tag{22-1}$$

where C_e = effluent BOD
C_i = influent BOD
r = ratio of recirculated flow to waste flow
D = depth of filter
K = experimental constant (0.49 for high-rate filters to 0.57 for low-rate)

The original equation presented above was considered to be valid for BOD removals of 90 percent or less. The Velz equation has been modified to include the effects of factors such as specific surface area, temperature, and flow rate. A variety of *modified* Velz equations have been suggested. Perhaps the most complex is that proposed by Eckenfelder:[6]

$$C_e = \left(\frac{C_i + rC_e}{1 + r}\right)e^{-K[A/(1+r)Q]^n D a_v^{1+m}} \tag{22-2}$$

where A is the filter plan area, a_v is the specific surface, and K, n, and m are experimental coefficients which vary with the medium, quantity of biological growth, and, to a degree, flow. A somewhat simpler form,

$$C_e = \left(\frac{C_i + rC_e}{1 + r}\right)e^{-K(1.035)^{(T-20)}D/(Q/A)^n} \tag{22-3}$$

is more commonly used. This equation is applicable to a single-stage filter. For the second stage of a two-stage system, the equation is modified to

$$C'_e = \left(\frac{C_e + rC'_e}{1 + r}\right)e^{-K(1.035)^{(T-20)}D(C_e/C_i)/(Q/A)^n} \tag{22-4}$$

in which C'_e is the effluent BOD from the second stage and the other terms are as given above. The values of K and n in these expressions vary somewhat with operating conditions and the specific medium used. Values reported in the literature may be based on different systems of units, hence care must be exercised in applying the equation. With filter area in m^2, flow in m^3/min, and depth in m, typical values of K and n for domestic sewage are 0.02 and 0.5, respectively.

Example 22-1 Calculate the effluent BOD of a two-stage trickling filter. Each stage has an area of 430 m^2, a depth of 2 m, and a recirculation rate of 125 percent of the flow. The flow is 3.15 m^3/min and the influent BOD is 170 mg/L following primary treatment. Also find the effluent BOD if there were only one filter of twice the area.

Solution
For the first stage, assuming $T = 20°C$,

$$C_e = \left(\frac{170 + 1.25C_e}{2.25}\right)e^{-0.02(2)/(3.15/415)^{0.5}}$$

$$= 73 \text{ mg/L}$$

For the second stage,

$$C'_e = \left(\frac{73 + 1.25C'_e}{2.25}\right)e^{-0.02(2)(0.429)/(3.15/415)^{0.5}}$$

$$= 49 \text{ mg/L}$$

For the single filter,

$$C_e = \left(\frac{170 + 1.25C_e}{2.25}\right)e^{-0.02(2)/(3.15/830)^{0.5}}$$

$$= 56 \text{ mg/L}$$

The improved efficiency provided by a two-stage system over a single-stage process of equal volume is typical of biological processes. It can be shown that the optimum configuration consists of an infinite number of infinitely small stages, which is not a practical alternative in trickling filter systems. Few trickling filter plants contain more than two stages.

The *NRC formula*[3] is based on data collected at military bases within the United States during World War II. It is useful in circumstances which approximate those under which the data were collected. Since most of the bases were in the south, the NRC formula predicts efficiencies which are higher than those achievable in northern areas. For a single-stage filter or the first stage of a two-stage system, the NRC equation is

$$\frac{C_i - C_e}{C_i} = \frac{1}{1 + 0.532(QC_i/VF)^{0.5}} \tag{22-5}$$

in which V is the filter volume in m^3, F is a recirculation factor given by

$$F = \frac{1 + r}{(1 + 0.1r)^2} \tag{22-6}$$

where r is the recirculation ratio, and the other terms are as defined above. For the second stage the equation is

$$\frac{C_e - C'_e}{C_e} = \frac{1}{1 + \dfrac{0.532}{1 - [(C_i - C_e)/C_i]}(QC_e/V'F')^{0.5}} \tag{22-7}$$

in which V' and F' are the volume and recirculation factor for the second stage and the other terms are as defined earlier.

Example 22-2 Recalculate the effluent quality of the system of the example above using the NRC equations.

Solution
For the first stage

$$F = \frac{1 + 1.25}{(1 + 0.125)^2} = 1.78$$

$$\frac{170 - C_e}{170} = \frac{1}{1 + 0.532\left(\frac{3.15 \times 170}{830 \times 1.78}\right)^{0.5}}$$

$$C'_e = 41 \text{ mg/L}$$

For the second stage r (and therefore F) is the same:

$$\frac{41 - C'_e}{41} = \frac{1}{1 + \frac{0.532}{1 - 0.757}\left(\frac{3.15 \times 41}{830 \times 1.78}\right)^{0.5}}$$

$$C'_e = 16 \text{ mg/L}$$

For the single filter,

$$\frac{170 - C_e}{170} = \frac{1}{1 + 0.532\left(\frac{3.15 \times 170}{1660 \times 1.78}\right)^{0.5}}$$

$$C_e = 31 \text{ mg/L}$$

As in the earlier example, the two-stage system is more efficient than a single unit.

Rankin[4] has developed a number of empirical formulas (based on the Ten States' Standards[20]) which are presented in detail in Ref. 18 and are summarized in Table 22-2. The loading criteria referred to in the table are

I. High-rate processes with dosing rates between 9.4 and 280 m/d and organic loads, including recirculation, less than 1.75 kg/(m^3 · day)

II. High-rate processes with loadings, including recirculation, of between 1.75 and 2.75 kg/(m^3 · day). For loadings in excess of 2.75 kg/(m^3 · day), the Ten States' Standards assume removal of 1.75 kg/(m^3 · day).

III. First-stage units without intermediate clarification prior to the second stage.

TABLE 22-2
Formulas for effluent BOD of trickling filter processes

Flow diagram (Fig. 22-6)	Loading criterion		
	I	II	III
a, b, c, e, and *f*	$C_e = (C_i - C_e) \times \left(\frac{1 + r}{1 + 1.5r}\right)$	$C_e = (C_i - C_e) \times \left[\frac{1.78(1 + r)}{2.78 + 1.78r}\right]$	$C_e = (C_i - C_e) \times \left(\frac{1 + r}{2 + r}\right)$

For second-stage filters, the effluent BOD is given by the equation under III in Table 22-2, except that, if the final BOD is less than 30 mg/L, the efficiency of the second stage cannot exceed 50 percent.

Example 22-3 Apply the appropriate Rankin formulas to the data of the example problems above.

Solution

In order to select an appropriate equation, an estimate must be made of the organic loading. Assuming the effluent BOD from the first stage will be about 50 mg/L,

$$\text{Organic load} = 170(3.15) + 50(3.15)(1.25) = 732 \text{ g/min} = 1054 \text{ kg/day}$$

For a filter volume of 830 m^3, the unit organic loading rate is 1.27 kg/($m^3 \cdot$ day). Therefore the loading criterion is I, and,

$$C_e = (170 - C_e)\left[\frac{1 + 1.25}{1 + 1.5(1.25)}\right]$$

$$C_e = 75 \text{ mg/L}$$

Checking the loading rate, we find it to be 1.44 kg/($m^3 \cdot$ day), which confirms that the loading condition is I. For the second stage, the effluent BOD is given by the equation under III,

$$C'_e = (75 - C'_e)\left(\frac{1 + 1.25}{2 + 1.25}\right)$$

$$C'_e = 31 \text{ mg/L}$$

The Ten States' Standards were developed from data from the area around the Great Lakes. These equations should thus be applicable in colder climates. The modified Velz equation includes a temperature correction and should thus be useful in any region. The temperature is the temperature of the waste, not of the air, but a trickling filter is a reasonably effective heat exchange device, hence the waste temperature may approach the air temperature, particularly when high recirculation rates are used.

Alternative design procedures may employ experimental data obtained for the particular medium being used. The designer must be certain that the experimental conditions are sufficiently similar to those of the proposed plant to justify use of the data. It is not uncommon to conduct pilot studies before designing major new facilities, particularly when new materials or plant configurations are under consideration. Procedures for the experimental evaluation of trickling filter processes may be found in Ref. 21.

The sizing of trickling filter processes is based on the application of the procedures discussed above. In design, as opposed to the analysis examples above, there are many possible solutions. In general, the designer will know the characteristics of the waste and the required characteristics of the effluent—C_i and C_e or C'_e in the equations above. The problem becomes one of finding a combination of area, volume, and recirculation rate which will provide the required degree of treatment. Increasing the recirculation rate will reduce the required volume up to a point, but the incremental benefit of recirculation decreases as the rate increases. The maximum useful recirculation rate predicted by the NRC formula is about 800 percent, and rates that high are not commonly used.

Example 22-4 Design a two-stage trickling filter to produce an effluent BOD of 30 mg/L. The design influent is 5 m³/min, the influent BOD is 160 mg/L, and the minimum waste temperature is expected to be 10°C. The modified Velz equation is to be used with $K = 0.03$ and $n = 0.5$.

Solution
The design equations are

$$C_e = \left(\frac{160 + C_e r}{1 + r}\right) e^{-0.03D(1.035)^{-10}/(5/A)^{0.5}}$$

and

$$30 = \left(\frac{C_e + 30r}{1 + r}\right) - e^{-0.03D(C_e/160)(1.035)^{-10}/(5/A)^{0.5}}$$

It is not necessary that the areas, recirculation rates, and depths be identical in the two filters, hence there are more possible solutions than will be developed here. Selecting a depth of 2 m and a recirculation rate of 200 percent, by trial and error one may find that each filter must have an area of 982 m².

Changing the depth or the recirculation rate will change the area. Permitting the two filters to have different areas will permit a reduction in the total area required. In an actual design, one would evaluate a range of alternatives in order to find an optimum configuration which minimizes capital and operating cost.

The distribution systems used in trickling filters are intended to spread the water more or less uniformly over the surface of the medium on which the bacterial growth is maintained. Both rotary and fixed systems are used. *Rotary distributors* consist of a hollow arm, pivoted at the center, and fitted with nozzles so sized and placed that they provide an essentially uniform flow per unit area. The arm may be driven by the liquid discharge or may be turned by an electric motor. Hydraulic drive is quite common and has the advantage of adjusting the rotor speed to the flow. A disadvantage is that the head required to drive the rotor increases as the square of the flow, and, with the variations which occur in wastewater flow, extreme changes in head may occur at the rotor supply. This problem can be minimized by flow equalization (Chap. 21), by provision of multiple arms—four, six, or more—which receive flow in steps as the head increases, and by automatically adjusting the recirculation rate to minimize flow variations. All of these techniques may be combined in some

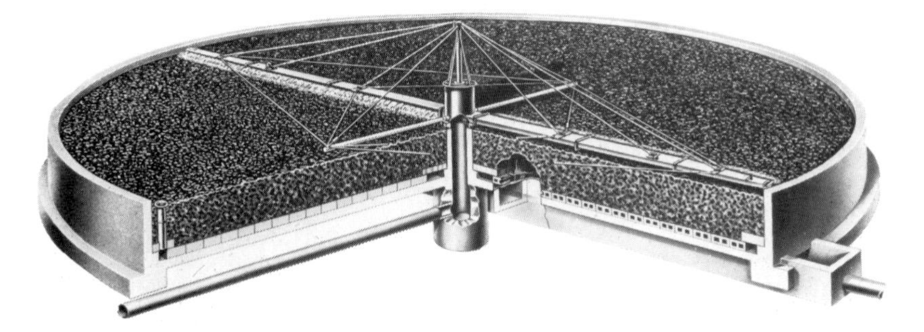

FIGURE 22-8
Rotary distributor trickling filter. (*Courtesy Door-Oliver Inc.*)

plants. The speed of rotation should be at least 6 revolutions per hour so that no part of the filter will have an opportunity to dry out. A standard rotary distributor is shown in Fig. 22-8.

Fixed-nozzle systems such as that shown in Fig. 22-4 are less expensive than rotary distributors and function equally well. Although most older trickling filters employ an intermittent dosing pattern of the sort given by a rotary distributor, many laboratory models of trickling filter systems employ continuous flow and there is no reason why intermittent dosing is necessary on coarse media with large void spaces. In fact, in some circumstances, continuous dosing has been shown to produce a superior effluent.[22]

The head loss in distribution systems may be calculated by using the principles of hydraulics presented in Chap. 3. Rotary distributors are supplied through a pipe which is either under the filter or supported on columns within it. Rotors are heavy structures which must be provided with bearings and watertight mechanical seals. This heavy, moving system is subject to a variety of mechanical problems which can be avoided by using fixed nozzles supplied by a pipe network.

The *underdrains* in trickling filters serve to support the filter medium, carry away the liquid effluent and sloughed solids, and distribute air through the bed. The underdrains are usually laid on a slope of about 1 percent toward a common collection point or channel and are sized so that they flow no more than half full at a velocity of at least 0.6 to 0.9 m/s (2 to 3 ft/s) at design flow (including recirculation). A variety of commercially available underdrains are shown in Fig. 22-9. Air will be transported through the filter and underdrains by gravity except under unusual (and transient) conditions. When the air is approximately 3.4°C warmer than the sewage, air flow will be relatively low, but will increase with any change in either temperature. In warm weather, air flow is downward and in cold weather upward through the bed. Forced ventilation is furnished by some designers, but is not necessary provided the underdrains are sized as described above and the following conditions are met:

1. Ventilation stacks or manholes with open grated covers should be installed at either end of the central collection channel.
2. Branch collecting channels on large filters should have ventilating stacks at the edge of the filter. Such a stack may be seen at the left end of the section of Fig. 22-8.
3. The open area in the top of the filter blocks (Fig. 22-9) should be at least 15 percent of the filter plan area.
4. The area of the open grating in ventilating stacks should be at least 0.4 percent of the filter plan area.

The filter medium may be contained in reinforced concrete basins as in Fig. 22-8 or in light corrugated metal or plastic structures as in Fig. 22-4. Rock filters are usually contained in concrete basins, while sheet plastic media are often housed in lighter structures. Rock media may occasionally need to be flooded to correct operational problems (see below) and thus require a water-tight tank. Plastic media, primarily because of the higher hydraulic loadings which are used, are not subject to these operational problems and thus offer some economy in construction of their containers.

22-4 Design of Rotating Biological Contactors

Although theoretical models for the RBC process similar to those for trickling filters have been developed,[17] the actual design process remains empirical.[23]

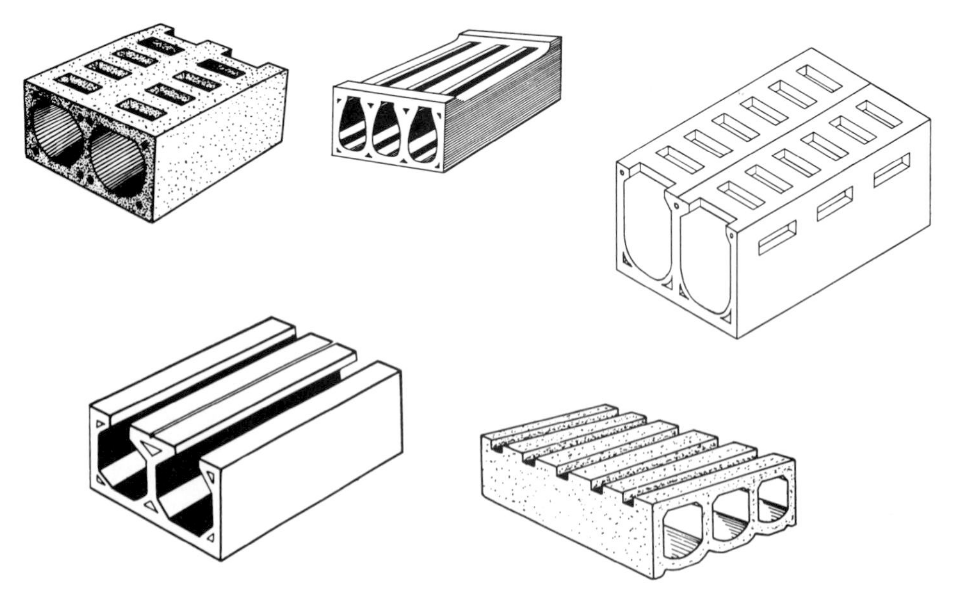

FIGURE 22-9
Blocks used in underdrain system of trickling filters.

Hydraulic loading rates in treating ordinary settled domestic sewage are 0.04 to 0.06 m/day and *organic loading rates* 0.05 to 0.06 kg BOD/($m^2 \cdot$ day) based on the surface area available for growth. In treating weaker wastewater, the hydraulic loading rate may be increased considerably. Designs may be prepared directly from manufacturers' design curves or by application of an equation of the form

$$Q(C_i - C_e) = PA\frac{C_e}{K_s + C_e} \tag{22-8}$$

for a single-stage system, where A is the total disk area covered by biological growth, K_s and P are experimental constants, and the other terms are as defined above. For multistage systems the equation becomes

$$Q(C_i - C_{en}) = PA\sum_{i=1}^{n}\frac{C_{ei}}{K_s + C_{ei}} \tag{22-9}$$

where C_{ei} is the effluent BOD of the nth stage and A is the area per stage. Solution of this equation requires a trial-and-error approach.

Example 22-5 Determine the required disk area for a rotating biological contactor intended to treat a wastewater flow of 8000 m^3/day with a BOD of 175 mg/L following primary treatment. The effluent BOD must be reduced to 20 mg/L. A treatability study has shown P to be 0.055 mg/($mm^2 \cdot$ day) and K_s to be 100 mg/L. Also find the required area if the system is divided into two stages.

Solution

For the single-stage system,

$$8 \times 10^6(175 - 20) = A(0.055)\frac{20}{100 + 20}$$

$$A = 1.35 \times 10^{11}\ \text{mm}^2 = 135{,}000\ \text{m}^2$$

Commercially available systems provide about 9300 m^2 of area on a shaft about 7.5 m long. This system would require 15 such shafts and would occupy a basin approximately 7.5 by 60 m in size. For the two-stage system, we will begin by assuming that the total area will be reduced (based on the results of the trickling filter analyses above). Assuming the area per stage is 50,000 m^2,

$$8 \times 10^6(175 - C_{e1}) = 0.055(50 \times 10^9)\frac{C_{e1}}{100 + C_{e1}}$$

and

$$8 \times 10^6(175 - C_{e2}) = 0.055(50 \times 10^9)\left[\frac{C_{e1}}{100 + C_{e1}} + \frac{C_{e2}}{100 + C_{e2}}\right]$$

Solving the first equation gives $C_{e1} = 54$ mg/L, which, when substituted in the second, yields $C_{e2} = 14$ mg/L.

The area can be reduced somewhat. A second trial at $A = 45{,}000$ m^2 yields $C_{e1} = 60$ mg/L and $C_{e2} = 16$ mg/L.

A third trial at $A = 40{,}000\ m^2$ yields $C_{e1} = 66$ mg/L and $C_{e2} = 20$ mg/L, which is the desired result.

The use of two stages in the example above reduces the total area from 135,000 to 80,000 m^2—a reduction of over 40 percent. The cost of the process would not be reduced by the same percentage, of course, but some economy would be possible. It has been reported[23] that RBC systems are insensitive to temperature above about 13°C. Below that point, the rate decreases by 5 percent for each 1° drop in temperature. The drop in rate requires a commensurate increase in disk area if equal treatment is to be achieved.

As noted earlier, RBC processes have been subject to some mechanical failures occasioned by the accumulation of heavy biological growths which, with the accompanying moisture content, overload and/or imbalance the shafts. Air-drive systems, in which air released at the bottom of the basin causes the shaft to turn, may provide some improvement in operation[24] although not a complete elimination of the problem.[19] The air also improves the dissolved oxygen balance at the head end of the plant and will help to control odors, which have been a problem at some installations. It is likely that, with normal domestic sewage, the aeration provided by the rotation of the disks is not adequate to meet the oxygen demand at the head of the plant. The disks are usually protected by a roof, since heavy rains may strip off the biological growth and hail may damage the plastic itself.

22-5 Design of Fluidized-Bed Systems

Fluidized beds represent a relatively new technology with few full-scale examples in existence. The important design considerations are maintenance of a hydraulic loading rate adequate to fluidize the bed, provision of an adequate detention time, and supply of adequate oxygen in aerobic systems.

The required velocity of fluidization may be calculated by using the procedures of Chap. 10, but it must be recognized that the characteristics of the suspension will change as the biological growth develops. Use of a basin shaped like the frustrum of a cone will permit substantial variation in bed properties without producing washout of the medium.

The required detention time depends on the concentration of biological solids in the basin. In this respect, fluidized bed systems are similar to activated sludge systems and perhaps might be better considered as a modification of that process. Levels of bacterial solids in excess of 20,000 mg/L have been reported, and, at such values, detention times on the order of 30 min are adequate for treating settled domestic sewage. The detention time and the upflow velocity required to fluidize the bed lead to very deep narrow basins or require recirculation of the flow. Recirculation is the most practical solution in full-scale systems.

Except in anaerobic fluidized-bed processes, supplying oxygen has been very difficult. The process has the capability to stabilize perhaps 300 mg/L of

BOD per hour, depending on the actual concentration of bacteria. Carrying out this stabilization requires between 300 and 400 mg/L of oxygen per hour, which is beyond the capacity of most available systems. Some ingenious techniques using high pressures and pure oxygen have been used, but this process at present must be considered experimental.

Anaerobic fluidized-bed processes have been developed and applied to the treatment of strong industrial wastes. The supply of oxygen, of course, is not a problem in such a system and satisfactory results have been reported at short detention times. This procedure is not useful, by itself, in treating domestic wastewater since the effluent quality from anaerobic processes will not meet secondary treatment standards. It might, however, be useful as the first stage in a two-stage anaerobic-aerobic process. Reducing the BOD in an anaerobic stage would make it easier to satisfy the oxygen demand in the second, aerobic, stage. Fluidized-bed systems offer some very real possibilities of improved waste treatment and will certainly undergo further development.

22-6 Operational Problems of Attached Growth Processes

The major operational problems of *trickling filters* are associated with cold weather operation. Efficiency in high-rate filters is reduced by approximately 30 percent per 10°C, and freezing may cause partial plugging of the filter medium and overloading of the remaining open area. In northern climates fiberglass covers have been employed to prevent ice formation. Covers have the additional advantage of helping to contain odors which may be produced in routine operation.

Psychoda alternata, or filter flies, breed in the slime layer in trickling filters and can create nuisance conditions in and near treatment plants. High hydraulic loading rates and maintenance of a thin biological film assist in washing the fly larvae from the filter before they can mature. The larvae, which look like small worms to the naked eye, are easily removed from the flow in secondary clarifiers which are provided with skimmers. It is important that the filter be uniformly dosed, since areas of low flow may serve as breeding zones. Flooding the filter for 24 h will drown the larvae. If it is anticipated that flooding may be necessary, the filter container must be designed to permit this. Plastic media trickling filters, because of the relatively smooth surface and high hydraulic rate typically used, seldom support large populations of flies.

In some areas, snails have created problems in rock filters. The snails feed on the slime growth, which is probably not harmful in itself. The difficulty lies in the snail shells which remain behind when the snails die and which can gradually fill the void spaces of the bed, interfering with the flow of both water and air. Removing the shells requires removing the medium, which is very expensive and time-consuming. The problem can be controlled, if it is known to be present, by periodically flooding the bed for several days. The snails will drown and as they decay the gases produced will buoy the shells to the surface where they can be skimmed by hand.

Odors in trickling filters are produced by anaerobic activity within the slime layer. The odors are reduced by high recirculation rates which thin the film and supply additional oxygen. In plants which are close to residential areas, masking perfumes are sometimes used. It is not always clear that the perfume or the combination of perfume and the odor offers an improvement over the original condition. Covers will help to contain the odor and the contained gases can be passed through a scrubber before being released.

RBC systems, like trickling filters, are affected by low temperatures. In northern areas the disks may be installed within heated buildings. This is more feasible for the RBC than for the trickling filter, since the former occupies substantially less land area. As mentioned earlier, some RBC systems have had problems associated with heavy and/or unbalanced growth and with production of odors. Mechanical problems have included motor, shaft, and bearing failures. Air-driven systems have been observed to lope and even to stop turning as a result of shaft imbalance. Turning the shafts at a higher speed will help to thin the film but substantially increases the energy costs, as does installation of an air-drive system.

22-7 Clarifier Design for Attached Growth Processes

Clarifiers following ordinary trickling filters and RBC processes are designed to remove relatively large particles of sloughed bacterial slime or humus. No thickening or hindered settling occurs, hence design criteria are based on particle size and density.

Surface overflow rates are 25 to 33 m/day [600 to 800 gal/(ft^2 · day)] at average flow and should not exceed 50 m/day [1200 gal/(ft^2 · day)] at peak flow. Weir loading rates and retention times are similar to those used in primary clarifiers. Recirculated flow is included when the clarifier is sized if it actually passes through the basins, as it does in flow patterns *a, c, c′, e, h, i, j,* and *m* in Fig. 22-6. It should also be noted that recirculated flow which passes through the primary clarifier will affect the design of that system. Sludge return is generally neglected in such calculations, since it is relatively small in trickling filter systems.

It has been suggested that, in plastic-media trickling filters employing very high recirculation rates, the mode of operation is more similar to that of a mechanically aerated activated sludge system than to an attached growth process. In such systems the designer may wish to consider application of the design standards presented below for activated sludge clarifiers.

Clarifier mechanisms used for attached growth processes are similar to those used in primary systems. Skimming is sometimes omitted, but this is a questionable economy. The additional cost of skimmers is quite small and floating solids may appear in final clarifiers following these processes as a result of gas production in the sludge layer in the bottom of the basin. If skimming is not provided, the floating solids may produce a deterioration in effluent quality.

The *quantity of waste sludge* produced in attached growth processes is somewhat less than that in suspended growth processes operated at similar loadings, since cell mass is lost as a result of anaerobic endogenous metabolism in the inner portion of the slime layer. Depending on hydraulic and organic loading rates, volatile suspended solids (VSS) production may be expected to range from 0.2 to 0.5 kg VSS per kilogram BOD removed, with the lower production in lightly loaded processes. The moisture content of the settled solids ranges from 90 to 98 percent, with lower moisture content on low-rate processes. Sludge is usually returned to the plant influent wet well and reseparated in the primary clarifier. The return flow is negligible with respect to the raw waste flow and does not affect the design of the primary basins. The mixed primary and secondary solids require further processing, as discussed in Chap. 23.

22-8 Suspended Growth Processes

Suspended growth processes maintain an adequate biological mass in suspension within the reactor by employing either natural or mechanical mixing. In most processes, the required volume is reduced by returning bacteria from a secondary clarifier in order to maintain a high solids concentration. Suspended growth processes include activated sludge and its various modifications, oxidation ponds, and sludge digestion systems.

The theory of suspended growth processes is well developed and generally accepted.[25,26] The basic factor in design, control, and operation of suspended growth systems is the *mean cell residence time* or *sludge age* θ_c defined by

$$\theta_c = \frac{X}{(\Delta X/\Delta t)} \tag{22-10}$$

in which X is the total microbial mass in the reactor and $\Delta X/\Delta t$ is the total quantity of solids withdrawn daily, including solids deliberately wasted and those in the effluent.

Two particular values of θ_c are of significance in design of biological processes. θ_c^m is defined as the lowest value of θ_c at which operation is possible. At retention times less than θ_c^m, organisms are removed more quickly than they are synthesized, hence failure will occur.

θ_c^d is the design value of θ_c and must be significantly greater than θ_c^m. The ratio, θ_c^d/θ_c^m, gives the safety factor of the system. Safety factors depend on the anticipated variability of the waste strength and flow and should never be less than 4. In standard processes safety factors of 20 or more are not unusual (Table 22-3).

For a system operating at equilibrium, the quantity of solids produced must equal that lost. The quantity produced per day is given by

$$\mu = \frac{\hat{\mu} S}{K_s + S} - k_d \tag{22-11}$$

TABLE 22-3
Typical process factors for activated sludge systems

Process	Normal loading per day		θ, days	θ_c^d days	Safety factor	r	Air supplied, $m^3/kg\ BOD_5$
	$kg\ BOD_5/m^3$	$kg\ BOD_5/kg$ MLVSS*					
Extended aeration	0.32	0.05–0.20	0.80–1.25	14–∞	≥ 70	0.50–1.00	90–125
Conventional activated sludge	0.56	0.20–0.50	0.25–0.30	4–14	20–70	0.15–0.30	45–90
Tapered aeration	0.56	0.20–0.50	0.25–0.30	4–14	20–70	0.15–0.30	45–90
Step aeration	0.80	0.20–0.50	0.20–0.30	4–14	20–70	0.20–0.50	45–90
Contact stabilization	1.12	0.20–0.50	0.01–0.04	4–14	20–75	0.50–1.00	45–90
Short-term aeration (high-rate activated sludge)	1.6–6.4	0.50–3.50	0.10–0.15	0.8–4	4–20	1.00–5.00	25–45

* MLVSS = mixed liquor volatile suspended solids

where μ = specific growth rate (increase in mass per unit mass per unit time)
$\hat{\mu}$ = maximum specific growth rate
S = concentration of rate-limiting nutrient surrounding microorganisms
K_s = constant equal to concentration of rate-limiting nutrient when rate of nutrient removal is one-half maximum rate
k_d = decay coefficient, mass per unit mass per unit time (reflects endogenous burn up of cell mass)

Application of mass balance equations and Eqs. (22-10) and (22-11) to specific process configurations permits general solutions for effluent quality in terms of experimental constants and sludge age.[26]

The *completely mixed process model* presented below is applicable to activated sludge systems of any configuration. The model was specifically developed for completely mixed systems, but is conservative in its prediction of effluent quality for plug flow systems which are not otherwise limited. The explicit assumptions of the model (Fig. 22-10) are that all waste utilization occurs in the biological reactor and that the total biological mass in the system is equal to the biological mass in the reactor. These imply that the clarifier volume is small and that recycling of solids is continuous. The mean cell residence time for this system, by definition, is

$$\theta_c = \frac{X}{(\Delta X/\Delta t)} = \frac{xV}{Q_w x_r + (Q - Q_w)x_e} \tag{22-12}$$

where Q = wastewater flow
Q_w = waste sludge flow rate
x = mixed liquor volatile suspended solids concentration (MLVSS)
x_r = concentration of volatile suspended solids in clarifier underflow
x_e = volatile suspended solids concentration in the effluent

If solids are wasted from the reactor as shown in the dashed line, rather than from the clarifier, the equation becomes

$$\theta_c = \frac{xV}{Q_w x + (Q - Q_w)x_e} \tag{22-13}$$

If the system is operating properly, that is, if the clarifier is functioning, the bulk of the solids will be removed in the waste sludge rather than in the

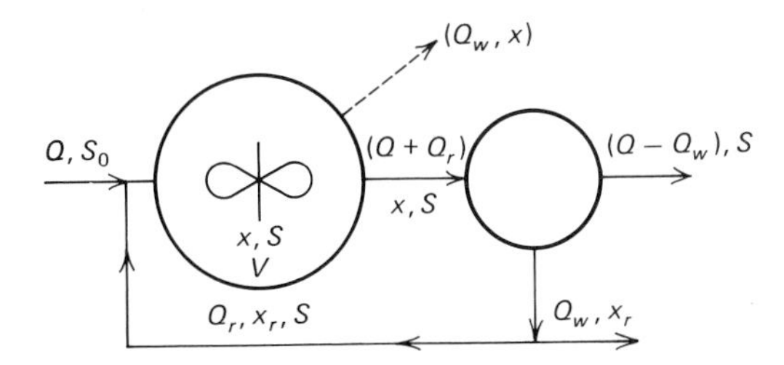

FIGURE 22-10
Completely mixed biological reactor with solids recycling.

effluent. θ_c can thus be controlled by varying Q_w and is not completely dependent on the reactor volume.

Lawrence and McCarty[26] have shown that the total microbial mass in the reactor is given by

$$xV = \frac{YQ(S_0 - S)\theta_c}{1 + k_d\theta_c} \tag{22-14}$$

in which Y is a growth yield coefficient relating cell yield to the mass of the rate-limiting nutrient metabolized. The ordinary circumstance in domestic waste treatment is that the rate-limiting nutrient is carbonaceous BOD; hence, in most cases, S_0 represents the influent BOD and S represents the effluent soluble BOD. The total effluent BOD will be greater than S, since the bacterial cells in the effluent (x_e) will also exert a BOD. The ultimate BOD of bacterial cell mass is approximately 1.15 mg/mg which yields a BOD_5 of about 0.77 mg/mg VSS.

When the required efficiency of treatment is known and the values of Y and k_d are established, it is possible to determine the required reactor volume and microbial concentration for various values of r and x_r.

Example 22-6 A wastewater with a flow of 10,000 m^3/day and a BOD of 160 mg/L is treated in an activated sludge process in which x is 2000 mg/L, V is 2500 m^3, Y is 0.65, and k_d is 0.05. The solids in the system are 80 percent volatile and the effluent suspended solids are estimated to be 30 mg/L. If the effluent BOD_5 is to be 30 mg/L, determine the required sludge age and the mass of solids wasted per day.

Solution

The BOD_5 of the effluent solids will be 30(0.80)(0.77) = 18 mg/L. The value of S therefore is

$$S = 30 - 18 = 12 \text{ mg/L}$$

Substituting in Eq. (22-14) gives

$$2000(2500 \times 10^3) = \frac{0.65(10{,}000 \times 10^3)\theta_c(160 - 12)}{1 + 0.05\theta_c}$$

$$\theta_c = 7 \text{ days}$$

The total mass of *volatile* solids removed per day is thus

$$\frac{2000 \times 2500 \times 10^3}{7} = 714 \text{ kg/day}$$

Since the solids are 80 percent volatile, the total mass removed per day will be 893 kg/day. Of this mass, the effluent will contain

$$30 \times 10{,}000 \times 10^3 = 300 \text{ kg/day}$$

The mass to be wasted is thus approximately 600 kg/day.

The solids wasted from the process will have a high moisture content (perhaps over 99 percent) and will require further treatment, as discussed in Chap. 23. In order to maintain the process, sludge must be returned to the aeration basin from the final clarifier. The return or recycle flow depends on the concentration of solids in the underflow of the clarifier. For most activated sludge processes, this is unlikely to exceed 2 percent (20,000 mg/L) and may well be less.

Example 22-7 Determine the required recycle rate for the process of the example above. The underflow concentration in the secondary clarifier is 15,000 mg/L.

Solution
A simple mass balance on the aeration basin yields

$$(15{,}000)(0.8)Q_r = 2000(Q_r + Q_r)$$

$$Q_r = 0.167Q = 1666 \text{ m}^3\text{/day}$$

In this example one should note the need to use either VSS or total suspended solids (TSS) for both solids concentrations. The underflow concentration is normally expressed as total solids, while x in Eq. (22-14) is volatile solids. The concentration of solids in the influent is neglected. This is justifiable on two bases: the concentration is low *and* the solids in the influent are not the bacterial solids which are needed to stabilize the waste.

The activated sludge process requires some means of supplying the oxygen required for stabilization of the waste. The quantity of oxygen can be calculated readily as the ultimate BOD of the waste less the ultimate BOD discharged in the effluent and in the waste solids flow. This may be expressed as

$$O_2 \text{ demand} = 1.47(S_0 - S)Q - 1.15(xV/\theta_c) \tag{22-15}$$

The factor 1.47 is a typical ratio of $BOD_{ultimate}$ to BOD_5, and the factor 1.15 is discussed above. Equation (22-15) will give the mass of oxygen required per unit time. This may be converted to *air required* by recognizing that air contains 23.2 percent oxygen by weight and, at standard temperature and pressure, has a density of 1.2 kg/m^3.

$$Q_{air} = \frac{O_2 \text{ demand}}{0.232(1.20)} = \frac{O_2 \text{ demand}}{0.278} \tag{22-16}$$

The efficiency of transfer of oxygen from air is always substantially less than 100 percent. The actual air required may be calculated by dividing the flow obtained from Eq. (22-16) by the efficiency of transfer expressed as a decimal.

Example 22-8 Calculate the air required for the preceding problem assuming the efficiency of transfer is 10 percent.

Solution

The oxygen demand is

$$O_2 \text{ demand} = 1.47(160 - 18)10{,}000 \times 10^3 - 1.15(714) \times 10^6 = 1266 \text{ kg/day}$$

The air flow required is

$$Q_{\text{air}} = \frac{1266}{0.278 \times 0.10} = 45{,}500 \text{ m}^3\text{/day} = 31.6 \text{ m}^3\text{/min}$$

Suspended growth processes without solids recycle are represented schematically as shown in Fig. 22-11. The effluent soluble BOD from such a process may be calculated from

$$S = \frac{K_s(1 + k_d\theta_c)}{Yk\theta_c - k_d\theta_c - 1} \tag{22-17}$$

in which the terms are as defined earlier. The concentration of solids in the reactor and in the effluent is given by

$$x = \frac{Y(S_0 - S)}{1 + k_d\theta_c} \tag{22-18}$$

From Eqs. (22-17) and (22-18) one may observe that S (the soluble effluent BOD) is not dependent upon S_0 (the influent BOD) but only on various rate constants and the solids retention time θ_c. Since no solids are returned, θ_c is equal to θ, the hydraulic retention time (V/Q). The effluent suspended solids concentration is, however, dependent on both S_0 and θ_c. Since the solids contribute to the total BOD, the effluent BOD does depend on the waste strength. Processes which may be represented by this mathematical model include oxidation ponds, whether aerobic, anaerobic, or facultative, and biological sludge digestion systems (Chap. 23).

Example 22-9 Determine the effluent BOD to be expected from an oxidation pond treating raw domestic sewage with a BOD_5 of 250 mg/L. Assume $Y = 0.65$, $K_s = 20$, $k = 5$, $k_d = 0.05$, and that the hydraulic detention time is 30 days.

Solution

The effluent soluble BOD is

$$S = \frac{20(1 + 0.05 \times 30)}{0.65(5)(30) - 0.05(30) - 1}$$

$$S = 0.5 \text{ mg/L}$$

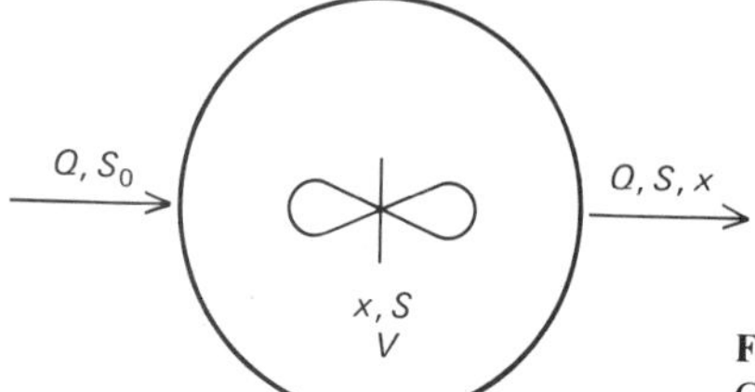

FIGURE 22-11
Completely mixed biological reactor without solids return.

The suspended solids concentration is

$$x = \frac{0.65(250 - 0.5)}{1 + 0.05(30)} = 65 \text{ mg/L}$$

The BOD_5 of the solids is estimated to be 0.77 times the VSS concentration. If the solids are 80 percent volatile, this yields a total BOD_5 of

$$BOD_5 = 0.5 + 0.77(0.8)(65) = 41 \text{ mg/L}$$

The actual measured BOD at any time might be higher or lower, depending on the instantaneous suspended solids concentration at the outlet. The latter is affected by quiescent or windy periods, algal blooms, etc.

22-9 Design of Suspended Growth Processes

In suspended growth, as with attached growth, there are many possible designs which will satisfy the requirements defined by the process equations. Examination of Eqs. (22-13) and (22-14), for example, indicates that maintenance of a higher value of x (the MLVSS) would permit a smaller reactor volume, that x is dependent on θ_c (and thus upon x_e, x_r, and Q_r), that the quantity of waste sludge produced which must be treated further depends on θ_c, that the oxygen demand depends on the quantity of waste sludge (and thus on θ_c), and, in short, that many solutions are possible. Bidstrup and Grady[27] have reported the development of an interactive menu-driven program for simulating both steady-state and transient behavior of completely mixed activated sludge processes. Another model is available[28] which is useful in simulating lagoon processes. Such simulations can be very helpful in gaining an understanding of the interactions of process variables. There are certain typical values of design parameters imposed, in part, by limitations on oxygen transfer and solids thickening. Such values are presented in Table 22-3. The major processes, including those listed in the table, are described below.

Conventional activated sludge represents the original configuration, consisting of a relatively long and narrow rectangular basin with air for oxygen supply and mixing being supplied through diffusers at the bottom of the tank. The flow passes to a clarifier, from the bottom of which solids are returned to the aeration basin. Excess solids are usually wasted from the sludge return line, although separate wasting from the aeration basin is possible and may be preferable. The returned solids are mixed with the incoming waste and pass through the basin in a plug flow fashion. Air is usually provided through porous diffusers. The high concentration of BOD and microbial solids at the head of the basin leads to rapid exertion of BOD and a high oxygen demand which may be difficult to meet. At the effluent end of the basin the air supplied may be in excess of the demand (Fig. 22-12). Because the plug flow regime offers little dilution of the incoming flow, this process is subject to upsets caused by shock loads and toxic materials.

The *tapered aeration process* attempts to match the oxygen supply to demand by introducing more air at the head end. This can be achieved by

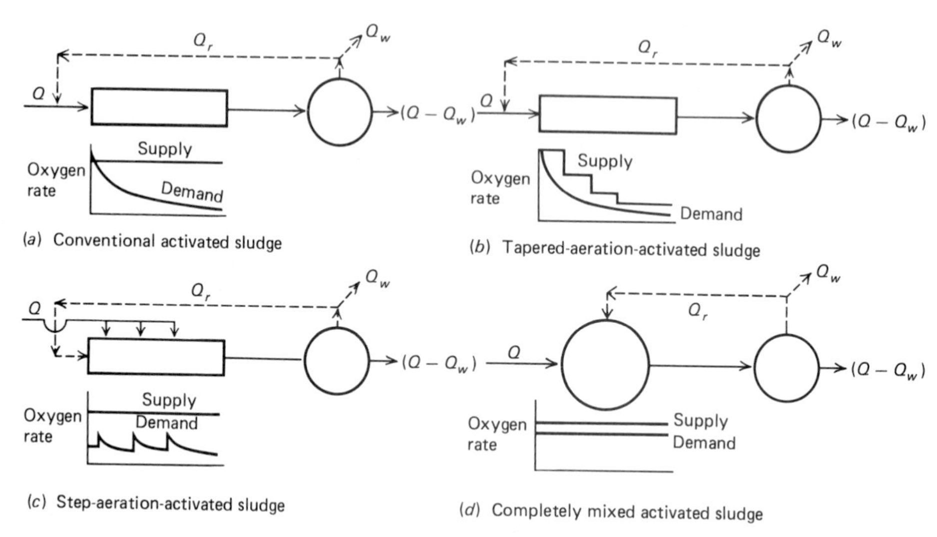

FIGURE 22-12
Activated sludge process configurations and effect on oxygen supply and demand.

varying the diffuser spacing. The process is otherwise the same as the conventional process and, like it, may be upset by shock loads or toxic materials.

The *step aeration process* distributes the waste flow to a number of points along the basin, thus avoiding the locally high oxygen demand encountered in the conventional and tapered aeration processes. The distribution of flow lessens the effect of peak hydraulic and organic loads and may provide sufficient dilution to protect the bacterial population against toxic materials.

The *completely mixed process* disperses the incoming waste and return sludge uniformly throughout the basin. The shape of the reactor is not important, provided it is conducive to good mixing. In such a process, the oxygen demand is uniform throughout the basin (Fig. 22-12). In actual basins of useful size, true complete mixing is difficult to achieve, but for practical purposes it can generally be approximated by careful selection of basin geometry and mixing and aeration equipment. The effect of peak organic and hydraulic loads is minimized in completely mixed systems, and toxic materials are usually diluted below their threshold concentration. In theory, the effluent of a completely mixed system is inferior to that of a plug flow process, but in practice the difference is not discernible.

Extended aeration is a completely mixed process operated at a long hydraulic detention time θ and high sludge age θ_c. The process is limited in application to small plants where its inefficiency is outweighed by its stability and simplicity of operation. Many extended aeration plants are prefabricated units ("package plants") which require little more than foundations and electrical and hydraulic connections. A typical extended aeration package plant is shown in Fig. 22-13. In selecting or specifying package plants the engineer

FIGURE 22-13
Extended aeration package plants. (*Courtesy Clow Corporation.*)

should give careful consideration to the quality and capacity of pumps, motors, and compressors as well as the stated capacity of the system.

Short-term aeration or *high-rate activated sludge* is a pretreatment process similar in application to a roughing filter. Retention times and sludge ages are low, which leads to a poor quality effluent and relatively high solids production. This process has potential application as the first stage of a two-stage process designed for biological nitrification (Chap. 24).

Contact stabilization takes advantage of the observed adsorptive properties of activated sludge. Returned sludge which has been aerated for stabilization of previously adsorbed organics is mixed with incoming wastewater for a brief period, perhaps 30 min. The mixed flow is then separated in a clarifier, the treated liquid is released, and the sludge with its burden of adsorbed organics is aerated for 3 to 6 h. During the aeration process the adsorbed organics are hydrolyzed and rereleased to the liquid prior to final stabilization. If the retention time in the contact stage is too long, rerelease may occur there with a consequent deterioration in effluent quality.

A significant reduction in required aeration capacity is possible in contact stabilization. If one assumes the sludge will thicken to 1 percent solids (10,000 mg/L) and that the concentration in the contact basin is 2500 mg/L, a simple calculation indicates that the total mixed volume (contact plus aeration) is only one-third that of a standard process if both employ an aeration period of 6 h. A schematic diagram of a contact stabilization process is illustrated in Fig. 22-14.

High-purity oxygen activated sludge systems have been developed in an attempt to permit easier matching of oxygen supply to oxygen demand and, perhaps, higher-rate processes through maintenance of higher concentrations

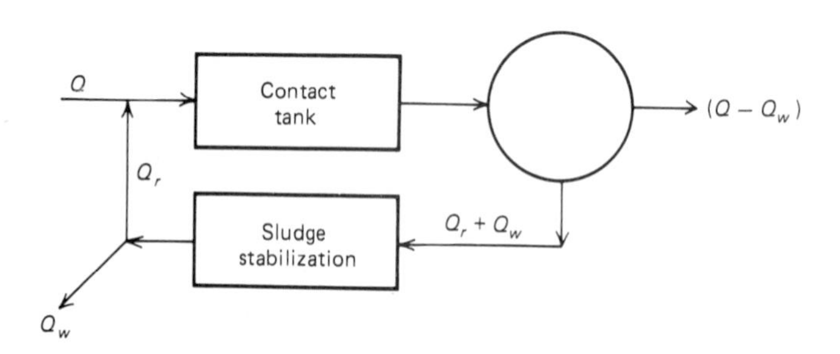

FIGURE 22-14
Contact stabilization process.

of biological solids. Two process configurations have been developed: closed reactors with a high-purity oxygen atmosphere (Fig. 22-15) and open reactors with fine bubble diffusion at the tank bottom (Fig. 22-16).

If it is possible to maintain higher mixed liquor solids concentrations, the oxygen processes will permit more rapid satisfaction of the oxygen demand. Maintaining high mixed liquor solids concentrations, however, is dependent on factors other than the source of oxygen. Specifically, it is dependent on the character of the solids, the design of the final clarifier, and the recycle rate. High-purity oxygen systems may have certain advantages in coping with unanticipated increases in organic loading[29] and in treating certain industrial wastes. Properly designed and operated air-activated sludge systems, however, can produce an effluent of comparable quality at comparable cost when treating domestic wastewater.[30]

Continuously fed, intermittently decanted processes represent a return to an older technology which has proved useful, particularly in small communities, although large systems have also been built.[31] In this design, wastewater is fed to a large basin and aerated without any discharge until the basin is full.

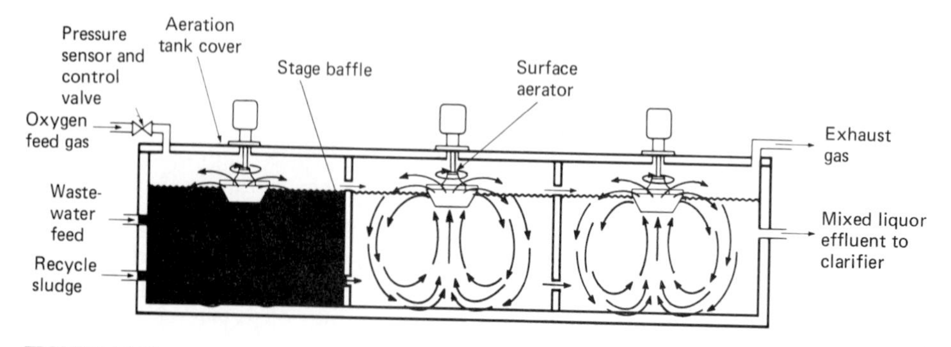

FIGURE 22-15
Closed-reactor high-purity oxygen system. (*Courtesy Union Carbide Corporation.*)

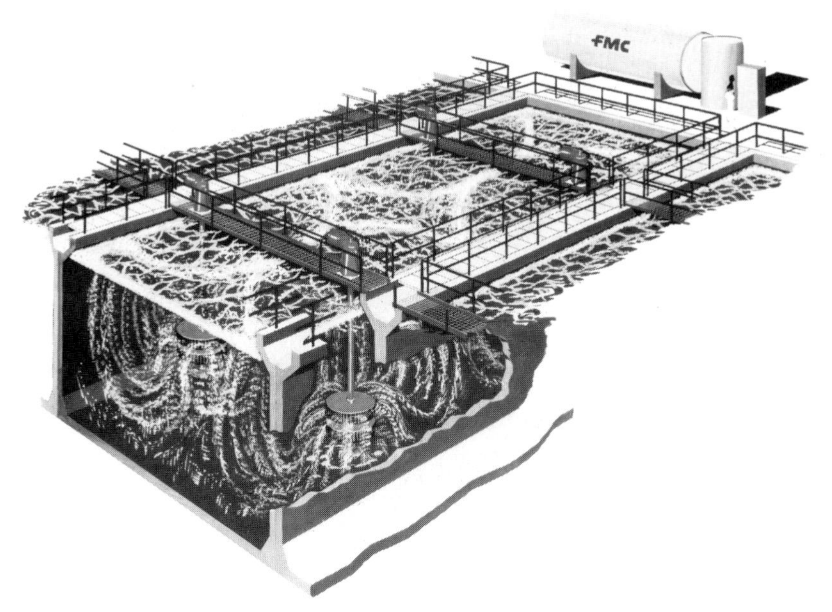

FIGURE 22-16
Open-reactor high-purity oxygen system. (*Courtesy FMC Corp.*)

The aeration is then shut off and the sludge is allowed to settle within the aeration basin itself. The supernatant, consisting of the treated waste, is then carefully drawn off without disturbing the settled sludge. The process is then begun again, and the basin is allowed to refill. Sludge is wasted from the process periodically—sometimes from the settled layer, more commonly from the basin during aeration. The sludge, like any waste sludge, requires further treatment.

Oxidation or *stabilization ponds* are a relatively low-cost treatment system which has been widely used, particularly in rural areas. The ponds may be considered to be completely mixed biological processes without solids return. The mixing is usually provided by natural processes (wind, heat, or fermentation) but may be augmented by mechanical or diffused aeration.

Aerobic ponds are generally constructed to operate at a depth between 1 and 1.5 m (3 to 5 ft). Shallower depths will encourage growth of rooted aquatic plants, while greater depth may interfere with mixing and with oxygen transfer from the surface.

Facultative ponds are sufficiently deep (more than 2 m) to provide separation of the basin into three horizontal strata. The pond depth inhibits mixing, hence organic solids which settle to the bottom will remain there and decay anaerobically. In the liquid above the sludge layer, there will be a zone in which facultative bacteria oxidize the incoming organics and the products of the anaerobic decomposition, while in the surface layer the pond will function in the same fashion as an aerobic pond.

Both aerobic and facultative ponds are biologically complex. The general reactions which occur are illustrated schematically in Fig. 22-17. Incoming organic matter is oxidized by bacteria to yield NH_3, CO_2, SO_4^{-2}, H_2O and the other end products of aerobic metabolism. These materials are then used by algae to produce more algal cells and, during daylight, oxygen which supplements that provided by wind action and is used by the bacteria to decompose the original waste. The symbiotic relationship between bacteria and algae leads to stabilization of the incoming waste as indicated in the example above, but may not yield the effluent BOD calculated from the model because of the presence of algal cells. The EPA and the states have permitted somewhat higher concentrations of suspended solids in the effluent of oxidation ponds than is permitted from other processes. These exceptions are limited to small communities and require a demonstration that no adverse environmental effects will result.

Design standards for oxidation ponds are specified by the individual states and vary with climate. Typical standards include embankment slopes (1:3 to 1:4), organic loading rate [2.2 to 5.5 g BOD/($m^2 \cdot$ day) or 20 to 50 lb/(acre $\cdot$ day), depending on climate], hydraulic detention time (60 to 120 days), and permissible seepage through the bottom (0 to 6 mm/day). In some climates it may be possible to operate oxidation ponds without discharge to surface waters.

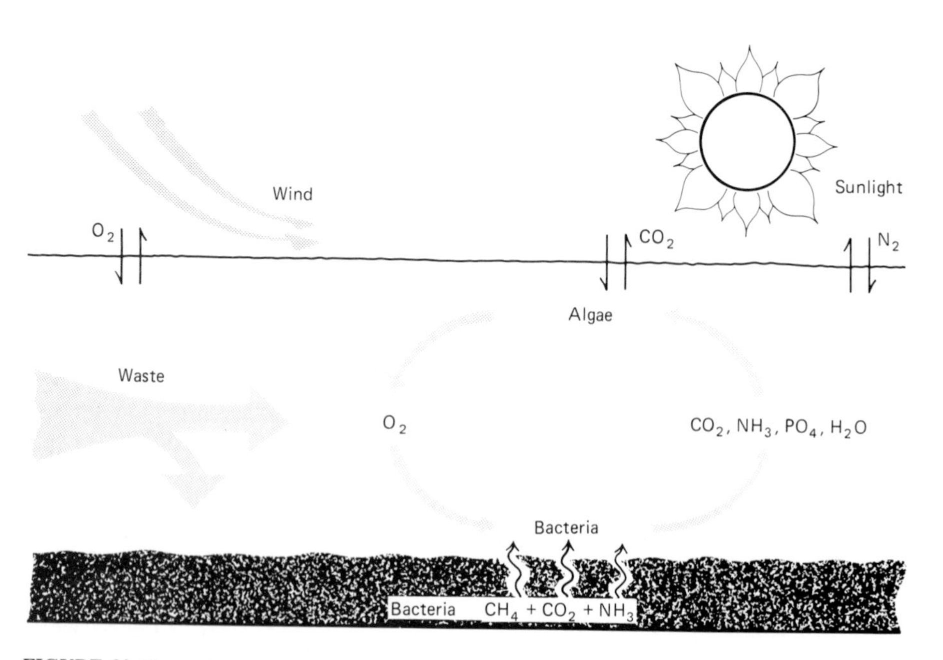

FIGURE 22-17
Oxidation pond schematic diagram.

22-10 Aeration and Mixing Techniques

Suspended growth processes require reasonably intense mixing in order to maintain the biological solids in suspension, disperse the waste through the basin, and provide the required oxygen for stabilization in aerobic systems. The transfer of gas into or from a liquid is a surface phenomenon, hence aeration equipment should maximize the interfacial area. This is most readily done by passing very fine bubbles of the gas through the liquid or very fine droplets of liquid through the gas. The rate of transfer is also affected by the partial pressure of the gas, thus techniques which increase the partial pressure of oxygen offer the potential of improved transfer.

Early aeration equipment employed fine-bubble diffusers made of porous ceramic materials. These devices provided good efficiency of transfer but were prone to plugging as a result of biological growth and contaminants in the air supply. Fine bubble diffusers with readily replaceable diffusion surfaces made of woven or porous plastic materials (Fig. 22-18) have also been used, generally connected to supply pipes which could be swung up or lifted out of the basin to simplify replacement of the surface.

Other devices such as that of Fig. 22-19 avoid the problem of plugging by using large holes which produce coarse bubbles which are then broken up by a rapidly rotating mechanical device called a *sparger*. Figure 22-20 illustrates a mechanism in which a stream of air and water is passed through a venturilike section in which the air is entrained in the liquid in the form of very fine bubbles. This device, like the mechanical sparger, is not prone to plugging and achieves its effectiveness by the use of mechanical energy.

Recently developed diffusers employ polymeric materials with very small openings providing fine bubbles (Fig. 22-21). These systems, unlike those used

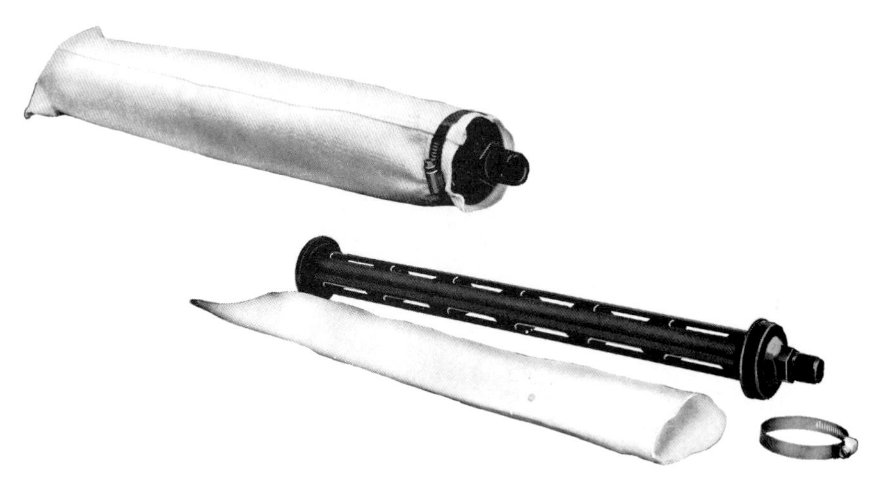

FIGURE 22-18
Typical porous sock diffuser.

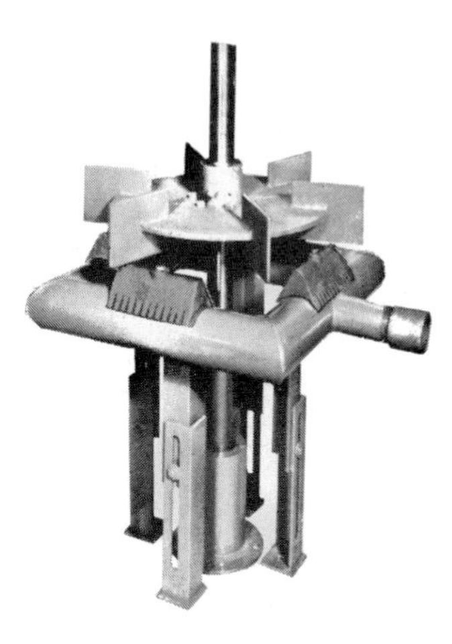

FIGURE 22-19
Mechanical sparger.

FIGURE 22-20
Jet aeration system. (*Courtesy Pentech Division, Houdaille Industries, Inc.*)

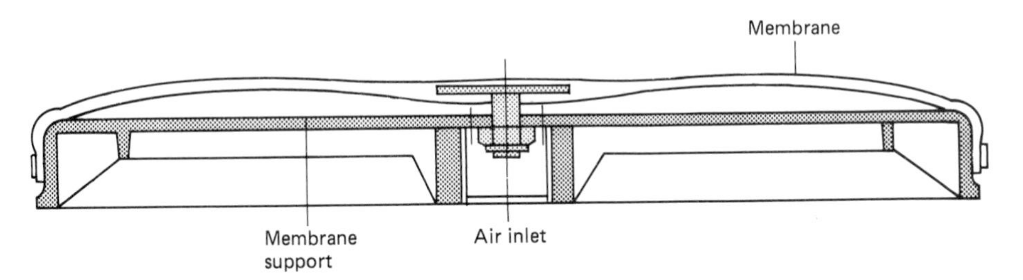

FIGURE 22-21
Fine bubble flexible diffuser.

in the past, seem to be relatively trouble-free in operation. Their effectiveness is a result of several factors. The air supplied to fine bubble diffusers is now filtered to remove dust which might cause plugging, and the diffusers themselves are flexible. When the air pressure is low, the openings are closed and the diffuser is watertight. As the pressure is increased, the diffuser expands and the air begins to pass through the openings. If the diffuser should plug, it will stretch a bit more until it is open again.

Fine bubble diffusers may have transfer efficiencies of up to 10 percent. At that level of efficiency, the air flow required to meet the oxygen demand in most biological systems is adequate to provide thorough mixing of the basin contents. A recently developed process uses thin synthetic polymer membranes through which the air passes by molecular diffusion.[32] In this technique, bubbles are not formed and there is no agitation of the liquid. Advantages claimed for the process are avoidance of foaming and stripping of volatile organics and the independence of oxygen partial pressure and depth of the tank. The process, however, will still require mixing, and, if this is not provided by the aeration technique, a supplementary system will be required.

Air-flow requirements can be calculated as illustrated in Example 22-8 of Art. 22-8. The flow for purposes of design is the maximum requirement during the design period and should not be based on average flow and strength. The air is delivered to the basins from a central location through an air pipe system. Techniques for determining the head loss in air pipe and fittings may be found in Ref. 18 and 33. Velocities range from 6 m/s (20 ft/s) in small (100-mm) pipe to 33 m/s (100 ft/s) in large (1520-mm) lines. Head losses in pipe headers are generally on the order of 100 to 200 mm (4 to 8 in) of water, while head losses through diffusers range to 400 or 500 mm (16 to 20 in) of water. Overall pressure loss usually approximates 25 percent of diffuser submergence in an economical design. The head loss through the diffuser must be large relative to the head loss in the pipe in order to assure reasonably uniform distribution of the flow. Submergence of diffusers is typically 5 to 6 m (16 to 20 ft). Deeper submergence improves gas transfer, since the partial pressure is increased and the bubble will have a longer contact time as it rises to the surface. Depths in excess of 5 to

6 m require compressors which are different in design (and cost) from those which are suitable at low pressures.

As an approximation, the friction factor of steel pipe carrying air may be taken to be

$$f = 0.029 \frac{D^{0.027}}{Q^{0.148}} \tag{22-19}$$

where D is the pipe diameter, m, and Q is the air flow, m³/min. The head loss in straight pipe can then be calculated from

$$H = f \frac{L}{D} h_v \tag{22-20}$$

where H = head loss, mm of water
L = pipe length, m
D = diameter, m
h_v = velocity head of air, mm of water

Alternatively, the head loss may be approximated from

$$H = 9.82 \times 10^{-8} \frac{fLTQ^2}{PD^5} \tag{22-21}$$

where T is the temperature, K, P is the pressure, atm, and the other terms are as previously defined. The absolute temperature may be estimated from the pressure rise at the compressor from

$$T_2 = T_1 \left(\frac{P_2}{P_1}\right)^{0.283} \tag{22-22}$$

Fittings in air distribution systems may be converted into equivalent lengths of straight pipe from

$$L = 55.4CD^{1.2} \tag{22-23}$$

where L = equivalent length, m
D = pipe diameter, m
C = factor from Table 22-4

TABLE 22-4
Resistance factors for air fittings

Fitting	C
Gate valves	0.25
Long-radius ell or run of standard tee	0.33
Medium-radius ell or run of tee reduced 25 percent	0.42
Standard ell or run of tee reduced 50 percent	0.67
Angle valve	0.90
Tee through side outlet	1.33
Globe valve	2.00

Example 22-10 Determine the head loss in a pipe header consisting of 305 m of 380-mm steel pipe which contains three long-radius ells. The air flow is 100 m^3/min at a gauge pressure of 0.55 atm. The ambient temperature is 30°C.

Solution
The equivalent length of the ells is

$$L = 3 \times 55.4 \times 0.33 \times (0.38)^{1.2} = 17 \text{ m}$$

The temperature at the compressor is

$$T_2 = (273 + 30)(1.55/1)^{0.283} = 343 \, K$$

The friction factor is approximately

$$f = 0.029 \frac{(0.38)^{0.027}}{(100)^{0.148}}$$

$$= 0.0143$$

and the head loss is

$$H = 9.82 \times 10^{-8} \frac{(0.0143)(305 + 17)(343)(100)^2}{(1.55)(0.380)^5} = 127 \text{ mm of water}$$

As air passes through a transmission line, the pressure will drop as a result of the friction losses. The pressure drop produces a decrease in density and thus an increase in velocity, since the mass flow is constant. This effect is offset to a degree by the decrease in temperature which also accompanies a drop in pressure and which produces an increase in density. Nonetheless, air being conveyed in closed conduits of constant size will move at increased velocity in the downstream direction. This factor is not significant in the relatively short lines used in most wastewater treatment plants, but could be important if the lines were long (1 km or more).

Compressors for use in diffused air systems may be either positive-displacement rotary lobe (Fig. 22-22) or centrifugal (Fig. 22-23) machines, depending on the specific application. For pressures up to about 7 kPa (10 lb/in^2) and flows above 15 m^3/min (5000 ft^3/min), centrifugal blowers are preferable. The flow of centrifugal blowers can be adjusted by throttling the inlet, which lowers the characteristic curve of the machine, producing a lower flow at a given head. Throttling the discharge side is not advisable, since this increases the discharge head. This also reduces the discharge, but may cause damage to the machine if the head is close to the shutoff value. In many wastewater applications, single-stage centrifugal blowers can provide the head required [up to about 6 m (20 ft) of water]. If higher pressures are required, multistage blowers capable of pressures up to 28 m (90 ft) of water are available.

Positive-displacement rotary lobe compressors are used primarily in small plants where air flow is low. The discharge of these machines can be altered only by varying their speed, which is a disadvantage. They are useful when pressures are over 50 kPa (7 lb/in^2) and flows are less than 15 m^3/min (5000 ft^3/

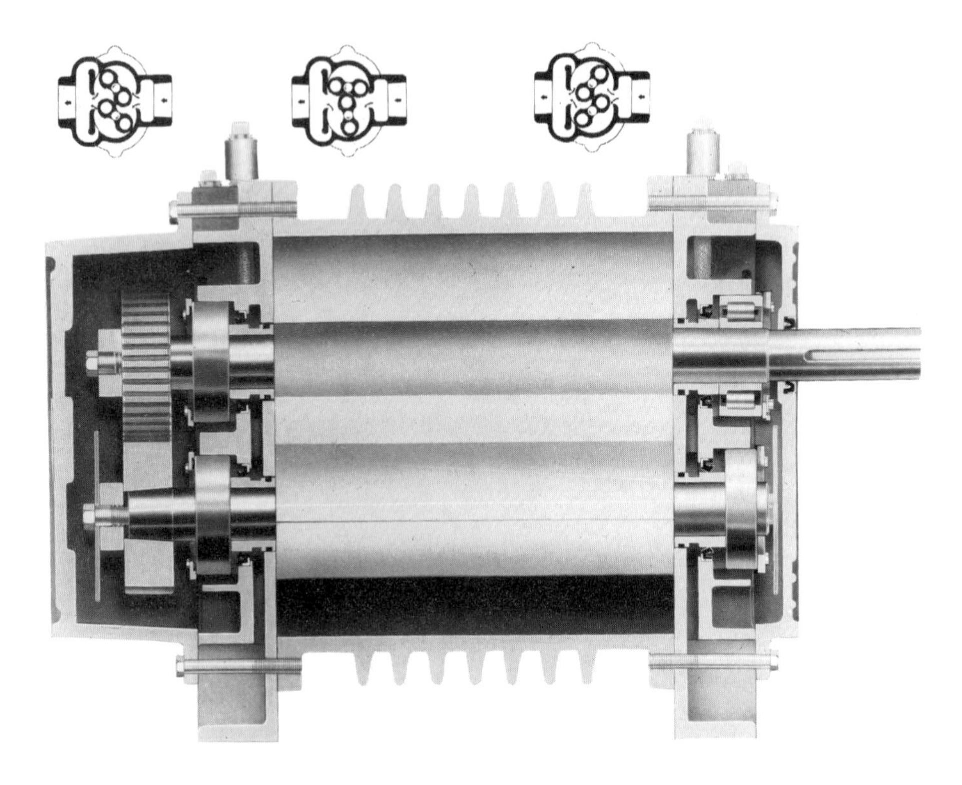

FIGURE 22-22
Rotary lobe compressor. (*Courtesy Dresser Industries, Inc.—Roots Blower Division.*)

min). Positive-displacement blowers should not be throttled, since this will cause overheating and overpressure and may result in damage to the machine.

Selection of compressors and design of compressor installations is at least as complicated as design of pump stations. On all but the smallest machines, sound suppressers are required on both the inlet and discharge. The inlet air is usually filtered to protect the machine and to minimize plugging of diffusers. The filters and sound suppressers produce additional head loss in the system, and these losses must be included in determining the pressure and power requirements of the compressor. Multiple units are usually installed in parallel with sufficient capacity to meet the maximum anticipated demand with the largest unit out of service.

Compressor power requirements may be estimated from the air flow, discharge and inlet pressures, and air temperature. Assuming adiabatic conditions,

$$P = \frac{wRT_1}{8.41e}\left[\left(\frac{P_2}{P_1}\right)^{0.283} - 1\right] \qquad (22\text{-}24)$$

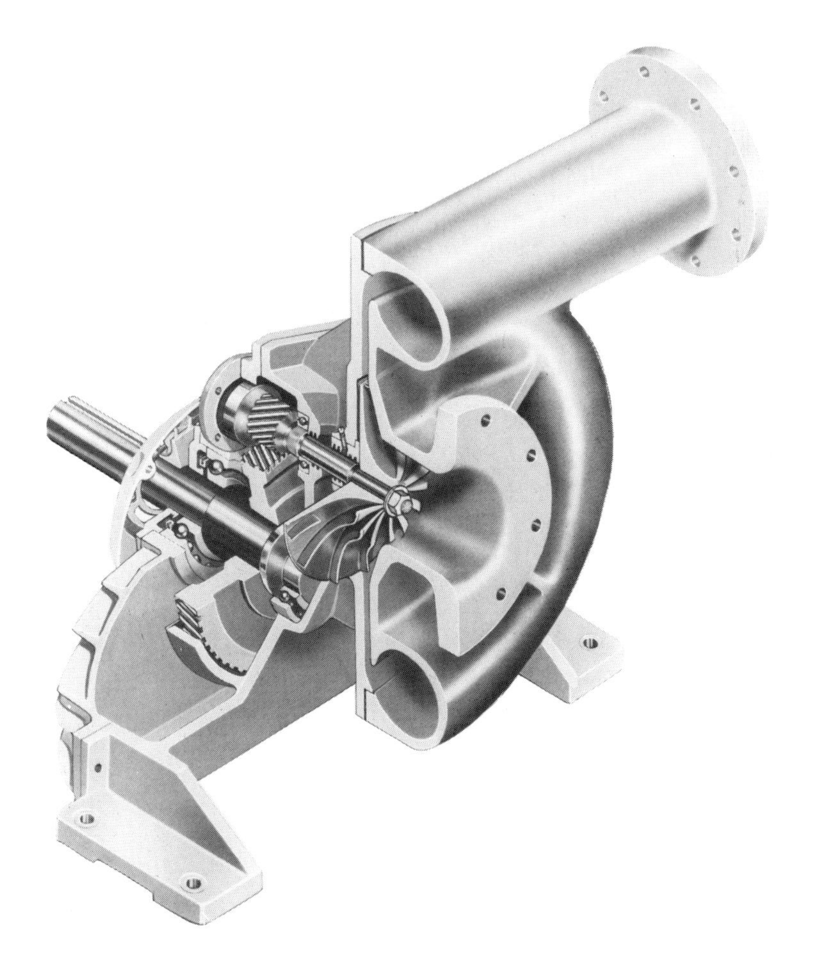

FIGURE 22-23
Centrifugal blower. (*Courtesy Dresser Industries, Inc.—Roots Blower Division.*)

where P = power required, kW
w = air mass flow, kg/s
R = gas constant (8.314)
T_1 = inlet temperature, K
P_1 = inlet pressure
P_2 = outlet pressure
e = efficiency of machine (usually 70 to 80 percent)

Example 22-11 Determine the power required to provide an air flow of 31.6 m^3/min at a discharge pressure of 1.7 atm. The inlet pressure (as a result of inlet filters and silencers) is 0.95 atm. The inlet temperature is 30°C and the efficiency of the machine is 75 percent.

Solution

The mass flow of the air is

$$w = 31.6 \times 1.2 = 37.9 \text{ kg/min} = 0.63 \text{ kg/s}$$

$$T_1 = 303 \text{ K}$$

$$P = \frac{(0.63)(8.314)(303)}{(8.41)(0.75)}\left[\left(\frac{1.7}{0.95}\right)^{0.283} - 1\right] = 45 \text{ kW}$$

A variety of *mechanical aeration* techniques are also used in suspended growth processes.

Surface aerators consist of motor-driven propellers mounted in either a fixed or a floating support. The propellers throw the liquid into the air, thus providing opportunity for oxygen transfer, and generate a flow in the basin which assists in mixing the contents. Oxygen transfer occurs both at the surface of the droplets and at the surface of the bulk solution. *High-speed* aerators such as that shown in Fig. 22-24 are driven by close-coupled motors and turn at the motor speed (usually 900 or 1800 r/min). *Low-speed* designs employing a transmission turn at 40 to 50 r/min (Fig. 22-25). They are consider-

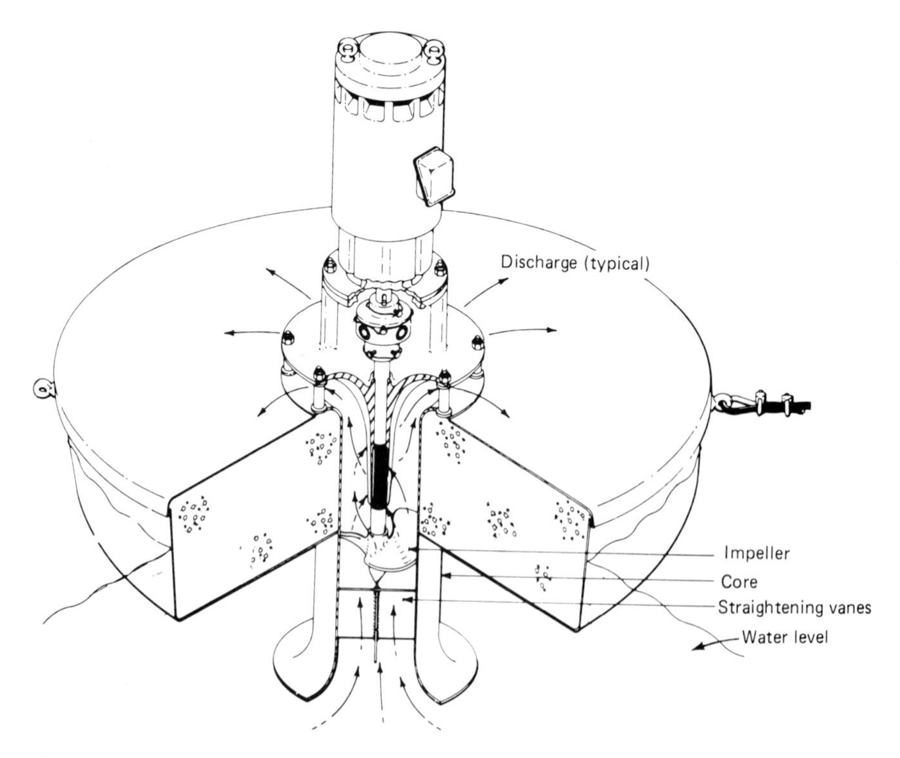

FIGURE 22-24
High-speed floating surface aerator. (*Courtesy Ashbrook-Simon-Hartley, Inc.*)

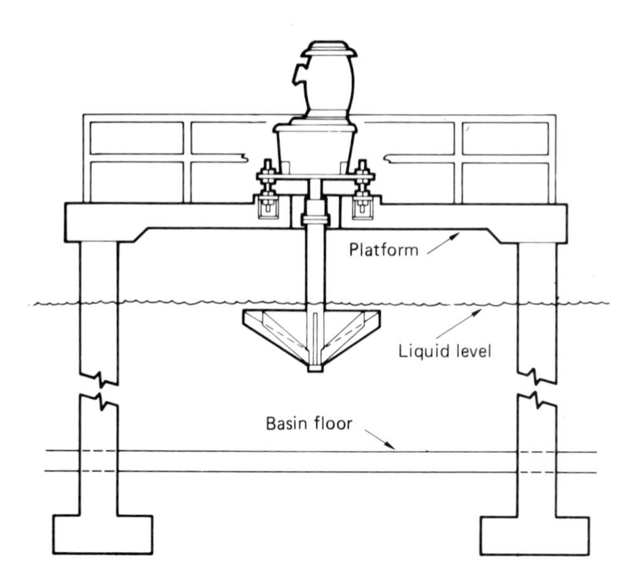

FIGURE 22-25
Low-speed fixed-platform surface aerator. (*Courtesy Ashbrook-Simon-Hartley, Inc.*)

ably more expensive than high speed units but are less subject to mechanical failure and more desirable from a process standpoint, since they produce less shearing of the biological floc. Finely divided biological solids tend to be difficult to remove and to thicken.

The effective mixing zone of surface aerators depends on the capacity of the unit and the design of the impeller. Specific information must be obtained from the manufacturer of the equipment which is selected, but the data of Table 22-5 may be useful as a rough guide.[18] Selection of mechanical aerators is based on consideration of both oxygen transfer and mixing requirements. The power required for mixing depends on the basin configuration and normally ranges from 13 to 26 kW/1000 m^3 of basin volume. Aeration requirements depend on the efficiency of the unit, the nature of the waste, the altitude (which affects the partial pressure of oxygen), the temperature, and the desired dissolved oxygen concentration. The interrelation of these factors is given by

$$M = M_0\left[\frac{\beta C_w - C_L}{9.17}(1.025)^{T-20}\,\alpha\right] \tag{22-25}$$

where M = rate of oxygen transfer, kg/MJ, under actual conditions
M_0 = manufacturer's rating of equipment at 20°C and 0 mg/L dissolved oxygen in clean water
β = salinity correction (1 for fresh water)
C_w = oxygen saturation concentration in water at given temperature and altitude
T = actual temperature of waste, °C
α = oxygen transfer correction for waste (usually 0.8 to 0.85)

TABLE 22-5
Surface aerator performance

Size, kW	Diameter of complete mixing, m	Diameter of oxygen dispersion, m
7.5	15	43
15	22	70
30	31	99
45	35	107
55	40	116
75	46	134
110	56	162

The solubility of oxygen at sea level is presented in App. 1. These values may be corrected for other altitudes (in meters) by using

$$C_w = C_T\left(1 - \frac{\text{altitude}}{9450}\right) \tag{22-26}$$

in which C_T is the saturation concentration at sea level at the temperature in question.

Example 22-12 Determine the power required to provide 1266 kg of O_2 per day to a waste treatment process located at an elevation of 500 m in which the desired minimum dissolved oxygen is 2.5 mg/L, $\alpha = 0.8$, and the maximum and minimum waste temperatures are 30 and 10°C, respectively. The manufacturer's reported transfer rate under standard conditions is 0.5 kg O_2/MJ.

Solution
From App. 1, $C_{10} = 11.2$ mg/L and $C_{30} = 7.6$ mg/L

$$C_w = C_T(1 - 500/9450) = 0.95C_T$$

At the lower temperature,

$$M = 0.5\left[\frac{(1)(0.95 \times 11.2) - 2.5}{9.17}(1.025)^{-10}(0.80)\right] = 0.28 \text{ kg/MJ}$$

At the higher temperature,

$$M = 0.5\left[\frac{(1)(0.95 \times 7.6) - 2.5}{9.17}(1.025)^{10}(0.80)\right] = 0.26 \text{ kg/MJ}$$

The energy required per day is thus

$$1266/0.26 = 4869 \text{ MJ}$$

and the connected power load is 56 kW.

The results of this calculation may be compared to the power required for a diffused air system above (45 kW). Both systems are intended to provide the same total oxygen supply and the diffused air system is seen to be somewhat more efficient.

Mixing and aeration is also effected in some systems by horizontally rotating devices which act, in a sense, as paddle wheels. Machines of this type are typically employed in relatively shallow basins, with the mechanical drive serving primarily to keep the liquid in motion at a velocity sufficient to prevent deposition of solids. A large fraction of the required oxygen is supplied by transfer through the free surface of the liquid rather than by aeration at the drive system. Systems employing this type of equipment (Figs. 22-26 and 22-27) in shallow basins are called *oxidation ditches* and are usually operated as extended aeration processes. Recommendations for unit size and power requirements are available from equipment manufacturers.

Any surface aerator may be adversely affected by ice formation during the winter months. Some manufacturers claim that their designs are less affected than others and some incorporate electrical heating of critical areas. The manufacturers' recommendations for cold weather operation should be followed to prevent loss of efficiency, damage to the machinery, and danger to the operators.

Aerated ponds are aerobic systems in which the natural oxygenation afforded by wind and algal action is supplemented by mechanical or diffused aeration. High-speed floating aerators are often used in this application, particularly in ponds which were initially not aerated but which have reached an

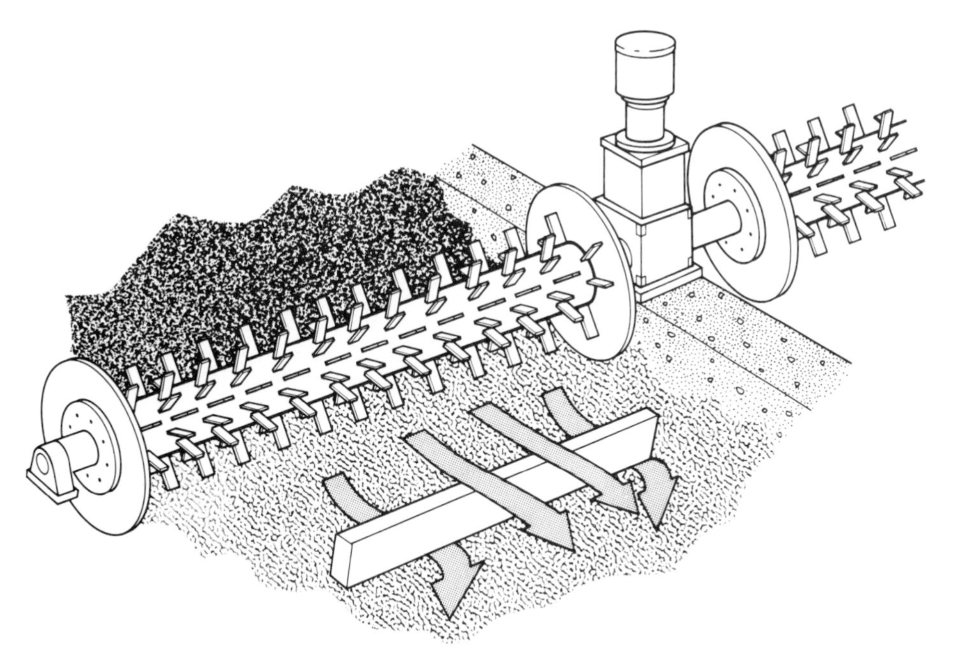

FIGURE 22-26
Horizontal brush surface aerator. (*Courtesy Passavant Corporation.*)

FIGURE 22-27
Horizontal surface aerator. (*Courtesy Envirex, a Rexnord Company.*)

overloaded condition. In shallow ponds the mixing currents generated by surface aeration may cause erosion of the bottom beneath the aerator. In order to prevent this, a flat plate called an *antierosion assembly* is mounted a short distance below the end of the intake of the aerator. This creates a more horizontal flow path and minimizes scour. In particularly deep ponds, the mixing may not reach the bottom. In such cases, the inlet may be extended by a *draft tube*. The circumstances under which these devices are required vary from machine to machine, but in general may be considered to be as represented in Fig. 22-28.[18]

In ponds which are originally designed to be aerated, both high- and low-speed aerators and static tube aerators are employed. The last consist of a cylinder containing a helical core (Fig. 22-29). Air is released from pipe distributors through a coarse diffuser at the bottom of the tubes. The upward flow is constrained to follow the helical path which extends its contact time and improves oxygen transfer. The air flow carries liquid with it and thus improves mixing of the basin. Simple perforated tube aerators have also been developed, but exhibit a tendency to plug and require considerable maintenance on that account.

Air requirements in aerated ponds may be calculated as for activated sludge processes. Power requirements, however, are usually dictated by mixing considerations (see Table 22-5), since organic loading rates are moderately low.

The procedures outlined above will permit tentative selection of aeration equipment and estimation of power requirements. Specification of aeration equipment should be based on performance in the actual system. The specification should address oxygen requirements in the system at a range of solids concentrations, dissolved oxygen concentrations, and temperatures as well as maximum power, mixing performance, noise, and other factors.[34]

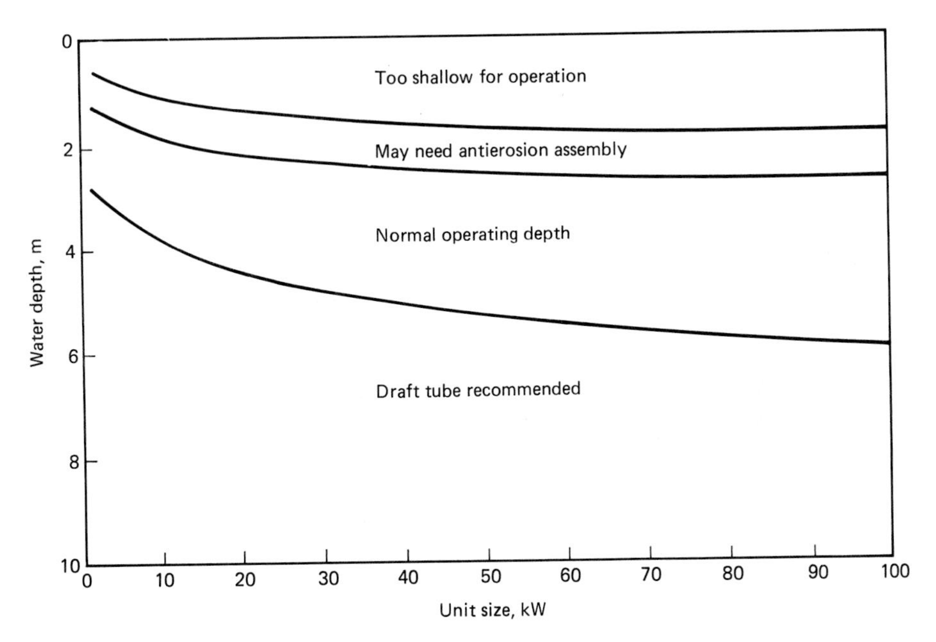

FIGURE 22-28
Application range of surface aerators.

22-11 Clarifier Design for Suspended Growth Processes

The design of the secondary clarifier is critically important to the operation of a suspended growth process. Not only do the solids which pass the clarifier contribute to the BOD of the effluent, but their loss may interfere with maintenance of the required sludge age in the biological reactor. The clarifier has a thickening function as well as one of clarification since a reasonable underflow density is necessary for solids recycle.

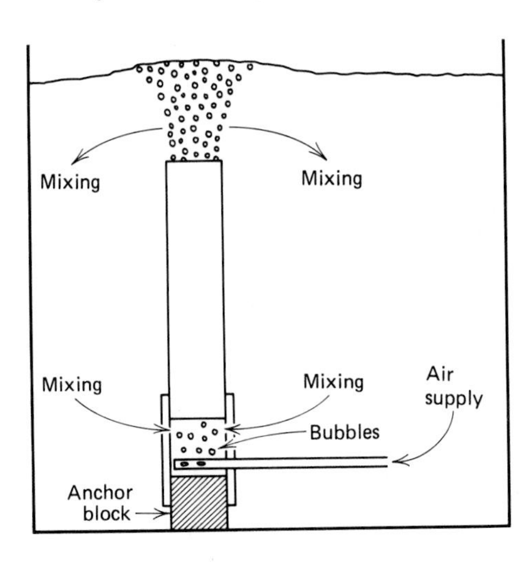

FIGURE 22-29
Static tube aerator.

The designer must consider the peak flow rate which is likely to enter the clarifier and its effect on surface overflow rate, weir loading rate, and solids loading rate. The sludge handling equipment should be sized to recirculate up to 100 percent Q to provide for short-term overloads.

Clarifiers may be circular or rectangular. The length of rectangular basins and the diameter of circular basins ordinarily does not exceed 10 times the depth, but dimensions are not critical. The important consideration is that the sludge collection system be of high capacity to ensure rapid removal in ordinary operation, high flow rates during unusual conditions, and capability to remove dense sludges which may accumulate during clarifier shutdown. Suction or siphon sludge removal systems are particularly suitable in secondary clarifiers and are available in configurations to fit both rectangular and circular basins.

Inlet baffles must be carefully designed since density currents are more pronounced in secondary than in primary clarifiers. Horizontal velocities are limited to about 0.5 m/min (1.5 ft/min) in rectangular basins and the annular inlet baffle in center-fed circular tanks should have a diameter equal to 15 to 20 percent that of the tank itself.

Solids loading is important in secondary clarifiers following suspended growth processes, since hindered settling is likely to occur and the settling velocity of discrete particles may not govern the basin design. The solids loading rate is expressed in kg/($m^2 \cdot$ h) of suspended solids and is equal to the mixed liquor suspended solids (MLSS) concentration times the flow (including recirculation) divided by the basin area:

$$\text{Solids loading rate} = \frac{\text{MLSS} \times (Q + Q_r)}{A_{\text{basin}}} \tag{22-27}$$

Typical solids loadings rates vary from 2.5 to 6.2 kg/($m^2 \cdot$ h) at average and peak loading conditions respectively. Design rates should be based on settling studies of the type described in Chap. 9. Somewhat simpler tests have recently been suggested[35] which may permit development of a settling flux curve from a single settling test.

Surface overflow rates should not exceed 57 m/day [1400 gal/($ft^2 \cdot$ day)], but will normally be less, since the solids loading rate ordinarily will govern except in processes with very low MLSS levels.

Weir loading rates should normally be kept below 370 m^2/day [30,000 gal/(ft $\cdot$ day)] for weirs located away from the tank periphery in water at least 3.5 m (12 ft) deep, and below 250 m^2/day [20,000 gal/(ft $\cdot$ day)] if located either against the wall, where density currents might turn upward, or in shallower basins. These values are based on peak flow rate. Loadings at average flow in small basins should be substantially less. The upward velocity in the vicinity of the weir should not exceed 3 to 5 m/h (10 to 15 ft/h). A great deal still remains to be learned about optimum design of secondary clarifiers[36] and standards may be expected to continue to change as additional information is developed.

Example 22-13 Design a secondary clarifier for an activated sludge process with a recycle rate of 30 percent, an MLSS concentration of 3000 mg/L, an anticipated peak flow of 10,000 m^3/day and an average flow of 3500 m^3/day. The basin is to be circular and the weir is along the wall.

Solution
The peak solids loading is

$$3 \text{ kg/m}^3 \times 10^4 \text{ m}^3\text{/day} \times 1.3 = 39{,}000 \text{ kg/day}$$

The average loading is

$$3 \times 3500 \times 1.3 = 13{,}650 \text{ kg/day}$$

The area required is 39,000/(24 × 6.2) = 262 m^2 at peak flow and 13,650/(24 × 2.5) = 228 m^2 at average flow. The peak flow governs, and A = 262 m^2 (2820 ft^2). A circular clarifier with this area will have a radius of 9.125 m and a perimeter of 57.4 m. The weir loading rate is

$$10{,}000/57.4 = 174 \text{ m/day at peak flow}$$

and

$$3500/57.4 = 61 \text{ m/day at average flow}$$

Both these values are below the peak rates specified above. With a depth of 3.5 m the retention time at peak flow will be 2.2 h.

The quantity of sludge produced in suspended growth processes is a function of the sludge age and may be calculated as discussed in the examples above. Not all the waste sludge must be treated, since a portion of the term $\Delta xV/\Delta t$ consists of the suspended solids in the effluent. The sludge from suspended growth processes is unlikely to thicken to more than 2 percent solids (20,000 mg/L), hence the volume which must be handled in the processes discussed in Chap. 23 will be large.

22-12 Operational Problems of Suspended Growth Processes

Operational problems of suspended growth processes generally hinge upon the inability to maintain the desired sludge age. Sludge may be lost from the clarifier as a result of bulking or floating, even when the clarifier is properly designed from a standpoint of solids loading.

Floating sludge results from denitrification (Chap. 24) and is associated with high sludge age and long solids retention time in the clarifier. In activated sludge processes with a sludge age in excess of 10 days and dissolved oxygen in excess of 2 mg/L, a degree of conversion of ammonia to nitrate will occur. If the sludge remains in the clarifier for too long its endogenous metabolism will exhaust the oxygen available and the bacteria will then reduce any nitrate present to nitrogen gas. The resulting bubbles may then buoy the sludge to the surface.

Floating sludge can be controlled by increasing the recirculation rate, thereby shortening the solids detention time in the clarifier, or by increasing the rate of solids wasting, thereby reducing the sludge age.

Bulking sludge results from the presence of filamentous microorganisms or from entrained water in individual cells. Filamentous growth consisting of fungi occurs in wastes deficient in nitrogen or other nutrients or at low pH. The problem can usually be corrected by providing the lacking nutrients and a neutral pH. The nutritional requirements of activated sludge are approximately as presented in Table 22-6. The values listed are not absolute minima, but amounts which are known to be adequate.

Poorly settling bacterial suspensions may include species such as *Beggiatoa* which are associated with low ratios of organic matter to cell mass (low *F*/*M* ratios) and low dissolved oxygen levels. The problem is less common in plug flow reactors (in which the organic load is high at the head end). This observation has led to a modification of the completely mixed activated sludge process in which a *selector* is added ahead of the completely mixed basin.[37–39] The selector consists of one or more chambers installed in series ahead of the aeration basin. The return sludge is mixed with the wastewater in these basins and aerated for a short period prior to entering the larger basin. The *F*/*M* ratio is high in these small basins, since no dilution with treated wastewater is provided. This condition favors the development of nonfilamentous species and thus "selects" the proper microbial population. A plug flow system, of course, provides a similar environment.

Instead of modifying the process as described above, bulking can often be minimized by reducing the sludge age. Clarifiers which have pronounced den-

TABLE 22-6
Approximate nutritional requirements in activated sludge (mg/mg of BOD)

Element	Concentration
Nitrogen	0.050
Phosphorus	0.016
Sulfur	0.004
Sodium	0.004
Potassium	0.003
Calcium	0.004
Magnesium	0.003
Iron	0.001
Molybdenum	Trace
Cobalt	Trace
Zinc	Trace
Copper	Trace

sity currents may permit short-circuiting with loss of fresh sludge and development of a mixed liquor with an average cell retention time which is different from (and longer than) that indicated by the sludge age. The actual sludge age should be kept in the range of 5 to 15 days to minimize the likelihood of bulking.

Bulking can be corrected in the short term by chlorination at a rate of about 0.5 to 1 percent of the sludge concentration. The organisms which are killed may appear in the plant effluent, producing a brief increase in BOD.

Stabilization ponds often have high concentrations of algal cells in their effluent. These cells may cause violations of both suspended solids and BOD limits. The cells can be removed by intermittent sand filtration, dissolved air flotation, coagulation and sedimentation, and filtration. None of these techniques are inexpensive, and their use would remove the cost advantage which lagoons offer to small communities.

A technique in which the effluent flows horizontally through a rock filter built in the form of a dike has demonstrated the capacity to reduce the cell mass substantially. The method is not filtration, but may, by providing flow through small, dark chambers, encourage settling of the algal cells. There has been no determination as yet with regard to the life of such rock barriers. It is possible that the void spaces may eventually fill. A variety of techniques for upgrading oxidation pond effluent may be found in Ref. 40.

22-13 Other Secondary Processes

Although most biological treatment processes may be characterized as either attached or suspended growth processes, there are some systems which combine features of both. As has been mentioned above, trickling filters constructed of sheet plastic media which are loaded at very high hydraulic rates exhibit many characteristics of mechanically aerated activated sludge processes. In a sense, the trickling filter could be considered as simply an aeration device, since most of the biological solids are suspended in the flow and the film on the medium is quite thin.

A proprietary process combines a horizontal redwood medium in a trickling filter stage with short-term aeration followed by clarification (Fig. 22-30). The settled solids from the clarifier are returned to the trickling filter together with recirculation taken from the bottom of the filter. This process is called *activated biofiltration* and is designed using curves and equations developed by the manufacturer. In recent studies of this process, it was found that the BOD removal obtained was less than the design prediction,[41] hence pilot studies are desirable prior to construction.

The concept of combining trickling filtration with an aeration or an activated sludge process has also been applied to systems employing other media in the filter and different flow patterns. Primary treatment is reportedly important to the stability of this process.[42] The bulk of the BOD removal appears to occur in the aeration process,[42,43] and the loading on the filter can be very

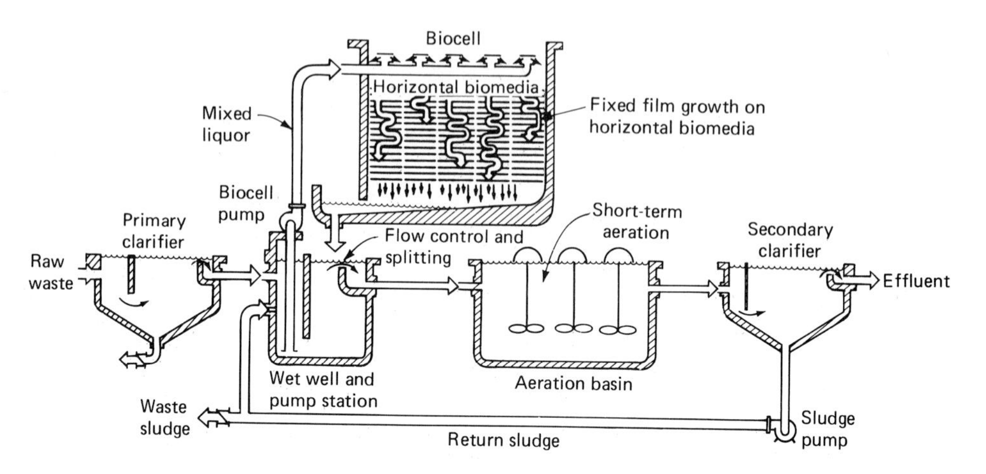

FIGURE 22-30
Activated biofilter process diagram. (*Courtesy Neptune Micrfloc.*)

high—up to nearly 1 kg/(m^3 · day) [60 lb/(1000 ft^3 · day)]—while still producing a high-quality effluent.[44]

Although the aeration basin is a very significant part of the process, its size can be quite small.[42] Detention times are typically on the order of 1 h. Clarifier overflow rates higher than those used for pure activated sludge processes yield a satisfactory effluent. Design is currently based on pilot studies, and the process is usually applied in expansions of existing trickling filter plants. Addition of short-term aeration after the filter may permit overloaded facilities to meet discharge permit limits.

The coupled trickling filter/activated sludge process appears to combine the best features of the two separate processes. Stability is very good, since the filter can accommodate substantial variations in loading. Aeration periods are short and the biological solids settle well.

Another proprietary process, developed in Japan, consists of an aeration basin in which woven plastic filaments called *ring lace* are suspended. The filaments serve as a support for development of attached growth, and the required oxygen is supplied by diffused air. This system appears to bear certain process similarities to aerated RBC systems, but has the apparent advantages of mechanical simplicity and better protection against environmental effects such as temperature extremes.

It is to be expected that further development of biological process technology will continue and that new systems will be regularly offered to the design engineer. In evaluating these new technologies, it is important to remember that the biological fundamentals are unchanged—the technique is simply the manner in which the natural process is applied.

22-14 Selection of Mechanical Equipment

Design of biological treatment systems involves sizing of treatment vessels and selection of process equipment. In some applications, it appears that these two tasks are intimately related, since manufacturers may claim that their equipment/process configuration offers substantial savings with regard to sizing of treatment vessels. Determining whether these claims are true requires very careful consideration of the manufacturer's data and examination of existing installations.

In considering the performance of existing treatment plants which are applying a new technology, one must recognize that such plants are almost certainly not operating at their design condition, but rather are receiving a lower flow and organic load. Satisfactory performance under such conditions is no guarantee of satisfactory performance at design load, nor is satisfactory performance of systems operated under laboratory conditions. Technologies such as high purity oxygen activated sludge and RBC systems were originally claimed (and largely accepted) as being inherently superior to older processes. As full-scale plants built in the last 20 years have approached design loadings, it has developed that the cost savings resulted at least in part from underdesign which was not evident at lower loads.

The design procedure is best separated into selection of a process, design of the process (reactor sizes, return flows, air/oxygen requirements, sludge production, etc.), and selection of process equipment. The process equipment consists of pumps, compressors, clarifier mechanisms, valves, diffusers, and other appurtenances which, typically, are available from a variety of manufacturers. Selection of a single manufacturer's equipment (except when there is a clear process need for a particular one-of-a-kind unit) leads to higher costs and can create legal problems for the engineer.

Process equipment should be described whenever possible by using a performance specification which establishes conditions of operation, required efficiency, durability, etc. Manufacturers are happy to provide "sample" specifications which may be useful to the designer in formulating requirements. These "samples" must be used with care, since it is very easy to incorporate some feature (material, manner of operation, dimensions) which can only be provided by a single supplier.

PROBLEMS

22-1. Determine the filter volume required according to the modified Velz equation in each of two identical filters if the waste in Example 22-4, Art. 22-3, is to be treated to yield a BOD of 30 mg/L. Use a recirculation ratio of 125 percent, a depth of 2 m, and a temperature of 25°C.

22-2. Repeat Prob. 22-1 using the NRC equation.

22-3. Write a computer program which will solve the modified Velz equations for the effluent BOD in a two-stage filter system. The input data should include temperature, influent BOD, the rate constants, and the volume, depth, and recirculation ratio for each unit.

22-4. Repeat Prob. 22-3 for the NRC equations. The input data in this case includes the influent BOD and the volume and recirculation ratio for each filter.

22-5. Using the program of Prob. 22-3, determine and plot the effect of distribution of total filter volume among the two units. That is, is it better to make the first filter larger than the second? The second bigger than the first?

22-6. Using the program of Prob. 22-3, determine the effect of varying the recirculation ratio on effluent BOD.

22-7. Repeat Prob. 22-5 using the program of Prob. 22-4.

22-8. Repeat Prob. 22-6 using the program of Prob. 22-4. Is there a maximum useful recirculation rate?

22-9. A wastewater flow of 2200 m^3/day with a BOD of 120 mg/L is to be treated in a trickling filter system to produce an effluent BOD of 25 mg/L. Design a system which will achieve this goal. Show that the system will work. Do you believe this is an economical design? Why did you choose this particular combination of volume, recirculation, number of units, etc.?

22-10. Determine the effluent BOD which will be produced by the design of Prob. 22-9 when the flow is 3300 m^3/day and the BOD is 140 mg/L. Is this result surprising?

22-11. Rework Example 22-5, Art. 22-4, using a three-stage rather than two-stage process.

22-12. Write a computer program to determine the area required in an RBC process as a function of influent BOD, rate constants, effluent BOD, and the number of stages.

22-13. Use the program of Prob. 22-12 to determine the relationship between the number of stages and the total area for influent BOD values from 180 to 120 and effluent BOD values from 20 to 30.

22-14. Design an RBC system to treat a waste flow of 12,500 m^3/day and to produce an effluent BOD of 25 mg/L when the influent BOD is 175 mg/L. Why did you chose this particular combination of stages?

22-15. Determine the effluent BOD from the process designed in Prob. 22-14 when the flow is 16,000 m^3/day and the influent BOD 150 mg/L.

22-16. Design an activated sludge system to produce an effluent BOD and suspended solids concentration of 25 mg/L when treating a waste flow of 15,000 m^3/day with a BOD of 160 mg/L. Make reasonable assumptions with regard to values of x_r and x. Find the basin volume, the recirculation rate, the quantity of waste sludge (mass and volume) produced per day, and the required air supply if the efficiency of transfer of O_2 is 10 percent. Compare your values to those in Table 22-3.

22-17. Calculate the effluent BOD produced by the process of Prob. 22-16 when the flow is 20,000 m^3/day and the BOD 140 mg/L. What does this result mean?

22-18. Determine the power required for surface aeration of a waste which has an oxygen demand of 1800 kg/day. The plant is at an elevation of 1000 m, the temperature range is 5 to 20°C, the required dissolved oxygen is 2 mg/L, α is 0.85, and M is 0.6 kg/MJ at standard test conditions.

22-19. Rework Prob. 22-18 for a diffused air system with a transfer efficiency of 9 percent, an inlet pressure of 0.95 atm, an outlet pressure of 1.70 atm, and a mechanical efficiency of 75 percent. The air temperature range is -5 to 30°C.

22-20. Calculate and plot the effluent soluble BOD and effluent total BOD produced by an oxidation pond treating domestic sewage with a BOD of 200 mg/L over a range of 0.5 to 60 days. Use enough values of time to define the shape of the curves. What seems to be the optimum retention time?

REFERENCES

1. Steven E. Harris et al., "Intermittent Sand Filtration for Upgrading Waste Stabilization Pond Effluents" *Journal of Water Pollution Control Federation*, **49**:83, 1977.
2. Khurshed J. Mistry and David M. Himmelblau, "Stochastic Analysis of Trickling Filter," *Journal of Environmental Engineering Division, American Society of Civil Engineers*, **101**:EE3:333, 1975.
3. "Sewage Treatment at Military Installations," *Sewage Works Journal*, **18**:5:787, 1946.
4. R. S. Rankin, "Evaluation of the Performance of Biological Beds," *Transactions of American Society of Civil Engineers*, **120**:823, 1955.
5. C. J. Velz, "A Basic Law for the Performance of Biological Beds," *Sewage Works Journal*, **20**:4, 1948.
6. Wesley W. Eckenfelder, Jr., "Trickling Filter Design and Performance," *Transactions of American Society of Civil Engineers*, **128**, 1963.
7. Bruce E. Jank and W. Ronald Drynan, "Substrate Removal Mechanism of Trickling Filters," *Journal of Environmental Engineering Division, American Society of Civil Engineers*, **99**:EE3:187, 1973.
8. W. F. Ames, V. C. Behn, and W. C. Collins, "Transient Operation of the Trickling Filter," *Journal of Sanitary Engineering Division, American Society of Civil Engineers*, **88**:SA3:21, 1962.
9. Bruce E. Logan, Slawomir W. Hermanowicz, and Denny S. Parker, "Engineering Implications of a New Trickling Filter Model," *Journal of Water Pollution Control Federation*, **59**:1017, 1987.
10. Bernard Atkinson and John A. Howell, "Slime Holdup, Influent BOD and Mass Transfer in Trickling Filters," *Journal of Environmental Engineering Division, American Society of Civil Engineers*, **101**:EE4:585, 1975.
11. Kenneth Williamson and Perry L. McCarty, "A Model of Substrate Utilization by Bacterial Films," *Journal of Water Pollution Control Federation*, **48**:1:9, 1976.
12. Terry L. Johnson and Gayle P. Van Durme, "Design and Evaluation of Biofilter Treatment Systems," *Journal of Water Pollution Control Federation*, **59**:1043, 1987.
13. James A. Mueller, "Oxygen Theory in Biological Treatment Plant Design—Discussion," *Journal of Environmental Engineering Division, American Society of Civil Engineers*, **99**:EE3:381, 1973.
14. "Filtering Materials for Sewage Treatment Plants," Manual of Practice No. 13, American Society of Civil Engineers, New York.
15. Tyler Richards and Debra Reinhart, "Evaluation of Plastic Media in Trickling Filters," *Journal of Water Pollution Control Federation*, **58**:774, 1986.
16. John R. Harrison and Glen T. Daigger, "A Comparison of Trickling Filter Media," *Journal of Water Pollution Control Federation*, **59**:679, 1987.
17. Edward D. Schroeder, *Water and Wastewater Treatment*, McGraw-Hill, New York, 1977.
18. "Wastewater Treatment Plant Design," Manual of Practice No. 36, American Society of Civil Engineers, New York, 1977.
19. W. A. Sack et al., "Operation of Air Drive Rotating Biological Contactors," *Journal of Water Pollution Control Federation*, **58**:1050, 1986.
20. Great Lakes–Upper Mississippi River Board of State Sanitary Engineers, *Recommended Standards for Sewage Works*, Health Education Service, Albany, N.Y., 1971.
21. Larry W. Benefield and Clifford W. Randall, *Biological Process Design for Wastewater Treatment*, Prentice-Hall, Englewood Cliffs, N.J., 1980.
22. Echol E. Cook and Leonard Crane, "Effects of Dosing Rates on Trickling Filter Performance," *Journal of Water Pollution Control Federation*, **48**:2723, 1976.

23. Ronald L. Antonie, *Fixed Biological Surfaces—Wastewater Treatment*, CRC Press, Cleveland, 1976.
24. Rao Y. Surampali and E. Robert Baumann, "Supplemental Aeration Enhanced Nitrification in a Secondary RBC Plant," *Journal of Water Pollution Control Federation*, **61**:200, 1989.
25. D. Jenkins and W. E. Garrison, "Control of Activated Sludge by Mean Cell Residence Time," *Journal of Water Pollution Control Federation*, **40**:1905, 1968.
26. Alonzo W. Lawrence and Perry L. McCarty, "Unified Basis for Biological Treatment Design and Operation," *Journal of Sanitary Engineering Division, American Society of Civil Engineers*, **96**:SA3:757, 1970.
27. Stephen M. Bidstrup and C. P. Leslie Grady, Jr., "SSSP—Simulation of Single Sludge Processes," *Journal of Water Pollution Control Federation*, **60**:351, 1988.
28. R. W. Sackellares, W. A. Barkley, and R. D. Hill, "Development of a Dynamic Aerated Lagoon Model," *Journal of Water Pollution Control Federation*, **59**:877, 1987.
29. Denny S. Parker and M. Steve Merrill, "Oxygen and Air Activated Sludge: Another View," *Journal of Water Pollution Control Federation*, **48**:2511, 1976.
30. A. A. Kalinske, "Comparison of Air and Oxygen Activated Sludge Systems," *Journal of Water Pollution Control Federation*, **48**:2472, 1976.
31. Jeppe S. Nielsen and Murray D. Thompson, "Operating Experiences at a Large Continuously Fed, Intermittently Decanted Activated Sludge Plant," *Journal of Water Pollution Control Federation*, **60**:199, 1988.
32. Pierre Côté et al., "Bubble-free Aeration Using Membranes: Process Analysis," *Journal of Water Pollution Control Federation*, **60**:1986, 1988.
33. Metcalf and Eddy, Inc., *Wastewater Engineering: Treatment/Disposal/Reuse*, 2d ed., McGraw-Hill, New York, 1979.
34. John R. Stukenberg, Valery N. Wahbeh, and Ross E. McKinney, "Experiences in Evaluating and Specifying Aeration Equipment," *Journal of Water Pollution Control Federation*, **49**:66, 1977.
35. Eric J. Wahlberg and Thomas M. Keinath, "Development of Settling Flux Curves Using SVI," *Journal of Water Pollution Control Federation*, **60**:2095, 1988.
36. Rudy J. Tekippe and Jon H. Bender, "Activated Sludge Clarifiers: Design Requirements and Research Priorities," *Journal of Water Pollution Control Federation*, **59**:865, 1987.
37. Glen T. Daigger, Millard H. Robbins, Jr., and Brian R. Maxwell, "The Design of a Selector to Control Low F/M Filamentous Bulking," *Journal of Water Pollution Control Federation*, **57**:220, 1985.
38. Grace Novak, Glen Brown, and Allan Yee, "Effects of Feed Pattern and Dissolved Oxygen on Growth of Filamentous Bacteria," *Journal of Water Pollution Control Federation*, **58**:978, 1986.
39. Stephen R. Linne and Steven C. Chiesa, "Operational Variables Affecting Performance of the Selector-Complete Mix Activated Sludge Process," *Journal of Water Pollution Control Federation*, **59**:716, 1987.
40. *Design Manual—Municipal Wastewater Stabilization Ponds*, Environmental Protection Agency, 625/1-83-015, 1983.
41. Madan L. Arora and Margaret B. Umphres, "Evaluation of Activated Biofiltration and Activated Biofiltration/Activated Sludge Technologies," *Journal of Water Pollution Control Federation*, **59**:183, 1987.
42. Raymond N. Matasci, Christopher Kaempfer, and James A. Heidman, "Full-Scale Studies of the Trickling Filter/Solids Contact Process," *Journal of Water Pollution Control Federation*, **58**:1050, 1986.
43. Brooks W. Newbry, Glen T. Daigger, and Diane Taniguchi-Dennis, "Unit Process Tradeoffs for Combined Trickling Filter and Activated Sludge Process," *Journal of Water Pollution Control Federation*, **60**:1813, 1988.
44. R. N. Matasci, et al., "Trickling Filter/Solids Contact Performance with Rock Filters at High Organic Loading," *Journal of Water Pollution Control Federation*, **60**:68, 1988.

CHAPTER 23

SLUDGE TREATMENT AND DISPOSAL

23-1 Importance of Sludge Management

The bulk of the suspended solids which enter a wastewater treatment plant and the waste solids generated in biological treatment must be handled as sludge at some point in the treatment process. The character and amount of the solids depend on the number and type of industries within the community, the degree to which their wastes are pretreated before discharge to the public sewers, and, to some extent, the primary and secondary processes employed within the treatment plant. The choice of primary and secondary processes may thus be influenced by anticipated problems with sludge handling. Vesilind[1] has noted that solids handling accounts for 30 to 40 percent of the capital costs, 50 percent of the operating costs, and 90 percent of the operational problems at sewage treatment plants. In view of the presently proposed federal regulations dealing with disposal of sewage treatment plant residues,[2] it is unlikely that these percentages will be reduced in the future.

Over 7 million tons of dry sewage solids are produced in the United States each year.[3] This material is not disposed of as a dry solid, but contains varying amounts of moisture so that the total mass actually involved may well exceed 70 million tons per year. This represents a volume of over 63 million m^3 (83 million yd^3) per year—this in addition to other solid wastes which constitute a major management problem in themselves.

The sludge from municipal treatment plants is likely to contain microorganisms which may contribute to the transmission of disease as well as organic and inorganic contaminants which may be hazardous or toxic to humans or have detrimental effects on the environment in general. For these reasons the proper handling of this material is a very important component of the design and operation of a wastewater treatment plant.

23-2 Amount and Characteristics of Sludge

Sewage sludge consists of the organic and inorganic solids present in the raw waste and removed in the primary clarifier, plus organic solids generated in secondary treatment and removed in the secondary clarifier or in a separate thickening process. The inorganic fraction may be assumed to have a specific gravity of about 2.5, while the organic matter has a specific gravity of 1.01 to 1.06, depending on its source.

The quantity of solids in raw domestic wastewater is typically 90 g/day per capita, with the reported concentration of 100 to 350 mg/L depending largely on the flow. Communities which contain major industries may vary from this value.

Of the solids in the raw waste, approximately 60 percent may be expected to be removed in primary clarification. The remainder is either oxidized in biological treatment or incorporated in the biological mass. The biological solids generated in secondary treatment average 0.4 to 0.5 kg/kg BOD applied in attached growth processes. They depend on the sludge age in suspended growth processes and range from 0.2 to 1.0 kg/kg BOD applied.

Example 23-1 Estimate the solids production from a trickling filter plant treating a flow of 1000 m^3/day with a BOD of 210 mg/L and suspended solids of 260 mg/L. Assume that the primary clarifier removes 60 percent of the suspended solids and 30 percent of the BOD.

Solution
The primary will produce solids at the rate of

$$0.60(260)(1000 \times 10^3) = 156 \times 10^6 \text{ mg/day}$$

The secondary will receive a BOD load of

$$0.70(210)(1000 \times 10^3) = 147 \times 10^6 \text{ mg/day}$$

from which solids will be generated at the rate of

$$0.5(147 \times 10^6) = 74 \times 10^6 \text{ mg/day}$$

The total mass of *dry* solids will thus be approximately 230 kg/day.

The solids production will, of course, vary from day to day and, as noted earlier, will contain a greater or lesser amount of water, depending in part on the processes involved. Mixed trickling filter and primary sludges may range from 5 to 10 percent solids and mixed primary and waste activated sludge from 2 to 5 percent solids, while waste activated sludge alone may contain less than 1 percent solids under adverse circumstances.

The moisture associated with sewage solids is in part free and separable by sedimentation, in part trapped in the interstices of floc particles and separable by mechanical dewatering, in part held by capillary action and separable by compaction, and in part chemically bound within or without the bacterial cells

TABLE 23-1
Constituents of waste-activated sludge

Constituent	Percentage by weight
Free water	70–75
Floc water	20–25
Capillary water	1–2
Bound water	1–2
Solids	0.5–1.5

and separable only by destruction of the cells. The relative proportions of water and solids in waste-activated sludge are presented in Table 23-1.

The effect of moisture content on sludge volume is enormous, and sludge handling techniques are directed toward reducing the moisture content, and thereby the volume of the sludge. Some of the processes discussed in detail below may be expected, in addition, to provide some reduction in the dry mass.

Example 23-2 A wastewater treatment plant produces 1000 kg of dry solids per day at a moisture content of 95 percent. The solids are 70 percent volatile with a specific gravity of 1.05 and 30 percent nonvolatile with a specific gravity of 2.5. Determine the sludge volume

(*a*) as produced
(*b*) after an anaerobic digestion process which reduces the volatile solids by 50 percent and produces a sludge with a moisture content of 90 percent
(*c*) after a dewatering process which reduces the moisture content to 75 percent
(*d*) after a drying process which reduces the moisture content to 10 percent
(*e*) after incineration, which leaves only dry nonvolatile solids

Solution
The original mass of sludge is 1000/0.05 or 20,000 kg/day. 1000 kg is solid, of which 70 percent is volatile with specific gravity 1.05 and 30 percent nonvolatile with specific gravity 2.50. The original sludge thus contains:

Component	Mass, kg	Volume, L
Volatile solids	700	667
Nonvolatile solids	300	120
Water	19,000	19,000
Total	20,000	19,787

After digestion, the volatile solids are reduced to 350 kg and the moisture content to 90 percent. The dry mass is thus 650 kg and the total mass is 650/0.10 = 6500 kg.

Component	Mass, kg	Volume, L
Volatile solids	350	333
Nonvolatile solids	300	120
Water	5,850	5,850
Total	6,500	6,303

After dewatering, the total mass is 650/0.25 = 2600 kg.

Component	Mass, kg	Volume, L
Volatile solids	350	333
Nonvolatile solids	300	120
Water	1,950	1,950
Total	2,600	2,403

After drying, the total mass is 650/0.90 = 722 kg

Component	Mass, kg	Volume, L
Volatile solids	350	333
Nonvolatile solids	300	120
Water	72	72
Total	722	525

And after incineration,

Component	Mass, kg	Volume, L
Volatile solids	0	0
Nonvolatile solids	300	120
Water	0	0
Total	300	120

The volume after each process, expressed as a percentage of the original, is (*a*) 100, (*b*) 32, (*c*) 12, (*d*) 3, (*e*) 1.

Although all these processes would not be applied in such a sequence, the calculations give an indication of the effect of moisture content on the volume. The most common combination is probably digestion followed by dewatering, which reduces the volume in this example by 88 percent. One may also observe that the calculation of the volumes of the individual fractions does not yield results significantly different from those which would be obtained by assuming that the entire mass had the density of water until the moisture content falls below about 50 percent. The mass of sludge itself cannot be estimated with great accuracy, so the sort of detailed calculation illustrated here is seldom justified.

23-3 Sludge Conditioning

Sludge conditioning includes a variety of processes, some biological, some chemical, and some physical, which may be applied to favorably alter the physical and chemical characteristics of sludge to improve its dewaterability. The biological and thermal processes also provide some decrease (perhaps as much as 50 percent) in the total mass of solids.

Conditioning processes include aerobic and anaerobic sludge digestion, composting, chemical addition, and heat treatment. The selection of a conditioning or digestion process should be made on the basis of projected costs and ease of operation, not on consideration of past practice. Modern conditioning techniques are more stable and may be less expensive than digestion processes, which in any event produce a sludge which may, depending on the dewatering technique, require further chemical treatment.

23-4 Digestion Processes

Digestion processes may be modeled as suspended growth processes without solids return (Chap. 22). Solids return in these systems is ineffectual because of the high proportion of inert material and the likelihood that the active biological population will settle less readily than the other solids. The minimum solids retention time in digestion processes is strongly affected by temperature: θ_c^m ranges from about 2 days at 35°C to 10 days at 20°C. θ_c^d is usually larger by a factor of 2 to 4. Digestion provides some reduction in populations of disease-causing organisms which may be present in the sludge and reduces the total solids mass and improves dewaterability as well.

Anaerobic digestion processes offer certain advantages over aerobic systems. The anaerobic process requires no oxygen supply, and, while mixing is desirable, the intensity required is not particularly high. In addition, anaerobic processes produce methane, which can be used as a source of energy within the treatment plant. On the other hand, the liquid separated from the solids after anaerobic digestion is somewhat worse in quality than that from aerobic systems.

Modern anaerobic digesters are both heated and mixed. The temperature is normally maintained at close to 35°C and mixing is sufficient to completely intermix the contents once daily. Most digesters are mixed more thoroughly, but complete mixing in the sense in which it exists in an activated sludge process is not necessary. Mixing may be provided by recirculated gas from the head space above the mixed liquor to the bottom of the tank or by mechanically driven propellers mounted in draft tubes (Fig. 23-1). A modest amount of mixing is provided by withdrawal and return of sludge for heating, but this is not sufficient to maintain a uniform concentration in the digester. The addition of waste sludge should be relatively constant in order to avoid upsetting the process.[4] Addition should be made at least twice daily, and more frequent loading is desirable.

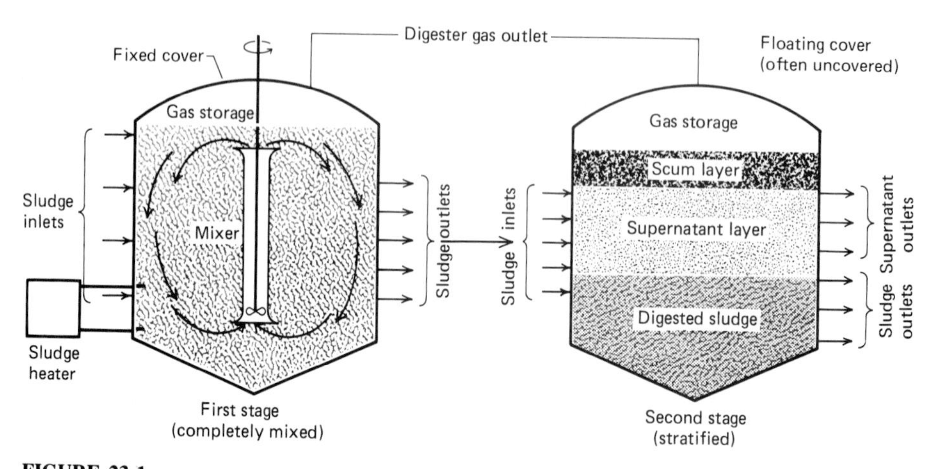

FIGURE 23-1
High-rate anaerobic digester. (*From Wastewater Engineering by Metcalf and Eddy, Inc. Copyright 1972 by McGraw-Hill, Inc. Used with permission of McGraw-Hill Book Company.*)

Heating has been provided in the past by circulating hot water through pipes within the digester. This has proven to be a poor technique, since the solids tend to cake on the warm surface and interfere with heat transfer to the liquid. A better method involves withdrawing a small flow from the digester and passing it through an external heat exchanger. The solids will still cake on the hot surfaces, but the external unit can be cleaned without draining the digester.

The amount of heat energy required depends on the temperature of the incoming sludge, the temperature desired in the digester, the temperature outside the digester, and the thermal conductivity of the digester walls, roof, and bottom. Procedures for calculating the required heat input may be found in Refs. 5 and 6.

In general, the methane produced supplies adequate heat energy to maintain the digester temperature. Under favorable circumstances it may do substantially more: there are plants in the United States in which the methane is used as a fuel in modified gasoline engines which drive generators or compressors. The cooling water from the engines serves to warm the digesters, and nearly all the energy needs of the plant are met.

The gases produced in anaerobic digestion consist of 60 to 70 percent methane by volume, the remainder being carbon dioxide and traces of ammonia, nitrogen, hydrogen sulfide, and hydrogen. Because of the contaminants (particularly the CO_2 and H_2S), it is not a very good fuel. The CO_2 is not combustible (and amounts to about 30 percent by volume), while the H_2S produces sulfuric acid when it is burned. Both these gases can be stripped by simple absorption techniques before the gas is used as a fuel.

Anaerobic digestion of sludge involves the biological breakdown of a portion of the solids. The dewaterability of sludge is primarily related to the particle size distribution, with poor dewatering being associated with a high percentage of small particles.[7,8] The destruction of these particles occurs in a two-stage process in which the complex organics in the waste solids are first broken down to simpler compounds, notably the short-chain volatile fatty acids. These intermediates are then further broken down, by a separate group of bacteria, to methane and carbon dioxide. The overall anaerobic process may be depicted schematically as shown in Fig. 23-2.

The intermediate formation of organic acids can lead to toxicity problems because of either depression of the pH or the cations associated with the acids. Additionally, heavy metals may exhibit toxic effects since they may be solubilized under the chemically reduced conditions which exist in anaerobic systems. The metals may be precipitated as sulfides by chemical addition if they are known to be present.[9] It is far preferable to exclude such materials by requiring pretreatment of industrial wastes, not only to protect the function of the digester, but also to ensure that the sludge can be disposed of without violation of EPA or state regulations.[2]

The obligate anaerobes which convert organic acids to methane and CO_2 are more sensitive to the surrounding environment than are the facultative organisms which effect the first step; hence toxic constituents will affect them first and cessation or slowing of their activity can lead to failure of the digester. It is not sufficient to monitor pH, since the alkalinity of anaerobic digesters is very high and substantial changes in acid concentration can occur with little or no change in pH. Regular measurement of volatile acids is desirable, but the normal function can often be monitored by observing the rate of gas production. Digester problems are always accompanied by a change in gas production or composition.

Design of anaerobic digesters is still largely empirical. Criteria for low-rate (unmixed) and high rate (mixed) systems are presented in Table 23-2.

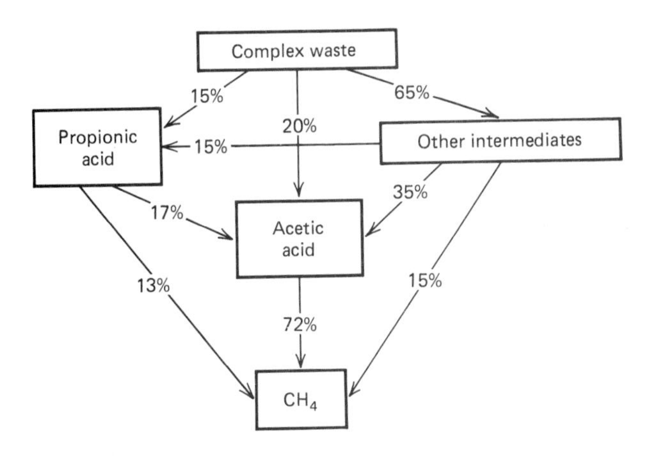

FIGURE 23-2
Metabolic pathways in anaerobic digestion.

TABLE 23-2
Anaerobic digestion design criteria

Design criterion	Process	
	High rate	Low rate
Solids retention time, days (θ_c)	10–20	30–60
Volatile solids loading (kg/m^3 per day)	2.4–6.4	0.06–1.6
Volume (m^3 per capita)		
Primary only	0.04–0.06	0.06–0.09
Primary and trickling filter	0.08–0.09	0.12–0.14
Primary and activated sludge	0.08–0.12	0.12–0.17
Digested solids concentration, %	4–6	4–6
Volatile solids reduction, %	50	60
Gas production (m^3/kg VSS added)	0.53	0.65
Methane content, %	65	65

Modern digesters are two-stage high-rate processes in which the first stage is heated and mixed and the second is quiescent and serves largely as a thickener for the digested sludge (Fig. 23-1). The critical factor in design is θ_c. The other factors in the table are artifacts which result from the typical characteristics and quantities of domestic sewage sludge.

Example 23-3 Sludge production in a "typical" plant treating domestic wastewater amounts to about 100 g/day per capita, of which 70 percent is volatile. The sludge is commonly 95 percent water. Calculate the volatile solids loading and the volume per capita in a digester with a detention time of 20 days.

Solution
The sludge mass per capita per day will be

$$100/0.05 = 2000 \text{ g}$$

and the volume approximately 2 L. At a 20 day detention time, the volume per capita is thus

$$20 \times 2 = 40 \text{ L} = 0.04 \text{ m}^3$$

If the sludge has a higher moisture content, the required volume will be greater. The volatile solids load will be

$$(0.10 \text{ kg})/(0.04 \text{ m}^3) = 2.5 \text{ kg/(m}^3 \cdot \text{day)}$$

The digested solids from anaerobic processes may be dewatered, without further treatment, by air-drying techniques. Chemical addition is usually required for satisfactory mechanical dewatering. Anaerobic processes are avoided by some designers because of their high first cost, reported susceptibility to biological upsets, and mechanical complexity. The choice in a given circumstance should be based on estimated costs and anticipated operational problems. Ordinary domestic sludges which do not contain significant quantities of industrial waste are readily digested in properly designed and operated anaerobic digesters.

Aerobic digestion processes are less susceptible to biological upsets than anaerobic systems. An aerobic process offers many different biological pathways and many different organisms which can be employed in the oxidation of complex organic materials. For this reason, aerobic systems are less likely to be upset by toxic substances.

Aerobic digestion produces volatile solids reductions comparable to those in anaerobic digestion,[10] has lower BOD in the liquid separated from the digested solids, fewer operational problems, and lower capital cost. Against these advantages must be counted the high energy cost for aeration and mixing, the lack of a useful product such as methane, and the difficulty of enhancing the process rate through heating. Heating is generally not feasible because raising the temperature reduces the solubility of oxygen, thus complicating oxygen transfer, and because mixed processes with long detention times tend to approach the air temperature. It is possible to provide a heated closed tank process through the use of high-purity oxygen, but this is only done in plants which already use high-purity oxygen for the secondary process. In such systems, a degree of heating may be provided by the fermentation reaction itself if the solids loading rate is sufficiently high.

Design criteria are presented in Table 23-3. The solids detention time for satisfactory dewatering should be kept in excess of 10 days and should be longer at low temperatures. If the sludge is digested in a high-shear regime such as that provided by high-speed aerators, or if it is permitted to become anaerobic after treatment, it will be difficult to dewater.[11] Properly handled, aerobically digested sludge can be dewatered by air-drying techniques. As with anaerobically digested sludge, chemical treatment is likely to be required prior to dewatering with mechanical systems.

Example 23-4 Design an aerobic digestion process for a plant which produces 750 kg of primary solids and 450 kg of waste activated sludge per day. The mixed sludge has a solids content of 4 percent.

TABLE 23-3
Aerobic digestion design criteria

Parameter	Value
Retention time θ_c^d	
Activated sludge only	15–20 days
Activated sludge plus primary	20–25 days
Air required (diffused air)	
Activated sludge only	20–35 L/(min · m^3)
Activated sludge plus primary	55–65 L/(min · m^3)
Power required (surface air)	0.02–0.03 kW/m^3
Solids loading	1.6–3.2 kg VSS/(m^3 · day)

Solution

The mass of the waste sludge is

$$\frac{750 + 450}{0.04} = 30{,}000 \text{ kg/day}$$

and its volume is approximately 30 m^3/day. For a retention time of 20 days, the volume is 600 m^3. The air required is in the neighborhood of 36 m^3/min. The solids loading, assuming the solids are 70 percent volatile, is

$$1200 \times 0.70/600 = 1.4 \text{ kg/(m}^3 \cdot \text{day)}$$

A higher loading would be possible at this sludge age only if the sludge were thickened before digestion.

23-5 Composting

Composting might be considered a disposal process, since it is possible, under some circumstances, to sell composted sewage sludge. As a general rule, however, the market for compost is less than the supply, hence most composted sludge is disposed of in sanitary landfills.

Composting of sewage sludge is less complex than composting of general municipal refuse, since it is relatively uniform in character and in particle size. The process involves aerobic thermophilic decomposition of a part of the organic content with a reduction in the number of pathogens. The sludge must be in a solid condition, with a moisture content less than 60 percent but more than 40 percent. This requires prior treatment or addition of a solid material such as wood chips or ground tire casings.[12] The solid material can be recovered, albeit not completely, by screening of the finished compost.

The optimum temperature is in the range of 55 to 65°C (130 to 150°F) and the optimum pH between 6 and 7.5. In general, pH control is neither feasible nor necessary. Temperature is affected by moisture content, aeration rate, mechanics of the process, and atmospheric conditions. The temperature will increase as a result of biological activity and can be reduced by increased mixing and/or aeration or by increased moisture content. Design procedures for composting may be found in Ref. 6. The major operational problem associated with composting is reportedly production of odors.[12,13]

Composting of sewage sludge will produce a reduction in volatile solids and will yield a product with a moisture content of 60 percent or less. For a raw sludge with a moisture content of 95 percent, the reduction in volume will be in the range of 95 percent. Although some revenue may be anticipated from sale of the compost, provided it does not violate EPA or state regulations for heavy metals or toxic organics, the ratio of revenue to cost in 1986 was estimated to range from \$2 to \$7 per \$100.[12] The major benefit of composting, then, must be seen as a reduction in volume. The presently proposed limits for contaminants in compost[2] are likely to complicate sale of this material.

23-6 Chemical Processes

Chemical conditioning may be applied to sludges which have been digested or to raw sludges. In the latter case, the objective may be reduction of nuisance potential (stabilization) as well as improvement in dewatering characteristics. Chemical conditioning processes include application of coagulants, elutriation, chlorination, heat treatment, freezing, and the addition of bulking agents.

The chemicals used in coagulation include the metallic and polymeric agents discussed in Chap. 9. The mechanism of chemical conditioning consists of neutralization of charge and formation of a polymeric bridge which incorporates the individual sludge particles into a lattice structure with sufficient rigidity and porosity to permit the water to drain. When stabilization is also provided (as with lime or chlorine) the additional mechanism involves disinfection, which prevents the decay of the organic material and resultant odor production.

The chemical dosages required are a function of pH, alkalinity, phosphate concentration, solids concentration, sludge storage time, and sludge makeup, i. e., whether primary, secondary, or a mixture of both.

Lime may be used alone as a conditioning or stabilizing agent. The dosages in the two cases are not identical. The quantity of lime which will permit dewatering is not sufficient for cessation of bacterial activity unless very intense mixing is provided.[6] For *stabilization*, dosages must be sufficient to raise the pH to 12 and maintain it at that level for at least 2 h and prevent the pH from falling below 11 for at least the time required to dispose of the sludge. Dosages range from 10 to 50 percent of the dry mass of solids in the sludge, being lowest for primary sludge and highest for waste activated sludge. If the pH is maintained above 12 for at least 2 h, bacterial reductions will be substantially better than those obtained in digestion processes.

Lime and ferric chloride are often used together in conditioning of sludges, particularly before pressure filtration. Typical dosages are presented in Table 23-4. Lime dosages, it may be noted, are substantially less in this

TABLE 23-4
Dosages of $FeCl_3$ and lime for sludge conditioning (percent of sludge concentration)

	Pretreatment					
	None		Digestion		Digestion plus elutriation	
Sludge	$FeCl_3$	CaO	$FeCl_3$	CaO	$FeCl_3$	CaO
Primary	1–2	6–8	1.5–3.5	6–10	2–4	
Primary and trickling filter	2–3	6–8	1.5–3.5	6–10	2–4	
Primary and activated sludge	1.5–2.5	7–9	1.5–4	6–12	2–4	
Activated sludge only	4–6					

process than in lime stabilization. Sludges which are chemically conditioned with lime or lime and $FeCl_3$ are generally not readily dewatered by air-drying techniques. Mechanical dewatering is commonly used in conjunction with these processes.

Example 23-5 Determine the required chemical dosages of lime and lime plus $FeCl_3$ for a raw primary sludge containing 8 percent solids. Assume the stabilization dose is 12 percent of the solids mass.

Solution

From Table 23-4, the lime and $FeCl_3$ dosages are estimated to be 7 and 1.5 percent of the sludge solids concentration, respectively. The dosages are thus

$$\text{Lime} = 0.07 \times 80{,}000 = 5600 \text{ mg/L}$$

$$FeCl_3 = 0.015 \times 80{,}000 = 1200 \text{ mg/l}$$

For lime alone, the dosage is

$$\text{Lime} = 0.12 \times 80{,}000 = 9600 \text{ mg/L}$$

Polymeric coagulants are widely used in conditioning of both raw and digested sludges prior to mechanical dewatering processes. Dosages, like the dosages of other conditioning chemicals, vary from sludge to sludge and from time to time. Typical values range from 0.02 to 0.75 percent of the solids concentration in the sludge, being lowest for primary sludge and highest for waste activated sludge. The dosage for the 8 percent solids sludge of the example above would be expected to be less than 40 mg/L, which indicates the substantial economies which may be available through the use of polymers.

An individual polymer, while generally effective, may not be capable of coagulating all the solids fractions within a particular sludge. Those fractions which are not coagulated will tend to build up in the plant, being recirculated in the water extracted from the sludge. In time, the polymer may become ineffective because of the increase in these materials. Occasional short-term changes in coagulant will usually remove these accumulations and result in better overall operation of the plant.

Elutriation is literally a washing of the sludge. Chemical conditioning with lime is dependent in part on the alkalinity; in fact, the lime dosages listed in Table 23-4 are required primarily to react with the alkalinity of the sludge in order to reduce the required dosage of ferric chloride. Sludges, particularly those which have been digested anaerobically, have alkalinities in the range of several thousand milligrams per liter, and this can be reduced by washing (elutriating) the sludge with water of low alkalinity, such as the effluent of the treatment plant. The relative water flow depends on the process configuration and the relative alkalinities of sludge and wash water. Typical flow rates are 4 to 5 times the sludge flow with a mixing time of about 1 min followed by settling in a gravity thickener (Art. 23-7) for 3 to 4 h. Surface loading rates in a thickener after elutriation are on the order of 25 m/day [600 gal/($ft^2 \cdot$ day)], with solids loading rates of 1.5 to 2 kg/($m^2 \cdot$ hr) [7 to 10 lb/($ft^2 \cdot$ day)].[5]

Elutriation can separate fine solids from the sludge and recirculate them through the plant. This may eventually result in their loss in the effluent and will create operational problems throughout the treatment process. The cost of treating the wash water and rethickening the sludge must be weighed against the cost of the chemicals saved. The widespread use of polyelectrolytes for sludge conditioning has reduced the need for elutriation, since no benefit is obtained by elutriating sludge prior to coagulation with polyelectrolytes.

Chlorine, in very high concentration, has been used as a stabilizing and conditioning agent. The chlorine stabilization process is proprietary and is marketed under the trademark Purifax. The chlorine demand is high, ranging to as much as 1200 mg/L for a sludge with 1 percent solids. The dosage increases with solids content in a more or less linear fashion and is higher for waste activated sludge than for primary sludge.

Chlorinated sludges can be dewatered by air-drying methods or in mechanical processes. The sludge and the liquid which drains from it are very acid (pH $\approx$ 2), and heavy metals present in the solids are readily leached under these conditions. Lime sludges, on the other hand, can be expected to complex the heavy metals as carbonates and hydroxides of very limited solubility.

Thermal conditioning actually destroys the biological cells in sewage sludge, permitting a degree of moisture release not achieved in other conditioning processes. The heat-treated sludge can be dewatered by air drying, and solids concentrations of 40 to 50 percent are routinely obtained in mechanical dewatering, ordinarily without the use of supplementary chemicals. A schematic of the Zimpro thermal sludge conditioning process is presented in Fig. 23-3. The sludge solids are mixed with air and heated to 180 to 200°C. The

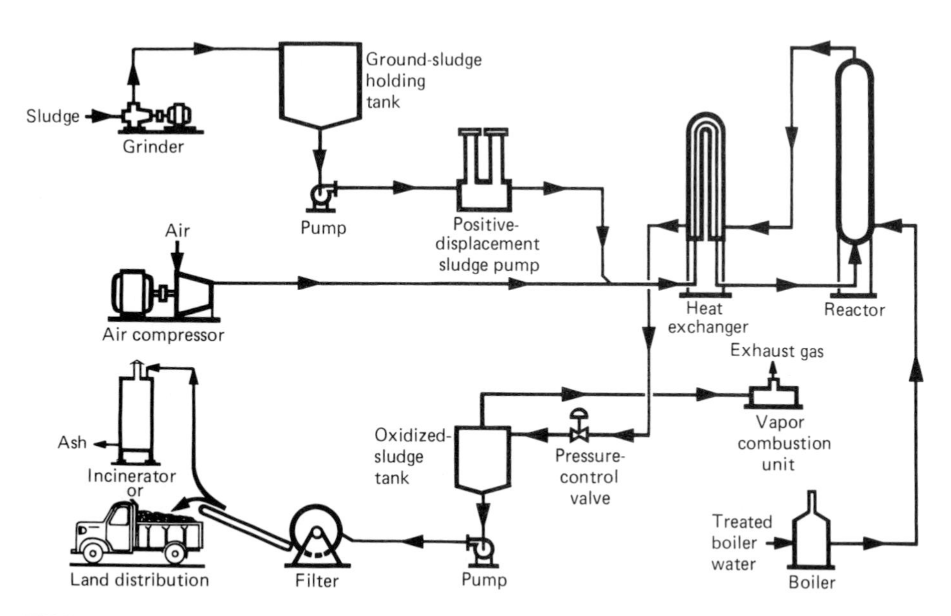

FIGURE 23-3
Thermal sludge conditioning process schematic. (*Courtesy Zimpro, Inc.*)

retention time in the process is 20 to 30 min at pressures on the order of 1500 kPa (200 lb/in^2).

Heat treatment produces a sterile sludge, and the process may be applied to wastes that contain toxic materials or are subject to wide variations in quality. Negative aspects of the process include the production of heavily contaminated gaseous and liquid process streams, which must be treated further, and high capital and operating costs. The COD of the liquid stream is about 0.6 mg/L per mg/L of solids; in effect, this returns to the biological process the bulk of the BOD removed by sludge wasting. The design capacity of aeration basins or other biological processes must be increased by as much as 15 percent to compensate for this return.[1]

Freezing under natural conditions will convert an undrainable jellylike sludge into a granular material that drains immediately when it thaws. In order for the freezing technique to be effective, the sludge must freeze throughout its depth; thus the depth is limited by the anticipated duration of the freezing period and the temperature differential below freezing. A relation between freezing depth, temperature differential, and time has been suggested:[14]

$$x = 2.04(\Delta T \times t)^{0.5} \qquad (23\text{-}1)$$

where x = depth of freezing, cm
ΔT = difference between freezing temperature and actual temperature
t = duration of freeze, days

Following a freeze-thaw cycle, a layer of sludge up to 2 m thick will drain within minutes on a sand bed to the point at which it can be removed by solids-handling equipment.[15]

Experiments with artificial freezing have not been completely successful. The sludge must freeze slowly if compaction of biological solids is to be avoided. Controlled evaporation of butane which has been mixed with the sludge has shown some promise of providing a workable process.[6]

Other conditioning processes have been developed that use materials such as fly ash, pulverized coal, and newspaper pulp. The product of such treatment is a relatively dry sludge after mechanical dewatering (40 to 50 percent solids), but the sludge volume may be only slightly decreased because of the large volumes of the additives.

23-7 Thickening

Thickening is done by a variety of concentration techniques in which relatively thin sludges are increased in solids content in order to reduce the total volume and thus the size of subsequent units. As has been shown above, a tremendous decrease in volume can be achieved through a modest increase in solids content. Thickening a waste sludge, for example, from 3 to 6 percent solids will reduce its volume by 50 percent—and permit a digester, if used, to be half as large, since sludge age is the critical design parameter.

Gravity thickeners are similar to sedimentation basins but may incorpo-

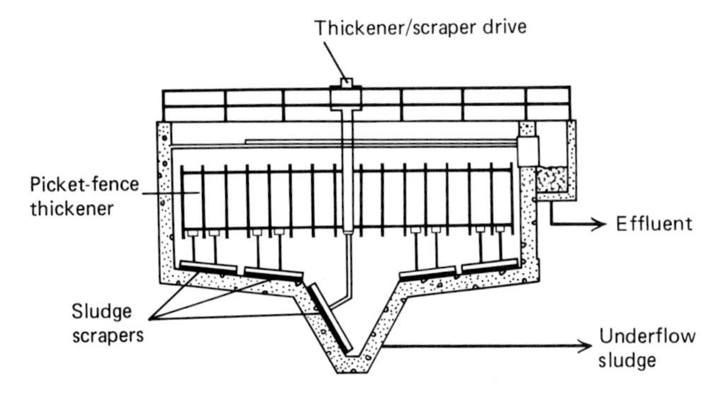

FIGURE 23-4
Gravity sludge thickener.

rate vertical pickets on the scraping mechanism (Fig. 23-4) which serve to gently agitate the sludge and aid in release of trapped gases and water.

The underflow concentration is controlled by the height of the sludge blanket in the basin. In general, the higher the blanket is, the thicker the solids will be. Excessive solids retention times, however, may lead to anaerobic activity, production of gases, and buoying of solids to the surface. Anaerobic activity may change the characteristics of aerobically digested sludge or waste activated sludge and interfere with subsequent dewatering techniques.

The design of gravity thickeners should be based on settling tests such as those described in Chap. 9 and Ref. 1. Pilot testing is desirable as well, but does not assure successful performance of a full-scale installation. In the absence of an existing plant, design may be based on data such as that in Table 23-5. The

TABLE 23-5
Loading and performance of gravity thickeners

Type of sludge	Influent solids, %	Underflow, %	Loading, kg/(m² · h)
Primary	2–7	5–10	4–6
Primary + activated sludge	0.5–4	4–7	1–3.5
Primary + trickling filter	2–6	5–9	2.5–4
Primary + rotating biological contactor	2–6	5–8	2–3.5
Activated sludge	0.5–1.5	2–3	0.5–1.5
Extended aeration	0.2–1.0	2–3	1–1.5
Trickling filter	1–4	3–6	1.5–2
RBC	1–3.5	2–5	1.5–2
Activated sludge + trickling filter	0.5–2.5	2–4	0.5–1.5
Digested primary	8	12	5
Digested primary + activated sludge	4	8	3
Thermally conditioned primary	3–6	12–15	8–10
Thermally conditioned primary + activated sludge	3–6	8–15	6–9
Thermally conditioned activated sludge	0.5–1.5	6–10	4–6

surface overflow rate is typically 15 to 35 m/day [350 to 850 gal/(ft^2 · day)] and is provided by recirculation of secondary effluent, which helps to maintain aerobic conditions in the thickener. Detention times are on the order of 3 to 4 h.

Flotation thickeners offer significant advantages in thickening light sludges like waste activated sludge. Such sludges have a density very close to that of water and thus are readily buoyed to the surface. Waste activated sludge can usually be concentrated to at least 4 percent solids by this technique. Surface overflow rates in flotation thickeners range from 10 to 45 m/day [250 to 1000 gal/(ft^2 · day)] at retention times of 30 min to 1 h. Typical solids loading rates and anticipated float concentrations are presented in Table 23-6.

In order to avoid destruction of the biological floc, the sludge itself is ordinarily not pressurized. Rather, a recycle stream is exposed to air at relatively high pressure and is combined with the incoming flow at the entrance to the basin. The reduced pressure in the basin permits the air in the recycle stream to come out of solution and float the solids to the surface.

Flotation thickeners should be designed, when possible, on the basis of laboratory or pilot testing. The important design variables are the air/solids ratio, the basin surface area, the detention time, and the recycle ratio. Typical air/solids ratios range from 0.005 to 0.06, with 0.02 being most common for activated sludge thickeners. The air/solids ratio may be calculated from

$$\frac{A}{S} = \frac{S_a(fP - 1)R}{C_0} \tag{23-2}$$

where A/S = air/solids ratio
S_a = saturation concentration of air in waste flow at 1 atm (Table 23-7)
f = fraction of saturation actually achieved (normally 0.5 to 0.8)
P = pressure, atm (usually 3.5 to 5)
R = recycle rate (the ratio of pressurized flow to waste flow)
C_0 = concentration of solids in incoming sludge

The short hydraulic detention time and the introduction of air in flotation thickeners are both favorable from the standpoint of further handling of the thickened sludge. Gravity thickeners may require return of plant effluent to maintain aerobic conditions, and, even then, the long detention time is likely to produce a loss in dewaterability. A typical flotation unit is shown in Fig. 23-5.

TABLE 23-6
Loading and performance of flotation thickeners

Type of sludge	Influent solids, %	Float, %	Loading, kg/(m^2 · h)
Primary	2–7	8–10	8–25
Primary + activated sludge	0.6–1.3	4.5–6.5	3.5–5
Primary + trickling filter	0.5–2	4.5–6.5	6–12
Activated sludge	0.5–1	2.5–7	2.5–3.4

TABLE 23-7
Solubility of air in water

Temp, °C	S_a, mg/L
0	37.2
10	29.3
20	24.3
30	20.9

The liquid extracted from the sludge in thickeners (supernatant in gravity thickeners, subnatant or underflow in flotation thickeners) will contain finely divided suspended solids and have a fairly high BOD. It usually must be returned to the head of the treatment plant.

23-8 Dewatering

Dewatering processes may be usefully divided into air drying and mechanical methods,[16] although some air-drying techniques employ mechanical equipment. Air drying includes those methods in which moisture is removed by evaporation and gravity or induced drainage such as sand beds, vacuum-assisted beds, wedgewire beds, sludge lagoons, and paved beds.

Sand beds are the oldest sludge dewatering technique and consist of 150 to 300 mm (6 to 12 in) of coarse sand underlain by layers of graded gravel ranging from 3 to 6 mm (⅛ to ¼ in) at the top to 40 mm (¾ to 1½ in) at the

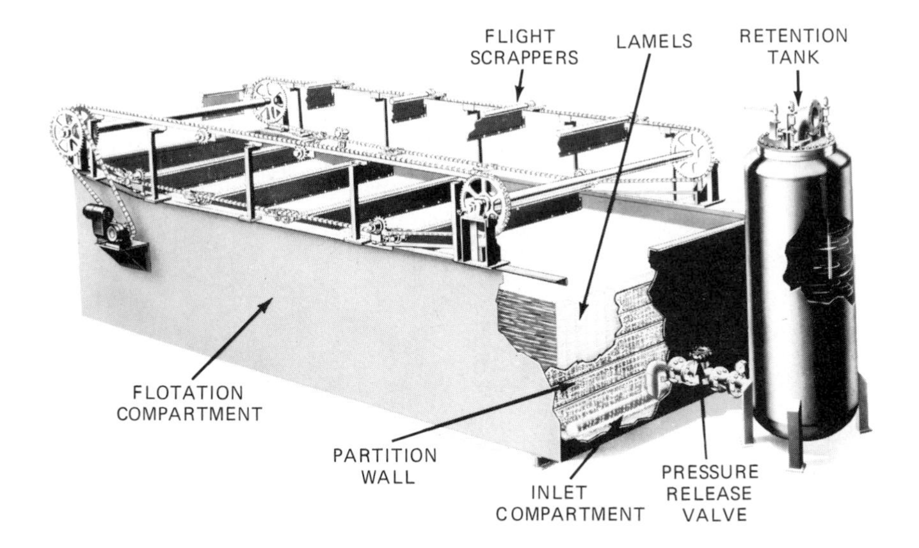

FIGURE 23-5
Flotation thickener. (*Courtesy Permutit Company, Inc.*)

bottom. The total gravel thickness is about 300 mm (1 ft). The bottom is usually natural earth graded to 100 or 150 mm (4 to 6 in) draintile placed on 6- to 9-m (20- to 30-ft) centers. Sidewalls and partitions between bed sections are of concrete or wooden planks and extend 300 to 400 mm (12 to 16 in) above the sand surface (Fig. 23-6).

The beds are 6 to 10 m (20 to 33 ft) wide and up to 40 m (130 ft) long. At least two beds are provided in even the smallest plants. Dewatering occurs as a result of drainage and evaporation and is very much affected by climate. Covering the beds with glass or translucent plastic has proven to be helpful in wet climates. Typical loading rates for digested sludges range from 60 to 200 $kg/(m^2 \cdot year)$. The higher rates apply to primary sludges in dry climates and the lower to mixtures of primary and waste activated sludge.

The beds are operated by filling them with digested sludge to a depth of 200 to 300 mm (8 to 12 in). Drying takes from a few weeks to a few months, depending on the climate and the season. After dewatering, the sludge solids content will range from 25 to 35 percent and the volume will have been reduced by 80 to 85 percent. The dried sludge can be removed from the bed with light mechanical equipment such as front-end loaders. A small amount of sand is lost in each drying cycle and the bed must be refilled periodically.

Freeze-assisted drying takes advantage of low winter temperatures to accelerate the dewatering process. As noted above, freezing separates the water from the solids and the water will drain very quickly when the frozen mass thaws. The feasibility of freezing sludge in a particular area can be related to the depth of frost penetration.[16] For purposes of preliminary assessment, the total depth of sludge which could be frozen in 75-mm (3-in) layers can be calculated from

$$Y = 1.76F_p - 101 \tag{23-3}$$

in which Y is the total depth of sludge and F_p is the maximum depth of frost penetration, both in cm. In this process the sludge is applied in 75-mm layers. When the layer is frozen, another layer is applied so that the total depth of frozen sludge may fill the bed by the end of the winter. It is very important that each layer be completely frozen before the next is applied. Any snow which falls should be swept from the surface, since it will otherwise serve to insulate the sludge.

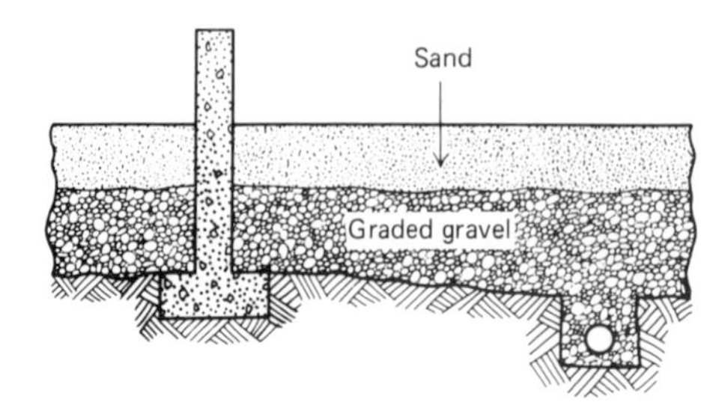

FIGURE 23-6
Section of a sludge drying bed.

Vacuum-assisted drying beds are constructed on an underdrain of porous rigid plates. A vacuum is applied to the underdrain, extracting the free water from the sludge. The time required to dewater conditioned sludges is about 1 day with solids loadings on the order of 10 kg/m^2 per cycle. After removal of the dewatered solids, the surface of the bed is washed with high-pressure sprays. The dewatered sludge can be moved by solids-handling equipment and, under optimum conditions, may approach a solids content of 35 percent, although a more normal expectation for this process is about 20 percent.

Wedgewire beds are similar in concept to the vacuum-assisted drying beds described above, with the medium consisting of a septum with wedge-shaped slots about 0.25 mm (0.01 in) wide. The bed is initially filled with water to a level above the wire screen. Polymer conditioned sludge is then added and, after a brief holding period, is allowed to drain through the screen.

The wedgewire process is proprietary, hence design information should be obtained from the manufacturer. Typical sludge loadings are reported to be 2 to 5 kg/m^2 per cycle, and the cycle time can be as little as 24 h. The sludge cake after 24 h has a solids content in the range of 10 percent. This can be handled with shovels, but is still plastic and further dewatering (longer cycle times) may be desirable.

Sludge lagoons are used both for storage and drying of waste solids. When dewatering is desired, the solids are placed in the lagoon after stabilization. Over a period of months, solids will tend to settle to the bottom and a crust will form on the surface. The crust is periodically broken up and removed, and the supernatant above the thickened solids is withdrawn and returned to treatment. The period required to fill the lagoon typically exceeds a year, and the solids content of the dewatered material reportedly ranges from 15 to 40 percent. Removal of the crust is critical to proper operation since, if left, it will prevent further evaporation of liquid.

Paved drying beds, like sludge lagoons, depend on evaporation for dewatering of the applied solids. The sludge is applied to a depth of about 300 mm (12 in) and is periodically mixed by a tractor mounted machine which is driven through the bed. The mixing breaks up any crust which forms and exposes wet surfaces to the atmosphere. Decanting of supernatant is also employed, as in sludge lagoons. Loadings employed in relatively dry climates have ranged from about 125 to 250 kg/(m^2 · year). Solids contents of 40 to 50 percent can be achieved within 30 to 40 days in dry climates.

Example 23-6 Estimate the land area required to dewater a digested municipal sludge which is produced at the rate of 1000 kg/day of dry solids at a moisture content of 95 percent. Evaluate a range of air-drying techniques.

Solution

The design criteria for the several processes range from a high of 10 kg/m^2 per cycle to a low of 60 kg/(m^2 · year). Assuming that the vacuum-assisted drying beds can be operated 200 cycles per year, their area would be

$$(1000 \times 365)/(200 \times 10) = 183 \text{ m}^2 \text{ (1970 ft}^2\text{)}$$

The sand drying beds would require an area of

$$1000 \times 365/60 = 6083 \text{ m}^2 \text{ (1.5 acres)}$$

Mechanical dewatering processes include belt filters, pressure filters, vacuum filters, centrifuges, and a number of one-of-a-kind proprietary devices. They are used in circumstances in which air-drying techniques require too much land area, odors associated with open processes are objectionable, or weather conditions do not permit their use.

Belt filters have become the most popular sludge dewatering device in new installations. Their capital cost and operating cost are generally less than for other mechanical techniques, and they produce a sludge which can be handled as a solid. These machines are made in a variety of configurations consisting of one or more endless woven belts which pass over and around a number of cylinders (Fig. 23-7). In most designs, the chemically conditioned sludge is fed to an open belt surface on which gravity drainage occurs as the belt moves forward. At the end of the gravity drainage section, provided the sludge has been properly conditioned, the solids content should be 10 percent or more. At this moisture content it behaves as a solid and can be subjected to pressure, shear, and vacuum in subsequent sections of the machine. A second belt is usually brought down toward the moving sludge at the end of the gravity zone and gradually applies pressure to the solids, squeezing out additional moisture. As the paired belts move around the rollers, their speed relative to each other varies and the sludge mass is sheared, aiding in the release of free

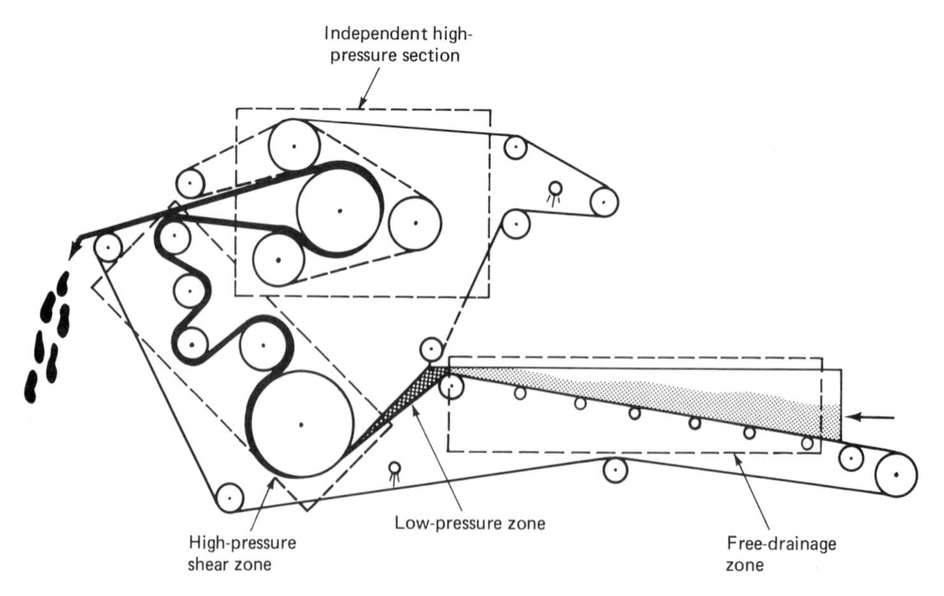

FIGURE 23-7
Schematic of belt filter.

water. The pressure gradually increases through the machine, and, in some designs, a partial vacuum applied. The belts then separate, and the sludge cake is dislodged by scraping the belt or by passing it around a very small radius roller. The product can be expected to have a solids content in the range of 12 to perhaps 40 percent, 20 percent being typical. Polymer dosages and filter loading rates which have been used for sludges of different types are presented in Table 23-8. Problems with belt filters are primarily associated with belt tracking and belt life. Large machines may have a belt width of 3 m, and it is difficult to keep the belt on the rollers as loading conditions vary during operation. Belt life is typically between 1000 and 2000 hours,[17] but may be much less if the sludge contains any large objects which can puncture the fabric.

Pressure filters consist, in most instances, of a filter press similar to that shown in Fig. 23-8. The filter support plates hold a woven filter medium against which the chemically conditioned sludge is pumped. Pumping is continued until the flow virtually ceases and the pressure is in the range of 4 to 12 mPa (600 to 1750 lb/in^2). The pressure is maintained for from 1 to 3 h. The frame is then opened, the plates are separated, and the cake is removed. The entire cycle can be automated, including dislodging of the dried cake with air. The solids content of the cake is often as high as 50 percent, but 30 to 40 percent is more likely. For very high solids content, addition of filter aids such as fly ash is required. Chemical conditioners for this process in the past have been largely limited to lime and ferric chloride; however, newer polymers have been applied successfully in some plants (Table 23-9).

Vacuum filters were, for many years, the only available alternative to sludge drying beds. In new installations they are not competitive with belt filters, but, since they are used in many existing plants, a description of their use will be presented. Chemically conditioned sludge is fed to a basin in which a large rotating drum covered with a filter medium is immersed. As the drum rotates, sludge is drawn against the filter and, as it leaves the liquid, water is drawn through the medium, leaving the solids behind on the surface. Further rotation carries the solids to a removal mechanism which may involve scraping the medium with a doctor blade, passing it around a small radius roller, or both. The filter medium may be of cloth or metal. The solids content of the product may range from as little as 5 percent for unthickened waste activated sludge to as much as 40 percent for thermally conditioned sludges. Chemical conditioning is generally with lime and ferric chloride. Polymers have generally not been applied successfully on vacuum filters. A typical vacuum filter is illustrated in Fig. 23-9, and reported performance on different sludges is presented in Table 23-10.

Centrifuges may be used as thickening devices for activated sludge or as a dewatering process for digested or chemically conditioned sludges. A typical solid-bowl horizontal centrifuge is illustrated in Fig. 23-10 . The screw conveyer or scroll rotates at a speed slightly higher than that of the bowl and thus carries the solids through the device and up the ramp to the sludge cake outlet.

TABLE 23-8
Loading and performance of belt filters

Sludge	Feed solids, %	Solids loading rate, kg/h per meter belt width	Polymer dose, g/kg	Cake solids, %
Raw:				
Primary	3–10	360–680	1–5	28–44
Waste activated sludge	0.5–4	45–230	1–10	20–35
Primary + waste activated sludge	3–6	180–590	1–10	20–35
Primary + trickling filter	3–6	180–590	2–8	20–40
Anaerobically digested:				
Primary	3–10	360–590	1–5	25–36
Waste activated sludge	3–4	40–135	2–10	12–22
Primary plus waste activated sludge	3–9	180–680	2–8	18–44
Aerobically digested:				
Primary + waste activated sludge	1–3	90–230	2–8	12–20
Primary + trickling filter	4–8	135–230	2–8	12–30
Thermally conditioned:				
Primary + waste activated sludge	4–8	290–910	0	25–50

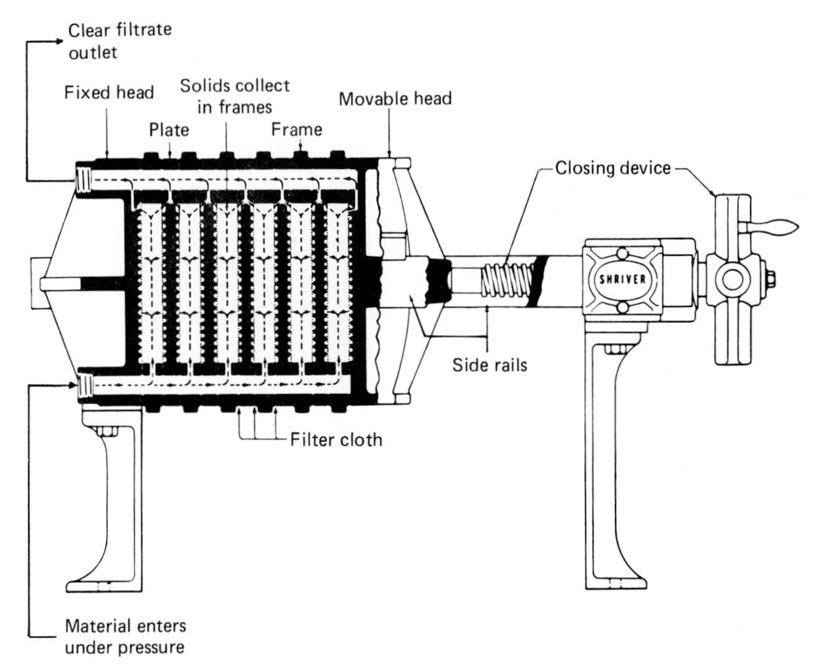

FIGURE 23-8
Filter press. (*Courtesy Shriver Division, Envirotech Corporation.*)

The variables subject to control include the bowl diameter, length, and speed; the ramp slope and length; the pool depth; the scroll speed and pitch; the feed point of sludge and chemicals; the retention time; and the sludge conditioning. Vesilind[1] presents a complete discussion of the interrelation of these variables and their effect on solids recovery and moisture content. Recovery of fine particles and light sludges can be improved by reducing the clearance between the scroll and ramp, sometimes by precoating with gypsum. In gen-

TABLE 23-9
Typical performance of pressure filters

Sludge	Conditioner dosage, %			Cake solids, %	Cycle time, min
	$FeCl_3$	Lime	Polymer		
Raw	1.5–12	15–30		36	70–330
			1–2.2	31	50–105
Anaerobic	3–15	12–43		38	52–168
			0.4–0.9	35	60–240

FIGURE 23-9
Vacuum filter. (*Courtesy Envirex, a Rexnord Company.*)

TABLE 23-10
Typical performance of vacuum filters

	Chemical conditioner			
	$FeCl_3$ plus CaO		Polyelectrolyte	
Type of sludge	**Yield, kg/(m² · h)**	**% solids**	**Yield, kg/(m² · h)**	**% solids**
Primary	20–60	25–38	40–50	25–38
Digested primary	20–40	25–32	35–40	25–32
Elutriated and digested primary	20–40	25–32		
Primary plus trickling filter	20–30	20–30	40–30	20–30
Digested primary plus trickling filter	20–25	20–30		
Elutriated and digested primary plus trickling filter	20–25	20–30		
Primary plus activated sludge	20–25	16–25	20–25	16–25
Digested primary plus activated sludge	20–25	14–22	17–30	14–22
Primary plus high purity oxygen activated sludge	25–30	20–28	20–30	20–28
Activated sludge alone	12–17	10–15	10–15	10–15

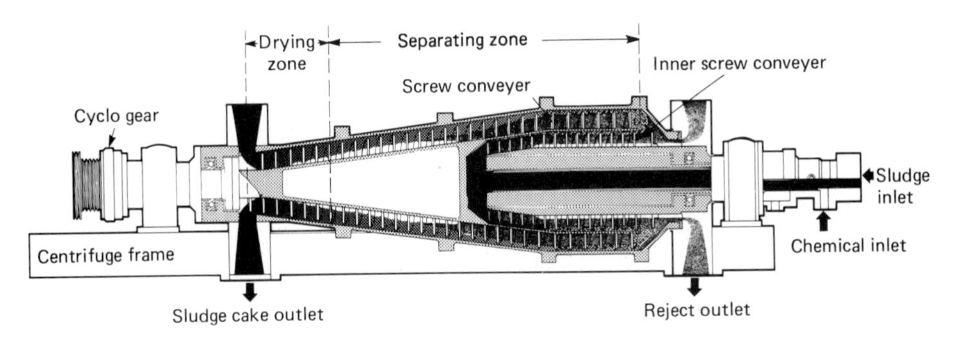

FIGURE 23-10
Horizontal solid-bowl centrifuge. (*Courtesy Environmental Division, Ingersoll-Rand Co.*)

eral, increased solids recovery in centrifuges is associated with increased moisture content in the dewatered sludge. The effect of any process change may be seen in a plot like that of Fig. 23-11. The machine must operate along the curve at whatever combination of recovery and solids concentration is acceptable. The only way the operator can move off the machine curve is by changing the characteristics of the sludge by physical, chemical, or biological conditioning.

Typical solids recoveries in newer centrifuges generally exceed 90 percent. Solids content depends on the source and degree of pretreatment and ranges from 12 to over 30 percent (Table 23-11). Centrifuges are comparable to belt filters in most respects other than solids recovery. Belt filters are generally superior in that respect.

A number of proprietary devices which employ specialized combinations of belt, press, and/or centrifuge technology have been developed in recent

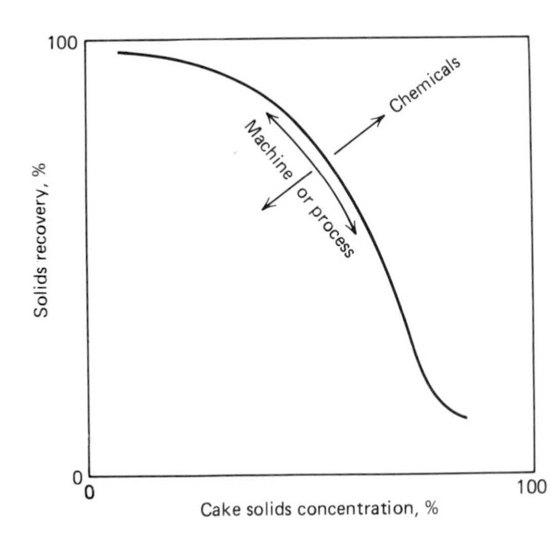

FIGURE 23-11
General solids recovery versus cake solids for a centrifuge. (*From Treatment and Disposal of Wastewater Sludge by P.A. Vesilind, Ann Arbor Science Publishers, Inc. 1974. Used with permission of Ann Arbor Science Publishers.*)

TABLE 23-11
Typical performance of centrifuge dewatering systems

Sludge	Cake solids, %	Solids recovery, %	Polymer, g/kg
Raw primary	28–34	90–95	1–2
Anaerobically digested primary	26–32	90–95	2–3
Raw waste activated sludge	14–18	90–95	6–10
Anaerobically digested waste activated sludge	14–18	90–95	6–10
Raw (primary + waste activated sludge)	18–25	90–95	3–7
Anaerobically digested (primary + waste activated sludge)	17–24	90–95	3–8
Extended aeration or aerobically digested sludge	12–16	90–95	6–10

years. The desirability of these machines for specific applications must be evaluated by the design engineer. In general, the manufacturers claim performance superior to that of other systems. The actual performance and relative costs should be established by parallel pilot testing of competing systems.

23-9 Drying and Combustion

The process of volume reduction begun in dewatering can be continued through other techniques which will reduce the moisture content to as little as 5 to 10 percent. Some incineration processes can even reduce the sludge to a moisture-free ash. The cost of these processes is justifiable only when disposal sites are very limited and haul distances are long or if the dried product can be sold as a soil amendment or fertilizer.

Flash drying consists of dispersing the sludge in a stream of hot, dry air and then separating the solids in a cyclone. The solids, depending on their makeup, may be utilized to provide all or part of the heat required for drying. In this case the process becomes a partial or complete incineration technique. Figure 23-12 is a schematic diagram of such a system.

Rotary dryers are steel drums turning at 5 to 8 r/min that are supplied with a mixture of mechanically dewatered and previously dried sludge which is exposed to a gas stream at a temperature of about 650 °C (1200 °F) for 20 to 60 min. This process, like flash drying, requires air pollution control equipment to recover fine particles and control odors.

Solvent extraction, which has been used in water treatment plant sludge handling (Chap. 11), can also be applied to wastewater sludges. Drying is more economical in a thermal sense with this process since the mixture of triethylamine and water has a lower latent heat than water alone. The process equipment is closed, hence no pollution control equipment should ordinarily be required. This process is proprietary and is still experimental.

The *Carver-Greenfield process* is a proprietary system which has been used successfully on a variety of sludges. The process employs a multiple effect

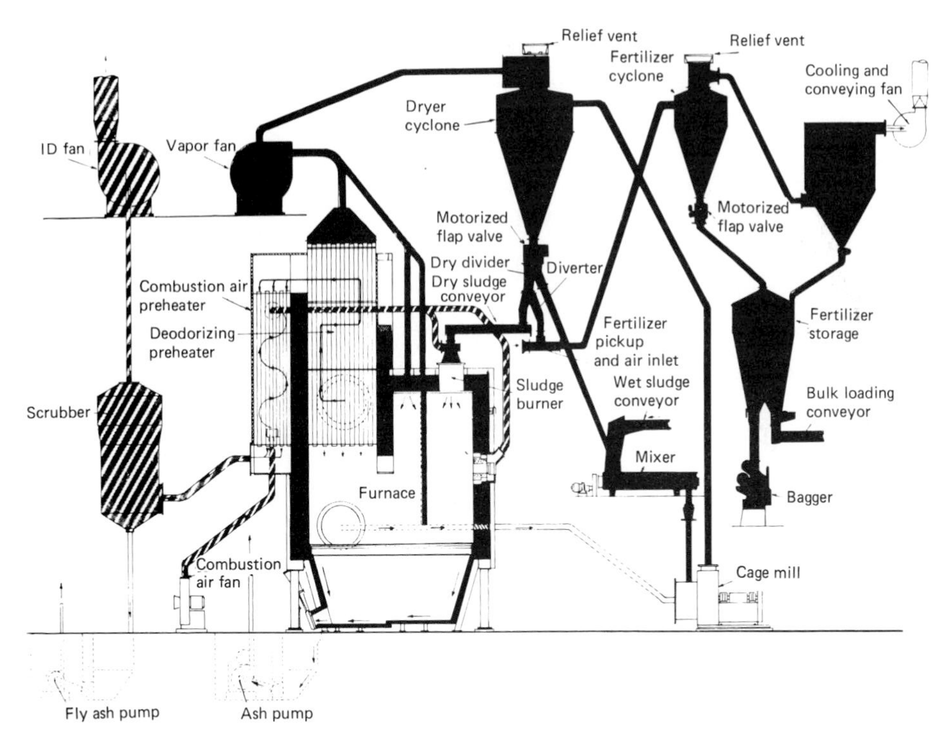

FIGURE 23-12
Flash-drying incineration system. (*Courtesy Combustion Engineering, Inc.*)

evaporator to separate water from a sludge which has first been mixed with oil. The oil is then separated by centrifugation and is returned for reuse. Liquid and gas process streams may require treatment. No particulate air contamination is likely.

Multiple-hearth incinerators (Fig. 23-13) are the most widely used sludge incineration device. Sludge enters at the top and is windrowed across each hearth, falling then to the next. As it passes through the furnace, it is dried and its temperature rises to the ignition point. The gas temperature in the furnace ranges from 550°C at the top (drying) zone to 1000°C in the middle (combustion) zone and 350°C at the bottom, where the ash is removed. Supplementary fuel is required to start the process and may be required in routine operation, depending on the moisture content and fuel value of the sludge. Major problems in operation include ineffective sludge dewatering and oversizing of the furnace,[18] both of which increase the heat required.

Multiple-hearth furnaces can be operated as *starved air processes* in which provision of less than the stoichiometric quantity of air produces incomplete combustion. Advantages claimed for this process include higher loading rates, reduced fuel consumption, more stable operation, easier control, and reduced air emissions.[6]

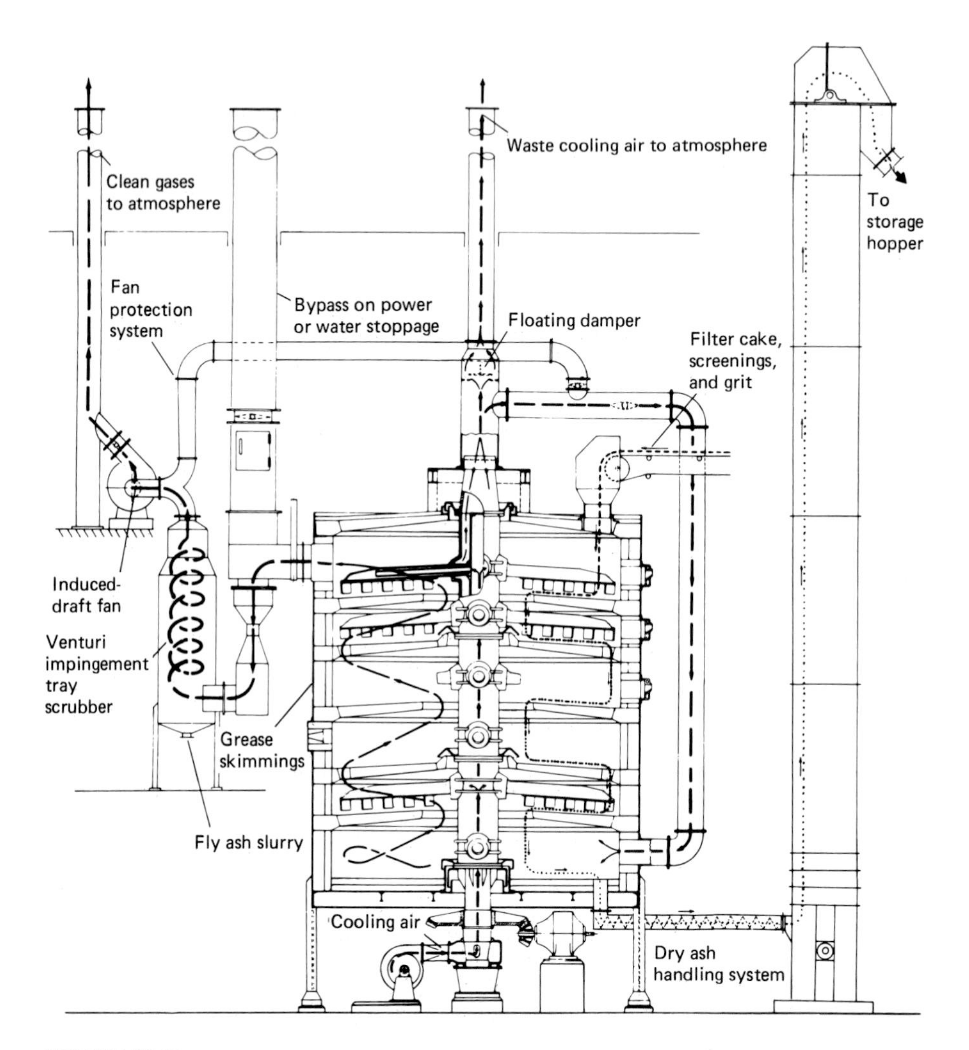

FIGURE 23-13
Multiple-hearth incinerator. (*Courtesy Nichols Engineering and Research Corporation, a subsidiary of Neptune International Corporation.*)

Fluidized-bed incinerators (Fig. 23-14) contain a bed of sand which is fluidized by the upward flow of injected sludge and air. Ash is carried out by the air flow and is removed by air pollution control systems. The sand is preheated to about 800°C prior to injection of the sludge. Depending on the fuel and moisture content of the sludge, auxiliary fuel may be required during routine operation. The sludge must be wet enough that it does not dry instantly on injection and plug the screw conveyer. An advantage of the fluidized-bed system lies in the ability of the sand to conserve heat during periods when the system is not in use. Start-up after short periods of disuse may not require supplementary fuel.

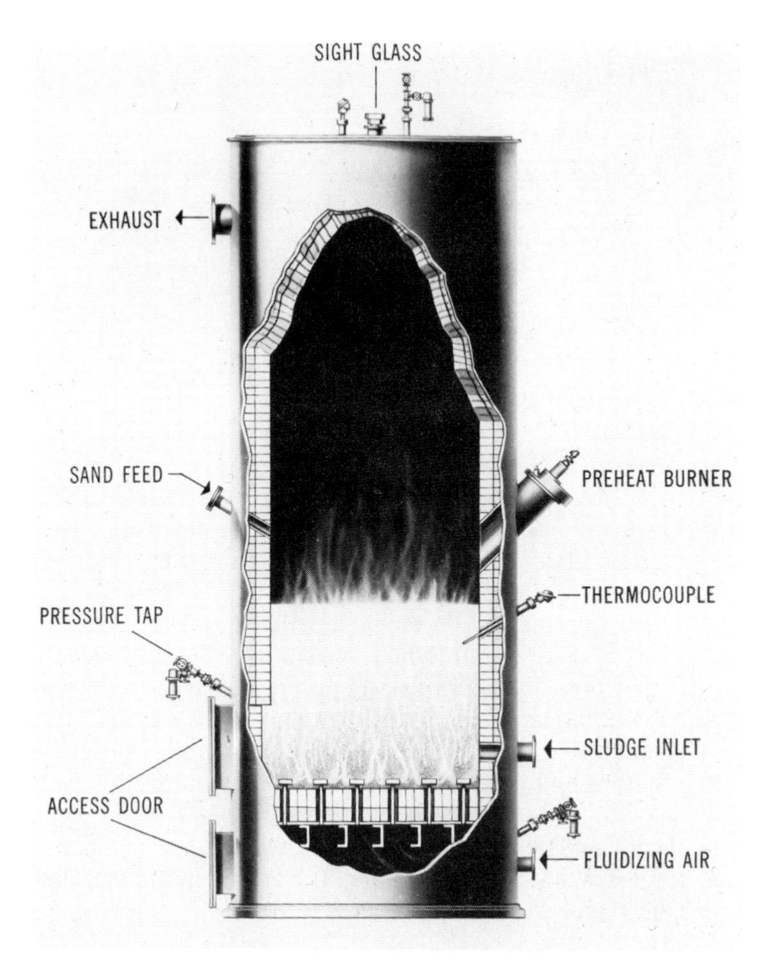

FIGURE 23-14
Fluidized-bed incinerator. (*Courtesy Dorr-Oliver, Inc.*)

Electric furnaces contain a moving woven wire belt that carries the sludge beneath infrared heating elements which heat it to the ignition point. If the sludge has a sufficiently low moisture content and a sufficiently high fuel value, the heating elements may be shut down after the process has been started. The electric furnace is less expensive than other incinerators and lends itself to intermittent operation, since a large mass does not need to be preheated. A schematic diagram of an electric furnace is presented in Fig. 23-15.

Wet oxidation is a process schematically similar to that of Fig. 23-3. Typical temperatures range from 180 to 270°C, with the optimum reported to be 230 °C.[19] Pressures vary from 7 to 20 mPa (1000 to 3000 lb/in^2). Under these conditions, the sludge is oxidized with production of a clear liquid stream which is lower in BOD than that from the thermal sludge conditioning process

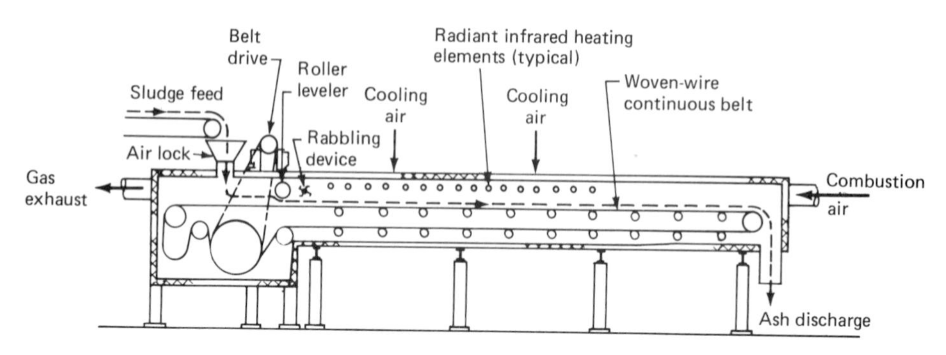

FIGURE 23-15
Schematic diagram of electric furnace.

and an ash which dewaters readily to 40 to 50 percent solids in any dewatering system. Reductions in volatile solids of 80 to 90 percent can be expected,[20] together with reductions in volume of over 96 percent after dewatering. Heavy metals in the sludge (other than copper) are concentrated in the solid residue.[19,20]

Pyrolysis is a destructive distillation technique that produces a carbon residue and a variety of gas and liquid products which may be economically recoverable. A process of this sort has been applied to sewage sludge with production of 0.2 to 0.3 g of oil per gram of volatile solids in the sludge.[21] The benefits of such a process include reduction in waste volume as well as production of a potentially useful product. Further development of this technology is likely.

In addition to the techniques discussed above, there are a number of proprietary thermal processes which are available. These techniques and their applicability to domestic sludges are discussed in Ref. 6.

23-10 Hazardous Constituents of Sewage Sludges

Sewage sludges may be expected to contain, to some degree, all of the contaminants present in the raw waste. Many of these materials are toxic and/or hazardous, and some of them are concentrated in the solid fraction by wastewater treatment. Those contaminants which have been detected in concentrations high enough to be of concern to EPA include arsenic, beryllium, cadmium, chromium, copper, lead, mercury, molybdenum, nickel, selenium, and zinc as well as 16 organic compounds.

The concentration of these materials in sewage sludge can be reduced by strict enforcement of industrial pretreatment standards, and it is very much in the interest of the municipality to do so. High concentrations of heavy metals or priority pollutants will restrict the options available with regard to ultimate disposal of the solid residue.

23-11 Management and Disposal of Residues

Whatever processes are used in handling the solids from sewage treatment, a residue will remain which must be disposed of. This material may range from raw solids at a moisture content of over 95 percent to incinerator ash.

The handling of waste solids and the ultimate disposal technique are very much interrelated. The optimum degree of volume reduction depends in large part on the distance to the disposal site, and the method of achieving the volume reduction is influenced by the proposed disposal technique. At present most of the sewage sludge in the United States is buried in sanitary landfills. About 16 percent is applied to land, slightly over 20 percent is incinerated, less than 10 percent is marketed after drying or composting, less than 6 percent is dumped in the ocean, and minor amounts are disposed of on-site at the treatment plant.

Ocean disposal has been more and more restricted in recent years. The current site used by New York and some other northeastern communities is 106 mi offshore, a distance that creates some very real economic incentives for development of alternative techniques. Congress has prohibited ocean disposal after January 1, 1992 but similar deadlines have come and gone in the past without result. The 106-mi haul is more likely to be effective.

Land disposal of sludges, like land disposal of wastewater, is subject to biological, chemical, and physical constraints. Disease transmission, odors, heavy metal accumulation, and social and esthetic problems must all be considered before land application is selected. The EPA presently distinguishes between application to agricultural land and nonagricultural land. Nonagricultural applications are expected to be limited by concentrations of specific contaminants. If prescribed levels are exceeded, application will be prohibited.

For agricultural applications, the limit of contaminants is based on total loading rather than concentration, with additional restrictions on indicators of disease-causing organisms. From a management point of view, application to agricultural land is far from perfect. Application is limited to certain seasons and the fertilizer value of the sludge is limited. Wet sludge can be applied by injection (Fig. 23-16), provided the rate is not so great that water is visible on the surface within an hour. Dewatered solids can be applied by spreading with ordinary farm equipment.

Marketing and distributing of dried or composted sludge is likely to be restricted to some degree. The regulations proposed by the EPA will limit the concentrations of specific contaminants in such products. Labeling will also be required which lists the concentrations of contaminants which might create health or environmental problems and suggests appropriate methods for application. Incidental products made with sludge as a constituent (such as bricks and concrete blocks) may also be regulated.

On-site disposal will be regulated depending on the degree to which the public has access to the site. The major concern in this instance is disease

FIGURE 23-16
Soil-injection sludge disposal. (*Courtesy Big Wheels, Inc.*)

transmission; hence those sludges treated with processes which give high microbial kills are more suitable for this method. Lime treatment, chlorination, or thermal processes should produce residues meeting the proposed standards.

Sanitary landfill disposal will probably remain the preferred technique for sewage sludge disposal. The present solid waste regulations require containment and groundwater monitoring features which should be adequate to ensure that the environment is protected. Dewatered, dried, composted, or incinerated sludges may be landfilled directly, while wet sludges, with careful handling, can be spread on the working face and mixed with other refuse. References 16 and 22 summarize current techniques for sewage sludge management and give case studies which offer a rationale for selection of processes and equipment.

PROBLEMS

23-1. Estimate the quantity of solids produced in an activated sludge plant with a flow of 5500 m^3/day and BOD and suspended solids equal to 200 mg/L. The primary removes 30 percent of the BOD and 55 percent of the suspended solids. The activated sludge plant has MLSS concentration of 2500 mg/L, a volume of 1200 m^3, and is operated at a sludge age of 8 days. Assume the mixed solids have a moisture content of 96 percent.

23-2. The solids of Prob. 23-1 are digested aerobically, which reduces the total solid dry mass by 30 percent. The digested sludge is then thickened in a flotation thickener to a solids content of 6 percent. Determine the volume after these steps and compare it to the volume of Prob. 23-1.

23-3. As an alternative to the procedure of Prob. 23-2, consider thickening the sludge before the digestion process. The thickened sludge would be at 6 percent solids, and a 30 percent reduction in solids would occur in digestion. What would be the final sludge volume in this case? What advantages (if any) would this procedure have over that of Prob. 23-2?

23-4. A digested sludge has a dry mass of 1200 kg/day and a solids content of 4 percent. Estimate the required size of a belt filter. Select the filter(s) so that all the solids can be dewatered in 8 h under normal circumstances and so that at least two machines are provided.

23-5. What depth of sludge could be frozen in an area where the minimum frost depth is 600 mm (2 ft)? How much would this add to the capacity of drying beds which could otherwise be used only 8 months per year at a loading rate of 10 kg/m^2 per month?

23-6. Determine appropriate dimensions for a gravity thickener used to thicken a primary sludge with a moisture content of 96 percent and a dry solids mass of 800 kg/day. What solids content would you expect to achieve in this process?

REFERENCES

1. P. Aarne Vesilind, *Treatment and Disposal of Wastewater Sludge*, Ann Arbor Science, Ann Arbor, Mich., 1974.
2. "Review of EPA Sewage Sludge Technical Regulations," *Journal of Water Pollution Control Federation*, **61**:1206, 1989.
3. Dan Morse, "Sludge in the Nineties," *Civil Engineering*, **59**:8:47, 1989.
4. Terence J. McGhee, "Volatile Acid Variation in Batch Fed Anaerobic Digesters," *Water and Sewage Works*, **117**:5:130, 1971.
5. Metcalf and Eddy, Inc., *Wastewater Engineering: Treatment, Disposal, Reuse*, 2d ed., McGraw-Hill, New York, 1979.
6. *Process Design Manual for Sludge Treatment and Disposal*, EPA 625/1-79-011, Environmental Protection Agency, Washington, D.C., 1979.
7. Desmond F. Lawler et al., "Anaerobic Digestion: Effects on Particle Size and Dewaterability," *Journal of Water Pollution Control Federation*, **58**:1107, 1986.
8. William R. Knocke and Terry L. Zentkovich, "Effects of Mean Cell Residence Time and Particle Size Distribution on Activated Sludge Vacuum Dewatering Characteristics," *Journal of Water Pollution Control Federation*, **58**:1118, 1986.
9. Terry M. Regan and Mercer M. Peters, "Heavy Metals in Digesters: Failure and Cure," *Journal of Water Pollution Control Federation*, **42**:1832, 1970.
10. Rajagopal Krishnamoorthy and Raymond C. Loehr, "Aerobic Sludge Stabilization—Factors Affecting Kinetics," *Journal of Environmental Engineering Division, American Society of Civil Engineers*, **115**:EE2:283, 1989.
11. C. W. Randall, J. K. Turpin, and P. H. King, "Activated Sludge Dewatering: Factors Affecting Drainability," *Journal of Water Pollution Control Federation*, **43**:102, 1971.
12. A. H. Benedict, Eliot Epstein, and John N. English, "Municipal Sludge Composting Technology Evaluation," *Journal of Water Pollution Control Federation*, **58**:279, 1986.
13. M. S. Finstein, F. C. Miller, and P. F. Strom, "Monitoring and Evaluating Composting Performance," *Journal of Water Pollution Control Federation*, **58**:272, 1986.

14. S. Reed, J. Bouzoun, and W. Medding, "A Rational Method for Sludge Dewatering via Freezing," *Journal of Water Pollution Control Federation*, **58**:911, 1986.
15. C. James Martel, "Dewaterability of Freeze-Thaw Conditioned Sludges," *Journal of Water Pollution Control Federation*, **61**:237, 1989.
16. *Design Manual—Dewatering Municipal Sludges*, EPA/625/1-87/014, Environmental Protection Agency, 1987.
17. "Belt Filter Press Dewatering of Wastewater Sludge," *Journal of Environmental Engineering Division, American Society of Civil Engineers*, **114**:EE5:991, 1988.
18. Walter G. Gilbert et al., "Improved Design and Operational Practices for Municipal Sludge Incinerators," *Journal of Water Pollution Control Federation*, **59**:939, 1987.
19. A. A. Friedman et al., "Characteristics of Residues from Wet Air Oxidation of Anaerobic Sludges," *Journal of Water Pollution Control Federation*, **60**:1971, 1988.
20. Y. C. Wu et al., "Wet Air Oxidation of Anaerobically Digested Sludge," *Journal of Water Pollution Control Federation*, **59**:39, 1987.
21. Kun M. Lee et al., "Conversion of Municipal Sludge to Oil," *Journal of Water Pollution Control Federation*, **59**:884, 1987.
22. P. Aarne Vesilind, Gerald C. Hartman, and Elizabeth T. Skene, *Sludge Management and Disposal for the Practicing Engineer*, Lewis Publishers, Inc., Chelsea, Mich., 1986.

CHAPTER 24

ADVANCED WASTEWATER TREATMENT

24-1 Purpose of Advanced Wastewater Treatment

It is possible to treat wastewater to whatever degree may be desired—to convert sewage into potable or even chemically pure water.[1-4] Advanced wastewater treatment encompasses those techniques which are applied in order to improve the quality of wastewater beyond that usually achieved in secondary treatment.

The soluble BOD of a secondary effluent is only a few milligrams per liter and, with proper design and operation, can easily be kept at that level. The total BOD, however, is also affected by the suspended solids, and reduction of suspended solids to levels less than 20 mg/L is difficult using sedimentation alone. *Suspended solids removal* is one application of advanced waste treatment and is probably the most common. Other advanced waste treatment techniques are directed toward reduction in *ammonia, organic nitrogen, total nitrogen, phosphorus, refractory organics*, and *dissolved solids*. This chapter discusses the principles of advanced waste treatment, but does not present detailed design procedures.

24-2 Suspended Solids Removal

The suspended solids in treated wastewater are in part colloidal and in part discrete, ranging in size from 10^{-3} to 100 μm. None of this material settles readily or it would be removed in secondary clarification. The techniques which have been applied to reduction of suspended solids in secondary effluents include chemical coagulation followed by sedimentation or dissolved air flotation, microscreens, diatomaceous earth filters, granular media filters, and ultrafiltration.

Chemical coagulation of water is described in Chap. 9, and the principles presented there are applicable in wastewater treatment as well. The coagulants used include lime, alum, iron salts, and polymers. Metallic coagulants also have a role in the removal of phosphorus, and lime is employed for pH adjustment in some ammonia-removal methods. Judicious combination of treatment techniques may thus permit the chemical addition to serve two or more purposes.

Solids removal after coagulation can be effected in solids contact processes or in simple sedimentation basins. In order to obtain very low suspended solids levels, some type of filtration is required. Coagulation and sedimentation of secondary effluent yields suspended solids concentrations between 3 and 25 mg/L.[5] Dissolved air flotation of the coagulated wastewater can be expected to yield a somewhat better effluent than simple sedimentation processes.

Microscreens (Fig. 24-1) consist of motor-driven drums that rotate about a horizontal axis in a basin which collects the filtrate. The drum surface is covered by a fine screen with openings ranging from 23 to 60 μm. Particles which are larger than the openings are retained by the screen as the wastewater flows from the inside of the drum to the surrounding tank. As the drum rotates,

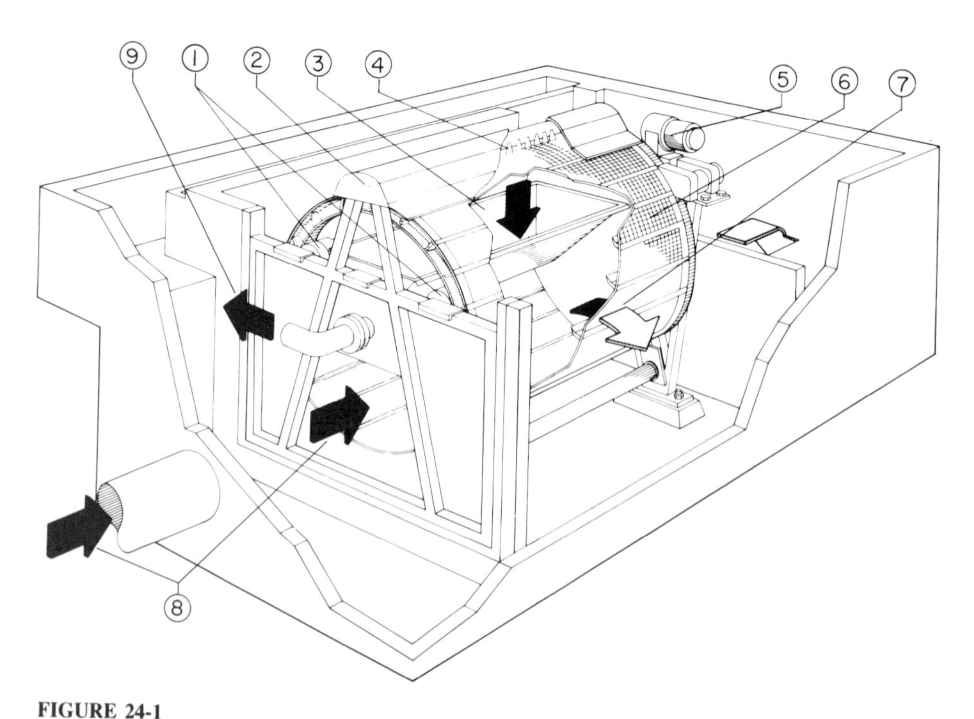

FIGURE 24-1
Microstrainer. (1) Drum support wheels; (2) drum lift; (3) screenings trough; (4) backwash spray; (5) drive; (6) grid; (7) effluent; (8) influent; (9) screenings return. (*Courtesy Zurn Industries, Inc., Envirosystems Division.*)

the screened solids are carried to the top, where they are cleaned from the screen by a high-velocity backwash spray. The backwash is collected in a trough and recirculated through the plant. Design parameters include mesh size, submergence, allowable head loss, and rotational speed.

Effluent suspended solids from microstrainers following activated sludge and trickling filter plants can be expected to range from 2 to 21 mg/L,[6,7] representing a 43 to 85 percent reduction from the level in the process influent. Typical hydraulic loadings are 0.06 to 0.44 m/min [1.5 to 10 gal/($ft^2 \cdot$ min)] on the submerged area. Backwash flow is ordinarily constant and ranges up to 5 percent of the product water. About 50 percent of this actually passes through the screen and must be recycled. The remainder runs down the outside of the screen and mixes with the filtrate.

Diatomaceous earth filters are described in Chap. 10. In applications to filtration of secondary treatment plant effluent, the precoat amounts to about 0.5 kg/m^2 and the body coat to 5 to 6 mg/L per unit of influent turbidity. Acceptable operation is reported to be possible only with influent suspended solids less than 13 mg/L,[5] and process costs are high. The effluent is comparable in turbidity to potable water, but is not suitable for drinking.

Granular media filters include the several systems discussed in Chap. 10. Filter configurations which would constitute a cross-connection in waterworks practice can be used advantageously in wastewater filtration. Upflow and biflow filters may be particularly useful in filtering waters containing high suspended solids concentrations. In addition, filters which have been specifically developed for wastewater use and which employ continuous cleaning or backwash, such as pulsed bed and traveling bridge designs, may be more economical than standard water treatment configurations.

There are some significant differences between filtration of treated wastewater and potable water. Flow rates in water plants are generally constant over extended periods of time because of the use of both upstream and downstream storage. When the flow does change, it is changed deliberately, and adjustments can be made in the number of units in service, in pretreatment, etc. In wastewater plants, on the other hand, the flow is seldom controllable by the operator and fluctuations in flow rate will affect the upstream treatment processes. In particular, the suspended solids concentration in the effluent of secondary clarifiers generally increases with increased flow. Thus, at peak flow rates, the filters will experience not only the worst flow but also the worst influent quality. For this reason, design must be based on peak flow rather than the average flow conditions which would be used in water plant design.

Design considerations include selection of an optimum filter configuration, including medium, size, and gradation, and appropriate filtration and backwash rates. The filtrate quality is generally improved with decreasing medium size and increasing depth, although the improvement is not great at low filtration rates. Media are typically coarser in wastewater than in water filtration. Effective sizes are generally greater than 1 mm. Filtration rates range from 0.06 to 0.5 m/min [1.5 to 12 gal/($ft^2 \cdot$ min)] with effluent suspended solids from 1 to 10 mg/L.[1,8] This represents a reduction of 20 to 95 percent from the concentration in the filter influent. Experimental rates up to 1.3 m/min [32 gal/(ft^2 · min)] have been applied with suspended solids reductions of 50 percent.[9] Reference 5 presents a complete discussion of design and economic factors which should be considered in selection of wastewater filtration systems.

The improvement provided by filtration of secondary effluents depends on the character of the solids, which in turn is influenced by the sludge age in suspended growth processes and the recycle rate in attached growth processes. Short sludge ages and high recycle rates adversely affect the filterability of the solids. The addition of appropriate coagulants prior to filtration or filtration of the effluent of a coagulation process will yield final turbidities similar to those in drinking water. Without further treatment, however, the water will not meet drinking water standards in other respects.

Ultrafiltration is similar in operation to reverse osmosis (Chap. 11) save that the membrane is coarser and the pressure lower. The process removes finely divided rather than dissolved solids, but has openings sufficiently small to remove colloidal particles as well. Ultrafiltration will remove all of the suspended solids and virtually 100 percent of the BOD, COD, and TOC remaining in treated wastewater. The liquid flux rate in long-term operation is about 0.33 m/day [8 gal/(ft^2 · day)] at a pressure of 170 kPa (25 lb/in^2). Design variables include the membrane area, configuration, and material, and the pressure applied. The membranes have a useful life of about 6 months in this application.

24-3 Nitrogen Removal

Nitrogen in wastewater is found as organic nitrogen, ammonia, and nitrate. Minor amounts of nitrite may also occur. When present as organic nitrogen or ammonia, nitrogen exerts an *oxygen demand* in accordance with

$$NH_4^+ + 2O_2 \rightarrow NO_3^- + 2H^+ + H_2O \tag{24-1}$$

Each milligram of ammonia (as N) exerts an oxygen demand of about 4.6 mg, if nitrogen uptake by bacterial synthesis is neglected. It is apparent from this that the nitrogenous oxygen demand of a waste can be a large fraction of the total oxygen demand since total Kjeldahl nitrogen (TKN), consisting of ammonium plus organic nitrogen, in wastewater amounts to 20 to 30 mg/L on average.

Example 24-1 A wastewater has a BOD of 220 mg/L and a TKN concentration of 30 mg/L. Estimate the total oxygen demand of the waste after primary treatment and the percentage of this contributed by nitrogen.

Solution

Assuming that primary treatment will remove 30 percent of the BOD, the ultimate carbonaceous oxygen demand will be approximately

$$L = 1.47(0.7)(220) = 225 \text{ mg/L}$$

The nitrogenous demand, neglecting removal in primary and secondary sludges, will be

$$4.6 \times 30 = 140 \text{ mg/L}$$

The nitrogenous demand thus contributes about 38 percent of the total and amounts to over 60 percent of the carbonaceous BOD.

Nitrogen in the form of ammonia is also toxic to some fish, its effect being dependent in part on the concentrations of carbon dioxide and dissolved oxygen, the pH, and the temperature as well as the concentration of ammonia.[10] Toxic levels as low as 0.01 mg/L have been reported.[11]

Most nitrogen which enters the environment will eventually be oxidized to nitrate, in which state it may be used as a nutrient by microscopic and macroscopic plants. Excess quantities of nitrogen in any form can thus contribute to eutrophication of surface waters. In addition, nitrate itself is a pollutant with respect to public water supplies.

Depending on the point of discharge and the applicable water quality standard, publicly owned treatment plants may have (1) no limit on nitrogen discharges, (2) a limit on ammonia or TKN or both, or (3) a limit on total nitrogen. Nitrogen can be removed and/or altered in form by both biological and chemical techniques. A number of methods which have been successfully applied are discussed below. The effect of a variety of secondary and advanced waste treatment processes on the nitrogen content of wastewater is summarized in Table 24-1.

Biological removal techniques include assimilation and nitrification-denitrification. The first step in the latter process, nitrification, is adequate to meet some water quality limitations since the nitrogenous oxygen demand is satisfied and the ammonia (which might be toxic) is converted to nitrate.

Biological assimilation occurs to a certain extent in any biological treatment process, since all living cells contain protein and therefore will take up nitrogen from solution. *Algal cultures* incorporate nitrogen, carbon dioxide, and trace nutrients to form algal cell mass. The algae may also produce sufficient oxygen for stabilization of the organic matter present in the raw waste, although this process (called *activated algae*) has not been demonstrated in a full-scale application. *Bacterial assimilation* can be encouraged by providing carbonaceous matter in a quantity commensurate with the nitrogen content (BOD:N ratio approximately 100:5). The disadvantages of bacterial assimilation include the cost of the carbon source—usually methanol or glucose—and the cost of handling the large quantities of sludge produced. The sludge cannot be digested or thermally treated, since this would result in re-release of the nitrogen in the cell mass. Assimilation processes, in themselves, are generally not practical, but some assimilation will occur in any biological treatment process. Assimilation can be maximized by operating at short sludge age, thus maximizing solids production.

TABLE 24-1
Effect of various treatment processes on nitrogen compounds

Treatment process	Effect on constituent			Removal of total nitrogen entering process, percent*
	Organic N	NH_3/NH_4^+	NO_3^-	
Conventional treatment processes				
Primary	10–20% removed	No effect	No effect	5–10
Secondary	15–25% removed† Urea → NH_3/NH_4^-	< 10% removed	Nil	10–20
Advanced wastewater treatment processes				
Filtration‡	30–95% removed	Nil	Nil	20–40
Carbon sorption	30–50% removed	Nil	Nil	10–20
Electrodialysis	100% of suspended organic N removed	40% removed	40% removed	35–45
Reverse osmosis	100% of suspended organic N removed	85% removed	85% removed	80–90
Chemical coagulation‡	50–70% removed	Nil	Nil	20–30
Land application				
Irrigation	→ NH_3/NH_4^+	→ NO_3^- → Plant N	→ Plant N	40–90
Inflation/percolation	→ NH_3/NH_4^+	→ NO_3^-	→ N_2	0–50

TABLE 24-1
Effect of various treatment processes on nitrogen compounds *(Continued)*

Treatment process	Effect on constituent			Removal of total nitrogen entering process, percent*
	Organic N	NH_3/NH_4^+	NO_3^-	
Major nitrogen removal processes				
Nitrification	Limited effect	$\rightarrow NO_3^-$	No effect	5–10
Denitrification	No effect	No effect	80–98% removed	70–95
Breakpoint chlorination	Uncertain	90–100% removed	No effect	80–95
Selective ion exchange for ammonium	Some removal, uncertain	90–97% removed	No effect	80–95
Ammonia stripping	No effect	60–95% removed	No effect	50–90
Other nitrogen removal processes				
Selective ion exchange for nitrate	Nil	Nil	75–90% removed	70–90
Oxidation ponds	Partial transformation to NH/NH_4^+	Partial removal by stripping	Partial removal by nitrification-denitrification	20–90
Algae stripping	Partial transformation to NH_3/NH_4^+	$\rightarrow$ Cells	$\rightarrow$ Cells	50–80
Bacterial assimilation	No effect	40–70% removed	Limited effect	30–70

* Will depend on the fraction of influent nitrogen for which the process is effective, which may depend on other processes in the treatment plant.

† Soluble organic nitrogen, in the form of urea and amino acids, is substantially reduced by secondary treatment.

‡ May be used to remove particulate organic carbon in plants where ammonia or nitrate are removed by other processes.

Example 24-2 A wastewater has a BOD of 160 mg/L and a nitrogen content of 20 mg/L after primary treatment. Estimate the nitrogen removed in a biological process which reduces the soluble BOD to 5 mg/L and produces 0.5 kg of waste solids per kilogram of BOD removed. Also estimate the supplementary dose of methanol (BOD = 1 g/g methanol) required for assimilation of the remaining nitrogen.

Solution

The BOD incorporated in the sludge amounts to $0.5 \times (160 - 5) = 77.5$ mg/L. The nitrogen removed by assimilation will thus be $0.05(77.5) = 4$ mg/L, or 20 percent.

To remove the additional 16 mg/L will require addition of 320 mg/L of BOD or 320 mg/L of methanol, assuming all the methanol is incorporated in new cell mass. Since this is not likely, additional methanol would be required.

Nitrification is a biological process in which ammonia is converted to nitrite and nitrite to nitrate by the species *Nitrosomonas* and *Nitrobacter*, respectively. The nitrification process can be modeled mathematically in a fashion similar to that shown for carbonaceous BOD removal in Chap. 22.

Complete design procedures are presented in Ref. 10. The process is dependent on sludge age, as are other biological systems, but is also strongly affected by temperature, dissolved oxygen concentration, final ammonium concentration, and pH. The reaction is enhanced at higher pH, higher temperature, and higher dissolved oxygen levels. The minimum dissolved oxygen for satisfactory nitrification in suspended growth processes is at least 2 mg/L at peak flow and load, although under some circumstances nitrification will occur at lower levels.

The optimum pH range for nitrification has been identified as 7.2 to 8.0.[10] The effect of pH is important, since, as can be observed from Eq. (24-1), nitrification results in production of free acid which will tend to depress the pH. As a rough guideline, one may assume that nitrification will reduce the alkalinity by about 7.2 mg/L as $CaCO_3$ per mg/L NH_3 as N. This is particularly significant in closed reactors such as those used in some high purity oxygen systems, since in these there is also an increase in CO_2 concentration which further depresses the pH.

The effect of *temperature* has been studied extensively, and it has been concluded that nitrification is more strongly inhibited at low temperature than is carbonaceous BOD removal.[12] Nitrification rates at 10°C are only about one-quarter those at 30°C even in attached growth systems,[10] and the latter are less affected by temperature than are suspended growth processes. Safety factors for suspended growth processes should be at least 2.5 unless the wastewater flow has been equalized. The nitrification process is particularly susceptible to variations in waste strength and flow and the required safety factor is also affected by the desired effluent ammonia concentration.

Nitrification can be obtained in *separate processes* after secondary treatment or in *combined processes* in which both carbonaceous and nitrogenous oxygen demand are satisfied. In combined processes the ratio of BOD to TKN is greater than 5, while in separate processes the ratio in the second stage is less than 3. Reducing the ratio to 3 does not require a high degree of treatment in the first stage, hence heavily loaded processes such as high rate activated sludge or roughing filters may be used. Any aerobic biological system, with proper design, can be used to achieve nitrification. In addition to activated sludge, rock-media trickling filters,[13] plastic-media trickling filters,[14] and rotating biological contactors[15] have been used. In all processes, the key to satisfactory operation lies in reduced organic load and high availability of oxygen.

Denitrification is a biological process which can be applied to nitrified wastewater in order to convert nitrate to nitrogen. The process is anoxic, with the nitrate serving as the electron acceptor for the oxidation of organic material. Denitrification will occur to a limited degree in some aerobic processes, but most research activity has been directed toward processes specifically designed for nitrate reduction. Using methanol as a carbon source, St. Amant and McCarty[16] developed the following equation for methanol demand:

$$C_m = 2.47(NO_3) + 1.53(NO_2) + 0.87(DO) \qquad (24\text{-}2)$$

where C_m = methanol required, mg/L
NO_2 and NO_3 = concentrations of nitrite and nitrate, mg/L as N
DO = dissolved oxygen concentration, mg/L.

Example 24-3 Calculate the methanol dosage required for denitrification of a waste which contains 18 mg/L NO_3 and 2 mg/L NO_2 as N and 2.5 mg/L DO.

Solution

Applying the experimental equation

$$C_m = 2.47(18) + 1.53(2) + 0.87(2.5) = 50 \text{ mg/L}$$

This may be compared with the dosage of methanol calculated above for an assimilation process.

A number of processes in which wastewater serves as the carbon source have been developed,[17,18] and industrial wastes have also been employed. Denitrification can be effected in packed bed or fluidized-bed reactors with roughly comparable efficiency,[19,20] and the packed beds can also serve as a filtration system if the medium is sufficiently fine.[21] A variety of process configurations and their advantages and disadvantages are presented in Table 24-2.

Chemical nitrogen-removal processes generally involve converting the nitrogen to a form which is of limited solubility in water, particularly the gases nitrogen (N_2) and ammonia (NH_3). The processes of major interest include breakpoint chlorination, ion exchange, and air stripping. The last process may be applied to the wastewater itself or to the regenerant waste stream from the ammonium-ion-exchange system.

Breakpoint chlorination converts 90 to 95 percent of the ammonia in wastewater to nitrogen gas. The remaining nitrogen is in the form of nitrogen trichloride or nitrate.[22] The overall reaction is

TABLE 24-2
Comparison of denitrification alternatives

System type	Advantages	Disadvantages
Suspended growth using methanol following a nitrification stage	Denitrification rapid, small structures required Demonstrated stability of operation Few limitations in treatment sequence options Excess methanol oxidation step can be easily incorporated Each process in the system can be separately optimized High degree of nitrogen removal possible	Methanol required Stability of operation linked to clarifier for biomass return Greater number of unit processes required for nitrification-denitrification than in combined systems
Attached growth (column) using methanol following a nitrification stage	Denitrification rapid, small structures required Demonstrated stability of operation Stability not linked to clarifier as organisms on media Few limitations in treatment sequence options High degree of nitrogen removal possible Each process in the system can be separately optimized	Methanol required Excess methanol oxidation process not easily incorporated Greater number of unit processes required for nitrification-denitrification than in combined system

$$NH_4^+ + 1.5HOCl \rightarrow 0.5N_2 + 1.5H_2O + 2.5H^+ + 1.5Cl^- \qquad (24\text{-}3)$$

which indicates a weight ratio of chlorine to nitrogen of 7.6:1. In practice, the dosage ranges from 8:1 to 10:1, with the amount required being influenced by the presence of reduced chemicals, both inorganic and organic, and by the pH.

The residual chlorine remaining after breakpoint chlorination is likely to exceed that which is permissible in the effluent. *Dechlorination* can be effected with sulfur dioxide:

$$SO_2 + HOCL + H_2O \rightarrow Cl^- + SO_4^{2-} + 3H^+ \qquad (24\text{-}4)$$

or activated carbon:

$$C + 2HOCL \rightarrow CO_2 + 2H^+ + 2CL^- \qquad (24\text{-}5)$$

Practical dosages of SO_2 range from 0.9 to 1.0 mg/mg Cl_2. Activated carbon has a chlorine reduction capacity of 370 to 2400 kg/m^3, depending on the hydraulic application rate and the type of carbon.

TABLE 24-2
Comparison of denitrification alternatives *(Continued)*

System type	Advantages	Disadvantages
Combined carbon oxidation-nitrification-denitrification in suspended growth reactor using endogenous carbon source	No methanol required Lesser number of unit processes required	Denitrification rates very low; very large structures required Lower nitrogen removal than in methanol-based system Stability of operation linked to clarifier for biomass return Treatment sequence options limited when both N and P removal required No protection provided for nitrifiers against toxicants Difficult to optimize nitrification and denitrification separately
Combined carbon oxidation-nitrification-denitrification in suspended growth reactor using wastewater carbon source	No methanol required Lesser number of unit processes required	Denitrification rates low; large structures required Lower nitrogen removal than in methanol-based system Stability of operation linked to clarifier for biomass return Tendency for development of sludge bulking Treatment sequence options limited when both N and P removal required No protection provided for nitrifiers against toxicants Difficult to optimize nitrification and denitrification separately

Ion exchange for ammonium removal employs a natural zeolite, *clinoptilolite*, which is relatively specific for ammonium. The beds are typically 1.5 m (5 ft) deep and are loaded at rates of 12 to 30 m/h [5 to 12 gal/(ft^2 · min)], which amounts to 7.5 to 20 bed volumes per hour. It is the loading rate which is most important. The depth serves only as a protection against short circuiting. The total volume treated between regeneration cycles depends on the ammonium concentration in the waste and the permissible concentration in the effluent. At influent concentrations of 15 to 17 mg/L (typical of domestic wastewater), 100 to 120 bed volumes can be treated before reaching an effluent concentration of 1 mg/L. This represents a 13- to 16-h cycle at a loading rate of 12 m/h. Design procedures for sizing clinoptilolite beds and for selecting the process configuration are presented in Ref. 10.

Regeneration of the ion exchange medium may be effected with alkaline or neutral sodium or calcium salts. Lime or sodium hydroxide will yield a waste regenerant stream which contains NH_4OH, which can be readily removed by stripping as discussed below. The use of alkaline regenerants, however, may precipitate divalent metals within the exchange medium. These precipitates may coat the clinoptilolite, reducing its effectiveness, and eventually may plug the bed.

Sodium chloride can also be used as a regenerant, although larger volumes (up to 25 to 30 bed volumes) are required. This represents a concentration factor of only 5 or 6 to 1 with respect to the original waste. The regenerant, however, can then be dosed with sodium hydroxide, elevating the pH and permitting stripping of the ammonia. Any metallic precipitates can be separated by sedimentation prior to the stripping process. The stripped regenerant can then be reused in the process.

Stripping of waste regenerant can be effected with air at an air/liquid ratio of 1100 to 2200 m^3/m^3 or with steam at rates of about 1.8 kg/m^3. Electrolytic destruction of the ammonia in the regenerant has also been evaluated, but has substantially higher energy costs than does stripping.

The waste itself can be stripped of ammonia if it is at the requisite pH (10.8 to 11.5) and adequate air is provided. The air/liquid ratio for the waste is higher than for the concentrated regenerant and ranges from 2200 to 6000 m^3/m^3, depending on the temperature and the pH of the liquid. The feasibility of stripping the waste itself depends on whether the necessary pH can be achieved at moderate cost. The dosage of alkali required is dependent primarily on the alkalinity of the waste, and stripping of the bulk flow is generally practical only in conjunction with lime precipitation of phosphorus (Art. 24-4). In this circumstance, the pH required is produced in conjunction with the phosphate removal.

The stripped ammonia, which is carried from the process by the air stream, can be released to the atmosphere or recovered by bubbling the air through an acid solution. The ammonia will dissolve in the solution, forming the ammonium salt of the acid. This solution may have economic value as a fertilizer or industrial chemical.

24-4 Phosphorus Removal

Phosphorus, like nitrogen, is a nutrient of microscopic and macroscopic plants, and thus can contribute to the eutrophication of surface waters. It is important to recognize, however, that the concentration of phosphorus necessary to support an algal bloom is only 0.005 to 0.05 mg/L as P, and that this level may be exceeded from natural sources in many surface waters. In such circumstances, treatment of the waste to remove phosphorus will not prevent algal growth.

Phosphorus may be removed biologically and chemically. In some instances, chemicals may be added to biological reactors instead of being used in separate processes while, in others, biologically concentrated phosphorus may be chemically precipitated.

Bacteria and algae require phosphorus for their metabolic activities in a ratio of about 1 part of phosphorus per 100 parts of carbon, hence some phosphorus removal can be anticipated in any biological process. The removal, however, ordinarily does not exceed 20 to 40 percent. The *activated algae* and the bacterial assimilation processes discussed above have the potential of removing additional phosphorus, but, if stoichiometric ratios are maintained, it would be necessary to add both carbon *and* nitrogen to the assimilation process. The resulting mass of waste sludge from either would be perhaps 5 times that produced in an ordinary plant.

Some treatment plants remove far more phosphorus than the theoretical bacterial requirement, and this phenomenon has been termed *luxury uptake*.[23] The mechanism of luxury uptake has been demonstrated to consist of cellular uptake rather than biologically mediated precipitation, and the cellular uptake has been reliably induced in laboratory experiments by manipulation of the relative concentrations of potassium and sodium.[24] It now appears, however, that the mechanism responsible for optimization of phosphate removal in full-scale treatment plants is quite different, resulting from selection of a particular bacterial population which has the property of storing large quantities of phosphorus within the cell.[25]

A novel biological-chemical process utilizes an activated sludge system and an anaerobic cell in which phosphorus taken up in the aerobic basin is released to the liquid phase, producing a concentrated phosphate stream and a phosphate-deficient sludge which is returned to the aeration basin, where the uptake of phosphorus is repeated. The concentrated phosphate stream is treated chemically to precipitate the phosphorus, which is then removed from the liquid in the primary clarifier. The advantage of this process lies in the biological concentration provided, since, as noted below, chemical dosages for phosphate precipitation are relatively insensitive to phosphate concentration. A schematic of this process is presented in Fig. 24-2.

Improved removal without any chemical addition can be obtained in similar processes which employ an anoxic or anaerobic zone prior to aeration or with the fill and draw activated sludge process described in Chap. 22. When operated to maximize phosphate removal, this system (sometimes called a *sequencing batch reactor*), is operated with an extra step in which the sludge is allowed to remain quiescent for a period after the treated liquid is withdrawn.[26] It appears that under the anoxic or anaerobic conditions in such processes, the fermentation of heterotrophic bacteria produces short-chain organics which can be metabolized by bacteria that have reserves of high-energy phosphate. This favors the development of such populations, which then are available to remove phosphorus in the subsequent aerobic stages. A number of proprietary processes have been developed which take advantage of this phenomenon and which are able to reduce the phosphorus content to the level of about 1 mg/L with no chemical addition. The operation of this type of system is severely

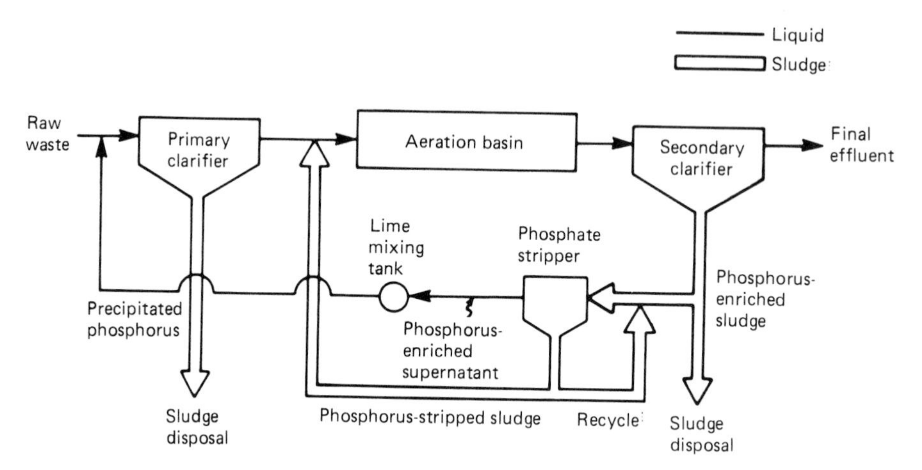

FIGURE 24-2
Biological-chemical phosphate-removal process. (*From Operation of Full-Scale Biological Phosphorus Removal Plant by G. V. Levin, G. J. Topol, and A. G. Tarnay, in Journal of Water Pollution Control Federation, 47, 1975. Copyright 1975, Water Pollution Control Federation. Used with permission of the Water Pollution Control Federation.*)

affected by the presence of nitrate, which permits other bacterial populations to metabolize the products of the anoxic fermentation, thus removing the advantage of the phosphate storing species and limiting their growth. Some sequencing batch reactor processes include a separate denitrification step in an attempt to reduce this problem.

Chemical phosphorus removal can be effected by precipitation in a variety of forms, the least expensive ordinarily being as aluminum, iron, or calcium salts. Phosphorus in wastewater may exist as *organic phosphate*, *polyphosphate*, or *orthophosphate*, the last consisting of four different ionic forms. For simplicity, the phosphorus is considered to be present as phosphate ion (PO_4^{-3}) and chemical reactions are presented on that basis. The actual reactions in real systems are more complex, hence chemical dosage calculations based on stoichiometric considerations will give only rough approximations of the actual requirements.

Aluminum phosphate can be precipitated in accord with

$$Al^{+3} + PO_4^{-3} \rightarrow AlPO_4 \tag{24-6}$$

The aluminum may be provided from alum [$Al_2(SO_4)_3 \cdot xH_2O$)] or sodium aluminate ($Na_2O \cdot AL_2O_3 \cdot xH_2O$). Alum, as noted in Chap. 9, reacts with water to produce free hydrogen ion, and thus will lower the pH of the waste. Sodium aluminate, on the other hand, is alkaline and will raise the pH. Since the solubility of aluminum phosphate is a minimum at pH 6, judicious use of the appropriate chemical or combination of chemicals may reduce the required dosage. The aluminum added will also combine with the alkalinity of the waste as described in Chap. 9. This will increase the chemical dosage somewhat.

TABLE 24-3
Aluminum/phosphate ratios for chemical precipitation

Percent P removal	Al:P Mole ratio	Al:P Weight ratio
75	1.38:1	1.2:1
85	1.72:1	1.5:1
95	2.30:1	2.0:1

Estimated ratios of aluminum to phosphate required for various degrees of removal are presented in Table 24-3.[27] Note that Eq. (24-6) indicates that the theoretical mole ratio of aluminum to phosphorus is 1:1.

Ferric and *ferrous phosphate* can be precipitated according to

$$Fe^{+3} + PO_4^{-3} \rightarrow FePO_4 \tag{24-7}$$

and

$$3Fe^{+2} + 2PO_4^{-3} \rightarrow Fe_3(PO_4)_2 \tag{24-8}$$

The reaction in Eq. (24-8) is possible, but does not appear to represent what actually happens, since iron/phosphate ratios are independent of the valance of the iron. Iron salts also react with the alkalinity to reduce the pH. The optimum pH for phosphorus removal with iron is 4 to 5 for the ferric form and 7 to 8 for the ferrous form. The iron/phosphorus ratio may be expected to vary with removal as shown in Table 24-4.

Hydroxyapatite may be precipitated by the reaction among lime, water, and phosphate:

$$5CaO + 5H_2O + 3PO_4^{-3} \rightarrow Ca_5(OH)(PO_4)_3 + 9OH^- \tag{24-9}$$

The precipitation reaction is pH-dependent and the lime will, of course, react with other chemical species as well. The lime dosage is primarily dictated by side reactions with alkalinity, hardness, and heavy metals rather than by the

TABLE 24-4
Iron/phosphate ratios for chemical precipitation

Percent P removal	Fe:P Mole ratio	Fe:P Weight ratio
70	0.67:1	1.2:1
80	0.89:1	1.6:1
90	1.28:1	2.3:1
95	1.67:1	3.0:1

phosphate concentration. The pH required for substantial phosphate removal is about 9. Removal increases with increased pH.

Chemical addition for phosphate removal can be made in the primary clarifier, in secondary biological processes, or in the final clarifier. The choice of point of addition depends on the chemical and other processes in the plant that may be affected. Addition in the primary is generally advantageous in that removals in the range of 80 percent will leave relatively low concentrations of phosphate which can be scavenged by subsequent biological processes. The efficiency of biological systems in removal of low levels of pollutants is far better than that of chemical systems. Addition in the primary will also improve the efficiency of that process with regard to BOD and suspended solids removal, permitting the secondary system to be smaller.

If stripping of ammonia is to be practiced, addition of lime in the final clarifier is more logical, since this will permit raising the pH for subsequent stripping while precipitating phosphorus. Nitrification processes, on the other hand, might be enhanced by addition of alum in the primary. This would serve to reduce both the phosphate concentration and the BOD load on a combined nitrification system.

Chemical addition in the biological process itself has also been evaluated and demonstrated to yield removal of phosphate and to assist in coagulation and separation of biological solids.[28,29] Phosphorus has been removed in this fashion in both trickling filter plants and activated sludge processes; however, dependable removal is not assured in trickling filter plants.[27]

Example 24-4 Compare the chemical dosages of alum required to remove 95 percent of the influent phosphate in a flow with a BOD of 200 mg/L and PO_4^{-3} of 10 mg/L if the chemical is added in the primary or in the secondary process.

Solution

Addition of alum in the primary may be expected to increase the BOD removal to at least 50 percent (Chap. 21). The influent BOD to the secondary will thus be about 100 mg/L. Biological removal will amount to about 1 mg/L per 100 mg/L of BOD. Thus, the influent to the biological process can contain 1.5 mg/L and the effluent will meet the 95 percent removal standard. Removal required in the primary is thus 8.5 mg/L or 85 percent. The chemical dosage, from Table 24-3, is $1.5 \times 8.5 = 12.75$ mg/L Al^{+3}. This would require about 140 mg/L of alum. Removal in the secondary clarifier will occur after biological treatment. The influent to that process, after ordinary primary treatment, will have a BOD of 140 mg/L and PO_4^{-3} of 10 mg/L. Removal by assimilation will leave 8.6 mg/L. To reduce this to 0.5 mg/L (94 percent) will require a chemical dosage close to $2 \times 8.1 = 16.2$ mg/L Al^{+3}, or 180 mg/L of alum.

24-5 Refractory Organics

Refractory organics are those which are not removed in biological systems operating under ordinary conditions of sludge age, liquid retention time, etc. The concentration of refractory organics in domestic wastewater may be roughly approximated as the difference between BOD and COD or TOC. The materials which contribute to this difference include organic molecules which are limited in solubility and those which contain resonant ring structures. Examples of such substances include lignin, cellulose, and other polysaccharides, fats, phenolic compounds, etc.

Removal of refractory organics can generally be provided by *adsorption* on activated carbon. The application of activated carbon to the treatment of potable water is presented in Chap. 11, together with a general discussion of the principles and chemistry of adsorption. The application of this technology in wastewater treatment is somewhat different because of the higher concentrations which are being removed.

Wastewater adsorption systems generally employ granular activated carbon held in a closed container through which the wastewater flow is passed. The beds may be operated in either upflow or downflow mode. Although the downflow process appears to offer a simple means of combining adsorption and filtration, this is not the case. If the bed is operated as a filter, it will need to be backwashed regularly. When it is backwashed, the spent carbon, being denser, will migrate to the bottom. In that position it will be in contact with treated water, and desorption is likely to occur. Downflow units must be taken out of service for removal and replacement of spent carbon. An upflow system, on the other hand, if loaded at appropriate hydraulic rates, can be operated as a fluidized bed in which the spent carbon will move progressively toward the bottom, maintaining a countercurrent operation in which the adsorptive capacity of the carbon is most fully utilized. The spent carbon can be readily removed and replaced on a continuous or semicontinuous basis. An upflow carbon column is illustrated in Fig. 24-3.

Regeneration of spent carbon can be effected by washing with organic solvents, mineral acid, or caustic; by steam; or by dry heat. To a certain degree, activated carbon can also be regenerated by biological growth on its surface. It has been observed that biological regeneration of activated carbon can enhance the removal even of nondegradable compounds.[30] This phenomenon can be attributed to the freeing of reactive sites occupied by degradable compounds, which increases the ability of the carbon to accept other compounds.

The regeneration techniques which employ liquids produce a liquid waste which must be handled further. The *thermal regeneration* process, on the other hand, produces only gases and a quantity of ash and fines. The thermal process includes three steps: drying at 100°C, pyrolysis of adsorbates at less than 800°C, and oxidation of decomposed adsorbates at temperatures in excess of 800°C. Carbon losses result from ignition in the furnace, mechanical attrition, and loss of fines. In addition, some loss of adsorptive capacity will result from incomplete removal of strongly adsorbed compounds. The net loss is likely to be about 10 percent per cycle from all causes, but may be substantially greater for some carbons.

Handling of wet carbon slurries is quite difficult. The material is abrasive, contributes to corrosion, and is very difficult to keep in motion in either pipes or open channels. Supplementary fluidization with air may be helpful in wet

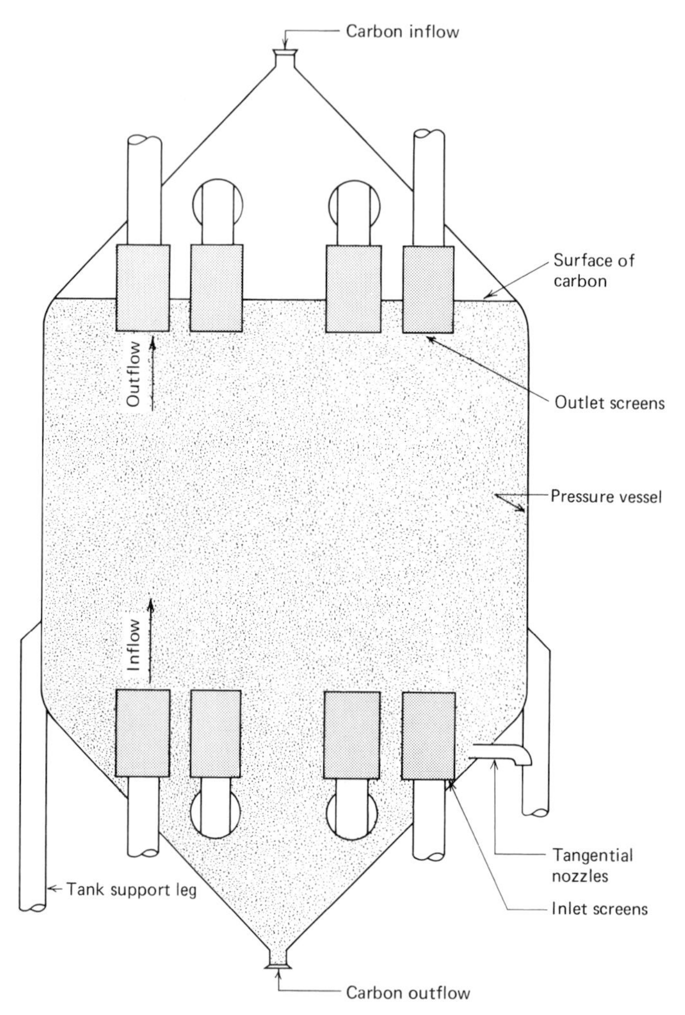

FIGURE 24-3
Upflow carbon column. (*From Process Design Manual for Carbon Adsorption, Environmental Protection Agency, 1971.*)

carbon conveyance systems. Materials of construction and handling techniques are discussed in Ref. 31. Unit dimensions are generally based on a ratio of depth to diameter of 2:1 to 10:1. Hydraulic detention time is 15 to 40 min based on the empty bed volume. Design criteria in each case, however, including criteria for selection of the carbon, should be based on pilot studies with the actual waste.

An interesting application of activated carbon to wastewater treatment is illustrated in Fig. 24-4. The process employs powdered activated carbon which serves both as an adsorbent and as a surface for biological growth. Mixed liquor suspended solids in this process can be maintained at levels of 1 to 1.5 percent. The waste solids (biological and carbon) are passed through a low-

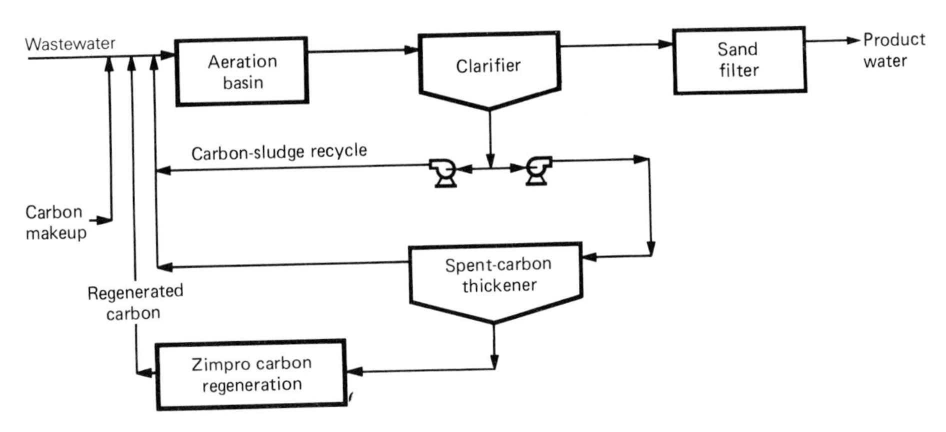

FIGURE 24-4
Biophysical wastewater treatment. (*Courtesy Zimpro, Inc.*)

pressure wet oxidation process (Chap. 23) in which the carbon is regenerated and the biological solids are destroyed. The waste solids stream from this process is relatively small, is inert, and is easily dewatered. Other combined adsorption-biological treatment processes usually depend on biologically mediated regeneration of the carbon, which tends to be less complete.

24-6 Dissolved Solids

Dissolved mineral species such as nitrate, nitrite, selenium, arsenic, and heavy metals may cause health problems, while others such as chloride or hardness may be esthetically or economically undesirable. Materials which are in true solution may be removed, in general, by the processes discussed in Chap. 11, but some specialized techniques have been developed for wastewater treatment.

Metals may be removed by ion exchange, precipitation, reverse osmosis, etc. Polyelectrolytes with a soluble fraction precipitated by metal have been used to remove heavy metals from solution and coagulate them in a single step,[32] while reverse osmosis has been applied to removal of dissolved organics as well as inorganics.

In most respects, the effluent of a well-designed and well-operated secondary treatment plant is comparable to, and in some ways, superior to, many surface waters. The water treatment techniques discussed in Chaps. 9, 10, and 11 can ordinarily be applied with satisfactory results.

24-7 Experiences with Advanced Waste Treatment

Advanced waste treatment processes have been applied in the United States only since about 1970; hence the body of information dealing with their performance is somewhat limited. Suspended solids removal by direct filtration through granular media can generally be depended on to reduce suspended solids concentrations to less than 20 mg/L, with average levels of about 10 mg/L. Ordinarily, this would be associated with a BOD of about the same level or less. Filtration processes which have been developed specifically for use on wastewater provide equal degrees of treatment and may have some operating advantages.

Nitrification process kinetics are well understood, and, with proper design, nitrification of domestic sewage can be dependably achieved. Denitrification processes are somewhat more difficult in that careful operation is required to balance the organic load with the nitrate concentration so that the nitrate is reduced and all of the BOD is removed. A nitrification process designed for a future peak load will work without modification at lesser loads. A denitrification system, on the other hand, must be constantly adjusted as wastewater strength and flow vary.

Ammonia-stripping processes are very sensitive to air temperature and, because of the high air/liquid ratio required, are prone to freezing during cold weather. Additionally, stripping processes applied to the entire wastewater flow often result in precipitation of metallic salts in the aeration tower. Chlorination is readily adjusted to flow and waste strength variations, but chlorination of wastewater, as discussed in Chap. 25, is likely to result in production of chlorinated hydrocarbons.

Precipitation of phosphorus with lime, alum, or iron salts is relatively dependable, although careful process control is necessary. The major problem with chemical processes has been the large quantities of sludge produced. These processes substantially increase sludge mass and volume as a result of side reactions with the alkalinity and other dissolved materials in the flow. Treatment plants designed for the biological removal of phosphorus through luxury uptake have generally functioned satisfactorily, although it may be observed that some of the existing installations have wastewater characteristics which particularly favor such processes.[25]

Activated carbon treatment of wastewater has generally not been satisfactory when compared with the expectations of the period from 1960 to 1970. Carbon-handling systems have been very troublesome and some plants have been literally incapable of conveying the carbon from the adsorbers to the regeneration furnace and back. Carbon loss has often been far greater than design estimates, with resulting higher costs. It is possible to operate adsorption systems successfully, but domestic wastewater treatment plants offer few examples of such performance.

REFERENCES

1. Hillel I. Shuval, *Water Renovation and Reuse*, Academic Press, New York, 1977.
2. M. T. Gillies (ed.), *Potable Water from Wastewater*, Noyes Data Corporation, Park Ridge, N.J., 1980.

3. Gordon Culp, George Wesner, and Robert Williams, *Wastewater Reuse and Recycling Technology*, Noyes Data Corporation, Park Ridge, N.J., 1980.
4. E. Joe Middlebrooks (ed.), *Water Reuse*, Ann Arbor Science, Ann Arbor, Mich., 1982.
5. *Process Design Manual for Suspended Solids Removal*, Environmental Protection Agency, Washington, D.C., 1975.
6. Forrest O. Mixon, "Filterability Index and Microscreener Design," *Journal of Water Pollution Control Federation*, **42**:1944, 1970.
7. Bart Lynam, Gregory Ettelt, and Timothy McAloon, "Tertiary Treatment at Metro Chicago by Means of Rapid Sand Filtration and Microstrainers," *Journal of Water Pollution Control Federation*, **41**:247, 1969.
8. P. G. Ripley and G. L. Lamb, "Filtration of Effluent from a Biological-Chemical System," *Water and Sewage Works*, **12**:2:67, 1973.
9. R. I. Nebolsine, I. Poushine, and C. Y. Fan, *Ultra-High Rate Filtration of Activated Sludge Plant Effluent*, Environmental Protection Agency, Washington, D.C., 1973.
10. *Process Design Manual for Nitrogen Control*, Environmental Protection Agency, Washington, D.C., 1975.
11. *Nitrogenous Compounds in the Environment*, Environmental Protection Agency, Washington, D.C., 1973.
12. P. M. Sutton et al., "Efficacy of Biological Nitrification," *Journal of Water Pollution Control Federation*, **47**:2665, 1975.
13. Denny S. Parker and Tyler Richards, "Nitrification in Trickling Filters," *Journal of Water Pollution Control Federation*, **58**:896, 1986.
14. Robert W. Okey and Orris E. Albertson, "Evidence for Oxygen-Limiting Conditions during Tertiary Fixed Film Nitrification," *Journal of Water Pollution Control Federation*, **61**:510, 1989.
15. Rao Y. Surampali and E. Robert Baumann, "Supplemental Aeration Enhanced Nitrification in a Secondary RBC Plant," *Journal of Water Pollution Control Federation*, **61**:200, 1989.
16. P. St. Amant and P. L. McCarty, "Treatment of High Nitrate Waters," *Journal of American Water Works Association*, **61**:659, 1969.
17. J. L. Barnard, "Cut P and N without Chemicals," *Water and Wastes Engineering*, **14**:7:33, 1974.
18. N. F. Matsche and G. Spatzierer, "Austrian Plant Knocks Out Nitrogen," *Water and Sewage Works*, **122**:1:18, 1975.
19. J. S. Jeris and R. W. Owens, "Pilot-Scale High-Rate Denitrification," *Journal of Water Pollution Control Federation*, **47**:2043, 1975.
20. R. P. Michael and W. J. Jewell, "Optimization of the Denitrification Process," *Journal of Environmental Engineering Division, American Society of Civil Engineers*, **101**:EE4:643, 1975.
21. John N. English et al., "Denitrification in Granular Carbon and Sand Columns," *Journal of Water Pollution Control Federation*, **46**:28, 1974.
22. T. A. Pressley et al., "Ammonia Removal by Breakpoint Chlorination," *Environmental Science and Technology*, **6**:662, 1972.
23. William F. Milbury, Donald McCauley, and Charles H. Hawthorne, "Operation of Conventional Activated Sludge for Maximum Phosphorus Removal," *Journal of Water Pollution Control Federation*, **43**:1890, 1971.
24. Judith B. Carberry and Mark W. Tenney, "Luxury Uptake of Phosphorus by Activated Sludge," *Journal of Water Pollution Control Federation*, **45**:2444, 1973.
25. *Design Manual—Phosphorus Removal*, Environmental Protection Agency, Cincinnati, 1987.
26. L. H. Ketchum et al., "A Comparison of Biological and Chemical Phosphorus Removals in Continuous and Sequencing Batch Reactors," *Journal of Water Pollution Control Association*, **59**:13, 1987.
27. *Process Design Manual for Phosphorus Removal*, Environmental Protection Agency, Washington, D.C., 1976.

28. D. A. Long and J. B. Nesbitt, "Removal of Soluble Phosphorus in an Activated Sludge Plant," *Journal of Water Pollution Control Federation*, **47**:170, 1975.
29. Richard F. Unz and Judith A. Davis, "Microbiology of Combined Chemical-Biological Treatment," *Journal of Water Pollution Control Federation*, **47**:185, 1975.
30. Gerald E. Speitel, Jr., et al., "Biodegradation and Adsorption of a Bisolute Mixture in GAC Columns," *Journal of Water Pollution Control Federation*, **61**:221, 1989.
31. *Process Design Manual for Carbon Adsorption*, Environmental Protection Agency, Washington, D.C., 1971.
32. Robert E. Wing et al., "Heavy Metal Removal with Starch Xanthate-Cationic Polymer Complex," *Journal of Water Pollution Control Federation*, **46**:2043, 1974.

CHAPTER 25

MISCELLANEOUS WASTEWATER TREATMENT TECHNIQUES

25-1 Disinfection

The present policy of the EPA with regard to disinfection of wastewater is to evaluate the need for such treatment on a case-by-case basis.[1] This policy is based on the observed toxicity of chlorine to aquatic wildlife and the demonstrated production of chlorinated hydrocarbons by chlorination of wastewater, coupled with concern about disease transmission. Both the need for disinfection and the choice of an appropriate disinfection technique are based on the water quality standard of the receiving stream.

Treatment of domestic wastewater produces substantial reductions in the numbers of disease-causing microorganisms. Both total and fecal coliform counts are typically reduced by 2 orders of magnitude by secondary treatment, and direct filtration will provide a further reduction by a factor of 10. The effluent is hardly sterile, however. Total and fecal coliform counts may be expected to range from 10^4 to 10^5 and 10^3 to 10^4 per 100 mL respectively following filtration. In addition, the treated waste will contain both viruses and protozoans which are agents of disease. Removal of these species is substantially less effective on a percentage basis, and far fewer of these organisms are required to produce clinical illness. It is therefore likely that treated wastes which are discharged to waters used for contact recreation, supply of potable water, or propagation of shellfish will require disinfection, although the determination in each case is generally made by the state agency charged with administration of the National Pollution Discharge Elimination (NPDES) System.

The disinfection techniques which are considered potentially applicable to wastewater include chlorination, chlorination/dechlorination, bromine chloride, chlorine dioxide, ozone, and ultraviolet light. The use of these methods in water treatment is discussed in Chap. 11, and their applicability to wastewater is summarized in Table 25-1.[2]

TABLE 25-1
Applicability of disinfection techniques to wastewater (*after Ref. 2*)

Consideration	Cl_2	$Cl_2/deCl_2$	BrCl	ClO_2	O_3	Ultraviolet
Size of plant	All sizes	All sizes	All sizes	Small to medium	Medium to large	Small to medium
Applicable level of treatment prior to disinfection	All levels	All levels	Secondary	Secondary	Secondary	Secondary
Equipment reliability	Good	Fair to good	Unknown	Unknown	Fair to good	Fair to good
Process control	Well developed	Fairly well developed	Problematic	No experience	Developing	Developing
Relative complexity of technology	Simple to moderate	Moderate	Moderate	Moderate	Complex	Simple to moderate
Safety concerns	Yes	Yes	Yes	Yes	No	No
Transportation on site	Substantial	Substantial	Substantial	Substantial	Moderate	Minimal
Bactericidal	Good	Good	Good	Good	Good	Good
Virucidal	Poor	Poor	Fair to good	Good	Good	Good
Fish toxicity	Toxic	Nontoxic	Slight to moderate	Toxic	None expected	Nontoxic

Hazardous by-products	Yes	Yes	Yes	Yes	None expected	No
Persistent residual	Long	None	Short	Moderate	None	None
Contact time	Long	Long	Moderate	Moderate to long	Moderate	Short
Contributes dissolved oxygen	No	No	No	No	Yes	No
Reacts with ammonia	Yes	Yes	Yes	No	Yes (high pH only)	No
Color removal	Moderate	Moderate	Unknown	Yes	Yes	No
Increased dissolved solids	Yes	Yes	Yes	Yes	No	No
pH-dependent	Yes	Yes	Yes	No	Slight (high pH)	No
O&M sensitive	Minimal	Moderate	Moderate	Unknown	High	Moderate
Corrosive	Yes	Yes	Yes	Yes	Yes	No

Chlorine is a proven disinfectant, but the toxicity of chlorine residuals to aquatic life at levels below the detection limit of ordinary monitoring procedures argues against its use in cases where aquatic toxicity is of concern. Although dechlorination with sulfur dioxide or activated carbon is possible, there is no assurance that the residual following such treatment will in fact be zero.

Neither chlorine dioxide nor bromine chloride are presently used in disinfecting municipal wastewaters in the United States and experience in their use is quite limited.[2] These considerations lead to the conclusion that the preferred technology in many cases will be either ozonation or ultraviolet light.

Although neither ozonation nor ultraviolet disinfection is widespread at present, as of 1985 some 42 municipal treatment plants with flows ranging from about 0.3 to 330 m^3/min (0.1 to 125 mgd) employed ozone as a disinfectant. Of these, 12 were plants operating with high-purity oxygen activated sludge processes. Production of ozone from high-purity oxygen is more efficient than production from air, and the oxygen can be used in the biological process after the ozone has been reduced in disinfecting the waste. Over 50 systems employing ultraviolet light were in operation in the United States and Canada and an additional 30 systems were in various stages of development in 1986.[2] These systems were designed for flows ranging from 0.05 to 33 m^3/min (0.02 to 12.5 mgd) but the majority were for flows less than 2.5 m^3/min (1 mgd).

Reference 2 presents a detailed discussion of the factors which should be considered in selecting a disinfection technique and in design of the required facilities. There are circumstances in which chlorination is appropriate either with or without subsequent dechlorination, and a great many situations in which disinfection by any method is neither required nor desirable.

25-2 Odor Control

Odors at wastewater treatment plants are normally associated with the products of anaerobic decomposition: reduced short-chain organics and hydrogen sulfide. Since these compounds are fairly volatile, they are readily released to the air at any point where the waste is exposed to the atmosphere, particularly if there is turbulence at the surface. Odor problems are particularly common at the headworks of treatment plants where stale sewage is exposed to the atmosphere in grit removal facilities, screening or comminution operations, and primary clarifiers. Odor problems are also encountered in operation of many trickling filters as a result of the anaerobic activity which occurs in the lower levels of the attached layer.

To be controlled, odors must be contained. The first step in reducing the problem, therefore, generally involves covering the source areas with flat or domed covers. The contained space must then be positively vented to prevent buildup of toxic concentrations of gas. For areas where workers will need access, the flow should be sufficient to give at least one air change in 5 min.[3] The vented air must then be managed in some fashion to reduce its impact on the neighborhood. Possible techniques include dispersion through tall stacks which depend on dilution for reduction of odor levels; treatment with oxidizing agents such as chlorine, ozone, permanganate, or hydrogen peroxide; wet scrubbing; adsorption on activated carbon; combustion, either catalyzed or uncatalyzed; biological treatment by passage of the air through aeration basins or composting operations; or addition of masking agents.[4]

Odors which arise from overloaded biological treatment processes can be controlled only by modification of the system. For short-term problems, the addition of nitrate will provide supplemental oxygen and may prevent creation of anoxic conditions and their consequent odors. The nitrate may result in an increase in nitrogen in the plant effluent, although under these conditions it would ordinarily be reduced to nitrogen gas.

25-3 Industrial Wastes

Industrial process wastes, wash waters, and cooling waters are subject to discharge restrictions similar to those applied to municipal wastes. Individual industries have specific NPDES requirements which must be met and which depend on the particular processes and/or products and the size of the manufacturing plant. Industrial waste discharges may, of course, be further restricted because of failure to meet water quality standards in particular areas.

Industries may have the option of treating their wastes to meet the NPDES requirements or of discharging them, either as is or after pretreatment, to a municipal treatment plant. If the wastes are treated by the community, the industry must expect to pay its fair share of the cost of construction (including capacity for future expansion) and the cost of operation of the facilities.

The cost of treatment is generally assigned to parameters involved in plant design such as flow, BOD, and suspended solids. If the plant includes nitrogen or phosphorus removal or if a particular waste is unusually high in some substance, such as grease or floating solids, these other parameters might be included in apportioning the cost. When the total load of each relevant parameter has been estimated, the costs of construction are then allocated, insofar as possible. The cost of pumping would be charged to flow; the cost of primary clarification in part to flow and in part to suspended solids; the cost of secondary treatment in part to flow and in part to BOD; and the cost of solids handling in part to BOD and in part to suspended solids. Industrial users would then pay in proportion to their contribution to the chargeable parameters.

Despite the requirement for proportional cost recovery, it is often advantageous for industry to arrange for the community to treat its wastes. Industries, in general, are not interested in waste treatment and would prefer that the responsibility lie with someone else. Combining industrial and municipal wastes in a single plant will ordinarily provide some economy of scale, reducing the unit costs of all users. Combining wastes from different sources may improve the treatability of the mixture, since one waste may supply the nutritional deficiencies of another. Finally, materials which might be difficult to treat at their original concentrations may be diluted by other flows to levels which are amenable to treatment.

From the standpoint of the community, addition of industrial wastes offers all of the advantages cited above save in the matter of responsibility. Once the community permits industrial discharges to the common sewers it accepts responsibility for treating them to whatever degree may be necessary to meet state and federal requirements. *Pretreatment* or *exclusion* is required for wastes which might cause problems in the collection system, the treatment facilities, or the receiving stream. The characteristics of a variety of industrial

TABLE 25-2
Industrial wastes and treatment processes

Source	Characteristics	Treatment
Food industries:		
Breweries	High nitrogen and carbohydrates	Biological treatment, recovery, animal feed
Canneries	High solids and BOD	Screens, lagoons, irrigation
Dairies	High fat, protein, and carbohydrate	Biological treatment
Fish	High BOD, odor	Evaporation, burial, animal feed
Meat processing	High protein and fat (may be warm)	Screens, sedimentation, flotation, biological treatment
Pickles	High BOD, high or low pH, high solids	Screening, flow equalization, biological treatment
Soft drinks	High BOD and solids	Screening, biological treatment
Sugar	High carbohydrate	Lagoons, biological treatment
Chemical industries:		
Acids	Low pH, low organics	Neutralization
Detergents	High BOD and phosphates	Flotation, precipitation
Explosives	High organics and nitrogen, trinitrotoluene	Chlorination or TNT precipitation
Insecticides	High organics, toxic to biological systems	Dilution, adsorption, chlorination at high pH
Materials industries:		
Foundries	High solids (sand, clay, and coal)	Screens, drying
Oil	Dissolved solids, high BOD, odor, phenols	Injection, recovery of oils
Paper	Variable pH, high solids	Sedimentation, biological treatment
Plating	Acid, heavy metals, cyanide toxic to biological systems	Oxidation, reduction, precipitation, neutralization
Rubber	High BOD and solids, odor	Biological treatment
Steel	Low pH, phenols, high suspended solids	Neutralization, coagulation
Textiles	High pH, and BOD, high solids	Neutralization, precipitation, biological treatment

wastes are summarized in Table 25-2. Some of these discharges are explosive, flammable, or corrosive or contain high concentrations of grease or solids and may cause damage to the sewers or danger to the public. Others contain toxic materials which may interfere with treatment, cause violations of the municipal permit, or create danger to workers.

Many toxic materials and heavy metals have been found to be concentrated in sludge, and such contaminants may severely restrict the disposal options open to the community. Industrial dischargers to common sewers should be required to pretreat their wastes to the degree necessary to prevent such problems.

Industrial waste treatment is an extensive topic which encompasses procedures such as *plant surveys* to determine individual waste sources and strengths, *process changes* to reduce waste flow and/or strength, *housekeeping improvements*, *pretreatment* to meet sewer discharge or other standards, and, finally, *treatment* either in a separate industrial plant or jointly with the municipality.

Treatment processes are dependent on the nature of the waste and may be physical, chemical, biological, or a combination of the three. Wastes which contain primarily simple organic compounds, such as those arising in food processing industries, may be treated by the techniques used for domestic wastewater. Others, such as metal plating wastes, may be treated primarily by chemical processes such as oxidation, reduction, precipitation, and neutralization. Still others, such as refinery wastes, may be treated by physical-chemical and biological techniques. Industrial wastes are often deficient in trace materials and nutrients necessary for biological treatment. When joint treatment is not practical, these deficiencies must be supplied by chemical addition.

Since about 1970 there has been a growing interest in the development of specialized strains of bacteria and cell-free bacterial extracts designed to enhance the treatment of industrial wastes. No one other than the manufacturers of these products has reported lasting benefit from their use and independent investigators have not been able to demonstrate improved treatment.[5] Bacterial populations change very rapidly in response to environmental stimuli, and wastes, in general, will select a bacterial population capable of their metabolism. It is important to note, however, that the design criteria and rate constants presented in Chap. 22 for biological treatment processes apply to domestic wastewater, and specific industrial wastes may have quite different parameters. Pilot studies should be conducted before large investments are made in process equipment.

References 6, 7, and 8 summarize recent data dealing with treatment of specific industrial wastes. The *Journal of the Water Pollution Control Federation* and the *Proceedings of the Annual Industrial Waste Conferences* held at Purdue University are excellent sources of information on industrial waste treatment, and the EPA has periodically issued publications presenting the best available treatment for specific industries.

25-4 On-Site Wastewater Management Techniques

Rural areas and the outskirts of urban areas may have insufficient population to support sewer systems and central treatment, even with the use of vacuum or pressurized collection. In such areas it is likely that unsanitary disposal techniques will be used unless local regulations are both well thought out and scrupulously enforced.

Satisfactory techniques will ensure that water supplies, particularly shallow wells, are not contaminated; that flies and vermin have no access to excreta; that surface waters are not polluted by runoff; and that nuisance conditions such as odors are minimized. Acceptable systems, depending on circumstances, include septic tanks and subsurface percolation; extended aeration, alone or following a septic tank; mounds; and intermittent sand filters. In some areas without running water (primarily, but not exclusively, campgrounds), the pit privy is still used.

Pit privies still exist in large numbers and continue to be used in rural areas with limited running water. As illustrated in Fig. 25-1, a typical privy consists of a pit about 1 m (3 ft) square by 1.25 m (4 ft) deep, lined with rough

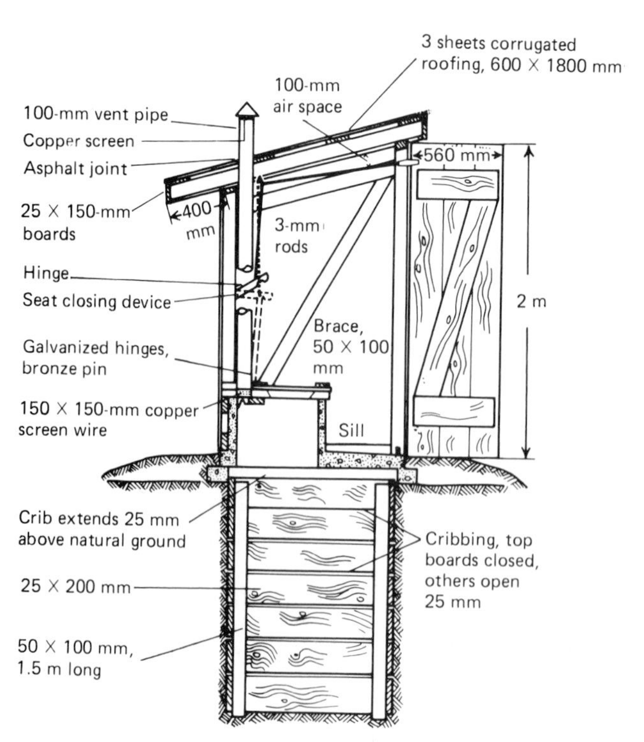

FIGURE 25-1
Concrete-slab pit privy.

boards on the sides and covered with a reinforced-concrete slab. A concrete riser supports the seat and a ventilator pipe conveys odors through the roof. The slab rests on a concrete curb to which the house is bolted. Earth is banked around the curb to prevent surface runoff from entering the pit. A privy of this type will serve an average family for about 10 years. Cleaning is not practical and a new pit must be dug when the old one is full. The house, slab, and curb can be moved to the new location.

Pit privies at campsites which receive heavy use are often lined with concrete and have an access door at the rear of the unit. This permits the contents to be removed and hauled to a municipal treatment plant or a suitable disposal site.

Septic tanks are primarily sedimentation basins, although a minor degree of solids destruction may occur as a result of anaerobic activity. Units are ordinarily sized to provide a 24-h retention time at average daily flow. Anticipated flows from various types of residential and public buildings are presented in Table 25-3.[9] These flows appear to be generous estimates in the light of more recent research results presented in Table 25-4.[10]

A typical septic tank is illustrated in Fig. 25-2. Modern septic tanks are manufactured of both concrete and fiberglass. The fiberglass tanks are cylindrical with domed ends and are placed with the axis of the cylinder horizontal. The inlet and outlet are baffled in order that floating material and grease will be retained. Heavy solids, including most organic solids, settle to the bottom where some biological activity may occur. This may not be beneficial, since gas production can interfere with the sedimentation function of the tank. The ratio of peak flow to average flow is very high in small septic tanks, and brief surges may disturb the solids in the bottom and carry them to the outlet. The effluent of a septic tank is offensive and potentially dangerous. The mean BOD concentration observed in a number of septic tank installations ranged from 120 to 270 mg/L, and the mean suspended solids from 44 to 69 mg/L.[11] Further treatment of septic tank effluent is required, either in an additional process or by subsurface disposal.

Aerobic treatment processes have been applied to single-family and other small flows. Available systems may employ a septic tank followed by aerobic treatment or aerobic treatment alone. The aerobic unit is usually an extended aeration system, but other small systems which utilize attached growth processes are also available. The latter employ more or less stationary media immersed in the wastewater, which is aerated by diffused air. In either type of process, a small compressor provides air for oxygen supply. In the extended aeration versions, solids recycle may be provided either by gravity or by an air-lift pump. These systems are installed below ground and their effluent may be discharged to a disposal field, a sand filter, or to surface drainage. The latter alternative is not desirable, since the effluent quality may not meet secondary standards under the best of circumstances, and there is no guarantee the unit will be properly maintained. Mean effluent BOD in a number of systems ranged from 38 to 57 mg/L with suspended solids of 45 to 64 mg/L.[11]

TABLE 25-3
Waste flow (*after Ref. 9*)

Type of establishment	Liters per person per day
Small swellings and cottages with seasonal occupancy	190
Single-family dwellings	280
Multiple-family dwellings (apartments)	225
Rooming houses	150
Boarding houses	190
Additional kitchen wastes for nonresident boarders	40
Hotels without private baths	190
Hotels with private baths (2 persons per room)	225
Restaurants (toilet and kitchen wastes per patron)	25–40
Restaurants (kitchen wastes per meal served)	10–12
Additional for bars and cocktail lounges	8
Tourist camps or trailer parks with central bathhouse	130
Tourist courts or mobile home parks with individual bath units	190
Resort camps (night and day) with limited plumbing	190
Luxury camps	380–570
Work or construction camps (semipermanent)	190
Day camps (no meals served)	55
Day schools without cafeterias, gymnasiums, or showers	55
Day schools with cafeterias, but no gymnasiums or showers	75
Day schools with cafeterias, gyms, and showers	95
Boarding schools	280–380
Day workers at schools and offices (per shift)	55
Hospitals	570–950+
Institutions other than hospitals	280–470
Factories (flow per person per shift, exclusive of industrial wastes)	60–130
Picnic parks with bathhouses, showers, and flush toilets	40
Picnic parks (toilets wastes only)	20
Swimming pools and bathhouses	40
Luxury residences and estates	380–570
Country clubs (per resident member)	380
Country clubs (per nonresident member present)	95
Motels (per bed space)	150
Motels with bath, toilet, and kitchen wastes	190
Drive-in theaters (per car space)	20
Movie theaters (per auditorium seat)	20
Airports (per passenger)	10–20
Self-service laundries (flow per wash, i.e., per customer)	190
Stores (per toilet room)	1500
Service stations (per vehicle served)	40

TABLE 25-4
Characteristics of rural household wastewater[10]

Average flow per capita	160 L/day
Peak flow per capita	272 L/day
BOD per capita	0.050 kg/day
Soluble solids per capita	0.035 kg/day

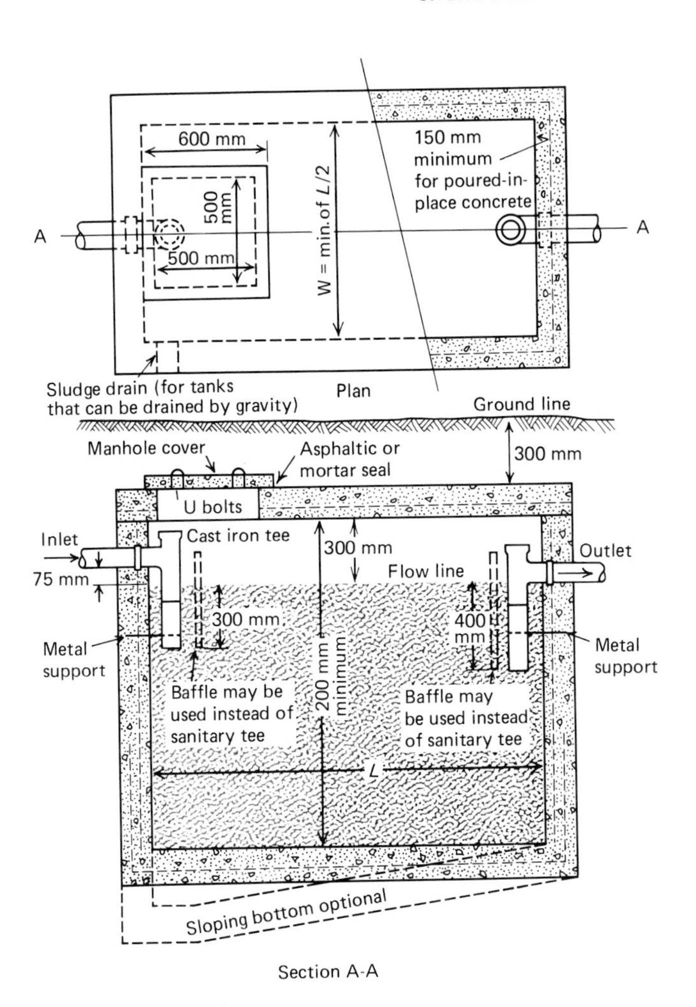

FIGURE 25-2
Residential septic tank.

The manufacturers of these processes also provide fabric filter bags and hypochlorination equipment which may be added to improve the effluent quality. The filters will reduce the suspended solids to levels less than 30 mg/L and hypochlorination will substantially reduce the bacterial population. The difficulty with this concept is the absence of any assurance of regular monitoring and maintenance. The filter bags must be changed regularly and the hypochlorite supply must be replenished. If these things are not done, indeed, even if the compressor fails, the wastewater will still continue to flow. Discharge to a subsurface disposal field appears to be a more dependable technique. Where soils are impermeable and aerobic units are used with discharge to surface drainage, a possible management technique is to require that all such units be maintained on a regular schedule either by the local government or by a single franchisee. This will help to ensure that they are operating as well as possible.

Subsurface disposal fields can serve as a point of further treatment and ultimate disposal for wastewater which has undergone some reduction in suspended solids and grease content in processes such as septic tanks or individual aerobic units. Many natural soils are suitable for such systems. Since the primary limitation lies in the long-term ability of the soil to transmit water, design is based on a standardized percolation test.[9]

The *percolation test* is conducted by boring a hole 100 mm (4 in) or more in diameter to the depth of the proposed disposal field (at least 500 mm). The sides of the hole are scratched and all loose earth is removed, after which 50 mm (2 in) of fine gravel or coarse sand is placed in the bottom. The hole is then filled with water to a depth of 300 mm (12 in), and that depth is maintained for at least 4 h and preferably overnight by adding water. If the hole holds water overnight, the depth is adjusted to about 150 mm (6 in) above the gravel and the drop in the water surface within 30 min is recorded. If the hole is empty in the morning, it is refilled to about 150 mm (6 in) above the gravel and the drop in water surface is recorded at 30-min intervals for 4 h, with water being added as necessary. The drop recorded in the last 30-min is used to determine the percolation rate. Table 25-5 may then be used to determine the required trench area. Alternatively, Eq. (25-1) will yield the flow which can be applied per unit area per day as a function of the percolation rate.

$$Q = \frac{204}{t^{0.5}} \tag{25-1}$$

where Q is the flow, L/(m² · day), and t is time, min, required for the water to fall 25 mm.

TABLE 25-5
Application rates for subsurface disposal

Percolation rate, time required for water to fall 25 mm (1 in), min	Maximum rate of waste application, L/(m² · day)
1 or less	204
2	143
3	118
4	102
5	90
10	65
15	53
30	37
45	33
60	24
> 60	Not suitable for percolation. Consider underground filter.

L/(m² · day) × 0.0245 = gal/(ft² · day)

If the subsidence rate is over 0.5 mm/min ($\frac{1}{4}$ in/min), a septic tank and disposal field will ordinarily prove to be a satisfactory system.[12] The disposal field is constructed of 100-mm (4-in) pipe, either short lengths of solid pipe laid with open joints or perforated plastic or fiber pipe. The slope of the distributing pipe is between 0.017 and 0.33 percent and individual lines are ordinarily 30 m (100 ft) or less in length. The pipe is placed in a ditch at least 500 mm (18 in) deep which has been excavated to a permeable stratum. The ditch is 300 to 900 mm (1 to 3 ft) wide and is backfilled to a depth of 300 to 400 mm (12 to 16 in) with gravel before the pipe is placed. An additional 50 mm (2 in) of gravel is placed over the pipe before the remainder of the trench is backfilled with soil. The total length of pipe depends on the trench width, since the product of these must equal the area obtained from the percolation rate. Laterals are laid about 2 m (6 ft) on centers.

Example 25-1 Determine the size of a septic tank and percolation field for a mobile home park which has 210 residents. The average percolation rate has been determined to be 6 mm/min.

Solution
From Table 25-2, the anticipated flow is 190 L/day per capita, or 39,900 L/day. The septic tank volume should thus be approximately 40 m^3. The percolation rate (time for the water to fall 25 mm) is $25/6 = 4.17$ min, which yields a hydraulic loading rate of 100 L/($m^2 \cdot$ day). The total trench area is thus 400 m^2, and the total length of the laterals 444 m if the trenches are 900 mm wide. Fifteen laterals, each 30 m long, placed 2 m on centers might be used. The area dedicated to the field would be approximately 30×30 or 900 m^2.

A variation of the disposal field utilizes a modified trench structure called a *capillary seepage trench*. In this system, the bottom and the lower part of the sides of the trench are sealed so that the water must flow horizontally in order to leave. This spreads the flow over a greater area and reduces the load in the immediate vicinity of the trench. The system is reported to yield a better effluent than the standard trench system.[13]

Mounds may be constructed above the surface of the ground in areas where soils are not suitable for subsurface disposal. The mound is constructed of imported pervious material over which the wastewater is distributed as in disposal fields. The mound depends on evapotranspiration for disposal of the bulk of the liquid. The remainder may be able to percolate through the soil below. Grass cover and proper shaping of the mound are important to ensure that rainfall will run off and that evapotranspiration of the wastewater will be maximized.

Sand filtration either on intermittent filters (Chap. 22) or buried filters may be required when soils are relatively impermeable. Loading rates on intermittent sand filters treating septic tank or aerobic unit effluent have ranged from 0.16 to 0.20 m/day [4 to 5 gal/($ft^2 \cdot$ day)].[14] The surface must be cleaned after 3 to 9 months, depending on the degree of pretreatment achieved.

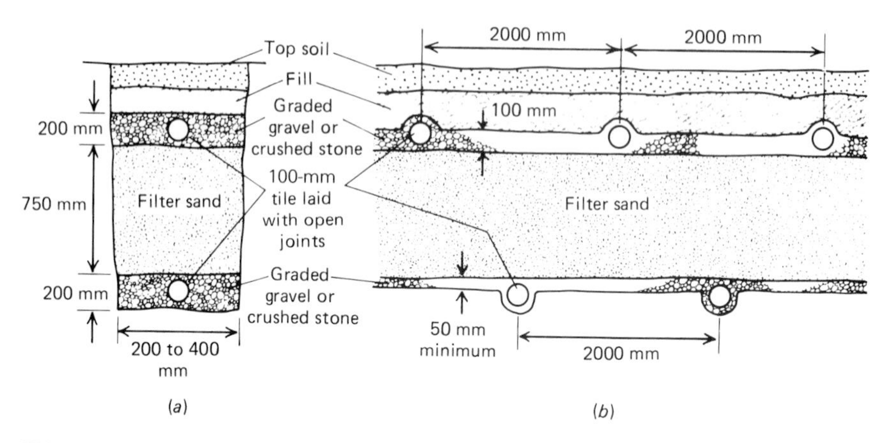

FIGURE 25-3
Subsurface filters. (*a*) Sand-filter trench; (*b*) sand filter.

Subsurface sand filters are installed in place of permeable material in a suitable excavation. A typical cross section is shown in Fig. 25-3. The sand should be relatively coarse (about 1 mm) and uniform to permit thorough ventilation. Loadings are about 0.04 m/day [1 gal/(ft^2 · day)]. The effluent from either filtration process must be collected and discharged to surface drainage. Depending on the point of discharge, disinfection may be required.

On-site wastewater treatment systems have the potential of providing an excellent effluent when they are properly designed and operated. Subsurface methods which will provide nitrification and denitrification adequate to remove up to 70 percent of the nitrogen content have been developed,[15] and small wetland systems can remove nearly all of the contaminants in ordinary wastewater.[16]

REFERENCES

1. D. Hubley et al., "Risk Assessment of Wastewater Disinfection," EPA-600/2-85/037, Environmental Protection Agency, Cincinnati, 1985.
2. *Design Manual—Municipal Wastewater Disinfection*, EPA/625/1-86/021, Environmental Protection Agency, Cincinnati, 1986.
3. Richard J. Pope and Jeffrey M. Lauria, "Odors: The Other Effluent," *Civil Engineering*, **59**:8:42, 1989.
4. Paul N. Cheremisinoff and Richard A. Young (eds.), *Industrial Odor Technology Assessment*, Ann Arbor Science, Ann Arbor, Mich., 1975.
5. Warren L. Jones and Edward D. Schroeder, "Use of Cell-Free Extracts for the Enhancement of Biological Wastewater Treatment," *Journal of Water Pollution Control Federation*, **61**:60, 1989.
6. W. Wesley Eckenfelder, Jr., *Industrial Water Pollution Control*, 2d ed., McGraw-Hill, New York, 1989.
7. Richard A. Conway and Richard D. Ross, *Handbook of Industrial Waste Disposal*, Van Nostrand Reinhold, New York, 1980.
8. Hardan Singh Azad (ed.), *Industrial Wastewater Management Handbook*, McGraw-Hill, New York, 1976.

9. "Manual of Septic Tank Practice," Publication No. 526, U.S. Public Health Service, Washington, D.C., 1957.
10. Robert Siegrist, Michael Witt, and William C. Boyle, "Characteristics of Rural Household Wastewater," *Journal of Environmental Engineering Division, American Society of Civil Engineers*, **102**:EE3:533, 1976.
11. Richard J. Otis and William C. Boyle, "Performance of Single Household Treatment Units," *Journal of Environmental Engineering Division, American Society of Civil Engineers*, **102:**EE1:175, 1976.
12. A. Ingham, "Discussion—Efficiency of a Septic Tile Field," *Journal of Water Pollution Control Federation*, **49**:335, 1977.
13. Brian E. Reed et al., "Improvements in Soil Adsorption Trench Design," *Journal of Environmental Engineering Division, American Society of Civil Engineers*, **115**:EE4:853, 1989.
14. David K. Sauer, William C. Boyle, and Richard J. Otis, "Intermittent Sand Filtration of Household Wastewater," *Journal of Environmental Engineering Division, American Society of Civil Engineers*, **102:**EE4:789, 1976.
15. Richard J. Piluk and Oliver J. Hao, "Evaluation of On-Site Waste Disposal System for Nitrogen Reduction," *Journal of Environmental Engineering Division, American Society of Civil Engineers*, **115:**EE4:725, 1989.
16. S. C. Reed et al., *Natural Systems for Waste Management and Treatment*, McGraw-Hill, New York, 1987.

CHAPTER 26

FINANCIAL CONSIDERATIONS

26-1 Cost Estimates

The development of engineering plans for water supply and wastewater treatment ordinarily involves economic evaluation of alternative schemes, selection of the least expensive satisfactory system, and development of a financing plan. The least expensive alternative frequently must be evaluated from several points of view: total cost, total cost to the utility, and total cost to the user. In some cases the federal or state government may provide a part of the capital cost while the local government or the public is responsible for other capital costs and for maintenance and operation. When such division of costs occurs, higher capital costs (paid by others) which reduce future operating costs (paid by users) may be selected even though the total cost is not a minimum.

The cost of engineering works of any kind is generally estimated on the basis of past costs for similar work. Unit costs for earthwork, pipe installation, reinforced concrete, piles, etc. are available from bid tabulations from other projects and may be projected to new construction.

In making such projections, it is important that the engineer be certain that the work is comparable in difficulty and that the costs are adjusted to account for inflation and other fluctuations in market prices. Adjustments of this sort are commonly made by using *cost indexes*. Examples of such indexes are the *Building Cost Index*, *Construction Cost Index*, *Materials Index*, *Skilled Labor Index*, and *Common Labor Index* published weekly in *Engineering News Record* for the United States. Additional values of these indexes on a city-by-city basis are published in the second issue of each month. The national indexes are based on a value of 100 in 1913. In 1989 their values were in the neighborhood of 4600 (construction), 2600 (building), 9300 (common labor), 4200 (skilled labor), and 1700 (materials). The individual city indexes for labor and materials are based on a value of 100 in 1967. Values for 1989 were in the neighborhood of 300 to 550. The EPA also publishes the *Sewage Treatment Plant Index*, which is useful in estimating costs.

Example 26-1 The cost of reinforced concrete in place in Philadelphia in July, 1989, was $750/m³. Estimate the cost of similar construction at a future date when the Construction Cost Index is 6150.

Solution
The Construction Cost Index for Philadelphia in July, 1989, was 5057. The estimated cost of the concrete in place at the future date is thus

$$750 \times 6150/5057 = \$912/\text{m}^3$$

Current costs of major equipment can be estimated in a similar fashion, but it is more usual to obtain estimates from manufacturers of such items as clarifier mechanisms, pumps, compressors, filter presses, chemical feeders, valves, and controls. A compilation of such costs, together with those for site preparation, construction, materials, and contingencies, permits preparation of reasonably accurate estimates for projects for which complete sets of plans are available.

Preliminary evaluation of alternatives (such as trickling filters versus activated sludge versus rotating biological disks) cannot be based on complete plans, since the cost of developing multiple detailed designs would be prohibitive. Rather, one tries to identify those features of the competing systems which would be different. If all would include identical primary clarifiers, the cost of this system is irrelevant to the comparison. If the sludge-handling systems would be different, then the cost of these must be included. The EPA periodically publishes graphic and computer-based data which facilitate comparison of different wastewater treatment systems.[1,2] The cost estimates provided by this sort of compilation are rough at best. Costs are presented on the basis of unit cost per volume of flow or other relevant design parameter such as mass of solids. The costs are developed for nominal average design parameters, which do not necessarily correspond to those which would be used in final design. In using such techniques, one must make sure that the values are current. Changes can easily occur in the relative costs of energy and materials which would alter the choice among the alternatives. The CAPDET model[2] is readily adjusted to the current local cost basis by introduction of costs for specific items from which other costs are then scaled. Such computer-based techniques are essential in studies dealing with large numbers of alternatives, but cannot be used as a basis for final cost estimates. In a study of a large metropolitan area in which there were over 200,000 alternatives with regard to number of plants, location of plants, and points of discharge, without considering process alternatives within the plants, CAPDET proved to be a very useful tool[3]. Desktop estimation techniques which were applied to the 20 least expensive alternatives selected by the program produced the same ranking, albeit at different cost levels. Since the object is selection of the least expensive solution, not precise estimation of its cost, the computer solution must be considered correct in this case.

A similar computer program designed for estimating costs of water treatment alternatives, WATMAN, has recently been extended to include plants designed for populations from 100 to 10,000 persons. The solution for cost is based on parametric equations in which the variables are peak and average daily flow. The user must identify the treatment trains which are required.[4]

Care must be taken in using computer-based estimation techniques to ensure that the results are reasonable. Complex programs often contain subtle errors which are not manifested during routine testing, but which may be important in applications to real problems. If the program gives answers which appear unlikely or impossible, other techniques must be used to check the solution.

26-2 Cost Comparisons

In selecting treatment or conveyance systems for water and wastewater, the goal is ordinarily to provide the desired level of service at minimum cost. The cost includes both capital costs for construction and recurring expenses for operation and maintenance. These must be combined in order to determine which alternative is least expensive. The principles of engineering economics are discussed elsewhere,[5] and only their direct application is presented here.

Annual costs such as operating expenses can be converted to present worth by multiplying by the *present worth factor* given by

$$(P/A, i, n) = \frac{(1 + i)^n - 1}{i(1 + i)^n} \tag{26-1}$$

in which i is the interest rate per period and n is the number of periods. The interest rate is usually a rate per year, and the value of n typically corresponds to the estimated life of the plant. Alternatively, capital costs may be converted to equivalent periodic costs by multiplying by the *capital recovery factor*

$$(A/P, i, n) = \frac{i(1 + i)^n}{(1 + i)^n - 1} \tag{26-2}$$

where the terms are as defined above. Values of these and other factors relating the value of money at one time to that at another are readily calculated and are tabulated in many handbooks and textbooks. The difficulty lies not in the calculation but in the selection of an appropriate life and interest rate. In projects financed by the federal government, the rate is usually established by the funding agency. Such rates are often based on political factors rather than fiscal principles and should not be adopted for other studies where their use is not required. If the project is to be financed by bonds (see below), the appropriate rate is the actual effective rate which will be paid by the issuer, including fees, discounts, and the nominal interest rate. This is not known exactly until the bonds are actually sold, but can be estimated with reasonable accuracy from recent similar issues.

The life of a project is never known until its use is abandoned. For purposes of cost comparisons, the appropriate life is that which was used as a basis for estimating design flows and loads. If the facility is expected to serve without expansion for 20 years, for example, that is the time which should be used in cost comparisons, even though it may be kept in operation beyond that time.

Not all alternatives have equal lives, yet comparisons must still be made on a rational basis. One might, for example, wish to compare the costs of a well field having an estimated life of 10 years with those of a reservoir with a life of 50 years. The reasons for selecting different design lives for such alternatives are discussed in Chap. 2.

Example 26-2 Compare the costs of a water supply system which will consist of either a reservoir with a design life of 50 years or a well field with a design life of 10 years. In either case the distribution system will be designed for ultimate development. The water treatment plants are not identical for the two alternatives, but both will be designed with a 15-year life. The actual cost of money is expected to be 9 percent per annum. The relevant costs are tabulated below. Costs which are common to both alternatives (disinfection, pumping, etc.) are not included.

	Capital cost	Annual O&M
Surface supply:		
Reservoir	415,000	1,800
Treatment	1,359,000	47,520
Total	1,774,000	49,320
Groundwater supply:		
Well field	200,000	5,000
Treatment	493,000	127,400
Total	693,000	132,400

Since the various alternatives have different lives, the comparison is most easily made on an annual cost basis. For the surface supply,

$$\begin{aligned}\text{Annual cost} &= 415{,}000(A/P,\ 9,\ 50) + 1{,}359{,}000(A/P,\ 9,\ 15) + 1800 + 47{,}520 \\ &= 415{,}000(0.09123) + 1{,}359{,}000(0.12406) + 1800 + 47{,}520 \\ &= \$255{,}778\end{aligned}$$

For the groundwater,

$$\begin{aligned}\text{Annual cost} &= 200{,}000(A/P,\ 9,\ 10) + 493{,}000(A/P,\ 9,\ 15) + 5000 + 127{,}400 \\ &= 200{,}000(0.15582) + 493{,}000(0.12406) + 5000 + 127{,}400 \\ &= \$224{,}726\end{aligned}$$

The comparison favors the groundwater supply. The fact that the surface supply will serve for 50 years rather than 10 has been taken into account in this calculation through the use of different interest factors for the two alternatives.

26-3 Optimization of Process Selection

The techniques described above may indicate that a particular sequence of treatment elements will be less expensive than other sequences, but will not establish the optimum proportion of the waste treatment load which should be borne by each stage, or the preferred selection among process variables for an individual system. For example, a primary clarifier and an activated sludge system will meet the ordinary secondary treatment standards for domestic wastewater, regardless of the primary clarifier size. As the surface overflow rate of the clarifier decreases, its cost will increase, but, up to a point, its efficiency will increase as well, which will reduce the load on the activated sludge process and, thus, the cost of that system. The activated sludge system can produce the required effluent quality if hydraulic detention time and solids return are manipulated to provide the required sludge age. Reduction of hydraulic detention time decreases the basin volume but increases the cost of recycling solids and may affect the stability of the process.

The determination of the minimum cost combination of such systems requires complete analysis of all liquid and solids-handling trains and the effects of changes in one process on the others. Middleton and Lawrence[6] have presented such an analysis for a process train including primary settling, activated sludge, final settling, gravity thickening, anaerobic digestion, vacuum filtration, incineration, and ultimate disposal of the sludge residue. The technique which they have developed is general and may be modified to include other processes.

Tarrer et al.[7] have prepared a computerized technique which allows optimization of the liquid treatment train, but not solids handling, for an activated sludge system. This technique is of limited practical application, since optimization of the liquid train alone will result in maximization of solids-handling costs. The technique is of interest, however, as a paradigm which might be used in development of a more complete simulation. The model, like that of Middleton and Lawrence, incorporates practical constraints and allows for uncertainty in waste parameters.

The design of the secondary clarifier has been found to exert a significant effect on optimum liquid treatment design;[6,7] hence approaches to optimization of secondary clarifier design are also necessary. Lee, Fan, and Takamatsu have presented such an analysis,[8] which indicates that under a given set of conditions there exists an optimal volume allocation in a multistage clarifier system and that a multistage clarifier is more efficient and stable than a single unit.

It is to be expected that further developments in system and unit process optimization will occur and that these simulations will simplify the engineer's task of selection among multiple alternatives.

26-4 Financing of Water Supply Systems

Water supply systems may be either privately or publicly owned. Privately owned water utilities are usually incorporated, but some small systems are partnerships or sole proprietorships.

Financing for corporations is provided by the sale of stocks and bonds and may be regulated by public utility commissions. Nonincorporated water companies must rely on the financial resources of the owners and whatever money may be borrowed using the assets of the company as collateral. Since these sources are more restricted than those of the bond market, such companies have traditionally been very limited in their ability to expand and to employ new technology. Recent regulations dealing with monitoring of priority pollutants and proposed mandatory treatment techniques may force these small utilities out of business.

The majority of the water supply utilities in the United States are publicly owned. There are both rural and municipal water supply districts which are administered either by the local government or by a separate elected or appointed governing board. Publicly owned utilities raise the capital necessary for construction through the sale of bonds which may be secured either by the revenues of the utility (called revenue bonds) or by all of the resources of the local government (called general obligation bonds). Most states impose a limit on the total debt backed by tax revenues which political subdivisions may assume. If the community is at or near its debt limit, revenue bonds may be the only means of raising money. Revenue bonds generally require the payment of higher interest rates than general obligation bonds.

When the proposed improvements to the water system benefit a limited group, a part of the cost may be raised through special assessments against the property served. Commonly, the cost of the distribution system serving new areas may be assessed against adjoining property—either at a charge based on frontage or at the actual cost of the main. Such charges are normally equitable, since the improvement will typically increase the value of adjoining property by more than the cost of the main.

The total amount of money to be raised is established on the basis of engineering estimates of the cost of construction plus engineering fees, legal fees, and a reserve for contingencies. The total is then divided among revenue bonds, general obligation bonds, and special assessments in whatever manner the debt limit, considerations of equity, and the local political climate may dictate.

When bonds are issued, the money to repay the principal and to pay the interest is normally generated by user fees, which are based on actual water consumption. A portion of the money may be derived from general property taxes, but a user fee is required in any event to generate funds for operation and maintenance, and this fee usually includes some money for debt service—even when the instrument is a general obligation bond. The distinction between general obligation and revenue bonds does not lie in how the community raises the money, but rather in the resources to which the bondholders may lay claim in the event of default. User fees typically must be adjusted periodically to accommodate new construction, changes in water consumption, and changes in operating and maintenance costs. The rates may be determined from the estimated costs of the debt service plus operation and maintenance costs.

Example 26-3 A small water supply system has financed the cost of its distribution system with general obligation bonds and real estate taxes. The treatment plant will cost \$3,500,000, which will be financed by 20-year revenue bonds. The bonds will have a nominal interest rate of 10 percent and will yield \$96.50 per \$100 of face value after payment of legal fees and brokers' commissions. The annual operation and maintenance costs of the plant and distribution system are estimated to be \$140,000 at a current production rate of 3800 m^3/day (1 mgd). Estimate the monthly bill of an average residential customer who uses 1600 L/day (425 gal/day).

Solution
The amount of the bond issue will be

$$\frac{3{,}500{,}000}{0.965} = \$3{,}627{,}000$$

The annual payment required for retirement of the debt is

$$3{,}627{,}000(A/P,\ 10,\ 20) = 3{,}627{,}000(0.11746) = \$426{,}000$$

The total revenue which must be raised per year is

$$426{,}000 + 140{,}000 = \$566{,}000$$

The unit cost of water supply is thus

$$\frac{566{,}000}{3800 \times 365} = \$0.408/\text{m}^3\ (\$1.55/1000\ \text{gal})$$

The monthly bill of the average residential user will be

$$1.6 \times 30 \times 0.408 = \$19.60$$

A number of political factors may modify the rather straightforward calculation presented above. Publicly owned waterworks may be operated at a profit, to reduce taxes, or may be operated at a loss by failure to perform routine maintenance. Users' bills may also be affected by provision of "free water" to schools and other public buildings. Such practices simply shift the cost from one segment of the population to another.

It is common practice to structure a rate system which will provide the required revenue while allocating the actual costs of service as equitably as possible. Charging users at a flat unit rate does not achieve this goal, since there are certain minimum costs associated with each connection to the system even if no water is used. In the past, many communities used declining rates in which successive increments of volume used per month were billed at lower costs. Such a scheme recognizes the economy of scale associated with increased production and sales, but has the flaw of encouraging use of a limited resource.

Modern rate structures are frequently based on a minimum bill which is paid by all customers and which provides a limited flow (perhaps 7500 L/month) at no additional cost. This is sometimes called a *lifeline rate* and is intended to ensure service to those with limited income. This charge should be based on the cost of meter reading, billing, collection and a part, at least, of the cost of the water. Use in excess of the minimum is billed at rates intended to generate the revenue necessary for debt service and operation of the system. These rates may increase with increasing use in order to discourage wasteful use and generation of high wastewater flows.

Careful record keeping is important if the community is to know the actual costs of providing water. In small cities it is not uncommon to have employees work on a variety of tasks such as animal control, street maintenance, snow removal, sewer repair, and water treatment during the course of a week. If their time and the equipment which is used is not billed to the appropriate activity, the apparent cost of water will be affected. Similarly, records of production are important to the establishment of rates. Knowing the cost without knowing how much water is produced is of little value. It should be possible to strike a rough balance between the total quantity of water billed to users and the total quantity produced in the plant. Differences will always exist, since water used for fire fighting and street cleaning is not metered, individual household meters, as they age, tend to underregister flow, and some leakage will occur in any system. If the differences between water produced and water sold exceed about 10 percent, a leak detection program or recalibration of meters may be financially justified.

26-5 Financing of Wastewater Systems

Collection and treatment of wastewater, historically, was not viewed as a business, hence it has become a public responsibility. Wastewater collection systems are owned by municipalities and most treatment facilities are also public property. In recent years there has been some development of "privatization" of wastewater treatment. In its initial stages this has taken two forms, the first of which involves turning over the operation of municipally owned facilities to a private contractor who provides the personnel required and is responsible for ensuring that the discharge permit is met. Under this type of contract, the energy and chemical costs may be paid by either the contractor or the community.

The second type of privatization system involves design, construction, and operation of a wastewater treatment plant by a contractor. Under this form of operation, all costs are typically paid by the contractor. With either system, the ownership and, generally, the maintenance and operation of the collection system remains with the community.

The advantage of privatization lies primarily in removing the day-to-day operation from political influence. Employees can be selected, promoted, or fired on the basis of their performance, and the potential for profit encourages economical operation. In at least one case the number of personnel required in a large treatment plant was reduced from over 20 to less than 10 with an improvement in the general level of operation and maintenance. Some large engineering firms have expanded into the area of wastewater treatment plant operation and offer their clients support services which would otherwise be very difficult to obtain. The decision to privatize wastewater treatment plant operation depends on the particular circumstances, but a number of small and large communities have selected this option in recent years. In particular, when new construction is required, privatization offers expanded service with no capital outlay on the part of the community.

Capital required for sewer systems is raised by frontage assessments, bonds, or both. In the past, the federal government has provided a substantial part of the cost of new sewerage systems through the EPA's construction grants program. This source of capital is no longer available, hence financing of new facilities is similar to that described above for water systems. Treatment facilities may be financed by either general obligation or revenue bonds, with revenue requirements being dictated by the total value of the issue, its term, and the effective interest rate. When treatment facilities are privately constructed and operated, the resulting capital costs are still paid by the users through their payments to the public agency which has entered into a contract with the operator.

Sewer charges are based principally on flow, but exceptions should be made for industrial users. If federal funds have been used in construction of the system, sewer charges for industrial contributors *must* be based on contribution of flow, BOD, suspended solids, grease, and other contaminants which may exceed normal domestic sewage concentrations in industrial wastes. The apportionment of charges is discussed in Art. 25-3.

Charges for domestic service are established, as for water treatment, on a cost per unit volume treated. This rate is then applied to each user's estimated sewage flow to determine the charge. Since individual residential wastewater flows are not measured, the charge may be based on a flat fee per residence, or on a flat fee per toilet, shower, bathtub, laundry facility, etc. It is more common, however, to base the charge on the water consumption since, as discussed in Chap. 2, nearly all the water which enters a home reappears in the form of wastewater. It may be more equitable to base the sewerage charge on water use in the winter, since a large part of that used in the summer may be applied to lawns or gardens and thus not enter the sewers.

REFERENCES

1. *The Cost Digest: Cost Summaries of Selected Environmental Control Technologies*, EPA-600/8-84-010, Environmental Protection Agency, Washington, D.C., 1984.
2. *Design of Wastewater Treatment Facilities—CAPDET Program User Guide*, EM 1110-2-501, part III, Department of the Army, Washington, D.C., 1980.
3. Terence J. McGhee, Parviz Mojgani, and Frank Vicidomina, "Use of EPA's CAPDET Program for Evaluation of Wastewater Treatment Alternatives," *Journal of Water Pollution Control Federation*, **55**:35, 1983.
4. Arun K. Deb and William G. Richards, "Evaluating the Economics of Alternative Technology for Small Water Systems," *Journal of American Water Works Association*, **75**:177, 1983.

5. Leland T. Blank and Anthony J. Tarquin, *Engineering Economy*, 3d ed., McGraw-Hill, New York, 1989.
6. Andrew C. Middleton and Alonzo W. Lawrence, "Least Cost Design of Activated Sludge Systems," *Journal of Water Pollution Control Federation*, **48**:889, 1976.
7. Arthur R. Tarrer et al., "Optimal Activated Sludge Design under Uncertainty," *Journal of Environmental Engineering Division, American Society of Civil Engineers*, **102**:EE3:657, 1976.
8. Chin R. Lee, L. T. Fan, and T. Takamatsu, "Optimization of Multistage Secondary Clarifier Design," *Journal of Water Pollution Control Federation*, **48**:2578, 1976.

APPENDIX 1

SATURATION VALUES OF DISSOLVED OXYGEN IN FRESHWATER AND SEAWATER

Saturation values of dissolved oxygen in fresh- and seawater exposed to an atmosphere containing 20.9 percent oxygen under a pressure of 760 mm of mercury*

(Calculated by G. C. Whipple and M. C. Whipple from measurements of C. J. J. Fox)†

Tempera-ture, °C	Dissolved oxygen (mg/L) for stated concentrations of chloride, mg/L					Difference per 100 mg/L chloride
	0	5000	10,000	15,000	20,000	
0	14.62	13.79	12.97	12.14	11.32	0.0165
1	14.23	13.41	12.61	11.82	11.03	0.0160
2	13.84	13.05	12.28	11.52	10.76	0.0154
3	13.48	12.72	11.98	11.24	10.50	0.0149
4	13.13	12.41	11.69	10.97	10.25	0.0144
5	12.80	12.09	11.39	10.70	10.01	0.0140
6	12.48	11.79	11.12	10.45	9.78	0.0135
7	12.17	11.51	10.85	10.21	9.57	0.0130
8	11.87	11.24	10.61	9.98	9.36	0.0125
9	11.59	10.97	10.36	9.76	9.17	0.0121
10	11.33	10.73	10.13	9.55	8.98	0.0118
11	11.08	10.49	9.92	9.35	8.80	0.0114
12	10.83	10.28	9.72	9.17	8.62	0.0110
13	10.60	10.05	9.52	8.98	8.46	0.0107
14	10.37	9.85	9.32	8.80	8.30	0.0104
15	10.15	9.65	9.14	8.63	8.14	0.0100
16	9.95	9.46	8.96	8.47	7.99	0.0098
17	9.74	9.26	8.78	8.30	7.84	0.0095
18	9.54	9.07	8.62	8.15	7.70	0.0092
19	9.35	8.89	8.45	8.00	7.56	0.0089
20	9.17	8.73	8.30	7.86	7.42	0.0088
21	8.99	8.57	8.14	7.71	7.28	0.0086
22	8.83	8.42	7.99	7.57	7.14	0.0084
23	8.68	8.27	7.85	7.43	7.00	0.0083
24	8.53	8.12	7.71	7.30	6.87	0.0083
25	8.38	7.96	7.56	7.15	6.74	0.0082
26	8.22	7.81	7.42	7.02	6.61	0.0080
27	8.07	7.67	7.28	6.88	6.49	0.0079
28	7.92	7.53	7.14	6.75	6.37	0.0078
29	7.77	7.39	7.00	6.62	6.25	0.0076
30	7.63	7.25	6.86	6.49	6.13	0.0075

* For other barometric pressures the solubilities vary approximately in proportion to the ratios of these pressures to the standard pressures.

† G. C. Whipple and M. C. Whipple, "Solubility of Oxygen in Sea Water," *Journal American Chemical Society*, **33**:362, 1911.

APPENDIX 2

RESISTANCE COEFFICIENTS OF VALVES AND FITTINGS

Reprinted from *Flow of Fluids through Valves, Fittings and Pipe*. Used with permission of Crane Co.

PIPE FRICTION DATA FOR CLEAN COMMERCIAL STEEL PIPE WITH FLOW IN ZONE OF COMPLETE TURBULENCE

Nominal Size	½"	¾"	1"	1¼"	1½"	2"	2½, 3"	4"	5"	6"	8-10"	12-16"	18-24"
Friction Factor (f_T)	.027	.025	.023	.022	.021	.019	.018	.017	.016	.015	.014	.013	.012

FORMULAS FOR CALCULATING "K" FACTORS*
FOR VALVES AND FITTINGS WITH REDUCED PORT

• **Formula 1**

$$K_2 = \frac{0.8\left(\sin\frac{\theta}{2}\right)(1-\beta^2)}{\beta^4} = \frac{K_1}{\beta^4}$$

• **Formula 2**

$$K_2 = \frac{0.5\,(1-\beta^2)\sqrt{\sin\frac{\theta}{2}}}{\beta^4} = \frac{K_1}{\beta^4}$$

• **Formula 3**

$$K_2 = \frac{2.6\left(\sin\frac{\theta}{2}\right)(1-\beta^2)^2}{\beta^4} = \frac{K_1}{\beta^4}$$

• **Formula 4**

$$K_2 = \frac{(1-\beta^2)^2}{\beta^4} = \frac{K_1}{\beta^4}$$

• **Formula 5**

$$K_2 = \frac{K_1}{\beta^4} + \text{Formula 1} + \text{Formula 3}$$

$$K_2 = \frac{K_1 + \sin\frac{\theta}{2}\left[0.8\,(1-\beta^2) + 2.6\,(1-\beta^2)^2\right]}{\beta^4}$$

• **Formula 6**

$$K_2 = \frac{K_1}{\beta^4} + \text{Formula 2} + \text{Formula 4}$$

$$K_2 = \frac{K_1 + 0.5\sqrt{\sin\frac{\theta}{2}}\,(1-\beta^2) + (1-\beta^2)^2}{\beta^4}$$

• **Formula 7**

$$K_2 = \frac{K_1}{\beta^4} + \beta\,(\text{Formula 2} + \text{Formula 4}) \text{ when } \theta = 180^\circ$$

$$K_2 = \frac{K_1 + \beta\left[0.5\,(1-\beta^2) + (1-\beta^2)^2\right]}{\beta^4}$$

$$\beta = \frac{d_1}{d_2}$$

$$\beta^2 = \left(\frac{d_1}{d_2}\right)^2 = \frac{a_1}{a_2}$$

Subscript 1 defines dimensions and coefficients with reference to the smaller diameter.

Subscript 2 refers to the larger diameter.

***Use "K" furnished by valve or fitting supplier when available.**

SUDDEN AND GRADUAL CONTRACTION

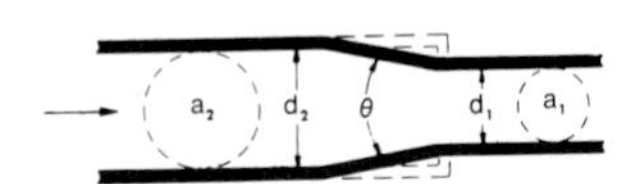

If: $\theta \leq 45^\circ$ .... K_2 = Formula 1

$45^\circ < \theta \leq 180^\circ$ ... K_2 = Formula 2

SUDDEN AND GRADUAL ENLARGEMENT

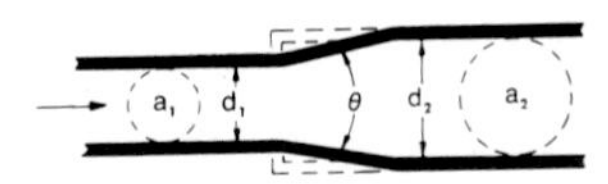

If: $\theta \leq 45^\circ$ .... K_2 = Formula 3

$45^\circ < \theta \leq 180^\circ$ ... K_2 = Formula 4

GATE VALVES
Wedge Disc, Double Disc, or Plug Type

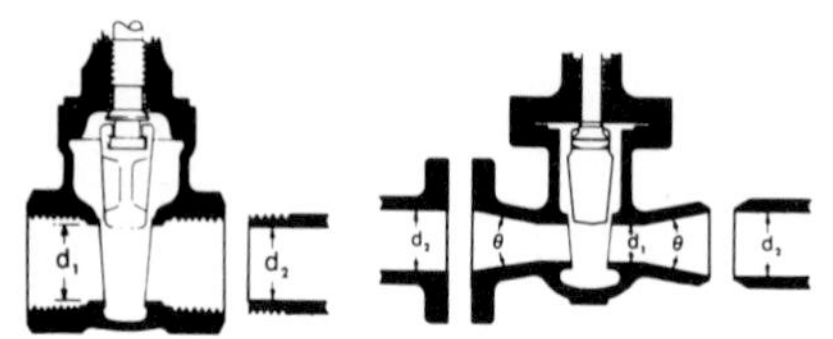

If: $\beta = 1,\ \theta = 0$ . . . . $K_1 = 8\,f_T$
$\beta < 1$ and $\theta \leq 45°$ . . . . K_2 = Formula 5
$\beta < 1$ and $45° < \theta \leq 180°$ . . . K_2 = Formula 6

GLOBE AND ANGLE VALVES

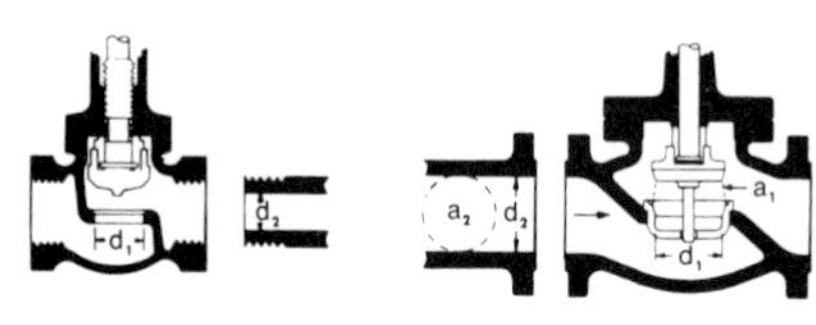

If: $\beta = 1 \ldots K_1 = 340\,f_T$

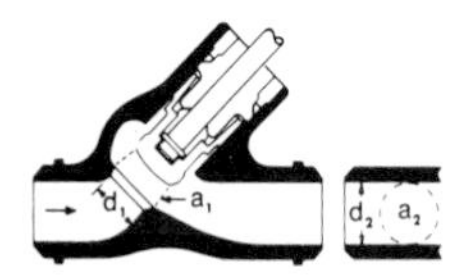

If: $\beta = 1 \ldots K_1 = 55\,f_T$

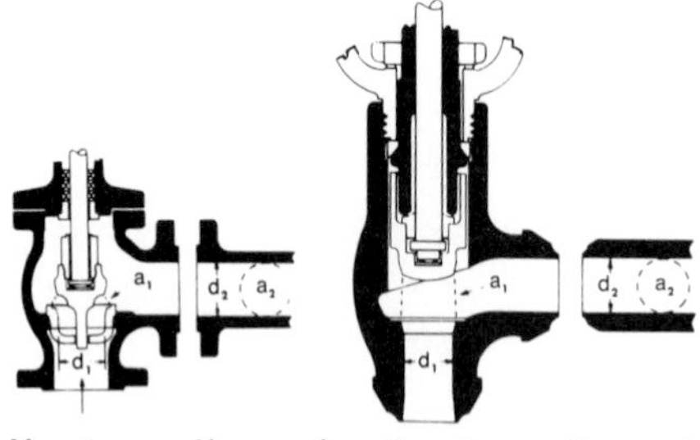

If: $\beta = 1 \ldots K_1 = 150\,f_T$ If: $\beta = 1 \ldots K_1 = 55\,f_T$

All globe and angle valves,
whether reduced seat or throttled,
If: $\beta < 1 \ldots K_2$ = Formula 7

SWING CHECK VALVES

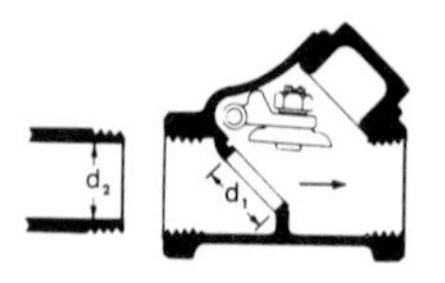

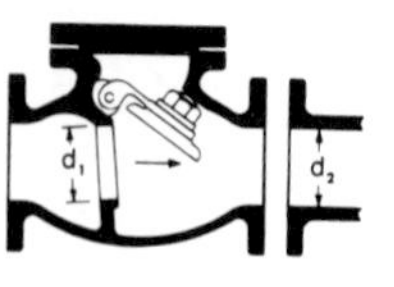

$K = 100\,f_T$

Minimum pipe velocity (fps) for full disc lift $= 35\sqrt{\bar{V}}$

$K = 50\,f_T$

Minimum pipe velocity (fps) for full disc lift $= 60\sqrt{\bar{V}}$ except U/L listed $= 100\sqrt{\bar{V}}$

LIFT CHECK VALVES

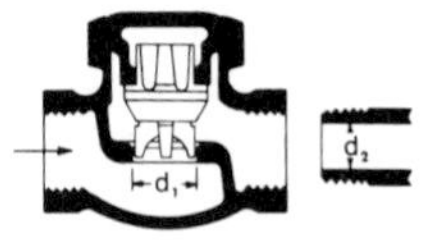

If: $\beta = 1 \ldots K_1 = 600\,f_T$
$\beta < 1 \ldots K_2$ = Formula 7

Minimum pipe velocity (fps) for full disc lift
$= 40\,\beta^2\sqrt{\bar{V}}$

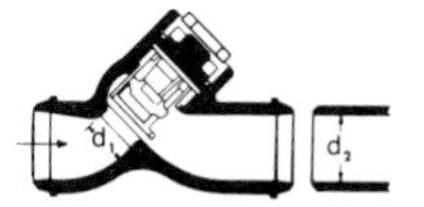

If: $\beta = 1 \ldots K_1 = 55\,f_T$
$\beta < 1 \ldots K_2$ = Formula 7

Minimum pipe velocity (fps) for full disc lift
$= 140\,\beta^2\sqrt{\bar{V}}$

TILTING DISC CHECK VALVES

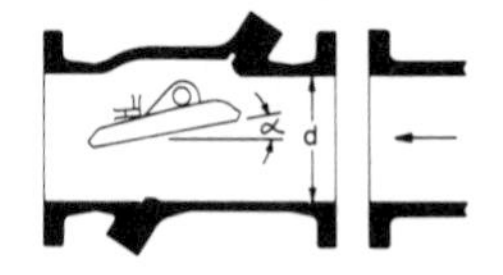

	$\alpha = 5°$	$\alpha = 15°$
Sizes 2 to 8″ . . . K =	$40\,f_T$	$120\,f_T$
Sizes 10 to 14″ . . . K =	$30\,f_T$	$90\,f_T$
Sizes 16 to 48″ . . . K =	$20\,f_T$	$60\,f_T$
Minimum pipe velocity (fps) for full disc lift =	$80\sqrt{\bar{V}}$	$30\sqrt{\bar{V}}$

STOP-CHECK VALVES (Globe and Angle Types)

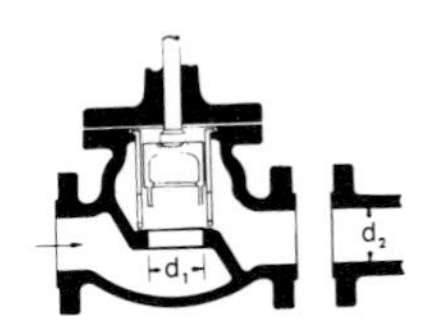

If:
$\beta = 1 \ldots K_1 = 400 f_T$
$\beta < 1 \ldots K_2 =$ Formula 7

Minimum pipe velocity for full disc lift
$= 55\,\beta^2 \sqrt{\overline{V}}$

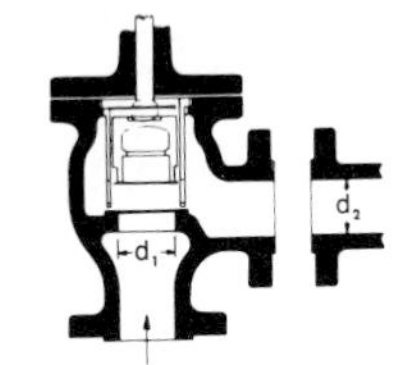

If:
$\beta = 1 \ldots K_1 = 200 f_T$
$\beta < 1 \ldots K_2 =$ Formula 7

Minimum pipe velocity for full disc lift
$= 75\,\beta^2 \sqrt{\overline{V}}$

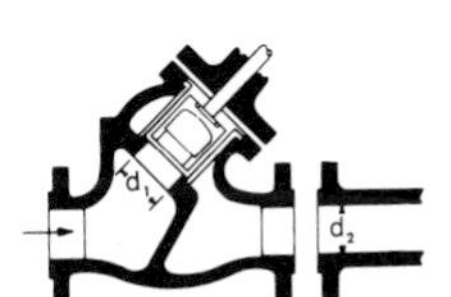

If:
$\beta = 1 \ldots K_1 = 300 f_T$
$\beta < 1 \ldots K_2 =$ Formula 7

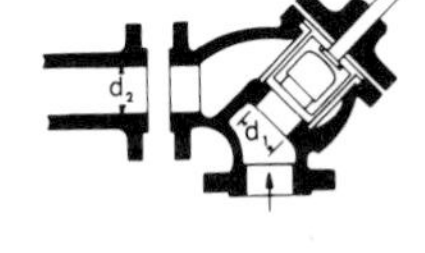

If:
$\beta = 1 \ldots K_1 = 350 f_T$
$\beta < 1 \ldots K_2 =$ Formula 7

Minimum pipe velocity (fps) for full disc lift
$= 60\,\beta^2 \sqrt{\overline{V}}$

If:
$\beta = 1 \ldots K_1 = 55 f_T$
$\beta < 1 \ldots K_2 =$ Formula 7

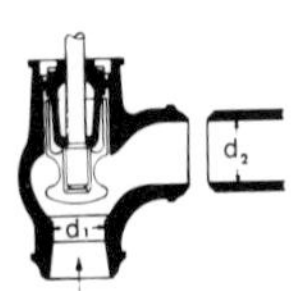

If:
$\beta = 1 \ldots K_1 = 55 f_T$
$\beta < 1 \ldots K_2 =$ Formula 7

Minimum pipe velocity (fps) for full disc lift
$= 140\,\beta^2 \sqrt{\overline{V}}$

FOOT VALVES WITH STRAINER

Poppet Disc | **Hinged Disc**

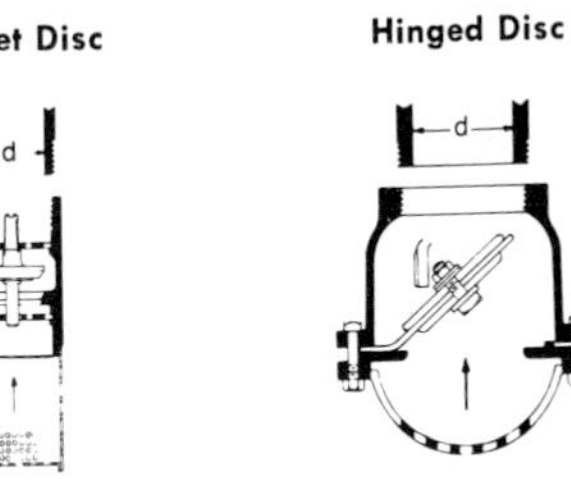

Poppet Disc: $K = 420 f_T$

Minimum pipe velocity (fps) for full disc lift
$= 15 \sqrt{\overline{V}}$

Hinged Disc: $K = 75 f_T$

Minimum pipe velocity (fps) for full disc lift
$= 35 \sqrt{\overline{V}}$

BALL VALVES

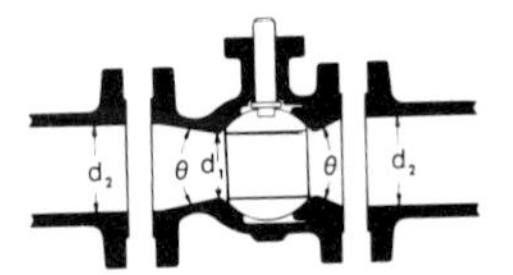

If: $\beta = 1,\ \theta = 0 \ldots\ldots\ldots\ldots K_1 = 3 f_T$
$\beta < 1$ and $\theta \leqq 45° \ldots\ldots\ldots K_2 =$ Formula 5
$\beta < 1$ and $45° < \theta \leqq 180° \ldots K_2 =$ Formula 6

BUTTERFLY VALVES

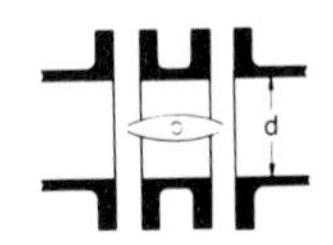

Sizes 2 to 8″... $K = 45 f_T$
Sizes 10 to 14″... $K = 35 f_T$
Sizes 16 to 24″... $K = 25 f_T$

PLUG VALVES AND COCKS

Straight-Way **3-Way**

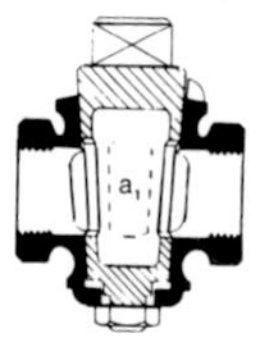

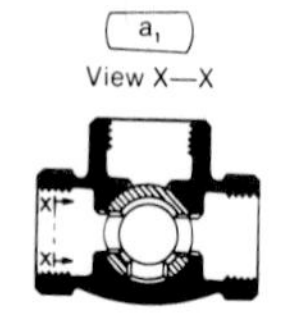

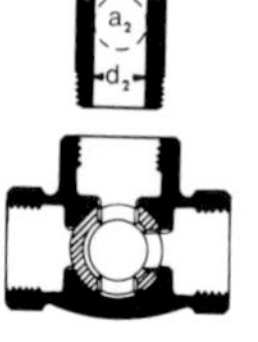

If: $\beta = 1$, $K_1 = 18\, f_T$

If: $\beta = 1$, $K_1 = 30\, f_T$

If: $\beta = 1$, $K_1 = 90\, f_T$

If: $\beta < 1 \ldots K_2 =$ Formula 6

MITRE BENDS

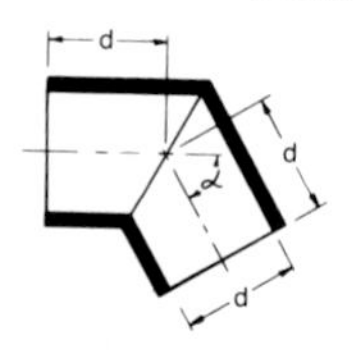

α	K
0°	2 f_T
15°	4 f_T
30°	8 f_T
45°	15 f_T
60°	25 f_T
75°	40 f_T
90°	60 f_T

90° PIPE BENDS AND FLANGED OR BUTT-WELDING 90° ELBOWS

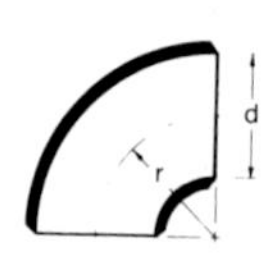

r/d	K	r/d	K
1	20 f_T	8	24 f_T
1.5	14 f_T	10	30 f_T
2	12 f_T	12	34 f_T
3	12 f_T	14	38 f_T
4	14 f_T	16	42 f_T
6	17 f_T	20	50 f_T

The resistance coefficient, K_B, for pipe bends other than 90° may be determined as follows:

$$K_B = (n-1)\left(0.25\,\pi f_T \frac{r}{d} + 0.5\,K\right) + K$$

n = number of 90° bends
K = resistance coefficient for one 90° bend (per table)

CLOSE PATTERN RETURN BENDS

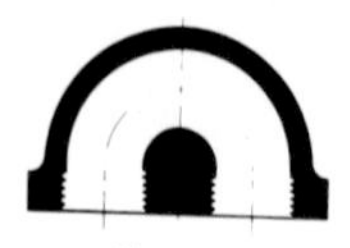

$K = 50\, f_T$

STANDARD ELBOWS

90° 45°

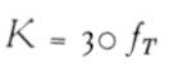

$K = 30\, f_T$ $K = 16\, f_T$

STANDARD TEES

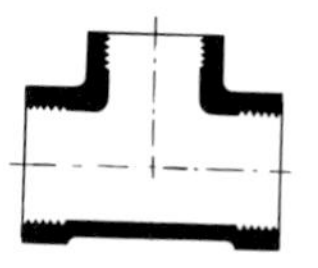

Flow thru run. . . . . . . $K = 20\, f_T$
Flow thru branch. . . . $K = 60\, f_T$

PIPE ENTRANCE

Inward Projecting

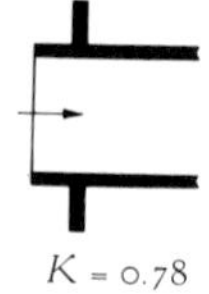

$K = 0.78$

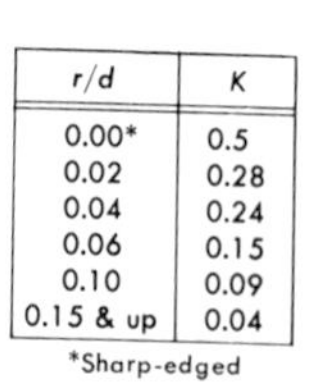

r/d	K
0.00*	0.5
0.02	0.28
0.04	0.24
0.06	0.15
0.10	0.09
0.15 & up	0.04

*Sharp-edged

Flush

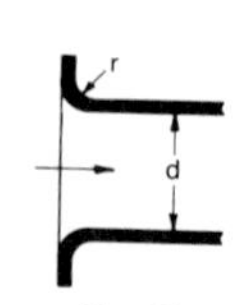

For K, see table

PIPE EXIT

Projecting **Sharp-Edged** **Rounded**

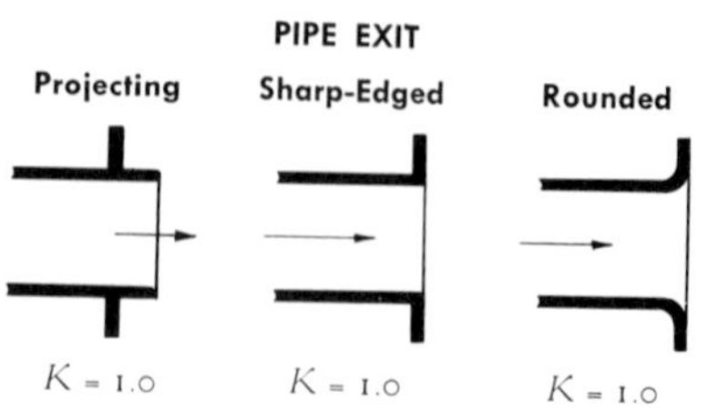

$K = 1.0$ $K = 1.0$ $K = 1.0$

APPENDIX 3

PHYSICAL PROPERTIES OF WATER

Physical properties of water at different temperatures

Temperature, °C	Density, g/cm³	Absolute viscosity, cP	Kinematic viscosity, cSt	Vapor pressure, kPa
0	0.99987	1.7921	1.7923	0.61
2	0.99997	1.6740	1.6741	0.71
4	1.00000	1.5676	1.5676	0.82
6	0.99997	1.4726	1.4726	0.94
8	0.99988	1.3872	1.3874	1.09
10	0.99973	1.3097	1.3101	1.23
12	0.99952	1.2390	1.2396	1.42
14	0.99927	1.1748	1.1756	1.61
16	0.99897	1.1156	1.1168	1.81
18	0.99862	1.0603	1.0618	2.02
20	0.99823	1.0087	1.0105	2.33
22	0.99780	0.9608	0.9629	2.66
24	0.99733	0.9161	0.9186	3.02
26	0.99681	0.8746	0.8774	3.38
28	0.99626	0.8363	0.8394	3.71
30	0.99568	0.8004	0.8039	4.24

g/cm³ × 62.42 = lb/ft³

Centipoise × 2.088×10^{-5} = lbf · s/ft²

Centistoke × 1.075×10^{-5} = ft²/s

Kilopascal × 0.145 = lb/in²

INDEX